FUNDAMENTALS OF
ROBOTICS
LINKING PERCEPTION TO ACTION

SERIES IN MACHINE PERCEPTION AND ARTIFICIAL INTELLIGENCE*

Editors: **H. Bunke** (Univ. Bern, Switzerland)
P. S. P. Wang (Northeastern Univ., USA)

*For the complete list of titles in this series, please write to the Publisher.

Series in Machine Perception and Artificial Intelligence

FUNDAMENTALS OF
ROBOTICS

LINKING PERCEPTION TO ACTION

Ming Xie

Singapore-MIT Alliance &
Nanyang Technological University, Singapore

World Scientific
New Jersey • London • Singapore • Hong Kong

Published by

World Scientific Publishing Co. Pte. Ltd.

5 Toh Tuck Link, Singapore 596224

USA office: Suite 202, 1060 Main Street, River Edge, NJ 07661

UK office: 57 Shelton Street, Covent Garden, London WC2H 9HE

Library of Congress Cataloging-in-Publication Data
Xie, M. (Min)
    Fundamentals of robotics : linking perception to action / Ming Xie.
      p. cm. -- (Series in machine perception and artificial intelligence ; v. 54)
    Includes bibliographical references and index.
    ISBN 981-238-313-1 -- ISBN 981-238-335-2 (pbk.)
    1. Robotics. I. Title. II. Series.

TJ211.X84 2003
629.8'92--dc21   2003043284

British Library Cataloguing-in-Publication Data
A catalogue record for this book is available from the British Library.

Printed by Fulsland Offset Printing (S) Pte Ltd, Singapore

Dedicated in memory of my dear mother.
To David and Sophie.

Preface

Purpose

The word "Robot" comes from the Czech word "Robota" which means "labor doing compulsory manual works without receiving any remuneration" or "to make things manually". Oxford dictionary defines "Robot" as "a machine resembling a human being and able to replicate certain human movements and functions automatically." These days, more and more robot is living up to its name. Gone is the time when robots are merely mechanisms attached to controls. Today's robots are a combination of manipulative, perceptive, communicative, and cognitive abilities. They can be seen welding heavy machinery or repairing nuclear power plants; they can be seen crawling through debris underwater or crawling across the craters on Mars; they can be seen de-mining military zones or cleaning windows on high buildings. Today's robots are capable of so many tasks. Yet, there is so much more on the horizon.

Tomorrow's robots, which includes the humanoid robot, will tutor children, work as tour guides, drive humans to and from work, do the family shopping. Tomorrow's robots will enhance lives in ways we never dreamed possible – Do not have time to attend the decisive meeting on Asian strategy? Let your robot go for you and make the decisions. Are not feeling well enough to go to the clinic? Let Dr. Robot come to you, make a diagnosis, and get you the necessary medicines for treatment. Do not have time to coach the soccer team this week? Let the robot do it for you.

Tomorrow's robots will be the most exciting and revolutionary thing to happen to the world since the invention of the automobile. It will change the way we work, play, think, and live. Because of this, nowadays robotics is one of the most dynamic fields of scientific research. These days, robotics is offered in almost every university in the world. Most mechanical engineering departments offer a similar course at both the undergraduate and graduate levels. And increasingly, many computer and electrical engineering departments are also offering it.

This book will guide you, the curious beginner, from yesterday to tomorrow. I will cover practical knowledge in understanding, developing, and using robots as versatile equipment to automate a variety of industrial processes or tasks. But, I will also discuss the possibilities we can look forward to when we are capable of creating a vision-guided, learning machine.

Contents

I have arranged the book, according to the following concerns:

- Focus:
  I focus a great deal on analysis, and control of today's robots. I also delve into the more recent developments being made with the humanoid robot.
- Educational:
  Whenever possible, I describe the philosophy and physics behind a certain topic, before discussing the concise mathematical descriptions.
- Systems Approach:
  The robot is a tightly-integrated entity of various engineering systems, ranging from mechanics, control, and information to perception and decision-making. Therefore, I use the systems approach to clearly organize and present the multiple facets of robotics.

I follow the motion-centric theme of robotics, because motion is a visible form of action which is intrinsically linked to perception. On the other hand, I intend to stick to the original meaning of robotics, namely: the study of the robot which is shaped like a human. It is my aim that this book provides a balanced coverage of various topics related to the development of robots. Whenever possible, I relate our discussion to the humanoid robot.

I have designed the flow of the chapters to be clear and logical, in order to ease the understanding of the motion-centric topics in robotics. I embrace both traditional and non-traditional fundamental concepts and principles of robotics, and the book is organized as follows:

In Chapter 1, I introduce the robot, from a philosophical point of view, as well as the theme of robotics, from the perspective of manufacturing and automation. Additionally, I illustrate the motion-centric topics in robotics. I link the directing concept of task-action-motion to perception in our discussion on artificial intelligence, and propose a new definition.

In Chapter 2, I review the mathematical description of a rigid body's motion. Motion is the unifying concept in robotics which links various topics together, such as generation of motion, modelling and analysis of motion, control of motion, and visual perception of motion. Here, I unify the mathematical notations related to geometrical entities under consideration in both kinematic analysis and visual perception of motion.

In Chapter 3, we study the pure mechanical aspect of a robot. We also cover, in detail, fundamental topics on mechanism and kinematics. I introduce a simple illustration on D-H parameters, as well as a new term, called the *simple open kinematic-chain*. This new term is useful for the kinematic modelling of the humanoid robot. Finally, I include a new scheme for solving inverse kinematics, called *discrete kinematic mapping*, which complements the classical solutions.

In Chapter 4, we study the electromechanical aspect of a robot. After I discuss the origin of the rigid body's motion, I introduce actuation elements at a conceptual level. I also introduce a new concept of the dynamic pair and discuss the solutions of kineto-dynamic couplings. I propose a new scheme on one-to-many coupling (i.e. a single actuator for all the degrees of freedom) in comparison with the traditional scheme of one-to-one coupling (i.e. one actuator per degree of freedom). In the latter part of the chapter, I focus on the fundamentals of establishing equations of motion, which relate force/torque to motion. I also mention the study of robot statics.

In Chapter 5, we study the control system of a robot. First, I introduce the basics of the control system. Then, I focus on the control and sensing elements of the system, especially on how to alter the direction of motion, how to regulate the electrical energy applied to the electric motors, and how to sense motion and force/torque variables. I discuss at an appropriate level the aspects of designing robot control algorithms in joint space, task space and image space.

In Chapter 6, we study the information system of a robot. Traditionally, this system aspect of the robot has been overlooked in robotics textbooks. Although an information system is not too important to an industrial robot, it is an indispensable subsystem of the humanoid robot. Here, we emphasize the basics of the information system, in terms of data processing, storage and communication. We study the fundamentals of computing platforms, micro-controllers and programming together with a conceptual description of multi-tasking. I also include discussions on various interfacing systems typically used by a micro-controller (i.e. I/O).

In Chapter 7, we study the visual sensory system of a robot. It is a common fallacy among novices that it is simple to form a vision system by putting optical lenses, cameras and computers together. However, the flow of information from the light rays to the digital images goes through various signal conversions. Thus, a primary concern of the robot's visual-sensory system is whether or not the loss of information undermines the image and vision computing by the robot's visual-perception system. I start this chapter with a study of the basic properties of light and human vision. Then, I focus on the detailed working principles underlying optical image-formation, electronic image-formation, and the mathematical modelling of the robot's visual sensory system. I also cover the important topics of CMOS-imaging sensors, CCD-imaging sensors, TV/Video standards, and image-processing hardware.

In Chapter 8, we study the visual-perception system of a robot. I keep to the motion-centric theme of robotics while focusing on the fundamentals underlying the process of inferring three-dimensional geometry (including posture and motion) from two-dimensional digital images. This process includes image processing, feature extraction, feature description and geometry measurement with monocular & binocular vision. When discussing image processing, I introduce the principle of dynamics resonance. It serves as the basis for explaining various edge-detection algorithms. I discuss in detail a probabilistic RCE neural network which addresses the issue of uniformity detection in images. In our study of binocular vision, I present two new solutions which cope with the difficult problem of binocular correspondence. Finally, I provide a theoretical foundation for further study of a robot's image-space control.

In Chapter 9, we study the decision-making system of a robot. Decision-making is indispensable in the development of autonomous robots. After I introduce the basics of decision-making, I discuss the fundamentals of task and action planning, at a conceptual level. However, I emphasize motion planning. We first study motion planning in task space, and introduce a new strategy, known as *backward-motion planning*. Then, we study image-guided motion planning in task space, discussing in detail, a unified approach, called *qualitative binocular vision*. I cover three important examples of image-guided autonomous behavior: a) hand-eye coordination,

b) head-eye coordination, and c) leg-eye coordination.

Acknowledgement

First of all, my sincere gratitude goes to:

- Professor Dhanjoo Gista (Ph.D, Stanford), one of the founders of the scientific discipline of bio-engineering, for his encouragement and initiative in persuading me to consolidate my teaching materials into this textbook. Without his moral support, this book would not have been possible.
- Professor Horst Bunke, the Series Editor for World Scientific Publisher, for his kindness in not only reviewing this book, but also including it in *Series in Machine Perception and Artificial Intelligence.*
- Mr. Ian Seldrup and Ms Tan Rok Ting, Editors at World Scientific Publisher & Imperial College Press, for providing me friendly editorial advice, and the final proof-reading of the manuscript.
- Dr. Hui Hui Lin (IBM) and his lovely wife, Jana, who have tirelessly helped in improving the readability of the manuscript and correcting errors.

I would also like to express my gratitude to:

- Nanyang Technological University and the Singapore-MIT Alliance, for providing me not only with a conducive environment to complete this book, but also generous research grants for research staff and various related projects.
- My research staff and research students for their assistance in graphic works and experimental results.

Finally, thanks to my wife and children for their love, support, and encouragement.

M. Xie
September, 2002

Note to Instructors

This book can serve as a textbook for a simple, one-semester course on robotics, provided you do not want to delve into vision, or a two-semester course which highlights vision as well. I have designed this textbook specifically for undergraduate- and graduate-level students in engineering. It can also serve as a supplementary text for students having an interest in robotics, vision, and artificial intelligence.

Contents

xviii

Chapter 1

Introduction to Robotics

1.1 Introduction

This chapter gives an overview of *robotics*, the engineering discipline which focuses on the study of robots. I will introduce relevant topics from the angle of manufacturing, which serves as a convincing basis to justify the usefulness and importance of robots in industry. For those who are actively undertaking research in the area of artificial intelligence, I also describe a framework from which to understand human intelligence. In the later part of the chapter, I discuss the major concerns of robotics, for the purpose of illustrating the simple, unifying theme, *motion*. An understanding of this theme would make the rest of the chapters easier to read.

1.2 Manufacturing

The word manufacture comes from the combination of two Latin words, *manus* (hand) and *factus* (to make). Thus, the literal meaning of manufacture is, "to make by hand" either directly, by producing handicrafts, or indirectly, by making use of re-programmable machine tools. Since ancient times, our ancestors exercised their creativity to the fullest, in making things by hand. This creativity led to the invention of *tools*, which made the process of "making things" much easier and more efficient. Most importantly, the discovery of *engineering materials*, such as metal, ceramics, and polymers, enlarged the scope of "things" which can be made by hand, with or without the help of tools. This, in turn, fuelled people's creativity in inventing various *processes* for making "things" of different complexity, property, and scale.

A direct consequence of the activity of "making things by hand" was

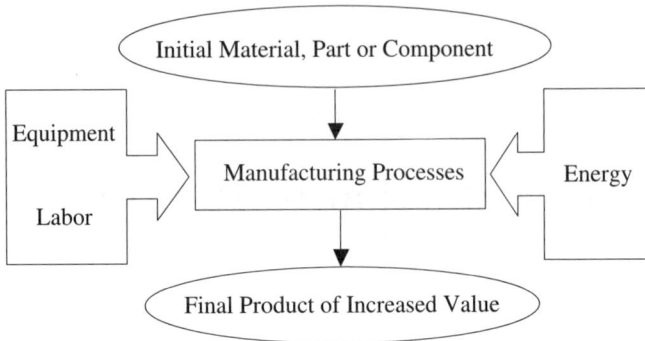

Fig. 1.1   Illustration of basic functional modules in manufacturing.

that craftsmen were able to produce a surplus of goods which far exceeded their needs. As a result, people began to exchange the surplus of one type of goods for the surplus of another type of goods. This led to the creation of commerce, which is a platform for stimulating the production of wealth for people and nations. With the advent and sophistication of finance and monetary systems, commerce has steadily reached a scale, which goes into a dimension far beyond geographical, social, and cultural boundaries. It is not exaggerating to say that today's commerce is the motor which drives all economic, social, and cultural activities. Regardless of the scale and dimension of commerce, the basic fundamentals, the exchange of goods or services, are still the same. Without *exchange*, there would be no commerce; without commerce, there would be no manufacturing.

If manufacturing is the art and science of "making things by hand," directly or indirectly, a formal definition of manufacturing can be stated, as follows:

Definition 1.1   Manufacturing is the application of processes which alters the geometry, property, and appearance of materials for the production of goods of increased value.

Refer to Fig. 1.1. A material, part or component is transformed from the initial state to the final state of product through the interaction among labor, equipment, material, and parts or components. This interaction results in energy consumption.

Growing commercial activities have undoubtedly pushed up the demand for product quality and manufacturing capacity, which is measured by the rate of output products over a fixed period of time. The production of goods in large volume has propelled manufacturing to evolve into a rigorous scientific discipline covering the following important aspects:

Products

Final products, produced by Manufacturers, can be classified into two categories: a) consumer products and b) capital products. The former are products purchased directly by consumers, such as personal computers, automobiles, TV, video recorders, household appliances, foods, beverages etc. The latter are products purchased by companies, such as industrial robots, mainframe computers, machine tools, construction equipment, processed materials, parts, devices, components etc. Capital products are used in manufacturing as facilities for the production of goods.

Materials

In addition to natural materials, such as wood, bamboo, stones, rocks, petroleum etc., the advent of engineering materials has undoubtedly enlarged the scope of things which can be manufactured. There are three basic categories of engineering materials, namely: metals, ceramics, and polymers. The combination of these three basic engineering materials forms another category called, *composite materials*, such as metal-ceramic composites, metal-polymer composites, and ceramic-polymer composites.

Processes

A manufacturing process consists of the interaction among labor, equipment, and input materials, parts, or components. A manufacturing process will change the geometry, property, or appearance of the input materials, parts, or components. Thus, the interaction in a manufacturing process will consume energy in the mechanical, thermal, chemical, or electrical domains.

Depending on the modes of interaction, a process can be either a serial process or a parallel process. A process is called a serial process if the interaction occurs locally on a material, part, or component while a process is called a parallel process if the interaction occurs globally across a material, part or component.

However, depending on the outcome of the interaction, a process can fall into one of the four categories:

(1) *Removal Process*:
This refers to the operation of removing certain portions of input materials. Examples include cutting, grinding, die stamping etc.

(2) *Addition Process*:
This refers to the operation of adding other materials to input materials, or joining parts or components together. Examples include painting, coating, assembly, 3D printing, welding, soldering etc.

(3) *Solidification Process*:
This refers to the operation of creating objects through the transformation of thermal states of materials. A typical example of the solidification process is injection molding.

(4) *Deformation Process*:
This refers to the operation of altering the shape of a material, part or component through the application of either mechanical or thermal energy. Typical examples include die forging, bending, rolling, heating etc.

Equipment

In order to satisfy the demand for high quality and high volume, it is necessary to use machine tools or automated equipment to operate manufacturing processes. Functionally, the role of manufacturing equipment is to control the interaction in a manufacturing process which will alter the geometry, property, or appearance of the initial material, part, or component.

Factories

The manufacturing industry consists of factories and organizations which produce or supply goods and services.

Refer to Fig. 1.2. A factory is typically composed of the following entities:

- *Production and Automation System*:
This includes the factory layout, process equipment, metrology equipment, material-handling equipment and labor.

- *Manufacturing Support System*:
This includes inbound logistics for ordering materials, parts, components and facilities. On the other hand, outbound logistics deals with distribution of the final products.

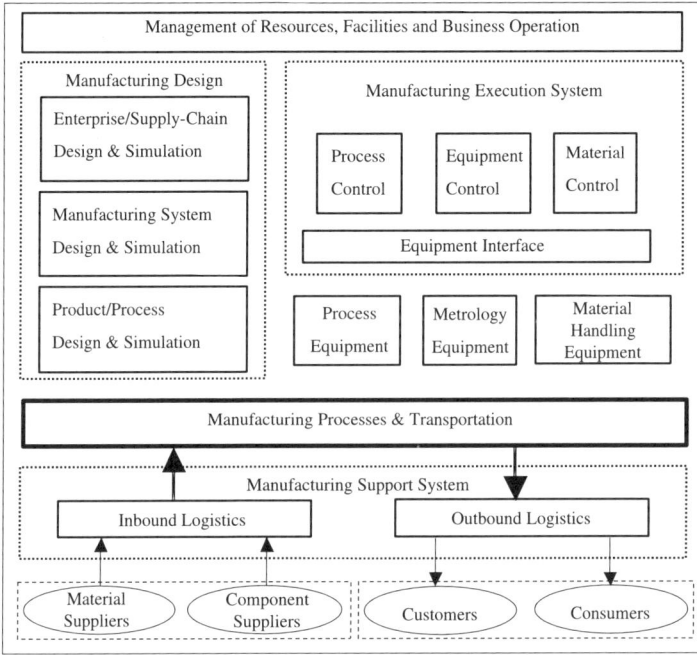

Fig. 1.2 Outline of a factory in manufacturing.

- *Manufacturing Execution System*:
  This includes process control, material flow control, equipment control and deployment of labor.
- *Manufacturing Design System*:
  This includes product design & simulation, process design & simulation and supply-chain design & simulation.
- *Enterprise Information and Management System*:
  This includes the management of resources, facilities and business operations.

1.3 Factory Automation

There are two ways to achieve high yields in manufacturing. The simplest, yet most expensive way is to increase the number of production lines. An alternative and more desirable way is to increase the rate of production in the existing production lines. It is possible to increase the production rate

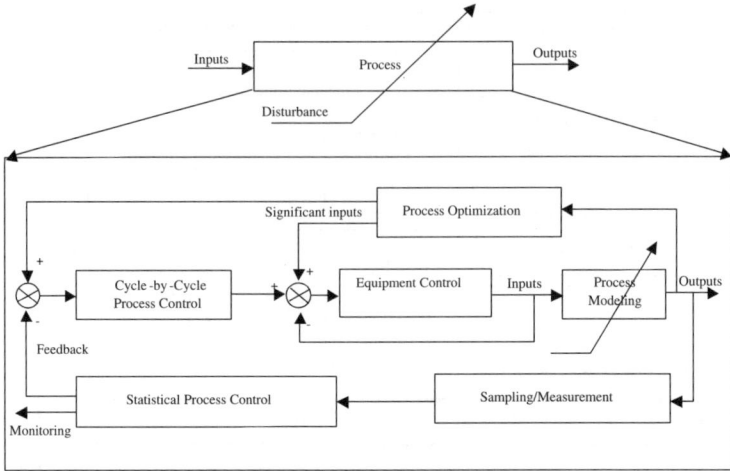

Fig. 1.3   Illustration of process optimization, process control, and equipment control in manufacturing.

by reducing the cycle time needed to produce a single part or product.

There are also two ways to reduce cycle time. The first approach is to improve the manufacturing process. The second approach is to automate the manufacturing process by using re-programmable and automatically controlled equipment.

As shown in Fig. 1.3, process optimization always starts with process modelling, which consists of establishing a mathematical description that relates a process's input $X = \{(X_1, X2, ..., X_n)\}$ to its output $Y = \{(Y_1, Y2, ..., Y_m)\}$. We can identify the subset of significant input variables by using the scientific method, called the *design of experiments*. If we regulate or control these significant input variables, the dynamics of a process can be optimized, in terms of stability, robustness, time responsiveness etc.

Therefore, one important aspect of factory automation is to automatically regulate or control the significant input variables of a process.

Refer to Fig. 1.3. Process control involves two cascaded control loops, which operate at different rates. The outer loop is the statistical process control, which aims at maximizing product quality. By definition, quality is inversely proportional to the output variation . Therefore, the quality control can be achieved by first monitoring the output variation of a process, and then automatically acting on a process's inputs for the purpose of minimizing this variation.

The inner loop, in a process control, is equipment control. As I mentioned before, the role of manufacturing equipment is to control the interaction in a process, which aims at altering in a controllable manner, the geometry, property, or appearance of the initial materials, parts, or components. Equipment control output is the direct input to a process. Therefore, the rate of equipment control depends on the dynamics of the process. In general, equipment control must operate in real-time. However, statistical-process control depends on sampling, measurement, and analysis of data. Its control cycle will be longer than that of equipment control.

1.3.1 *Automation and Robots*

As we discussed above, a manufacturing process consists of the interaction among labor, equipment, and materials, parts, or components. This interaction results in energy consumption in mechanical, thermal, chemical or electrical domains.

The aim of automation is to eliminate the direct involvement of labor in the process interaction. This is only achievable by setting up automatically-controlled equipment. In this way, the role of labor in the factory has shifted, from direct involvement to indirect programming and/or monitoring of automated equipment.

The most typical energy-driven interactions are those which convert electrical energy to mechanical energy. The manifestation of energy in the mechanical domain takes the form of motions which can be altered by using a mechanism. A mechanism is not a machine. An example of a familiar mechanism is the bicycle. A formal definition of mechanism is as follows:

Definition 1.2 A mechanism is a set of (mechanical) elements arranged in certain configurations for the purpose of transmitting motions in a pre-determined fashion.

Motion is the visible form of mechanical energy. An element which converts electrical energy into motion is called an *electric motor or actuator*. If a system includes at least one element intended for energy conversion, this system is called a *machine*. A typical example of a machine is a motorcycle or car. A formal definition of machine can be stated as follows:

Definition 1.3 A machine is a super-set of mechanism(s), and contains elements which supply energy to drive this mechanism(s).

In mathematics, motion is fully described by the parameters: position (p), velocity (v), and acceleration (a). These motion parameters (p, v, a) are important input variables for many manufacturing processes. A large family of manufacturing equipment among the variety which exists, is the one which supplies the motion required by a manufacturing process, such as: arc-welding, spray painting, assembly, cutting, polishing, deburring, milling, drilling etc. Of this class of equipment, an increasingly popular type is the industrial robot.

The English word *robot* was derived from the Czech word *robota*, meaning *forced workers*. The word *robot* became popular in 1921 because of a play named "Rossum's Universal Robots" by Czechoslovakian writer, Karel Kapek. In the play, a scientist called Rossum created human-like machines that revolted, killed their human masters, and took control of the world.

The American company, Unimation Inc., founded in 1962 by Joseph Engelberger and George Devol, was the first company to actually produce industrial robots.

From an engineering point of view, a robot is the embodiment of manipulative, perceptive, communicative, and cognitive abilities in an artificial body, which may or may not have a human shape. For example, industrial robots are merely a combination of an arm and a hand. According to the Robot Institute of America, the formal definition of industrial robot is as follows:

Definition 1.4 A robot is a programmable, multi-functional manipulator designed to move material, parts or specialized devices through variable programmed motions for the performance of a variety of tasks.

However, in view of the evolution of the robot into a sophisticated mechanical, electronic, controlling, informative, perceptive, and cognitive system, a new definition of robot is necessary:

Definition 1.5 A robot is the embodiment of manipulative, locomotive, perceptive, communicative and cognitive abilities in an artificial body, which may or may not have a human shape. It can advantageously be deployed as a tool, to *make things* in various environments.

Nowadays, the robot is not only gaining more popularity in industry, but also slowly entering society, in the form of humanoid or animal-like entertainment robots. Figs 1.4 − 1.7 show some prototypes developed by Japanese companies. Accordingly, it is necessary to concisely define the term *humanoid robot*. A formal definition of humanoid robot is as follows:

Fig. 1.4 PINO: An open development platform for a desktop humanoid robot. Photo by Author.

Definition 1.6 A *humanoid robot* is the embodiment of manipulative, locomotive, perceptive, communicative and cognitive abilities in an artificial body *similar to that of a human*, which possesses skills in executing motions with a certain degree of autonomy, and can be advantageously deployed as *agents* to *perform tasks* in various environments.

In this book, *robot* refers to both industrial and humanoid robots.

In the manufacturing industry, tasks or processes which can typically be accomplished by using robots include:

- Welding:
  This is the process of joining two work-pieces together by applying molten weld metal. For spot welding, the important motion parameter is position; for arc welding, an additional important motion parameter is the speed of travel.
- Cutting:
  This is the process of applying thermal or mechanical energy to cut a

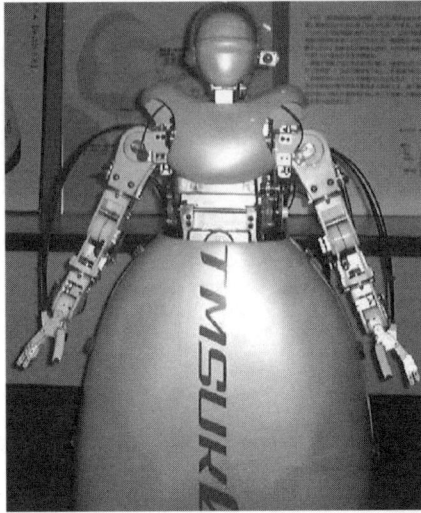

Fig. 1.5   TMSUK's humanoid robot prototype. Photo by Author.

work piece into a specific shape. In this process, the important motion parameters are position and velocity.

- Assembly:
  This is the process of either adding components to form a single entity, or affixing components to a base unit (e.g. to place components on a printed circuit board). In this process, the important motion parameter is position.

- Material Handling:
  This is the process of either packaging parts into a compartment (box) or loading/unloading parts to/from another station. In this process, position is an important motion parameter.

1.4   Impact of Industrial Robots

The industrial robot, a combination of arm and hand, can advantageously be deployed in the manufacturing industry to automate many processes, which have motion parameters as input. Obviously, automation using robots shifts the use of human labor from direct interaction with a process to various indirect interactions. These indirect interactions include process monitoring, process diagnostics, equipment setting, equipment program-

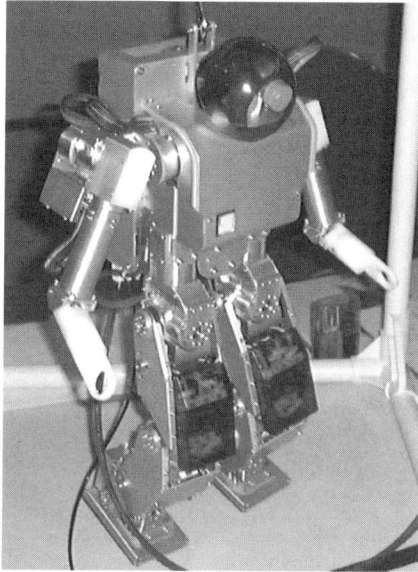

Fig. 1.6   FUJITSU's humanoid robot prototype. Photo by Author.

ming, the development of automation solutions etc.

Despite all these new job opportunities, one major concern in the manufacturing industry is that the proliferation of robots will cause displacement of human labor and eventually, unemployment. This is a real problem, faced by both developed and developing countries. There is no simple solution to this issue. However, it can be addressed at various levels:

- *Individual Level:*
  It must be clear that we are living in a changing environment. We are now witnessing the third wave of the industrial revolution: Information Technology. The tools for us to make things are constantly evolving with increased complexity and sophistication. Therefore, individuals must constantly learn new knowledge and skills in order to make use of the modern tools being deployed in the industry.
- *Industrial Level:*
  Perhaps, there should be an adjustment in the company goals . The initial goal of a company is to earn profits for the shareholders through the manufacturing and supply of goods and/or services. It is clear that a second dimension should be added to this initial goal. This second dimension is the social responsibility of the company to constantly shape

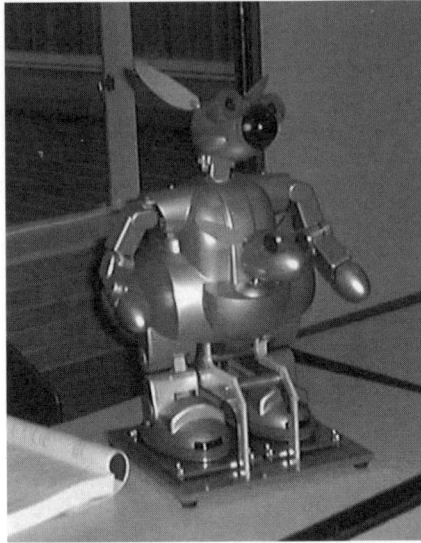

Fig. 1.7   SHARP's Animal-like entertainment robot prototype. Photo by Author.

the knowledge and skills of its employees into a pattern, which will keep them relevant to the current market condition and the evolution of the industry (or emergence of new industry).

- *Education Level*:
Perhaps, we should re-emphasize the true nature of education. It should be a process of developing brain power in mastering knowledge and acquiring skills, on top of understanding theories in a specific field of science and technology. The development of brain power should be considered more important than the memorization of theoretical facts and data. For an individual to be adaptable to a changing environment or job market, which is increasingly dominated by knowledge and skill, it is necessary that he/she adequately develop his/her brain power in philosophy, physics, mathematics, and computing.

- *Social Level*:
The deployment of robots in the manufacturing industry will undoubtedly increase the productivity and quality of goods. This, in turn, will generate more wealth for companies and society. We should make sure that the re-distribution of wealth, generated by the deployment of robots, is wisely regulated, in order to create more jobs and activities for human beings in other employment sectors (i.e. service, entertain-

ment, sports, arts, healthcare, education, research etc.)

Despite social concerns about the displacement of human labor, the wide use of robots in the manufacturing industry will certainly have positive impact, as well:

- *Productivity*:
  It is impossible to achieve a high production yield using human labor, because the biological system cannot deliver continuous physical effort without rest. There is no doubt that automation with robots will increase productivity.
- *Flexibility*:
  The winning characteristics in manufacturing today are volume, price, quality and service. Automation using robots will add a fifth characteristic: choice. In other words, the ability to supply customized products or services in a timely manner. A company which wishes to offer choices to its customers must have flexibility in configuring and operating its production line. Therefore, flexible automation or agile manufacturing is the key to success in today's manufacturing industry. The robot, being re-programmable equipment capable of executing various motions within a reasonably large working space, is undoubtedly one of the best types of equipment for flexible automation.
- *Quality of Products*:
  Humans are not capable of making things with a consistent degree of accuracy. For example, our vision does not make metric measurements when performing tasks. And, without accurate visual guidance, we cannot perform motions with any reasonable degree of accuracy or replication. Robots can not only execute accurate motions repeatedly, but are immune to the emotional states which affect human's performance.
- *Quality of Human Life*:
  The use of robots can free humans from doing dirty, dangerous, and difficult jobs.
- *Scientific Discipline*:
  The proliferation of robots in industry reveals the importance of *robotics*, the study of robots as a scientific discipline for education and research. Indeed, robotics is an important subject in engineering, and is being widely taught in any university having an engineering department.

1.5   Impact of Humanoid Robots

The enhanced locomotive ability of humanoid robots will increase the range of the manipulative function. With the tight integration of perceptive ability, a humanoid robot will gain a certain level of autonomy by interacting with human masters or executing tasks intelligently. The inclusion of cognitive ability will make a humanoid robot a physical agent, capable of self-developing its mental and physical capabilities through real-time interaction with the environment and human masters. There is no doubt that the emergence of humanoid robots will have great impact on many aspects of our modern society.

1.5.1   *Industry*

The enhanced and human-like locomotive ability of the humanoid robot makes it possible to deploy humanoid robots to places where humans are still working like machines. Within the manufacturing industry itself, it is foreseeable that the emergence of the humanoid robot will certainly automate more tasks, such as: maintenance, diagnostics, security etc. We can even imagine that an un-manned factory may become a reality one day. It is clear as well, that the humanoid robot will also benefit other industries. For example, construction of buildings or houses can be considered as an assembly process, which could automatically be completed by specialized teams of humanoid robots under the supervision of human masters.   In the healthcare industry, humanoid robots could provide great service in the rehabilitation of patients. Additionally, humanoid robots could be of assistance in hospitals for certain tasks (e.g. precision surgery).

1.5.2   *Society*

Up until today, the greatest consumer product has undoubtedly been the automobile. It provides us with unprecedented mobility, increases our sense of social status, and offers us high-quality, in-vehicle entertainment experiences among other things. Without the automobile, there would be no efficient transportation. Now, the question is: What will be the next great consumer product? If we look at the evolution of the computer as a consumer product, we may discover the following trends:

- The microprocessor is becoming smaller, while computational power is increasing.

- The microprocessor is consuming less electrical power, despite the increased computational power.
- The computer has more functions and better performance while the price is constantly decreasing.
- The computer is constantly improving its cognitive abilities: soft computing (i.e. computation with words rather than digital logic), speech recognition, auditory communication, visual communication, visual perception etc.
- The computer has the ability to express inner emotional states through visible motions (e.g. facial expressions).

The evolution of the computer coincides with the evolution of the robot, which is characterized by the following clear trends:

- The robot's arm is becoming smaller, yet it has increased manipulative ability.
- The robot's hand is becoming more compact, yet it has increased dexterity and sensing ability.
- The robot has a head, which may influence the design of future computer's monitor, and provide the means for facial expressions and perceptive sensory input and output.
- The robot has legs to perform human-like, biped walking.
- The robot is becoming intelligent. The incorporation of demanding computational power and cognitive abilities enables the robot to perform real-time interaction with its environment and humans.
- The robot is becoming the perfect embodiment of an artificial body with artificial intelligence.

The parallel development of the computer and robot industries, each of which has its own group of manufacturers with sound and profitable track records, will contribute to the emergence of the humanoid robot as the next great consumer product. Undoubtedly, this will bring new benefits to consumers, such as:

- *Robot-assisted Entertainment*:
  Computerized games are very popular. These are basically pre-programmed interactions between users and artificial creatures in a virtual world. With the humanoid robot, or an animal-like robot, computerized games will take on a new dimension: a pre-programmed interaction with artificial creatures in real space. Moreover, with the humanoid robot, certain animations or tours could also be automated

in places, such as museums, hotels, shopping centers, theme parks, film studios etc.

- *Robot-assisted Healthcare at Home*:

  Healthcare is expensive. One effective way to lower the cost is to increase the accessibility of healthcare services. This would be possible, if humanoid robot technology advanced to the stage where we could deploy humanoid robots to homes, both locally and in remote areas. By simply activating the appropriate pre-coded programs, a humanoid robot could instantaneously be configured as an "expert" in the necessary medical field. (e.g. dentist). Therefore, some pre-hospital diagnostics or treatments could be done at home or in a neighborhood community center, at a much lower cost.

  In addition, the humanoid robot would provide the means for a human medical expert to diagnose and treat a patient at a remote location, or even a remote village, through tele-presence. With the humanoid robot, it is possible for us to envision in the future the concept of a virtual hospital.

- *Robot-assisted Education at Home*:

  In our modern society, parents are heavily occupied with their professional activities. There is limited time for parents to contribute to the educational process of their children. In order to compensate for this deficiency, more and more parents rely on private tutors to complement schooling. In the future, an alternative to private tuition may be the use of humanoid robot-tutors, with pre-programmed and selectable knowledge and skills. One obvious advantage to using the humanoid robot as a tutor is the ability for the same humanoid robot to be configured as a tutor in different disciplines, as well as at different skill and knowledge levels. As a result, robot-assisted education is not only appropriate for children, but also relevant to the continuing education of adults. With the humanoid robot, it is possible to envision a virtual university for life-long learning at home.

- *Robot-assisted Tele-existence*:

  The human body is not only a collection of sub-systems for output functions, but also a collection of sub-systems for input of sensory information. The humanoid robot is also a complex system, having output and input functions, and is an ideal platform to extend the reach of a human's output functions, for example, to a remote location. It can act as a "complex sensor" placed at a remote location, feeding back information to humans who can then feel, experience, interpret, respond etc. This is what the commonly called *tele-existence*. With the humanoid robot, it may be possible one day for humans to travel or shop in a remote place without going anywhere. This type of tele-existence will certainly enrich our lives.

- *Mechanized Buddy*:
  More and more researchers in the community of robotics and artificial intelligence are studying developmental principles, underlying the self-development of mental and physical abilities through real-time interaction between the environment and humans. This research will have a direct impact on the manufacturing of smart artificial animals (e.g. SONY's AIBO, a type of mechanized dog). The animals of the future will possess certain degrees of autonomy to self-develop emotional, affective and cognitive functions through real-time interaction with humans. It will not be long before we will see the market flourishing with smart toys or animals.

1.5.3 *Space Exploration and Military*

The humanoid robot is an ideal platform for tele-presence or tele-existence. Because of this, it is easy for us to imagine its applications in space exploration and the military. It would be possible to dispatch a space shuttle commanded by humanoid robots, yet controlled by humans at a ground station on Earth. It would also be possible to assign humanoid robots to maneuver military land, air, or underwater vehicles in a battle field without risk of human casualty. In the battle against terrorism, humanoid robots could help to prevent, dissuade, and rescue.

1.5.4 *Education*

Today's engineering education emphasizes the integrative aspect of various engineering disciplines, or Mechatronics. Mechatronics originated in Japan in the 1970s and is gaining worldwide attention because of the importance of integrating different, but inter-related engineering disciplines. A formal definition of Mechatronics can be stated as follows:

Definition 1.7 Mechatronics is the study of the synergetic integration of physical systems with information technology and complex decision-making in the design, analysis and control of smart products and processes.

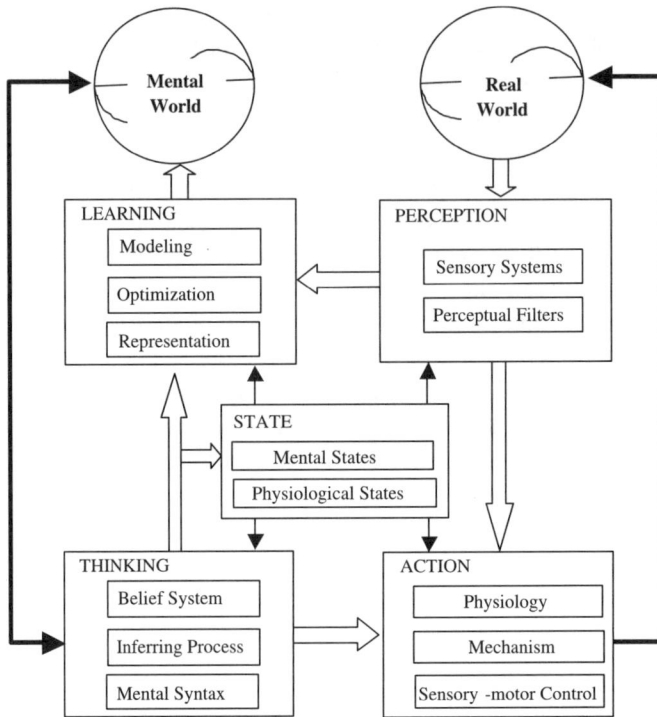

Fig. 1.8   A proposed framework for human intelligence which encompasses four constituent elements: perception, learning, thinking and action.

The humanoid robot, being a combination of mechanical, electronic, control, information, perceptive and cognitive systems, is an ideal platform to illustrate the essence of Mechatronics.

1.5.5   *Research*

Besides education, which propels research on the humanoid robot, two additional forces which simultaneously stimulate research in this fascinating field are: a) computational neuroscience and b) intelligent (physical) systems.

Computational neuroscience is the study of how the human brain plans and controls behaviors to achieve desired results (i.e. outcome). By definition, a "behavior" is a sequence of actions with a specific syntax (i.e. ordered arrangement). Human life is an unceasing journey of achieving desired results through actions which are decided as a result of our internal representations of the real world, our mental state-of-mind, and our physiological state. An important goal of computational neuroscience is to develop computational theories and models from an understanding of the brain's functions in these important areas. (See Fig. 1.8 for illustration).

- *Perception*:
  Humans have five sensory systems, namely: visual, auditory, kinesthetic, gustatory, and olfactory. These sensory systems supply input to the brain to build the personalized internal representations of the external world. In addition, mental and physical states affect the way the brain reacts to certain sensory input.

- *Learning*:
  The human brain has the ability to process filtered sensory data to derive structured representations which form our mental world, in terms of knowledge and skills. Generally speaking, *knowledge* describes the relationships between causes (e.g. stimuli, actions, transformations, conditions, constraints etc) and effects (e.g. results, facts, situations, conceptual symbols etc). However, *skill* describes the association of behaviors (i.e. the ordered sequence of actions) with results. Language is undoubtedly an important component in learning visual, auditory, and kinesthetic representations of the external real world. And, *Learning* is a process for us to gain not only knowledge, but also skills which result in actions. Hence, action is the source of results. We all have similar bodies and neurological systems. If we undertake the same actions under the same mental and physiological states, we can achieve similar results. This form of learning encourages us to effectively imitate successful people.

- *Thinking*:
  Human beings are capable of not only communicating with the real world, but also communicating with the internal, mental world. We all have our own mental syntax (i.e. ordered sequence of mental actions) which conditions our thinking process. Simply speaking, *thinking* is a process of associating causes with, or disassociating them from effects. The thinking process is dictated by our belief system, because *belief* is pre-formed and pre-organized predictions of achievable outcomes. Our belief system also determines the configuration of our mental and physiological states. For example, the belief that one is resourceful places a person in a different state-of-mind than the belief that one is

miserable.

- *Action*:

    The human body is capable of performing actions, driven by our sensory-motor systems. The ordered actions are behaviors acting on the real world for the achievement of desired results. The performance of our sensory-motor systems obviously depends on the body's mechanisms, as well as physical energy which is affected by our mental state. For instance, being vital, dynamic and excited certainly releases more physical energy than being depressed, scared or uninterested etc.

Intelligence and the ability to produce intended results are unique attributes associated with humans. Now, here comes the question of human intelligence. How do we define intelligence? From an engineering point of view, it is constructive to form an objective, precise definition of "intelligence" to prevent it from being misused. One possible definition of intelligence is as follows:

Definition 1.8   Intelligence is the ability to link perception to actions for the purpose of achieving an intended outcome. Intelligence is a measurable attribute, and is inversely proportional to the effort spent in achieving the intended goal.

The study of computational neuroscience not only helps us have a better understanding of the brain's functions, but also helps guide the engineering approaches in the development of artificial intelligence. However, human-inspired artificial intelligence must be tested by an artificial body. The humanoid robot is an ideal testing ground for computational theories or models derived from the study of the human brain's mechanisms in perception, learning, thinking, and action.

On the other hand, *intelligent (physical) system* is the study of computational principles for the development of perception, learning, decision-making and integration in an artificial body. An artificial body requires artificial intelligence in order to adapt to a changing environment for the purpose of better performing the assigned tasks. Undoubtedly, the humanoid robot is a perfect research platform for the study of the embodiment of artificial intelligence with an artificial body.

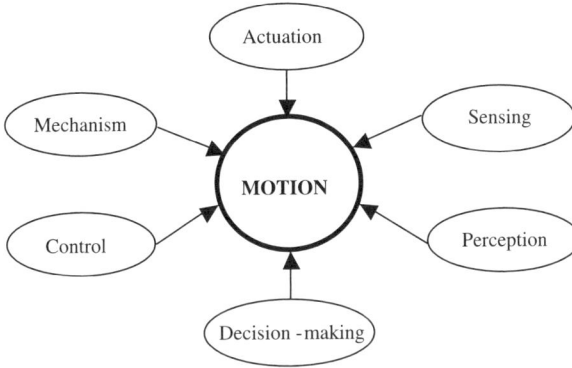

Fig. 1.9   A set of related topics in robotics with a motion-centric theme.

1.6   Issues in Robotics

Robotics, or the study of robots, is an engineering discipline. Functionally, a robot is a physical agent which is capable of executing motion for the achievement of tasks. A robot's degree of autonomy depends on its ability to perform the ordered sequence of perception, decision-making and action.

As we know, a robot's dynamics in motion execution is dictated by mechanical energy consumption in connection with kinematic constraints imposed by the robot's mechanisms. Literally, the definitions of kinematics and dynamics are as follows:

Definition 1.9   Kinematics is the study of motion without consideration of force and torque, while dynamics is the study of motion in relation to force and torque.

Therefore, the unifying concept in robotics is *motion*, being a visible form of action. As illustrated in Fig. 1.9, the major issues in robotics will logically include:

1.6.1   *Mechanism and Kinematics*

From a mechanical point of view, a mechanism is a set of linkages without an actuator. The purpose of a mechanism is to impose kinematic constraints on the types of motion the mechanism can deliver at a particular point. By default, this particular point is at the tip of an end-effector.

In general, a mechanism consists of joints and links. In robotics, a *link* is a rigid body inside a mechanism, while a *joint* is the point of intersection between any pair of adjacent links. Any changes in the relative geometry among the links will induce a specific type of motion. Therefore, it is important to study the relationship between the motion parameters of the linkages and the motion parameters of a particular point on the mechanism. This study is the object of *robot kinematics*. There are two problems with robot kinematics:

- How do we determine the motion parameters of a particular point on the mechanism from the knowledge of the motion parameters of the linkages? This is commonly called the *forward kinematics* problem.
- How do we determine the motion parameters of the linkages necessary to produce a desired set of motion parameters at a particular point on a mechanism? This is known as the *inverse kinematics* problem.

1.6.2   *Actuation Elements and Dynamics*

In the mechanical domain, any motion is produced by the conversion of mechanical energy. The study of the relationship between motion parameters and force/torque is the object of *robot dynamics*. This discipline aims at determining the equations of motion imposed by force and torque on a robot's mechanism.

Mechanical energy can be generated either through gravitational force or some other form of energy. Today, the most widely available source of energy is electrical energy. Hence, it is important to study the conversion from electrical energy to mechanical energy. This leads to the study of electric motors, which use the electro-magnetic principle to induce mechanical movement from the interaction of two independent magnetic fields. A magnetic field can be generated by either a permanent magnet or electromagnet.

Being able to convert electrical energy into mechanical energy is important but not sufficient. We must also be able to modulate the mechanical energy being produced. There are two ways to modulate the final output of mechanical energy. It is common at the output side to use a transmission mechanism which is a combination of reducers, tendons, gears and bearings. It is common at the input side to use a power amplifier to regulate the amount of electrical energy, which is converted into the corresponding mechanical energy.

1.6.3 *Sensing Elements*

From an engineering point of view, any system in a real environment is invariably subject to certain types of noise, uncertainty and disturbance. Therefore, no mathematical description of the input and output relationship is ever exact. Sensing is the only way to obtain the actual output values.

In robotics, sensing elements are used to convert physical quantities such as the motion parameters of actuators into corresponding electrical signals. Output from a sensing element provides feedback on the motion parameters regarding the linkages inside a robot's mechanism.

Besides the sensing of motion parameters, there are two other requirements for sensing: a) the measurement of the interaction force/torque between a robot's end-effector and its environment, and b) the motion parameters of a robot's workpieces and workspace. (This latter will be the object of study when we discuss the robot's visual-perception system).

1.6.4 *Control*

The robot's intrinsic dynamics are described by the relationship between the motion parameters of a robot's end-effector and the forces/torques applied to the robot's mechanism. Since a robot is designed to perform tasks through the execution of motions, the dynamics of a robot must closely (if not exactly) follow the desired dynamics imposed by a task. How to determine the desired dynamics from a given task will be studied under the topic of motion planning.

In general, it is impossible to design a robot which has intrinsic dynamics that meets the intended dynamics of any given task. Therefore, it is desirable to have a mean to alter the robot's intrinsic dynamics externally. In engineering, the discipline of automatic-feedback control is the study of methods and tools for the analysis and synthesis of system dynamics. The beauty of the automatic-control theory is its versatility. You can alter a system's intrinsic dynamics through the insertion of a control element (controller) in a closed-feedback loop so that the system's actual dynamics meet a wide range of specifications, in terms of stability, time responsiveness, and output accuracy.

1.6.5 *Information and Decision-Making*

Today, it would be inconceivable to attempt to build an automatic machine without a brain. The hardware aspect of a machine's brain can be as simple as a micro-controller, or a microprocessor having a certain level of computational power. The first function of a robot's brain is to perform computational tasks, such as sensory-data processing (including visual perception) and the mathematical computations underlying robot kinematics, robot dynamics, and robot control.

The second function of a robot's brain is to support not only the interaction between the robot and human master, but also the communication between the robot and the outside world. The third and most important function of a robot's brain is to implant computational algorithms of artificial intelligence. One important aspect of a robot's intelligence is the ability to plan tasks, actions and motions without human intervention.

In fact, research on developmental principles for acquiring perceptive, thinking and acting skills is aimed at developing a skill-based approach for the autonomous planning of task, action and motion. This is because associating a sequence of ordered actions/motions with the corresponding results is a skill which can be learned with the help of linguistic processing and programming.

1.6.6   *Visual Perception*

Vision is the most important sensory channel for humans (as well as animals). Without vision, our abilities to act and learn would be tremendously weakened. In a similar manner, vision plays a vital role in any machine which intends to perform autonomous motions, actions, tasks, or even behaviors.

In robotics, the performance of a task implies that a robot's actual dynamics closely follows the dynamics required by the task itself. The actual dynamics of a robot, manifested in the form of visible motions, can be measured in two ways. The first approach is to use sensors to measure the motion parameters of the linkages inside the robot's mechanism. Subsequently, it is possible to use forward kinematics to derive the motion parameters at a particular point (e.g. end-effector) on the mechanism. The second approach is to use artificial vision to measure the motion parameters at a particular point on the robot's mechanism. This ability to provide accurate visual feedback to the control loop of motion execution constitutes the first function of a robot's visual-perception system.

The second important function of a robot's visual perception system is to describe an action in two-dimensional (2-D) image space instead of expressing it in three-dimensional (3-D) task space. This is a crucial step

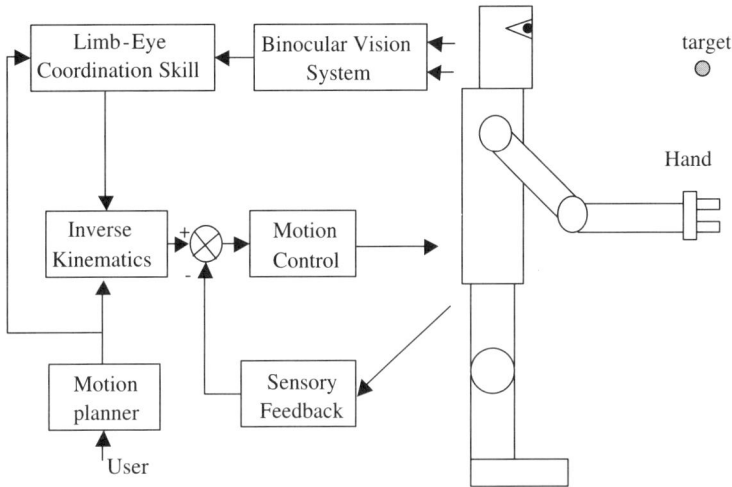

Fig. 1.10 A Framework for Limb-Eye Coordination Behavior.

towards achieving the autonomous execution of a given action.

The third important function of visual perception is to infer the 2-D and/or 3-D geometry of a robot's workpieces and workspace as a necessary step towards the robot's automated planning of an action/task. This is because the automatic generation of an action/task plan requires knowledge of the geometric models of the workspace and workpieces.

The fourth important function of visual perception is to provide visual input to support a robot's learning process.

Finally, if one considers image-based communication/interaction a special skill, it can be regarded as the fifth function of visual perception.

As we know, action is the source of results. With well-developed perception capability, the limb-eye coordination behavior, as illustrated in Fig. 1.10, will be achievable. Of these behaviors, the most impressive ones will be: a) vision-guided manipulation (i.e. hand-eye coordination), b) vision-guided positioning (head-eye coordination), and c) vision-guided locomotion (leg-eye coordination). Chapter 9 will not only cover well-known, pure engineering principles, but also report the latest findings on human-like, engineering approaches to limb-eye coordination behavior.

1.7   Exercises

(1) Describe two key factors which pushed manufacturers to adopt automation.
(2) What is a manufacturing process?
(3) Identify some critical parameters of an arc-welding process.
(4) What is a robot ? What is robotics?
(5) How do you shape the motions exhibited by a machine?
(6) What is kinematics? What is dynamics?
(7) What is behavior? What is intelligence?
(8) How do you address the issue of unemployment caused by increased automation?
(9) What is the possible impact the humanoid robot will have on industry and society?
(10) Explain the key topics in robotics.

1.8   Bibliography

(1) Bowyer, K. and L. Stark (2001). Special Issue on Undergraduate Education and Computer Vision, *International Journal of Pattern Recognition and Artificial Intelligence*, **15**, 5.
(2) Craig, J. (1986). *Introduction to Robotics: Mechanics and Control*, Addison-Wesley.
(3) DeVor, R. E., T. H. Chang and J. W. Sutherland (1992). *Statistical Quality Design and Control*, Prentice-Hall.
(4) Fu, K. S., R. C. Gonzalez and C. S. G. Lee (1987). *Robotics: Control, Sensing, Vision and Intelligence*, McGraw-Hill.
(5) Groover, M. P. (1996). *Fundamentals of Modern Manufacturing*, Prentice-Hall.
(6) Knoll, A., G. Bekey and T. C. Henderson (2001). Special Issue on Humanoid Robots, *robotics and Autonomous Systems*, **37**, 2-3.
(7) Rehg, J. A. (1997). *Introduction to robotics in CIM Systems*, Prentice-Hall.
(8) (2001). *Proceedings of International Conference on Humanoid Robots*, Tokyo, Japan.

Chapter 2

Motion of Rigid Bodies

2.1 Introduction

Motion is the unifying theme in robotics. Motion is the visible form of actions for the achievement of results in a real and physical world. From a physics point of view, a motion is the result of force or torque applied to a body. For example, linear acceleration is generated when force is applied to a body which is free to move in the direction of the force (Newton's Law). If a body is fixed at a particular point, an applied force will cause the body to rotate about an axis which goes through the point.

Therefore, it is important to understand the mathematical description of motion in terms of displacement, velocity and acceleration. This will also help as a sound basis for the study of some important topics in robotics, such as: motion generation (i.e. a robot's mechanism and actuation), motion planning & control (i.e. decision-making and control), and visual perception of motion. From a mechanical point of view, a robot is an assembly of links that are normally of a rigid body. (See Chapter 3.) Thus, in this chapter, we will restrict our discussion to the motion of a rigid body.

2.2 Cartesian Coordinate Systems

In order to describe the motions of a rigid body, we need to define two references: a) the time reference for velocity and acceleration, and b) the spatial reference for position and orientation. A time reference is described by the one-dimensional axis t. The values of time variable t depend on our common clock system which is unchangeable, and is applicable to all the real and physical systems on Earth.

A common approach for assigning the spatial reference to our physical three-dimensional (3-D) world is to define an orthogonal reference system formed by X, Y and Z axes. These three axes are mutually perpendicular to one another and intersect at a common point called *the origin*, and denoted by O. In a spatial reference system, an axis serves as a reference to measure the position of an object along it. In order to obtain a measurement with a physical meaning (in terms of meters, centimeters etc.), we must define a unit of measurement for each axis in the spatial reference system. This is normally done by associating a unit vector to each axis. This unit vector is called a *base vector*. Therefore, a Cartesian coordinate system can be defined as follows:

Definition 2.1   Any spatial reference system with base vectors that are mutually perpendicular to one another is called a *Cartesian coordinate system* (or coordinate system for short). Any space with a Cartesian coordinate system as its spatial reference system is called a *Cartesian Space*.

In a (Cartesian) coordinate system $O_0 - X_0Y_0Z_0$ having the assigned base vectors $(\vec{i}, \vec{j}, \vec{k})$ where

$$\vec{i} = \begin{pmatrix} 1 \\ 0 \\ 0 \end{pmatrix} \qquad \vec{j} = \begin{pmatrix} 0 \\ 1 \\ 0 \end{pmatrix} \qquad \vec{k} = \begin{pmatrix} 1 \\ 0 \\ 0 \end{pmatrix}$$

the position of point P on a rigid body (see Fig. 2.1) is represented by a vector $r = (x, y, z)^t$, where

- x is the distance to the origin, measured along the base vector $\vec{i}$,
- y is the distance to the origin, measured along the base vector $\vec{j}$,
- z is the distance to the origin, measured along the base vector $\vec{k}$.

These (x, y, z) are called the Cartesian coordinates (or coordinates for short). Accordingly, position vector r of point P can be written as:

$$r = x \bullet \vec{i} + y \bullet \vec{j} + z \bullet \vec{k}. \tag{2.1}$$

Interestingly, Eq. 2.1 can also be written in a matrix form, as follows:

$$r = [\vec{i},\ \vec{j},\ \vec{k}] \bullet \begin{pmatrix} x \\ y \\ z \end{pmatrix} = R \bullet r \tag{2.2}$$

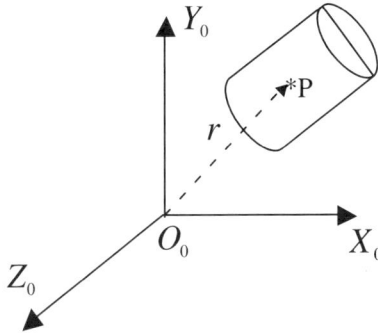

Fig. 2.1   A spatial reference system for a rigid body.

where

$$R = [\vec{i},\ \vec{j},\ \vec{k}] = \begin{pmatrix} 1 & 0 & 0 \\ 0 & 1 & 0 \\ 0 & 0 & 1 \end{pmatrix}.$$

The equation $r = R \bullet r$ means the mapping by matrix R, which is applied to coordinate system $O_0 - X_0 Y_0 Z_0$. This concept of mapping can be applied to any pair of coordinate systems in a Cartesian space. It is an important concept, which is simple and helpful in understanding robot kinematics.

2.3   Projective Coordinate Systems

We can manipulate a geometric object in a Cartesian space in different ways, such as translation, rotation, scaling, and projection. The projection of a geometric object along a set of straight lines has interesting properties and is useful in practice. One typical application of projection is the display of three-dimensional objects onto a two-dimensional screen. Another typical application is visual perception, which aims at inferring the three-dimensional geometry of an object from its two-dimensional projection onto an image plane.

Here, let me briefly introduce the concept of projective coordinates for the purpose of explaining the meaning of *homogenous coordinates*, which are useful for concisely describing general motion transformations by 4×4 matrices.

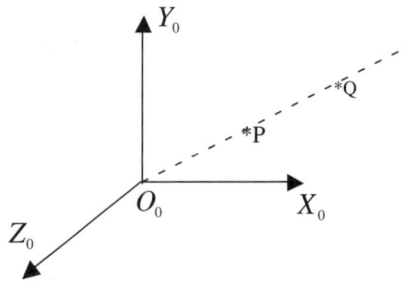

Fig. 2.2   Illustration of projective coordinate systems.

Let $P = (x, y, z)^t$ be a point in a spatial reference system (e.g. a three-dimensional Cartesian coordinate system). Refer to Fig. 2.2. We draw a straight line which passes through the origin and point P. It turns out that the coordinates of any arbitrary point Q on this straight line can be written as

$$Q = (k \bullet x, k \bullet y, k \bullet z)^t$$

where k is a scaling variable. When $k = 1$, $Q = P$. Then we can form a new reference system represented by (X, Y, Z, k). This new reference system is called the *projective coordinate system*. Accordingly, the coordinates $(X, Y, Z, k)^t$ are called the *projective coordinates*. The conversion from projective coordinates $(X, Y, Z, k)^t$ to Cartesian coordinates $(x, y, z)^t$ is done in the following way:

$$x = \frac{X}{k} \quad y = \frac{Y}{k} \quad z = \frac{Z}{k}. \tag{2.3}$$

When $k = 1$, the Cartesian coordinates can equivalently be represented by their projective coordinates. Thus, we call $(x, y, z, 1)^t$ the *equivalent projective coordinates* with respect to the corresponding Cartesian coordinates $(x, y, z)^t$. In robotics, element k is commonly called the *homogenous coordinate*. In the following, we will make use of the equivalent projective coordinates to uniformly describe translational and rotational motion transformations in a matrix form.

2.4   Translational Motions

In a Cartesian space, any complex motion of a rigid body can be treated as a combination of translation and rotation. The physical meaning of translational motion is motion along a straight line. Therefore, the orientation of a rigid object will remain unchanged if the motion is translational.

2.4.1 *Linear Displacement*

A translational motion can be described by a linear displacement vector, a linear velocity vector (if any), and a linear acceleration vector (if any). These entities are equal at all points on a rigid body.

It is worth noting that linear acceleration is the result of a force applied to a rigid body. Since motions at all points on a rigid body are equal, it is sufficient to choose one point as representative for the study of the motion caused by a force or torque. However, as one point in a three-dimensional Cartesian space does not represent any orientational information, it is necessary to assign two directional vectors to a point on a rigid body in order to represent the variations in orientation caused by a rotational motion. For the sake of consistency, it is common to assign, to each rigid body, an orthogonal coordinate system located at a chosen point. In this way, the study of a rigid body's motion becomes the study of the relative motion between its coordinate system and a reference coordinate system. In robotics, the orthogonal coordinate systems are commonly called *frames*.

Example 2.1   Fig. 2.3 shows the assignment of frame $O_1 - X_1 Y_1 Z_1$ to a rigid body. If this body undergoes a translational motion described by displacement vector $T = (t_x, t_y, t_z)^t$ within time interval $[t_1, t_2]$, the position vector of the origin O_1 at time instant t_2 will be

$$O_1(t_2) = O_1(t_1) + T. \qquad (2.4)$$

Thus, the origin of frame $O_1 - X_1 Y_1 Z_1$ can serve as a representative point on the rigid body for the study of its motion with respect to any other frame.

◇◇◇◇◇◇◇◇◇◇◇◇◇◇◇◇◇◇◇

With more than one frame in a Cartesian space, a point's position vector may have different values for its elements, depending on the choice of the reference frame. For the sake of clarity, when dealing with mathematic symbols, it is necessary to explicitly indicate the frame which a vector refers to. In robotics, there is no unique way to do this. One approach is to use a superscript before each symbol to indicate the reference frame. Another approach is to use a superscript after the symbol. Here, we adopt

(a) Before translation   (b) After translation

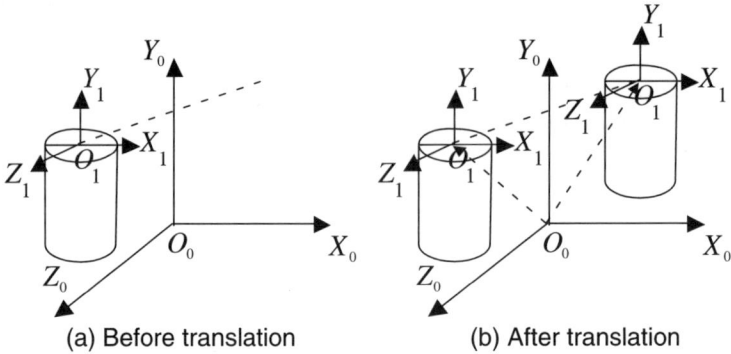

Fig. 2.3   Frame assignment to a rigid body, and an illustration of a translational motion.

the first approach. For example, ^0P means position vector P referenced to frame 0.

Example 2.2   Refer to Fig. 2.3. The position vector of origin O_1 with respect to frame 0 is denoted by $^0O_1$, while the same position vector with respect to its own frame 1 is denoted by $^1O_1$. Since $^1O_1 = (0,0,0)$ and this position vector with respect to its own frame remains unchanged (i.e. frame 1 is fixed on the rigid body), we have

$$^0O_1 = {}^1O_1 + {}^0O_1. \tag{2.5}$$

If we define $^0T_1 = {}^0O_1$, the above equation becomes

$$^0O_1 = {}^1O_1 + {}^0T_1. \tag{2.6}$$

◇◇◇◇◇◇◇◇◇◇◇◇◇◇◇◇◇◇◇

This example clearly shows that vector $^0T_1$ describes the relative distance (or displacement) between the origin of frame 0 and the origin of frame 1. In other words, it describes the translational motion of frame 1 with respect to frame 0, if we imagine that frame 1 initially coincides with frame 0 at time instant t_0. This vector is called the *translation vector*.

In general, the translational motion of frame j with respect to frame i can be denoted by ^iT_j.

Example 2.3   Refer to Fig. 2.4. We denote $^0T_1(t) = (t_x, t_y, t_Z)^t$ the translational motion transformation from frame 1 to frame 0, at time instant t. If the coordinates of point P on a rigid body with respect to its own frame (i.e. frame 1) are $^1P = (x_1, y_1, z_1)^t$, the position vector of the same point P with respect to frame 0 will be

$$^0P(t) = (x_0, y_0, z_0)^t = \;^1P + \;^0T_1(t).\qquad(2.7)$$

In Eq. 2.7, position vector ^1P is a constant vector because frame 1 is fixed on the rigid body.

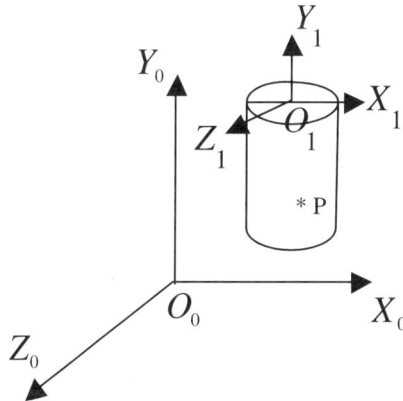

Fig. 2.4 Translational transformation from frame 1 to frame 0.

◇◇◇◇◇◇◇◇◇◇◇◇◇◇◇◇◇◇◇◇

Interestingly enough, if we use the equivalent projective coordinates to substitute for the Cartesian coordinates in the above example, Eq. 2.7 can be expressed in a matrix form as follows:

$$\begin{pmatrix} x_0 \\ y_0 \\ z_0 \\ 1 \end{pmatrix} = \begin{pmatrix} 1 & 0 & 0 & t_x \\ 0 & 1 & 0 & t_y \\ 0 & 0 & 1 & t_z \\ 0 & 0 & 0 & 1 \end{pmatrix} \bullet \begin{pmatrix} x_1 \\ y_1 \\ z_1 \\ 1 \end{pmatrix}\qquad(2.8)$$

or

$$^0P = \{^0M_1\} \bullet \{^1P\}\qquad(2.9)$$

where

$$
{}^0M_1 = \begin{pmatrix} 1 & 0 & 0 & t_x \\ 0 & 1 & 0 & t_y \\ 0 & 0 & 1 & t_z \\ 0 & 0 & 0 & 1 \end{pmatrix}.
$$

In fact, matrix ${}^0M_1$ in Eq. 2.9 is called the *homogenous motion transformation matrix*. It describes the motion transformation from frame 1 to frame 0. If we ignore the elements of the fourth column, the first three column vectors of ${}^0M_1$ are exactly the base vectors of frame 1 with respect to frame 0 while the last column vector is exactly the position vector of frame 1's origin, with respect to frame 0. These physical meanings remain valid for any complex motion between any pair of frames.

2.4.2   *Linear Velocity and Acceleration*

Displacement only describes the effect of motions at discrete time instants among the frames under consideration. However, the instantaneous variations of a rigid body's motion are described by its velocity and acceleration.

Let us consider a rigid body to which frame j is assigned with respect to frame i. If we denote ${}^iT_j(t) = (t_x, t_y, t_z)^t$ the linear displacement vector due to translational motion, the corresponding linear velocity and acceleration vectors will be

$$
\begin{cases} {}^iv_j(t) = \dfrac{d\{{}^iT_j(t)\}}{dt} = \left(\dfrac{dt_x}{dt}, \dfrac{dt_y}{dt}, \dfrac{dt_z}{dt} \right)^t \\[4mm] {}^ia_j(t) = \dfrac{d^2\{{}^iT_j(t)\}}{dt^2} = \left(\dfrac{d^2t_x}{dt^2}, \dfrac{d^2t_y}{dt^2}, \dfrac{d^2t_z}{dt^2} \right)^t. \end{cases} \tag{2.10}
$$

2.5   Rotational Motions

The physical meaning of a rotational motion is the rotation of a rigid body about any straight line called a *rotation axis*. This occurs when a force, or torque, is applied to a rigid body constrained by a rotation axis. A typical example of rotational motion is the rotation about one of the X, Y, and Z axes.

2.5.1 *Circular Displacement*

Rotational motion can be described by circular displacement, circular velocity, and if any, circular acceleration (the sum of tangential and centrifugal
accelerations). And, these entities are equal at all points on a rigid body.

Now, let us examine rotational motion without regard to the force,
or torque, which causes the motion. Fig. 2.5 shows a rigid body's frame
undergoing rotational motion about the Z axis from its initial location at
$(d, 0, 0)$ (referenced by frame 0) to a new location.

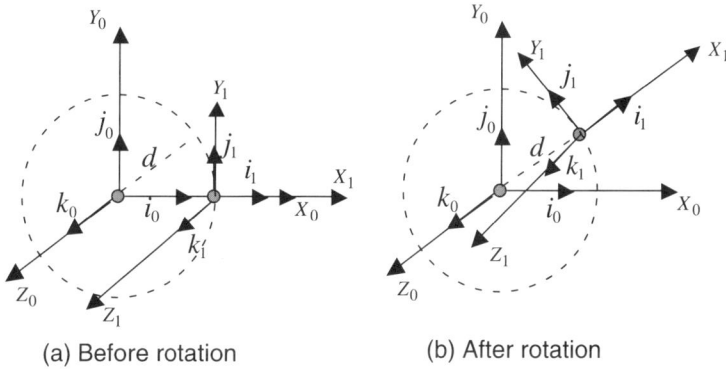

(a) Before rotation   (b) After rotation

Fig. 2.5   Illustration of a rigid body's rotational motion to which frame 1 is assigned.

Under the rotational motion, not only will the origin of a rigid body's
Frame change (except in the case when $d = 0$), but the orientation of
its base vectors will be as well (except in the case when one base vector
coincides with the rotation axis). Another important fact is that rotational
motion will not cause the rotation axis to change.

Consider time interval $[t_1, t_2]$. The base vectors and the origin's position
vector of frame 1 at time instant t_1 will be

$$\begin{cases} {}^0\vec{i}_1(t_1) = (1, 0, 0)^t \\ {}^0\vec{j}_1(t_1) = (0, 1, 0)^t \\ {}^0\vec{k}_1(t_1) = (0, 0, 1)^t \\ {}^0O_1(t_1) = (d, 0, 0)^t. \end{cases} \qquad (2.11)$$

If the rotational angle about the Z axis is denoted by θ at time instant

t_2, these vectors will become

$$\begin{cases} {}^0\vec{i}_1(t_2) = (\cos(\theta),\ \sin(\theta),\ 0)^t \\ {}^0\vec{j}_1(t_2) = (-\sin(\theta),\ \cos(\theta),\ 0)^t \\ {}^0\vec{k}_1(t_2) = (0,0,1)^t \\ {}^0O_1(t_2) = (d\bullet\cos(\theta),\ d\bullet\sin(\theta),\ 0)^t. \end{cases} \quad (2.12)$$

If we define matrix R as follows:

$$R = [{}^0\vec{i}_1(t_2),{}^0\vec{j}_1(t_2),{}^0\vec{k}_1(t_2)] = \begin{pmatrix} \cos(\theta) & -\sin(\theta) & 0 \\ \sin(\theta) & \cos(\theta) & 0 \\ 0 & 0 & 1 \end{pmatrix}, \quad (2.13)$$

Eq. 2.12 can be rewritten in the following matrix form:

$$\begin{cases} {}^0\vec{i}_1(t_2) = R \bullet \{{}^0\vec{i}_1(t_1)\} \\ {}^0\vec{j}_1(t_2) = R \bullet \{{}^0\vec{j}_1(t_1)\} \\ {}^0\vec{k}_1(t_2) = R \bullet \{{}^0\vec{k}_1(t_1)\} \\ {}^0O_1(t_2) = R \bullet \{{}^0O_1(t_1)\}. \end{cases} \quad (2.14)$$

Now, if we imagine that frame 1 performs a rotational motion which always starts from its initial configuration, and this initial configuration is where the frame's base vectors are respectively parallel to the base vectors of a reference frame (e.g. frame 0) at time instant t_0, then we have

$$\begin{cases} {}^0\vec{i}_1(t_1) = (1,0,0)^t = {}^1\vec{i}_1 \\ {}^0\vec{j}_1(t_1) = (0,1,0)^t = {}^1\vec{j}_1 \\ {}^0\vec{k}_1(t_1) = (0,0,1)^t = {}^1\vec{k}_1. \end{cases} \quad (2.15)$$

If we denote ${}^0R_1(t) = R$ and substitute Eq. 2.15 into Eq. 2.14, the base vectors of frame 1 at time instant t (i.e. substitute t_2 with t) can be expressed as follows:

$$\begin{cases} {}^0\vec{i}_1(t) = {}^0R_1(t) \bullet \{{}^1\vec{i}_1\} \\ {}^0\vec{j}_1(t) = {}^0R_1(t) \bullet \{{}^1\vec{j}_1\} \\ {}^0\vec{k}_1(t) = {}^0R_1(t) \bullet \{{}^1\vec{k}_1\}. \end{cases} \quad (2.16)$$

The physical meaning of matrix ${}^0R_1(t)$ is the rotational motion transformation at time instant t from frame 1 to frame 0. This matrix is commonly called a *rotation matrix*. A rotation matrix is a square and invertible matrix. It is interesting and useful to note that this matrix describes the orientation of frame 1 with respect to frame 0, because the three column vectors of ${}^0R_1(t)$ are the base vectors of frame 1 with respect to frame 0 at time instant t. Interestingly enough, the inverse of $^0R_1(t)$ describes the orientation of frame 0 with respect to frame 1. So, it is logical to denote the inverse of $^0R_1(t)$ with $^1R_0(t)$, that is,

$$^1R_0(t) = \{^0R_1(t)\}^{-1}.$$

Example 2.4 Refer to Fig. 2.6. We denote $^0R_1(t)$ the rotational motion transformation from frame 1 to frame 0 at time instant t. If the vector connecting origin O_1 to point P on a rigid body is $^1P = (x_1, y_1, z_1)^t$, the same position vector P with respect to frame 0 will be

$$^0P(t) = \{^0R_1(t)\} \bullet \{^1P\}. \tag{2.17}$$

And, the inverse of the above equation yields the expression for ^1P:

$$^1P = \{^0R_1(t)\}^{-1} \bullet \{^0P(t)\} = \{^1R_0(t)\} \bullet \{^0P(t)\}.$$

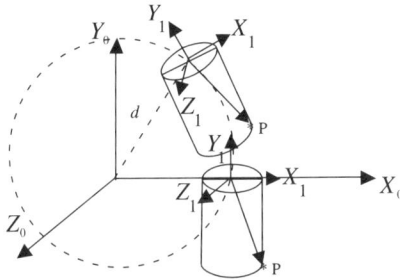

Fig. 2.6   Rotational transformation from frame 1 to frame 0.

◇◇◇◇◇◇◇◇◇◇◇◇◇◇◇◇◇◇◇◇

Now, if we substitute the equivalent projective coordinates for the Cartesian coordinates of point P, Eq. 2.17 can be written as follows:

$$^0P = {}^0M_1 \bullet \{^1P\} \tag{2.18}$$

with

$$^0M_1 = \begin{pmatrix} ^0R_1 & \vec{0}^t \\ \vec{0} & 1 \end{pmatrix}$$

(NOTE: $\vec{0} = (0, 0, 0)$).

From Eq. 2.9 and Eq. 2.18, we can see that the translational and rotational motions can uniformly be described by the homogenous motion transformation matrix M.

2.5.2   *Circular Velocity and Acceleration*

So far, we have examined the circular displacement of a rotational motion. It is important to understand the mathematical description of circular velocity and circular acceleration because these motion parameters are directly related to the force and torque applied to a rigid body.

Refer to Fig. 2.5. Assume that we choose the Z axis as the rotation axis. In reference frame 0, the position vector of frame 1's origin O_1 is

$$^0O_1 = d \bullet \{^0\vec{i_1}\}. \tag{2.19}$$

Then, the circular velocity vector at this origin O_1 with respect to reference frame 0 will be

$$^0v_1 = \frac{d\{^0O_1\}}{dt} = d \bullet \frac{d\{^0\vec{i_1}\}}{dt}. \tag{2.20}$$

The instantaneous variation of a vector's orientation is due to the instantaneous variation of the rotation angle θ. There will be no change in the vector's orientation if there is no instantaneous variation in the rotation angle θ. Since a vector's variation caused by rotational motion follows a circular path, it is in the tangential direction of the circular path and its magnitude of variation is equal to the instantaneous variation of the rotation angle. If we denote $\dot{\theta}$ (i.e. $\frac{d\theta}{dt}$) the instantaneous variation of the rotation angle, then we have

$$\frac{d\{^0\vec{i_1}\}}{dt} = \dot{\theta} \bullet \{^0\vec{j_1}\}. \tag{2.21}$$

Since $^0\vec{j_1} = \,^0\vec{k_1} \times \,^0\vec{i_1}$, substituting Eq. 2.21 into Eq.2.20 yields

$$^0v_1 = \frac{d\{^0O_1\}}{dt} = d \bullet \dot{\theta} \bullet \left(\{^0\vec{k_1}\} \times \{^0\vec{i_1}\} \right) = (\dot{\theta} \bullet \{^0\vec{k_1}\}) \times (d \bullet \{^0\vec{i_1}\}). \tag{2.22}$$

If we define

$$^0\omega_1 = \dot{\theta} \bullet \{^0\vec{k_1}\}$$

and recall that $^0O_1 = d \bullet \{^0\vec{i_1}\}$, Eq.2.22 can be rewritten as follows:

$$^0v_1 = \frac{d\{^0O_1\}}{dt} = {}^0\omega_1 \times \{^0O_1\}. \tag{2.23}$$

Now, it is important to understand the physical meaning of Eq. 2.23. In fact, vector $^0\omega_1$ is called the *angular velocity vector* of a rigid body to which frame 1 is assigned with respect to reference frame 0. This vector coincides with the rotation axis. The norm of $^0\omega_1$ is equal to the instantaneous variation of the rotation angle, denoted by $\dot{\theta}$, which is also called the *angular velocity* or *angular velocity variable* of a rigid body.

Eq. 2.23 also means that a rigid body's angular velocity will cause a *circular velocity* at any arbitrary point P on the rigid body. This point's circular velocity vector is equal to the cross product between the rigid body's angular velocity vector and the point's position vector. That is,

$$^iv_j = \frac{d\{^iP_j\}}{dt} = {}^i\omega_j \times \{^iP_j\}. \tag{2.24}$$

By further differentiating Eq. 2.23, the expression of the circular acceleration vector will be

$$^0a_1 = \frac{d\{^0\omega_1\}}{dt} \times^0 O_1 +^0 \omega_1 \times \frac{d\{^0O_1\}}{dt}. \tag{2.25}$$

By applying the following equalities

$$\begin{cases} \frac{d\{^0\omega_1\}}{dt} = \frac{d(\dot{\theta}\bullet^0\vec{k_1})}{dt} = \ddot{\theta}\bullet^0 \vec{k_1} + \dot{\theta} \bullet \frac{d\{^0\vec{k_1}\}}{dt} \\ \\ \frac{d\{^0O_1\}}{dt} = {}^0\omega_1 \times \{^0O_1\} \end{cases}$$

into Eq. 2.25, we obtain

$$^0a_1 = \left(\ddot{\theta}\bullet^0 \vec{k_1} + \dot{\theta} \bullet \frac{d\{^0\vec{k_1}\}}{dt} \right) \times^0 O_1 +^0 \omega_1 \times (^0\omega_1 \times^0 O_1). \tag{2.26}$$

Since a rigid body's rotational motion about an axis will not cause any change to the rotation axis, then we have

$$\frac{d\{^0\vec{k_1}\}}{dt} = 0.$$

By applying this result into Eq.2.26, the expression of the circular acceleration vector becomes

$$^0a_1 = (\ddot{\theta}\bullet^0 \vec{k_1}) \times^0 O_1 +^0 \omega_1 \times (^0\omega_1 \times^0 O_1). \tag{2.27}$$

If we define the following terms:

$$\begin{cases} {}^0\alpha_1 = \ddot{\theta} \bullet {}^0\vec{k}_1 \\[2mm] a_t = {}^0\alpha_1 \times {}^0 O_1 \\[2mm] a_n = {}^0\omega_1 \times ({}^0\omega_1 \times {}^0 O_1) \end{cases} \qquad (2.28)$$

then Eq. 2.27 can be rewritten as follows:

$${}^0 a_1 = {}^0\alpha_1 \times {}^0 O_1 + {}^0 \omega_1 \times ({}^0\omega_1 \times {}^0 O_1) = a_t + a_n \qquad (2.29)$$

where ${}^0\alpha_1$ is the rigid body's angular acceleration vector to which frame 1 is assigned with respect to frame 0. Vector ${}^0\alpha_1$ is parallel to the rotation axis and its norm is equal to the second-order derivative of the rotation angle, denoted by $\ddot{\theta}$. This is called a rigid body's *angular acceleration* or *angular acceleration variable*.

From Eq. 2.28, it is possible to prove the following equalities:

$$\begin{cases} a_t = {}^0\alpha_1 \times {}^0 O_1 = (\ddot{\theta} \bullet d) \bullet \{{}^0\vec{j}_1\} \\[2mm] a_n = {}^0\omega_1 \times ({}^0\omega_1 \times {}^0 O_1) = (\dot{\theta}^2 \bullet d) \bullet \{-{}^0\vec{i}_1\}, \end{cases} \qquad (2.30)$$

where a_t is in the direction of ${}^0\vec{j}_1$, and a_n is in the opposite direction of ${}^0\vec{i}_1$. In fact, a_t is called the *tangential acceleration vector* and a_n is called the *centrifugal acceleration vector*. From Eq. 2.29, we can see that the circular acceleration vector is the sum of the tangential acceleration and the centrifugal acceleration.

In general, a rigid body's angular acceleration ${}^i\alpha_j$ will generate circular acceleration at any point P in the following way:

$${}^i a_j = {}^i\alpha_j \times {}^i P_j + {}^i\omega_j \times ({}^i\omega_j \times {}^i P_j). \qquad (2.31)$$

2.6 Composite Motions

As I mentioned before, any complex motion can be treated as the combination of translational and rotational motions. With the help of frames, it is easy to study the composite motions among the frames.

2.6.1 *Homogenous Transformation*

For simplicity's sake, we commonly use the term *configuration* or *posture* to refer to the relative geometry of a frame with respect to another frame. The configuration of a frame, with respect to another frame, encapsulates the relative position and orientation of these two frames.

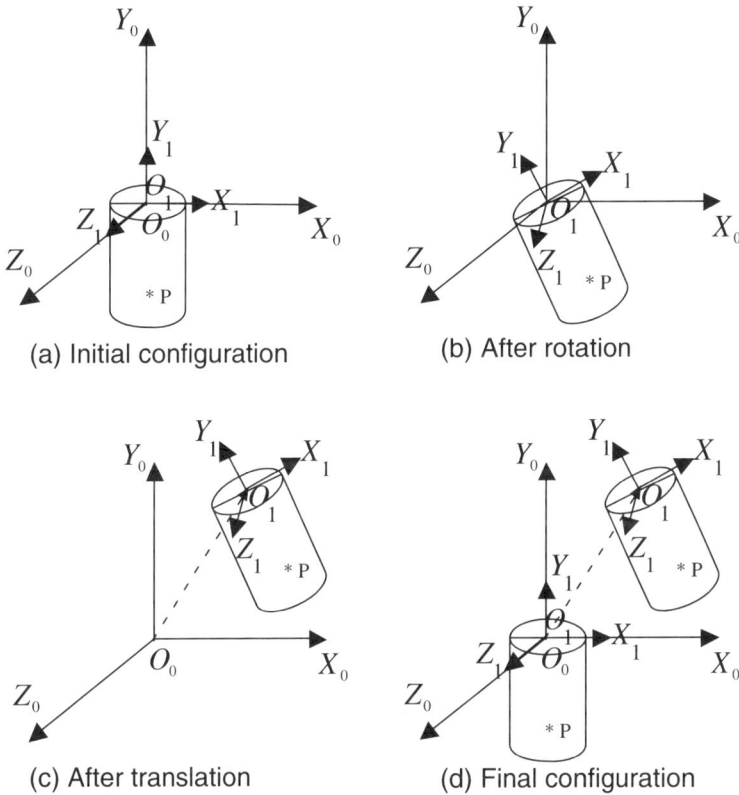

(a) Initial configuration

(b) After rotation

(c) After translation

(d) Final configuration

Fig. 2.7   Illustration of composite motions.

Fig. 2.7 illustrates the fact that it is possible to configure any pair of frames by applying a rotational motion followed by a translational motion to the initial configuration where the two frames coincide. In this example, frame 1 is assigned to a rigid body, and the reference frame is denoted by frame 0. Here, let us assume that the rotational motion occurs within time interval $[t_0, t_1]$ while the translational motion occurs within time interval

$[t_1, t_2]$. If we denote $^0R_1$ as frame 1's rotation matrix with respect to frame 0 in time interval $[t_0, t_1]$, point P's position vector at time-instant t_1 will be

$$^0P(t_1) = (x_1, y_1, z_1)^t = {}^0R_1 \bullet \{^1P\}, \tag{2.32}$$

where $^1P = (x_0, y_0, z_0)^t$ is point P's position vector with respect to frame 1. Since P is a point on the rigid body, its position vector remains unchanged, with respect to the frame assigned to the rigid body.

Now, let $^0T_1$ be the translational vector of frame 1 with respect to frame 0 within time interval $[t_1, t_2]$. Then, point P's position vector at time-instant t_2 will be

$$^0P(t_2) = (x_2, y_2, z_2)^t = {}^0P(t_1) + {}^0T_1. \tag{2.33}$$

The combination of Eq. 2.32 and Eq. 2.33 yields

$$^0P(t_2) = {}^0R_1 \bullet \{^1P\} + {}^0T_1. \tag{2.34}$$

Since rotation matrix R is formed by a set of orthogonal base vectors, its inverse exists. From Eq. 2.34, we can derive the expression for ^1P as follows:

$$^1P = \{^0R_1\}^{-1} \bullet \{^0P(t_2)\} - \{^0R_1\}^{-1} \bullet \{^0T_1\}. \tag{2.35}$$

If we substitute the equivalent projective coordinates of P for the Cartesian coordinates of P in Eq. 2.34 and Eq. 2.35, and substitute t_2 for time variable t, these two equations can be written in a compact matrix form as follows:

$$\begin{cases} {}^0P(t) = (x_0, y_0, z_0, 1)^t = {}^0M_1(t) \bullet \{^1P\} \\ {}^1P = (x_1, y_1, z_1, 1)^t = {}^1M_0(t) \bullet \{^0P(t)\} \end{cases} \tag{2.36}$$

with

$$^0M_1(t) = \begin{pmatrix} {}^0R_1 & {}^0T_1 \\ \vec{0} & 1 \end{pmatrix} \tag{2.37}$$

and

$$^1M_0(t) = \begin{pmatrix} \{^0R_1\}^{-1} & -\{^0R_1\}^{-1} \bullet^0 T_1 \\ \vec{0} & 1 \end{pmatrix}. \tag{2.38}$$

Matrix $^0M_1(t)$ describes mapping from the equivalent projective coordinates of a point expressed in frame 1 to the equivalent projective coordinates of the same point expressed in frame 0. And, matrix $^1M_0(t)$ describes inverse mapping by $^0M_1(t)$.

In general, matrix $^iM_j(t)$ describes mapping from frame j to frame i. An alternative interpretation of matrix $^iM_j(t)$ is that it represents both position (i.e. the origin) and orientation of frame j with respect to frame i. Mapping by this type of matrix is commonly called the *homogenous transformation* because it describes the homogenous transformation applied to the equivalent projective coordinates between two frames.

2.6.2 *Differential Homogenous Transformation*

Refer to Fig. 2.7. Assume that the rigid body, to which frame 1 is assigned, undergoes a continuous composite motion. If we denote $^0R_1(t)$ the continuous rotation transformation, and $^0T_1(t)$ the continuous translation transformation of frame 1 with respect to frame 0, point P's instantaneous coordinates with respect to frame 0 will be

$$^0P(t) = \ ^0R_1(t) \bullet^1 P + \ ^0T_1(t) \tag{2.39}$$

or in a matrix form:

$$^0P(t) =^0 M_1(t) \bullet \{^1P\}$$

with

$$^0M_1(t) = \begin{pmatrix} ^0R_1(t) & ^0T_1(t) \\ \vec{0} & 1 \end{pmatrix}$$

By differentiating Eq. 2.39, we obtain the expression for point P's velocity vector with respect to frame 0, that is,

$$^0v(t) = \frac{d(^0P(t))}{dt} = \frac{d(^0R_1(t))}{dt} \bullet^1 P + \frac{d(^0T_1(t))}{dt}. \tag{2.40}$$

Since a rotation matrix's column vectors are a frame's base vectors, we have

$$^0R_1(t) = \left(\{^0\vec{i}_1\}, \ \{^0\vec{j}_1\}, \ \{^0\vec{k}_1\} \right). \tag{2.41}$$

Differentiating the above rotation matrix with respect to time gives

$$\frac{d(^0R_1(t))}{dt} = \left(\frac{d\{^0\vec{i_1}\}}{dt}, \frac{d\{^0\vec{j_1}\}}{dt}, \frac{d\{^0\vec{k_1}\}}{dt} \right). \tag{2.42}$$

If the angular velocity vector of the rigid body, with respect to frame 0, is $^0\omega_1 = (\omega_x, \omega_y, \omega_z)^t$, we have

$$\left(\frac{d\{^0\vec{i_1}\}}{dt}, \frac{d\{^0\vec{j_1}\}}{dt}, \frac{d\{^0\vec{k_1}\}}{dt} \right) = {}^0\omega_1 \times \left(\{^0\vec{i_1}\}, \{^0\vec{j_1}\}, \{^0\vec{k_1}\} \right). \tag{2.43}$$

If $S(^0\omega_1)$ denotes the skew symmetric matrix of vector $^0\omega_1$, then we have

$$S(^0\omega_1) = \begin{pmatrix} 0 & -\omega_z & \omega_y \\ \omega_z & 0 & -\omega_x \\ -\omega_y & \omega_x & 0 \end{pmatrix}. \tag{2.44}$$

Accordingly, Eq. 2.43 can also be written, as follows:

$$\left(\frac{d\{^0\vec{i_1}\}}{dt}, \frac{d\{^0\vec{j_1}\}}{dt}, \frac{d\{^0\vec{k_1}\}}{dt} \right) = S(^0\omega_1) \bullet \left(\{^0\vec{i_1}\}, \{^0\vec{j_1}\}, \{^0\vec{k_1}\} \right). \tag{2.45}$$

By applying Eq. 2.45 and Eq. 2.42, Eq. 2.40 will become

$$^0v(t) = S(^0\omega_1) \bullet \left(\{^0\vec{i_1}\}, \{^0\vec{j_1}\}, \{^0\vec{k_1}\} \right) \bullet \{^1P\} + \frac{d(^0T_1(t))}{dt}$$

$$= S(^0\omega_1) \bullet \{^0R_1(t)\} \bullet \{^1P\} + \frac{d(^0T_1(t))}{dt} \tag{2.46}$$

$$= S(^0\omega_1) \bullet \{^0P(t)\} + \frac{d(^0T_1(t))}{dt}.$$

If we use the equivalent projective coordinates, the above equation can be concisely written as follows:

$$^0v(t) = \frac{d(^0M_1(t))}{dt} \bullet \{^1P\} \tag{2.47}$$

with

$$\frac{d(^0M_1)}{dt} = \begin{pmatrix} S(^0\omega_1) \bullet \{^0R_1(t)\} & \frac{d(^0T_1(t))}{dt} \\ \vec{0} & 1 \end{pmatrix}.$$

Therefore, matrix $\frac{d(^0M_1)}{dt}$ describes the differential homogenous transformation from frame 1 to frame 0. In other words, it determines how the velocity of point P in frame 1 is perceived from frame 0, even though point P has no motion with respect to frame 1.

Another useful result from the above development is the expression for the differential of the rotation matrix. That is,

$$\frac{d\{^0R_1(t)\}}{dt} = S(^0\omega_1) \bullet \{^0R_1(t)\} \tag{2.48}$$

where $^0\omega_1$ is frame 1's angular velocity vector, and $^0R_1(t)$ describes frame 1's orientation, with respect to frame 0, at time instant t.

From Eq. 2.48, we can derive one interesting property of the skew-symmetric matrix $S(^0\omega_1)$. That is,

$$A \bullet S(^0\omega_1) \bullet A^{-1} = S(A \bullet^0 \omega_1) \tag{2.49}$$

where A is a 3×3 matrix which has an inverse. The above expression is useful in the study of motion kinematics (i.e. to derive a compact expression of the Jacobian matrix). Proof of the above expression is as follows:

First, we compute the derivative of matrix $A \bullet^0 R_1$:

$$\frac{d(A\bullet^0R_1)}{dt} = \frac{d}{dt}\left(A \bullet \{^0\vec{i}_1\}, \; A \bullet \{^0\vec{j}_1\}, \; A \bullet \{^0\vec{k}_1\}\right)$$
$$= \left(A \bullet \frac{d\{^0\vec{i}_1\}}{dt}, \; A \bullet \frac{d\{^0\vec{j}_1\}}{dt}, \; A \bullet \frac{d\{^0\vec{k}_1\}}{dt}\right). \tag{2.50}$$

By applying the following equalities:

$$\begin{cases} A \bullet \frac{d\{^0\vec{i}_1\}}{dt} = A \bullet (^0\omega_1 \times^0 \vec{i}_1) = (A \bullet^0 \omega_1) \times (A \bullet^0 \vec{i}_1) \\[2mm] A \bullet \frac{d\{^0\vec{j}_1\}}{dt} = A \bullet (^0\omega_1 \times^0 \vec{j}_1) = (A \bullet^0 \omega_1) \times (A \bullet^0 \vec{j}_1) \\[2mm] A \bullet \frac{d\{^0\vec{k}_1\}}{dt} = A \bullet (^0\omega_1 \times^0 \vec{k}_1) = (A \bullet^0 \omega_1) \times (A \bullet^0 \vec{k}_1) \end{cases}$$

into Eq. 2.50, we have

$$\frac{d(A\bullet^0R_1)}{dt} = (A \bullet^0 \omega_1) \times \left(A \bullet \{^0\vec{i}_1\}, \; A \bullet \{^0\vec{j}_1\}, \; A \bullet \{^0\vec{k}_1\}\right)$$
$$= S(A \bullet^0 \omega_1) \bullet (A \bullet^0 R_1). \tag{2.51}$$

By applying the following equality

$$\frac{d(A \bullet {}^0 R_1)}{dt} = A \bullet \frac{d({}^0 R_1)}{dt} = A \bullet S({}^0 \omega_1) \bullet {}^0 R_1$$

into Eq. 2.51, we obtain

$$A \bullet S({}^0 \omega_1) \bullet {}^0 R_1 = S(A \bullet {}^0 \omega_1) \bullet (A \bullet {}^0 R_1). \tag{2.52}$$

By eliminating ${}^0 R_1$ from both sides, and multiplying A^{-1} to both sides, Eq. 2.52 finally becomes

$$A \bullet S({}^0 \omega_1) \bullet A^{-1} = S(A \bullet {}^0 \omega_1).$$

2.6.3 *Successive Elementary Translations*

The study of successive elementary translations and/or rotations is of a particular interest in robotics. First of all, it provides a better understanding of the minimum number of variables necessary to describe a pure translation or pure rotation. Secondly, it provides a sound basis for describing the relationship among a set of frames assigned to a set of rigid bodies which are connected (i.e. constrained) in a series, where each rigid body has only one degree of freedom (e.g. the links inside an arm manipulator). Elementary motion denotes the motion of a frame having only one independent (motion) parameter or degree of freedom (DOF).

Consider the translational motion between reference frame 0 and frame 1 (assigned to a rigid body in a Cartesian space). If we denote ${}^0 T_1(t) = (t_x, t_y, t_z)^t$, and the translation vector of frame 1 with respect to frame 0, it is clear that the translation vector ${}^0 T_1(t)$ can be decomposed in the following way:

$$\begin{pmatrix} t_x \\ t_y \\ t_z \end{pmatrix} = \begin{pmatrix} t_x \\ 0 \\ 0 \end{pmatrix} + \begin{pmatrix} 0 \\ t_y \\ 0 \end{pmatrix} + \begin{pmatrix} 0 \\ 0 \\ t_z \end{pmatrix}. \tag{2.53}$$

We can see that any arbitrary translation will be the sum of three elementary translations along the X, Y and Z axes. A graphic illustration of this fact is shown in Fig. 2.8. Therefore, we can conclude that the minimum number of variables to fully describe a translational motion in a Cartesian space is 3.

Fig. 2.8 shows the temporal series of frame 1 in a Cartesian space after three successive translations. However, we can also treat these as a series of spatially arranged frames which are consecutively supporting each other.

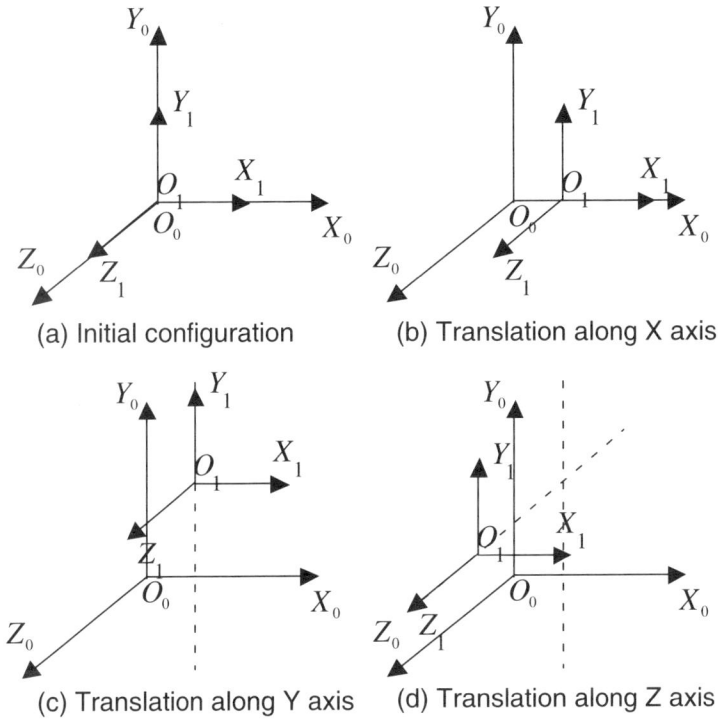

(a) Initial configuration

(b) Translation along X axis

(c) Translation along Y axis

(d) Translation along Z axis

Fig. 2.8   Illustration of three successive translations.

A typical scenario is the assignment of frames to links which are serially connected to form an arm manipulator. In fact, we can rename the temporal series of frame 1 in the following manner:

- Frame 1 after the translation along the X axis is renamed as frame i.
- Frame 1 after the translation along the Y axis is renamed as frame $i+1$.
- Frame 1 after the translation along the Z axis is renamed as frame $i+2$.

If we examine the consecutive translational motions among frames 0, i, $i+1$ and $i+2$, as shown in Fig. 2.8, the following result is straightforward:

$$\begin{cases} {}^0T_i = (t_x, 0, 0)^t = t_x \bullet \{{}^0\vec{i}_i\} \\ {}^iT_{i+1} = (0, t_y, 0)^t = t_y \bullet \{{}^i\vec{j}_{i+1}\} \\ {}^{i+1}T_{i+2} = (0, 0, t_z)^t = t_z \bullet \{{}^{i+1}\vec{k}_{i+2}\}, \end{cases} \tag{2.54}$$

where we explicitly emphasize the direction of translation ^0T_i (i.e. the base vector $\vec{i}$ of frame i), the direction of translation $^iT_{i+1}$ (i.e. the base vector $\vec{j}$ of frame $i+1$), and the direction of translation $^{i+1}T_{i+2}$ (i.e. the base vector $\vec{k}$ of frame $i+2$).

These expressions are helpful to better understand motion kinematics which will be studied in Chapter 3. For example, just as translational motion does not cause any changes in a base vector, it will also not cause any changes to a directional vector. Thus, differentiating Eq. 2.54 with respect to time yields

$$\begin{cases} \frac{d(^0T_i)}{dt} = \frac{dt_x}{dt} \bullet \{^0\vec{i}_i\} \\[2mm] \frac{d(^iT_{i+1})}{dt} = \frac{dt_y}{dt} \bullet \{^i\vec{j}_{i+1}\} \\[2mm] \frac{d(^{i+1}T_{i+2})}{dt} = \frac{dt_z}{dt} \bullet \{^{i+1}\vec{k}_{i+2}\}, \end{cases} \tag{2.55}$$

By applying Eq. 2.54 into Eq. 2.53, the translational motion transformation from frame $i+2$ to frame 0 can be expressed as follows:

$$^0T_{i+2} = {}^0T_i + {}^iT_{i+1} + {}^{i+1}T_{i+2}. \tag{2.56}$$

In general, Eq. 2.56 is valid for any set of spatially arranged frames. If we use the equivalent projective coordinates, Eq. 2.56 can be rewritten as follows:

$$^0M_{i+2} = \{^0M_i\} \bullet \{^iM_{i+1}\} \bullet \{^{i+1}M_{i+2}\} \tag{2.57}$$

with

$$^0M_{i+2} = \begin{pmatrix} 1 & 0 & 0 & t_x \\ 0 & 1 & 0 & t_y \\ 0 & 0 & 1 & t_z \\ 0 & 0 & 0 & 1 \end{pmatrix} \qquad ^0M_i = \begin{pmatrix} 1 & 0 & 0 & t_x \\ 0 & 1 & 0 & 0 \\ 0 & 0 & 1 & 0 \\ 0 & 0 & 0 & 1 \end{pmatrix}$$

$$^iM_{i+1} = \begin{pmatrix} 1 & 0 & 0 & 0 \\ 0 & 1 & 0 & t_y \\ 0 & 0 & 1 & 0 \\ 0 & 0 & 0 & 1 \end{pmatrix} \qquad ^{i+1}M_{i+2} = \begin{pmatrix} 1 & 0 & 0 & 0 \\ 0 & 1 & 0 & 0 \\ 0 & 0 & 1 & t_z \\ 0 & 0 & 0 & 1 \end{pmatrix}.$$

In robotics, the above equation is the exact expression for the forward kinematics of a Cartesian or XYZ robot having three links moving in the X, Y and Z directions respectively.

2.6.4 *Successive Elementary Rotations*

Fig. 2.9 shows three successive rotations of frame 1 in a Cartesian space referenced to frame 0. At the initial configuration, frame 1 coincides with frame 0. The three successive elementary rotations are as follows:

- The first elementary rotation is that of frame 1 (at its initial configuration) about its own X axis. This elementary rotation is fully determined by rotation angle θ_x.
- The second elementary rotation follows the first one, and it is that of frame 1 (at the current configuration) about its own Y axis. This elementary rotation is fully determined by rotation angle θ_y.
- The third elementary rotation follows the second one, and it is that of frame 1 (at the current configuration) about its own Z axis. This elementary rotation is fully determined by rotation angle θ_z.

After the first elementary rotation, frame 1's base vectors $(\vec{j}, \vec{k})$ remain inside frame 0's YZ plane. Clearly, one elementary rotation does not represent a general rotation. After the second elementary rotation, frame 1's base vector $\vec{j}$ still remains inside frame 0's YZ plane. Therefore, two consecutive elementary rotations are not sufficient to produce a general rotation. Only after three successive elementary rotations will frame 1 reach an arbitrary orientation with respect to frame 0. Hence, the minimum number of variables necessary to describe a general rotational motion between two frames in a Cartesian space is 3.

Similarly, a temporal series of frame 1 can also be treated as a series of spatially-arranged frames which are consecutively supporting each other. This can be done by renaming the frames in the following way:

- Frame 1, after the first elementary rotation, is renamed frame i.
- Frame 1, after the second elementary rotation, is renamed frame $i+1$.
- Frame 1, after the third elementary rotation, is renamed frame $i+2$.

Since each elementary rotation only depends on its corresponding rotation angle about one of the X, Y and Z axes, the orientation of frame 1 with respect to frame 0 after three successive elementary rotations can be expressed, as follows:

$$^0R_{i+2} = \{^0R_i\} \bullet \{^iR_{i+1}\} \bullet \{^{i+1}R_{i+2}\} \qquad (2.58)$$

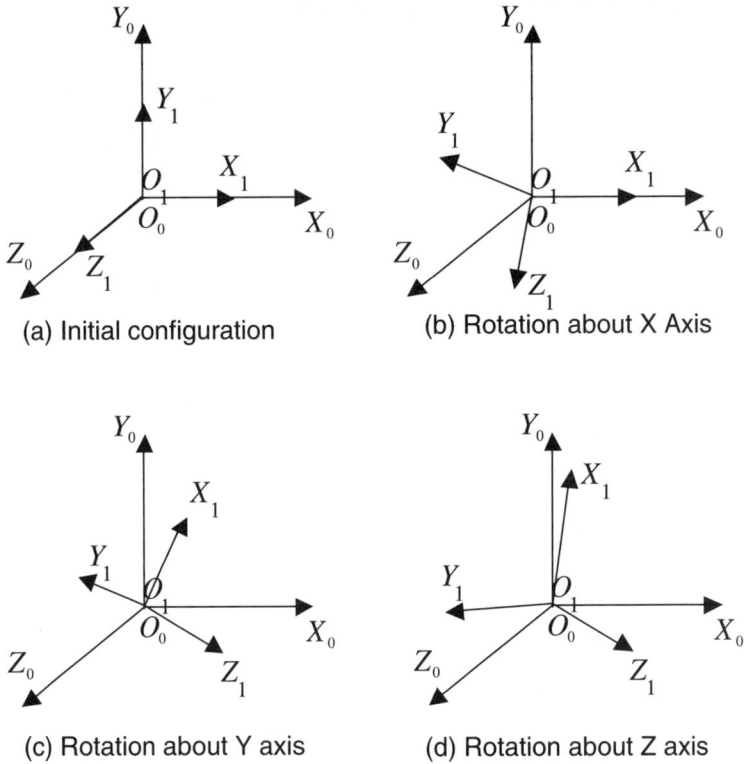

(a) Initial configuration

(b) Rotation about X Axis

(c) Rotation about Y axis

(d) Rotation about Z axis

Fig. 2.9   Illustration of three successive rotations.

with

$$^0R_i = \begin{pmatrix} 1 & 0 & 0 \\ 0 & \cos(\theta_x) & -\sin(\theta_x) \\ 0 & \sin(\theta_x) & \cos(\theta_x) \end{pmatrix}$$

$$^iR_{i+1} = \begin{pmatrix} \cos(\theta_y) & 0 & \sin(\theta_y) \\ 0 & 1 & 0 \\ -\sin(\theta_y) & 0 & \cos(\theta_y) \end{pmatrix}$$

$$^{i+1}R_{i+2} = \begin{pmatrix} \cos(\theta_z) & -\sin(\theta_z) & 0 \\ \sin(\theta_z) & \cos(\theta_z) & 0 \\ 0 & 0 & 1 \end{pmatrix}.$$

If we consider the equivalent projective coordinates, Eq. 2.58 can equivalently be written as follows:

$$^0M_{i+2} = \{^0M_i\} \bullet \{^iM_{i+1}\} \bullet \{^{i+1}M_{i+2}\} \tag{2.59}$$

with

$$^0M_i = \begin{pmatrix} ^0R_i & \vec{0}^t \\ \vec{0} & 1 \end{pmatrix}, \qquad ^iM_{i+1} = \begin{pmatrix} ^iR_{i+1} & \vec{0}^t \\ \vec{0} & 1 \end{pmatrix},$$

$$^{i+1}M_{i+2} = \begin{pmatrix} ^{i+1}R_{i+2} & \vec{0}^t \\ \vec{0} & 1 \end{pmatrix}.$$

In robotics, Eq. 2.59 describes the forward kinematics of a spherical joint having three degrees of freedom.

Imagine now that θ_x, θ_y, and θ_z can undergo instantaneous variations with respect to time. Differentiating Eq. 2.58 yields

$$\begin{aligned} \frac{d(^0R_{i+2})}{dt} &= S(^0\omega_{i+2}) \bullet \{^0R_{i+2}\} \\[2mm] &= \frac{d(^0R_i)}{dt} \bullet \{^iR_{i+1}\} \bullet \{^{i+1}R_{i+2}\} \\[2mm] &\quad + \{^0R_i\} \bullet \frac{d(^iR_{i+1})}{dt} \bullet \{^{i+1}R_{i+2}\} \\[2mm] &\quad + \{^0R_i\} \bullet \{^iR_{i+1}\} \bullet \frac{d(^{i+1}R_{i+2})}{dt}. \end{aligned} \tag{2.60}$$

By applying the following property of skew-symmetric matrix:

$$A \bullet S(^0\omega_1) \bullet A^{-1} = S(A \bullet^0 \omega_1)$$

we can derive the following equalities:

$$\begin{cases} \frac{d(^0R_i)}{dt} \bullet \{^iR_{i+1}\} \bullet \{^{i+1}R_{i+2}\} = S(^0\omega_i) \bullet \{^0R_{i+2}\} \\[3mm] \{^0R_i\} \bullet \frac{d(^iR_{i+1})}{dt} \bullet \{^{i+1}R_{i+2}\} = S(^0R_i \bullet^i \omega_{i+1}) \bullet \{^0R_{i+2}\} \\[3mm] \{^0R_i\} \bullet \{^iR_{i+1}\} \bullet \frac{d(^{i+1}R_{i+2})}{dt} = S(^0R_{i+1} \bullet^{i+1} \omega_{i+2}) \bullet \{^0R_{i+2}\}. \end{cases} \tag{2.61}$$

By applying Eq. 2.61 into Eq. 2.60, we finally obtain the following interesting and useful expression:

$$S(^0\omega_{i+2}) = S(^0\omega_i) + S(^0R_i \bullet^i \omega_{i+1}) + S(^0R_{i+1} \bullet^{i+1} \omega_{i+2}) \tag{2.62}$$

or

$$^0\omega_{i+2} = {}^0\omega_i + {}^0R_i \bullet \{^i\omega_{i+1}\} + {}^0R_{i+1} \bullet \{^{i+1}\omega_{i+2}\}. \qquad (2.63)$$

The physical meaning of the above expression will become clearer in Chapter 3. It simply means that frame $i + 2$'s angular velocity vector with respect to frame 0 is the sum of the angular velocity vectors of the intermediate frames between frame 0 and frame $i + 2$, all being expressed with respect to frame 0.

2.6.5   *Euler Angles*

Refer to Fig. 2.9, again. The first elementary rotation can choose X, Y or Z axes as its rotation axis. As for the second elementary rotation, it has only two axes to choose from, because its rotation axis should not be the same as the previous one. Similarly, the third elementary rotation also has two axes to choose from, because its rotation axis should not be the same as the previous one. As a result, we have a total of $3 \times 2 \times 2$ (or 12) possible combinations to form a sequence of three successive (elementary) rotations. Each combination has its own set of three rotation angles as the minimum number of variables to describe a general rotation. Therefore, we have 12 possible sets of rotation angles. They are:

$$(\theta_x\theta_y\theta_z) \ \ (\theta_x\theta_y\theta_x) \ \ (\theta_x\theta_z\theta_y) \ \ (\theta_x\theta_z\theta_x)$$

$$(\theta_y\theta_z\theta_x) \ \ (\theta_y\theta_z\theta_y) \ \ (\theta_y\theta_x\theta_y) \ \ (\theta_y\theta_x\theta_z) \qquad (2.64)$$

$$(\theta_z\theta_x\theta_y) \ \ (\theta_z\theta_x\theta_z) \ \ (\theta_z\theta_y\theta_x) \ \ (\theta_z\theta_y\theta_z).$$

These sets are commonly called *Euler Angles*.

Example 2.5   Refer to Fig. 2.9. Assume that the actual orientation of frame 1 is reached through three successive elementary rotations, as follows:

- The first elementary rotation is that of frame 1 about its Z axis at the current configuration (i.e. the initial configuration). The rotation angle is α.
- The second elementary rotation is that of frame 1 about its Y axis at the current configuration (i.e. after the first elementary rotation). The rotation angle is β.

- The third elementary rotation is that of frame 1 about its Z axis at the current configuration (i.e. after the second elementary rotation). The rotation angle is ϕ.

If frame 1 is renamed frame i after the first rotation, frame $i + 1$ after the second rotation, and frame $i + 2$ after the third rotation, the rotation matrix describing the orientation of frame 1 at its actual orientation with respect to frame 0 will be

$$^0R_1 = {}^0R_{i+2} = \{^0R_i\} \bullet \{^iR_{i+1}\} \bullet \{^{i+1}R_{i+2}\}$$

with

$$^0R_i = \begin{pmatrix} \cos(\alpha) & -\sin(\alpha) & 0 \\ \sin(\alpha) & \cos(\alpha) & 0 \\ 0 & 0 & 1 \end{pmatrix}$$

$$^iR_{i+1} = \begin{pmatrix} \cos(\beta) & 0 & \sin(\beta) \\ 0 & 1 & 0 \\ -\sin(\beta) & 0 & \cos(\beta) \end{pmatrix}$$

$$^{i+1}R_{i+2} = \begin{pmatrix} \cos(\phi) & -\sin(\phi) & 0 \\ \sin(\phi) & \cos(\phi) & 0 \\ 0 & 0 & 1 \end{pmatrix}.$$

In this example, the three angles (α, β, ϕ) are called ZYZ Euler Angles.

◇◇◇◇◇◇◇◇◇◇◇◇◇◇◇◇◇◇◇◇

2.6.6 *Equivalent Axis and Angle of Rotation*

Euler angles are the set of minimum angles which fully determine a frame's rotation matrix with respect to another frame. Each set has three angles which make three successive elementary rotations. Thus, the orientation of a frame, with respect to another frame, depends on three independent motion parameters even though the rotation matrix is a 3×3 matrix with 9 elements inside.

In robotics, it is necessary to interpolate the orientation of a frame from its initial orientation to an actual orientation so that a physical rigid body (e.g. the end-effector) can smoothly execute the rotational motion in real space. Since a set of Euler Angles contains three angles, called *angular variables*, the interpolation applied to three variables may take a different path or trajectory in a space defined by these three variables. Therefore, it is not easy to manipulate Euler angles for the purpose of determining smooth, intermediate orientations of a frame between its initial and actual orientations.

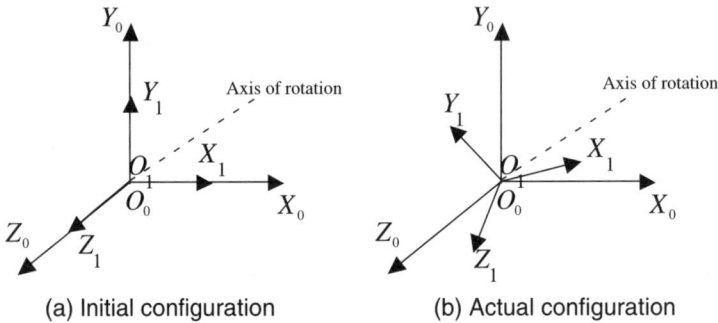

(a) Initial configuration   (b) Actual configuration

Fig. 2.10   Equivalent axis of rotation: any orientation of a frame, with respect to another frame, can be treated as a rotation about an equivalent axis from its initial configuration to its actual configuration.

Interestingly enough, there is an *equivalent axis of rotation* which brings a frame from its initial orientation to its actual orientation with a corresponding angle of rotation. The truth of this property can be verified in the following way:

Assume that such an equivalent axis of rotation exists, and is denoted by unit vector $r = (r_x, r_y, r_z)^t$. Based on the fact that a rotational motion will not cause any changes to the rotation axis itself, the following equality must be valid, if r is the equivalent axis of rotation for the rotational motion between frame 1 and frame 0:

$$^0R_1 \bullet \{^1r\} = \,^0r \tag{2.65}$$

with

$$^1r = \,^0r = (r_x, r_y, r_z)^t.$$

Now, imagine that there is an intermediate frame i which has the Z axis that coincides with the equivalent rotation axis r. (See Fig. 2.11.) And assume that the orientation of frame i with respect to frame 0, is determined by the YXZ Euler angles.

If ^0R_i denotes the orientation of frame i, with respect to frame 0, we

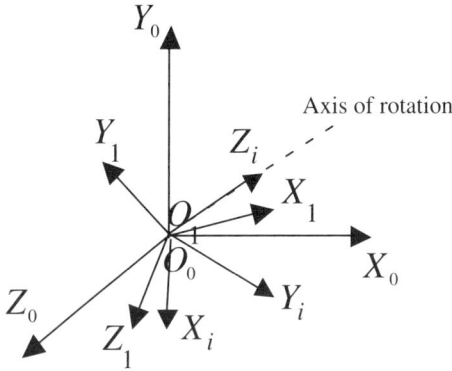

Fig. 2.11 An intermediate frame i which has the Z axis that coincides with the equivalent rotation axis.

have

$$^0R_i = R_y \bullet R_x \bullet R_z \tag{2.66}$$

with

$$R_y = \begin{pmatrix} \cos(\alpha) & 0 & \sin(\alpha) \\ 0 & 1 & 0 \\ -\sin(\alpha) & 0 & \cos(\alpha) \end{pmatrix}$$

$$R_x = \begin{pmatrix} 1 & 0 & 0 \\ 0 & \cos(\beta) & -\sin(\beta) \\ 0 & \sin(\beta) & \cos(\beta) \end{pmatrix}$$

$$R_z = \begin{pmatrix} \cos(\theta) & -\sin(\theta) & 0 \\ \sin(\theta) & \cos(\theta) & 0 \\ 0 & 0 & 1 \end{pmatrix}.$$

In the above expression, angles (α, β) must take the following values in order for the Z axis of frame i to coincide with the equivalent rotation axis after two successive elementary rotations about its Y and then, X axes:

$$\begin{cases} \alpha = \arccos\left(\dfrac{r_z}{\sqrt{r_x^2 + r_z^2}}\right) \\ \\ \beta = -\arccos\left(\sqrt{r_x^2 + r_z^2}\right). \end{cases}$$

When the equivalent rotation axis coincides with the Z axis of frame i, then

$$^1r = \,^0R_i \bullet \{^i\vec{k_i}\} = R_y \bullet R_x \bullet R_z \bullet \{^i\vec{k_i}\} \tag{2.67}$$

and

$$^0r = \,^0R_i \bullet^i \vec{k_i} = R_y \bullet R_x \bullet R_z \bullet \{^i\vec{k_i}\} \tag{2.68}$$

where $^i\vec{k_i} = (0,0,1)^t$.

Since the angle θ inside matrix R_z is unconstrained, if $\theta = 0$ (i.e. $R_z = I_{3\times3}$), Eq. 2.67 becomes

$$^1r = R_y \bullet R_x \bullet \{^i\vec{k_i}\} \tag{2.69}$$

Now, by applying Eq. 2.69 and Eq. 2.68 to Eq. 2.65, the following equality holds:

$$^0R_1 \bullet R_y \bullet R_x = R_y \bullet R_x \bullet R_z, \tag{2.70}$$

or more explicitly we have

$$^0R_1 = R_y \bullet R_x \bullet R_z \bullet R_x^{-1} \bullet R_y^{-1}. \tag{2.71}$$

The above equation proves the validity of the assumption that an equivalent rotation axis exists for any given orientation of a frame with respect to another frame. And, the rotation matrix can be formed by five successive elementary rotations which are fully determined by the rotation axis's unit vector and the rotation angle θ (i.e. the R_z matrix).

If $^0R_1$ is given as the input:

$$^0R_1 = \begin{pmatrix} r_{11} & r_{12} & r_{13} \\ r_{21} & r_{22} & r_{23} \\ r_{31} & r_{32} & r_{33} \end{pmatrix},$$

we can derive the solutions for r and θ. From Eq. 2.71, we can obtain

$$\theta = \arccos\left(\frac{r_{11} + r_{22} + r_{33} - 1}{2}\right) \tag{2.72}$$

and

$$r = \frac{1}{2\sin(\theta)} \bullet \begin{pmatrix} r_{32} - r_{23} \\ r_{13} - r_{31} \\ r_{21} - r_{12} \end{pmatrix}. \tag{2.73}$$

2.7 Summary

This chapter provides a review of the fundamentals of a rigid body's motions. The descriptions cater to the need in robotics, to precisely establish kinematic relationships among a set of kinematically-constrained rigid bodies. The concept of frame is introduced to summarize a rigid body's mathematical abstraction in space. This abstraction indicates the need to adopt frame-centric notations when dealing with the symbols which denote motion transformations. The frame-centric notations also serve as a sound basis to unify motion-related notations in robotics and computer vision.

Motion is caused by a force or torque applied to a rigid body which is free to move in certain way (i.e. straight path/trajectory, circular path/trajectory, or a combination of both). A motion can be described in terms of displacement, velocity and acceleration. All these motion parameters could be studied by working with the frames assigned to the rigid bodies in space. Interestingly enough, the relationship between any pair of frames is consistently described by a homogenous-motion transformation, which uniformly describes translational motion, rotational motion, or the combination of both.

In this chapter, we also discussed the concept of successive elementary motions. This helps to understand the minimum number of motion parameters necessary to describe any combination of translational and rotational motions. On the other hand, it also serves as a sound basis to justify the most advantageous way of assigning frames to kinematically constrained bodies (i.e. links). This will be discussed in more detail in Chapter 3.

2.8 Exercises

(1) In a Cartesian space referenced by a frame 0 a rigid body, to which frame 1 is assigned, is free to perform any composite motion. At time instant t, the rigid body reaches a configuration which is described by this homogenous motion transformation matrix:

$$^0M_1(t) = \begin{pmatrix} r_{11} & r_{12} & r_{13} & t_x \\ r_{21} & r_{22} & r_{23} & t_y \\ r_{31} & r_{32} & r_{33} & t_z \\ 0 & 0 & 0 & 1 \end{pmatrix}$$

What is the expression of the base vectors of frame 1 with respect to frame 0 at time instant t?

(2) Explain the meaning of *homogenous coordinate*.

(3) $^0R_1(t)$ describes the actual orientation of frame 1 with respect to frame 0 at time instant t. At this moment, the rigid body associated with frame 1 undergoes a rotational motion about the Z axis of frame 0. Assume that the angular rotation angle is θ. Prove the following expression:

$$\frac{d(^0R_1(t))}{dt} = \dot{\theta} \bullet S(^0\vec{k}_0) \bullet \{^0R_1(t)\}.$$

(4) We normally denote $S(v)$ the skew-symmetric matrix of vector v. Assume that vector v is a 3×1 vector which has an inverse. Prove the following equality:

$$R \bullet S(v) \bullet R^{-1} = S(R \bullet v).$$

(5) Prove the solutions in Eq. 2.63 and Eq. 2.64.

2.9   Bibliography

(1) Baruh, H. (1999). *Analytical Dynamics*, McGraw-Hill.

(2) Nakamura, Y. (1991). *Advanced Robotics: Redundancy and Optimization*, Addison-Wesley.

(3) Rogers, D. F. and J. A. Adams (1990). *Mathematical Elements for Computer Graphics*, McGraw-Hill.

Chapter 3

Mechanical System of Robots

3.1 Introduction

Action is the source of results. Thus, mechanical system is a part of any physical system, such as a robot. In general, a mechanical system consists of: a) a structure and b) a mechanism.

A structure is a set of elements with static couplings. There is no motion between any two adjacent elements in a structure. On the other hand, a mechanism is a set of elements with flexible couplings. The purpose of mechanism is to constrain types of motion produced by a mechanical system. Thus, it is necessary to study a robot's mechanism in order to better understand the robot's functions during motion execution.

Because it takes a lot of experience and knowledge to create a mechanism, we will study the robot mechanism at the basic level. However, in this chapter, we will focus on the principles and solutions underlying the robot mechanism's kinematic analysis.

3.2 Robot Mechanism

Fig. 3.1 shows a conceptual design of a humanoid robot. Fig. 3.2 shows an actual prototype of a single-motor-driven (SMD) arm manipulator with 20 degrees of freedom. From these two examples, we can see that a robot's mechanism is an assembly of elementary rigid bodies, called *links*, which are connected through entities called *joints*.

3.2.1 *Links*

A formal definition of a link can be stated as follows:

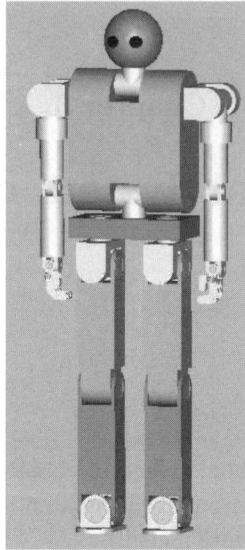

Fig. 3.1   A conceptual design of a humanoid robot.

Definition 3.1   A link is a rigid body with at least one particular point, called a *node*, which supports the attachment of other link(s).

According to the number of nodes each link possesses, we can divide the links into these common categories:

- Unary Link:
  This refers to a link with a single node. In Fig. 3.2, the big conic entity is a unary link (the virtual model offers a better view). The tip unit (i.e. tip link) of a finger assembly is also a typical unary link (see Fig. 3.3).
- Binary Link:
  This refers to a link with two nodes. The robot's arm manipulators, legs, and fingers are composed of serially-connected binary links (Figs. 3.1 and 3.2 show two examples with many binary links).
- Ternary Link:
  This refers to a link with three nodes. A typical example is the hip of the humanoid robot, shown in Fig. 3.2. One of the nodes in the hip link supports the connection to the upper body of the humanoid robot, while the other two nodes connect to the assemblies of the two legs in the humanoid robot.

<table>
<tr><td>(a) Virtual Model</td><td>(b) Real Prototype</td></tr>
</table>

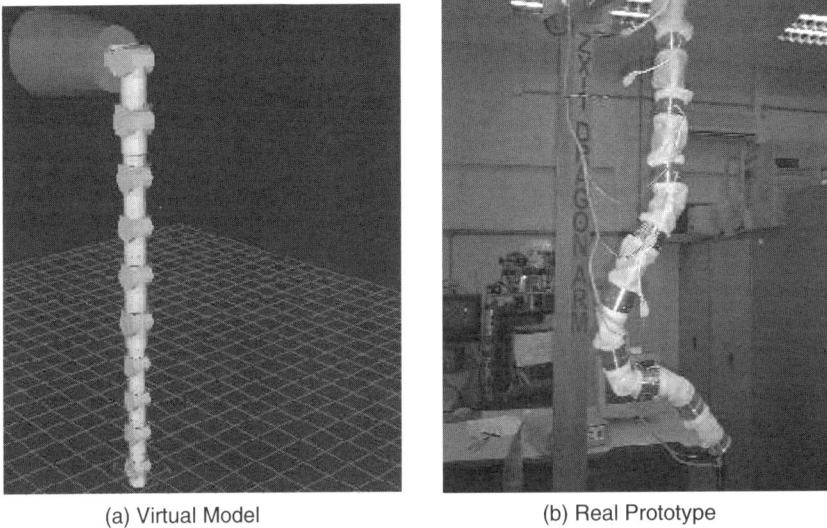

Fig. 3.2   The first prototype of the single-motor-driven robot arm manipulator: a) the virtual model and b) the actual prototype with 20 degrees of freedom, which are simultaneously and independently driven by a single motor located at the base (on top).

- Quaternary Link:
  This refers to a link with four nodes. A typical example is the upper body of the humanoid robot, shown in Fig. 3.2. The node on the top is connected to the head/neck assembly of the humanoid robot. The node at the bottom is connected to the hip assembly. The remaining two nodes, at the left and right, support the attachment of the assemblies of the left and right arms.

In practice, there are cases when a link needs more than four nodes. A typical example is the palm link inside a robot's hand. If a robot's hand has five fingers, then the palm must have 6 nodes: 5 nodes for the connection of the five fingers and one node for the attachment to the end-point of a robot's arm.

Example 3.1   Fig. 3.3 shows three examples of links: a) a binary link, b) a unary link, and c) a ternary link.

◇◇◇◇◇◇◇◇◇◇◇◇◇◇◇◇◇◇

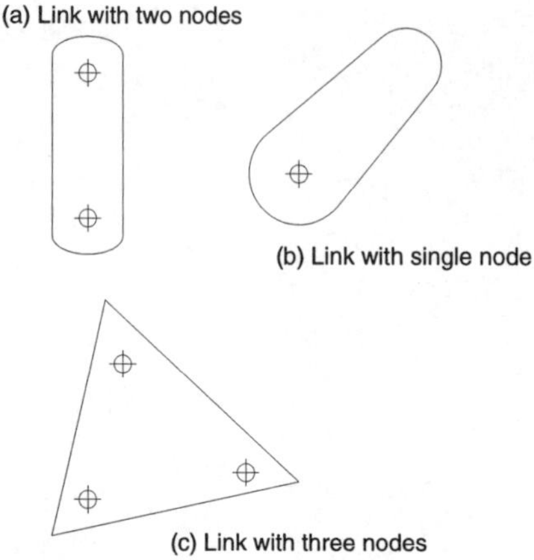

Fig. 3.3   Examples of unary, binary, and ternary links.

Example 3.2   Fig. 3.4 shows a conceptual design of a multiple-fingered hand. Notice one link with 6 nodes (the palm link), five unary links, and ten binary links (i.e. two links per finger with the first binary link of the thumb in the shape of a cross-cylinder).

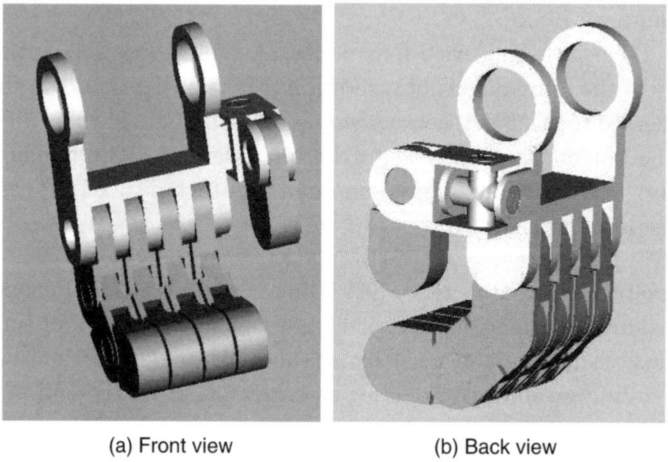

Fig. 3.4   A close-up view of the conceptual design of a multiple-fingered hand.

◇◇◇◇◇◇◇◇◇◇◇◇◇◇◇◇◇◇

3.2.2 *Joints*

A formal definition of joint is as follows:

Definition 3.2   A joint is the connection between two or more links at their nodes. It constrains the motions of the connected links.

If a joint connects only two links, the entity is also called a *kinematic pair*. Depending on the degrees of freedom (DOF) allowed for the kinematic pair, a joint can be classified as: one-DOF joint, two-DOF joint, three-DOF joint etc. In robotics, for the simplicity of kinematic and dynamic analysis, we use one-DOF joints because any joint with a higher order of DOFs can easily be treated as the combination of multiple one-DOF joints.

In a kinematic pair, one degree of freedom (DOF) needs one independent parameter, which uniquely determines the relative geometry (position and/or orientation) of the two links in the kinematic pair. Generally speaking, a mechanical system's degrees of freedom refer to the number of independent parameters, which uniquely determine its geometry (or configuration) in space and time.

In three-dimensional space, any complex motion with respect to a reference coordinate system can be treated as the combination of two basic motions: translation and rotation. Therefore, if a one-DOF joint imposes a translational motion, it is called a *prismatic joint*. If the constrained motion is rotational, it is called a *revolute joint*

Example 3.3   Fig. 3.5 shows a mechanism with one base, three links, and a robotic hand. The base has a slotted track so that Link 1 can undergo linear motion along it. The joint constraining the base and Link 1 is a prismatic joint. In a similar way, Link 1 also has a slotted track so that Link 2 can undergo linear motion along it. Thus, the joint constraining Link 1 and Link 2 is also a prismatic joint. Since Link 3 is a rod which is fixed inside the inner cylindrical tube of Link 2, Link 3 moves up and down along the axis of Link 2. Therefore, the joint constraining Link 2 and Link 3 is another prismatic joint.

We can see that the three moving directions of Links 1, 2, and 3 are mutually perpendicular to each other. A robot with such a mechanism is called a *Cartesian robot* or *XYZ robot*.

$$\diamond\diamond\diamond\diamond\diamond\diamond\diamond\diamond\diamond\diamond\diamond\diamond\diamond\diamond\diamond\diamond\diamond$$

Example 3.4   Fig. 3.6 shows a mechanism with one base and three links. Link 1 is fixed inside the inner cylindrical tube of the base, and can only

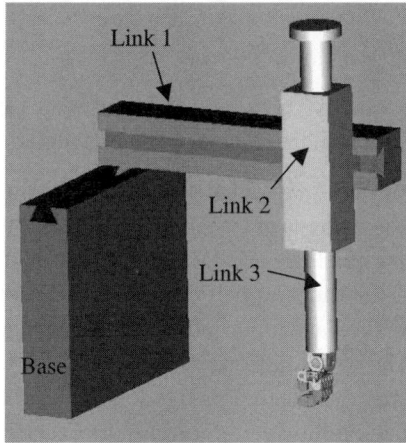

Fig. 3.5   A mechanism with three prismatic joints.

undergo a rotary motion about the axis of that tube. Hence, the joint constraining the base and Link 1 is a revolute joint. Now, look at the connection between Link 1 and Link 2. The fork-shaped node (attachment point) of Link 1 is joined using a pin to the corresponding node (attachment point) of Link 2. The joint constraining Link 1 and Link 2 is a revolute joint. In a similar way, the joint constraining Link 2 and Link 3 is also a revolute joint. A robot with such a mechanism is called an *articulated robot* or *RRR-type robot* (R stands for revolute).

◇◇◇◇◇◇◇◇◇◇◇◇◇◇◇◇◇◇

The above examples illustrate that robots may structurally be different. One way to differentiate between the various robots is to identify the first three joints in a robot's mechanism. Because a joint may be either prismatic or revolute, there are eight possible combinations. For example, an RRR-type robot means that the first three joints in the robot's mechanism are all revolute joints.

3.2.3   *Kinematic Chains*

As a robot is a physical agent which performs given tasks through the execution of motions, it should have a sufficient number of degrees of freedom in order to be flexible in a three-dimensional space where any free-moving rigid body exhibits six degrees of freedom: three for rotary motions and

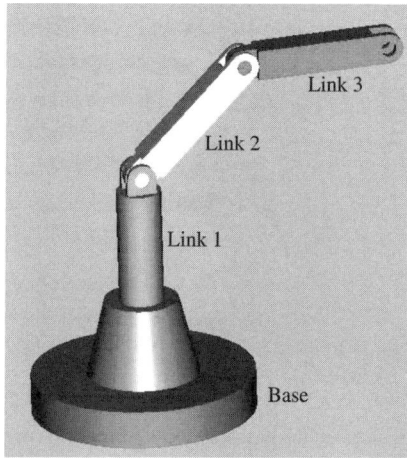

Fig. 3.6   A mechanism with three revolute joints.

three for translational motions. In general, a robot should have at least six degrees of freedom in order to be able to position a tool at a certain location. However, robots dedicated to specialized tasks may require more or less than six degrees of freedom. And, it is a challenge to figure this out at the design stage.

The common approach to arranging degrees of freedom is to form an ordered sequence of links connected in a series through the joints. This type of mechanism is called a *kinematic chain*. A formal definition of kinematic chain can be stated as follows:

Definition 3.3   A kinematic chain is an assembly of links connected in a series through joints, the output motion of which at any chosen point, only depends on the motion parameters of the joints.

A kinematic chain is said to be *open* if there is no joint connecting the first and last links inside the chain. Otherwise, it is called a *closed kinematic-chain*.

Example 3.5   Fig. 3.7 shows an example of a mechanism with one base link and six other links. These elements are all connected in a series to form an open kinematic-chain. In robotics, this mechanism is called an *articulated arm manipulator* in robotics. Refer to Fig. 3.4. It is easy to see that each finger in a robotic hand is also an open kinematic-chain.

◇◇◇◇◇◇◇◇◇◇◇◇◇◇◇◇◇◇◇

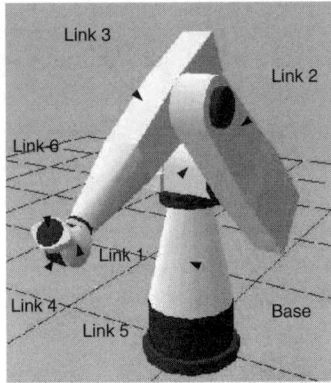

Fig. 3.7   An open kinematic-chain with one base and six links, connected in a series.

Example 3.6   Fig. 3.8 shows a conceptual design of a parallel robot mechanism. In this example, the circular disk at the top is a ternary link which is supported by three open kinematic-chains. The base links of these open kinematic-chains are fixed on a common ground plane, and are treated as a single ternary link. It is clear that this mechanism contains three closed kinematic-chains.

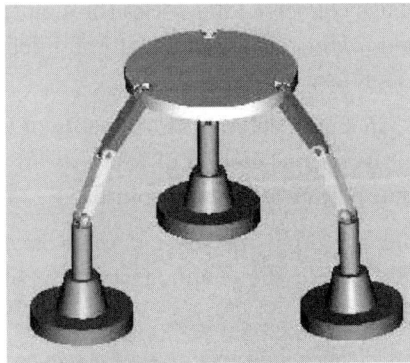

Fig. 3.8   An example of a parallel robot with three closed kinematic-chains.

◇◇◇◇◇◇◇◇◇◇◇◇◇◇◇◇◇

3.2.4   *Multiplicity of an Open Kinematic-Chain*

When we deal with an arm manipulator, we can treat the underlying mechanism as a *simple open kinematic-chain*. This is because the base link (i.e. the first link) and the end-effector link (i.e. the last link) are clearly indicated, and will remain unchanged. However, it is more complicated when dealing with a humanoid robot. Here, we use the term *simple open kinematic-chain* to differentiate it from the more general term *open kinematic-chain*. A definition of a simple open kinematic-chain can be stated as follows:

Definition 3.4   Any open kinematic-chain with designated base and end-effector links is called a simple open kinematic-chain.

Clearly, an open kinematic-chain can form different simple open kinematic-chains. As a result, it is necessary to emphasize the multiplicity of an open kinematic-chain in a humanoid robot's mechanism for at least two reasons:

(1) An open kinematic-chain inside a humanoid robot should not be treated as a simple open kinematic-chain. This is because the designations of the base and end-effector links depend on the types of activities the humanoid robot is performing.
(2) If we divide a humanoid robot's activity into a sequence of phases, an open kinematic-chain in a humanoid robot can advantageously be treated as different simple open kinematic-chains corresponding to different phases.

Example 3.7   Fig. 3.9 illustrates a simulated humanoid robot in a seated position. The right arm can be treated as a manipulator with its base link attached to the upper body. Similarly, the right leg can also be treated as a manipulator with its base link attached to the hip. In this example, the right hand and the right foot are both the end-effector links.

◇◇◇◇◇◇◇◇◇◇◇◇◇◇◇◇◇◇◇

Example 3.8   Fig. 3.10 illustrates a simulated humanoid robot in a standing position. For the activities performed by a humanoid robot in such a standing position, it is easy to study the motions if the right leg, the body, and the right arm/hand form a simple open kinematic-chain, like an arm manipulator. In this case, the right foot serves as the base link, and the right hand is the end-effector link.

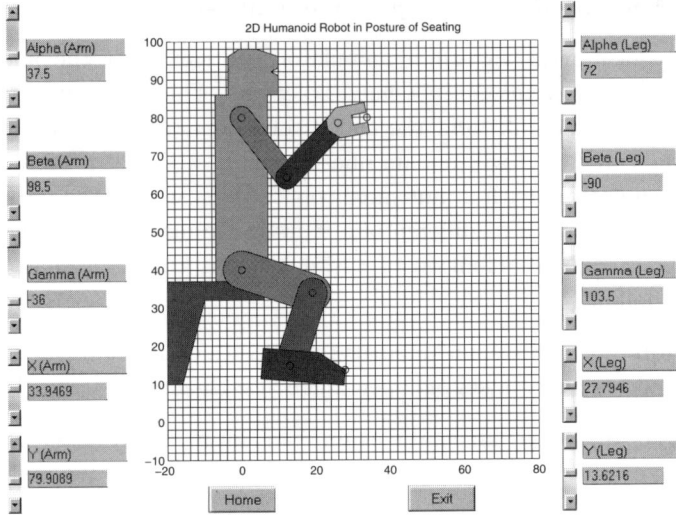

Fig. 3.9   A simulated humanoid robot in a seated position, where the right arm and the right leg can be treated as two independent manipulators.

◇◇◇◇◇◇◇◇◇◇◇◇◇◇◇◇◇

Example 3.9   A robot's human-like biped walking can be treated as a combination of two sequences of successive cycles: a) the sequence of the left leg's walking cycles and b) the sequence of the right leg's walking cycles. Each cycle has three successive phases. Fig. 3.11 shows the first two phases of the left leg's walking cycle (the third phase overlaps with the first phase of the right leg's walking cycle).

A cycle begins with the first phase. The first phase starts when the left heel elevates off the ground, and ends when the left toe elevates off the ground. The second phase starts at the end time-instant of the first phase, and ends when the left heel touches the ground again. Within the left leg's walking cycle, it is clear that we can consider the right foot to be the base link of the simple open kinematic-chain formed by the two legs.

◇◇◇◇◇◇◇◇◇◇◇◇◇◇◇◇◇

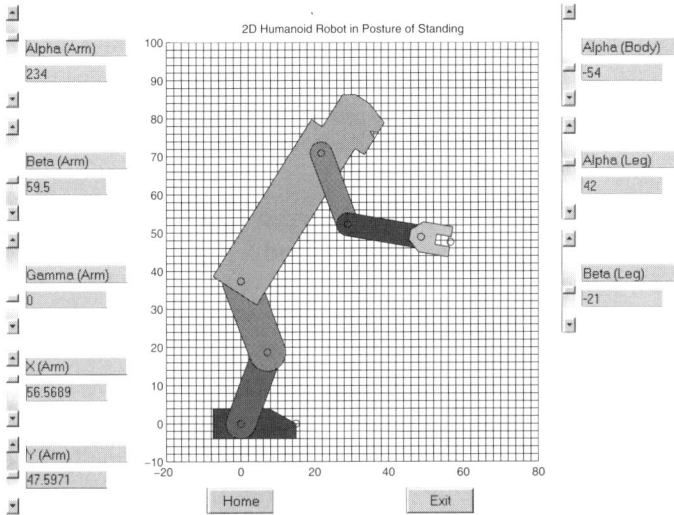

Fig. 3.10 A simulated humanoid robot in a standing position, where the right leg, the body, and the right arm/hand can advantageously be treated as a simple open kinematic-chain.

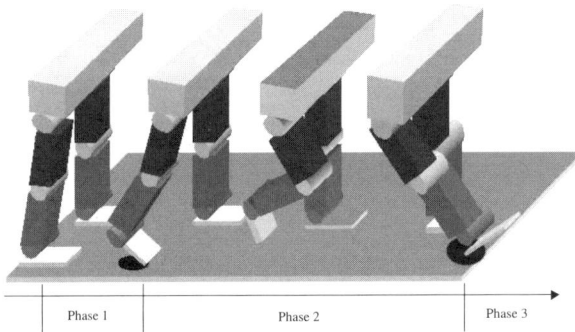

Fig. 3.11 Illustration of a simple open kinematic-chain for the study of the left leg's walking cycle when the right foot is treated as the base link.

3.3 Robot Kinematics

From a mechanical point of view, a robot is an assembly of a set of links where any pair of two adjacent links is connected by a one-DOF (Degree Of Freedom) joint. Every joint inside a robot's mechanism can be either a prismatic or revolute joint which imposes one degree of freedom and can be described by one input motion parameter. Since the purpose of a robot's mechanism is to shape the output motion at a particular point on the mechanism itself, it is important to understand the relationship between the robot mechanism's input motion parameters and its output motion parameters.

It is worth noting at this point, that the robot mechanism's output motion is specified and described with respect to a Cartesian space which is commonly called *task space* in robotics. And a three-dimensional Cartesian space can be represented by the orthogonal coordinate system $O - XYZ$, where a general motion involves six parameters: a) three for translation and b) three for rotation.

As for the robot mechanism's input motions of a robot's mechanism, a joint imposes one kinematic constraint on the corresponding kinematic-pair. The motion parameter of a one-DOF joint is normally represented by a variable which describes the relative position between the two links in the kinematic pair. The motion parameter of joint i is commonly called a *joint variable* and is denoted by q_i. The joint variable q_i refers to an angle if joint i is a revolute joint. Otherwise, it refers to a linear or circular displacement. All the joint variables in a mechanism form a vector of joint coordinates which define a space commonly called a *joint space*.

Therefore, the robot mechanism's kinematic analysis is simply the study of the mapping between task space and joint space. (See Fig. 3.12). Mapping from joint space to task space is called *forward kinematics*, while the reverse mapping (from task space to joint space) is called *inverse kinematics*.

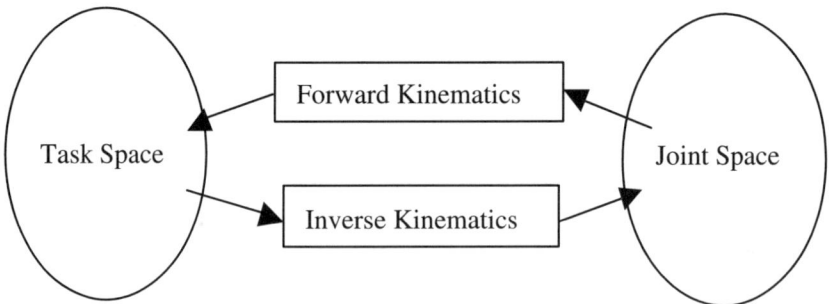

Fig. 3.12   An illustration of kinematic mapping.

3.3.1   *Kinematics of a Link*

It is a common practice in science and engineering to develop a systematic approach to describe any complex system. As for kinematic analysis, it is desirable to have a systematic method.

Since a robot's mechanism can be treated as the combination of a set of open kinematic-chains, it is important to develop a theoretical framework for the modelling and analysis of an open kinematic-chain. An open kinematic-chain consists of a set of serially connected links. Thus, the starting point for kinematic analysis is to study the kinematics of a link.

3.3.1.1   *Assignment of a Link's Frame*

As we discussed in Chapter 2, the study of motions among rigid bodies is equivalent to the study of motions among frames assigned to the rigid bodies under consideration. In general, motion transformation between any pair of frames involves six motion parameters: three for the elementary translations and three for the elementary rotations. In robotics, there is a simple method which only requires four motion parameters to represent the relative posture between two consecutive links. This method is called *Devanit-Hartenberg* (or DH) representation. The basic idea behind DH representation can be explained as follows:

At a time instant, an open kinematic-chain can be treated as a simple open kinematic-chain with a clear designation of its base and end-effector links. For the example shown in Fig.3.7, the open kinematic-chain has one base link and three binary links. We treat a unary link as a degenerated binary link, the two nodes of which are merged into one. For clarity's sake, the node of a binary link which is closer to the base link (when measured along the open kinematic-chain) is called the *proximal node*, while the other node is called the *distal node*. With reference to a link, the joint at the proximal node is called the *proximal joint* and the one at the distal node is called the *distal joint*.

For link i inside a simple open kinematic-chain, the question is: Where do we place frame i assigned to it? If we want to reduce the number of motion parameters, it is imperative to constrain a frame's orientation and origin as much as possible. Since two consecutive links form a kinematic pair (the corresponding joint has one degree of freedom), a simple solution is to constrain one of the X, Y, or Z axes of frame i by aligning it with the motion axis of the joint.

In robotics, a common way is to align the Z axis of a link's frame with the motion axis of one of its two joints. Now, the question is: Should we align the Z axis of frame i with the motion axis of the proximal joint or the distal joint? The obvious answer is to align it with the motion axis of the distal joint, as illustrated in Fig. 3.13, because the robot mechanism's output motion is usually at the distal joint of the end-effector link.

It is important to note that the end-effector link in a simple open kinematic-chain is physically driven by its proximal joint when an actuator applies force/torque to it. This fact is also true for the rest of the links inside a simple open kinematic-chain. For easy notation, the proximal joint of link i is called joint i. The designation of link i and joint i is also illustrated in Fig. 3.13.

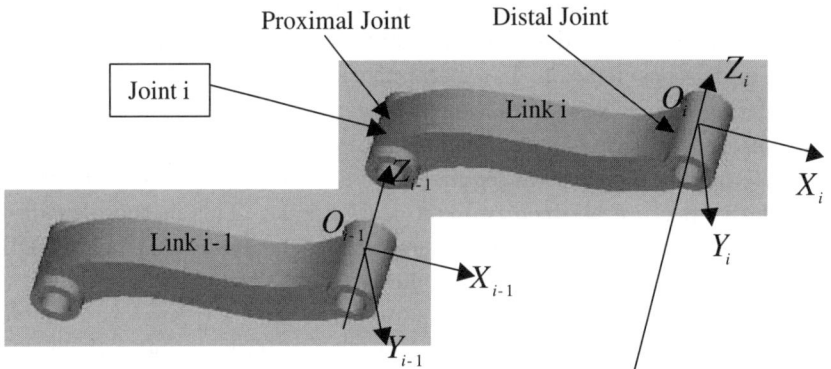

Fig. 3.13   Designation of link i and joint i, including frame assignment to link i.

Once we have imposed a constraint on the Z axis of a link's frame, it is also easy to impose a constraint on the link frame's origin along the Z axis. This can be done by placing the origin at the intersection point between the Z_i axis (its direction is represented by $^0\vec{k}_i$ with respect to frame 0) and the common normal vector from the Z_{i-1} axis to the Z_i axis. And, this common normal vector defines the X axis of frame i. If $^0\vec{k}_i$ and $^0\vec{k}_{i-1}$ (i.e. the direction vector of the Z_{i-1} axis with respect to frame 0) are parallel to each other (as shown in Fig. 3.13), the origin of frame i is chosen to be at the center of the distal joint.

Finally, the Y axis of frame i is determined by the right-hand Rule, so that a Cartesian coordinate system is formed for frame i.

Example 3.10   Fig. 3.14 shows a link which has non-parallel motion axes at the proximal and distal joints. In this case, it is easy to determine the origin of the frame assigned to the link. It is at the center of the distal joint.

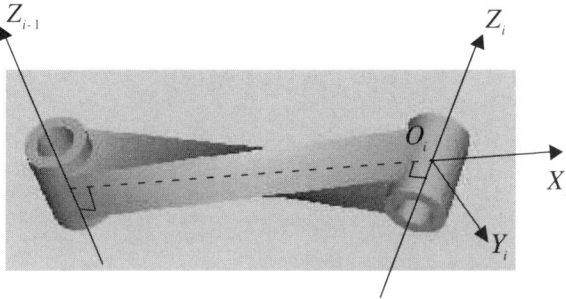

Fig. 3.14   A link which has non-parallel motion axes at the proximal and distal joints are not parallel.

◇◇◇◇◇◇◇◇◇◇◇◇◇◇◇◇◇◇◇

Example 3.11   Fig. 3.15 shows a link which has motion axes at the proximal and distal joints that are co-planar and intersect at a point. In this case, the origin of the frame assigned to the link is at this intersection point, and the X axis is determined by the cross-product of the Z_i and Z_{i-1} axes' base vectors:

$$^0\vec{i}_i = ^0\vec{k}_i \times ^0\vec{k}_{i-1}.$$

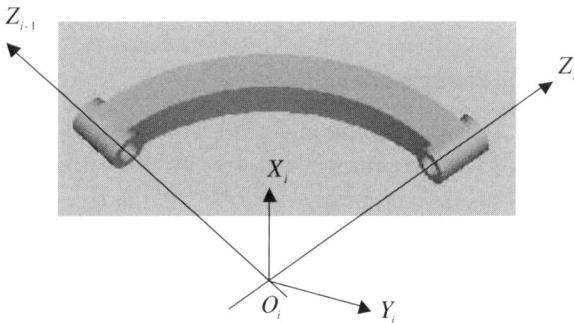

Fig. 3.15   A link which has motion axes at the proximal and distal joints that are co-planar and intersect at a point.

◇◇◇◇◇◇◇◇◇◇◇◇◇◇◇◇◇◇◇

Example 3.12   Fig. 3.16 shows a link which has motion axes at the proximal and distal joints that are in a general configuration. In this case, the common normal vector from the Z_{i-1} axis to the Z_i axis is

$$^0\vec{i}_i = \{^0\vec{k}_i\} \times \{^0\vec{k}_{i-1}\}.$$

This normal vector indicates the X axis of the frame assigned to the link.

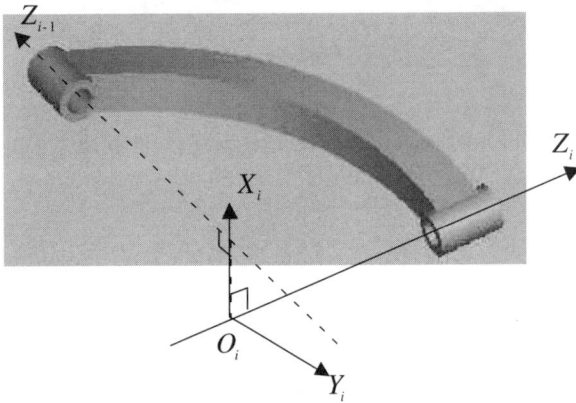

Fig. 3.16   A link which has motion axes at the proximal and distal joints that are in a general configuration.

◇◇◇◇◇◇◇◇◇◇◇◇◇◇◇◇◇◇◇

3.3.1.2   *Geometric Parameters of a Link*

In Fig. 3.17, we make use of two parallel planes to better illustrate the frame assignment to link i. It also shows in a clearer way that the minimum set of four parameters is necessary to fully describe the relative posture between the two consecutive links. Let us assume that the Z_{i-1} and Z_i axes are not collinear (i.e. not on the same line). In practice, there is no reason to design two consecutive links in such an arrangement.

If we know the common normal vector $^0\vec{i}_i$ (i.e. the X_i axis), we can define two parallel planes p_i and p_{i-1}, both using this vector as their normal vector. And, plane p_i contains the Z_i axis while plane p_{i-1} contains the

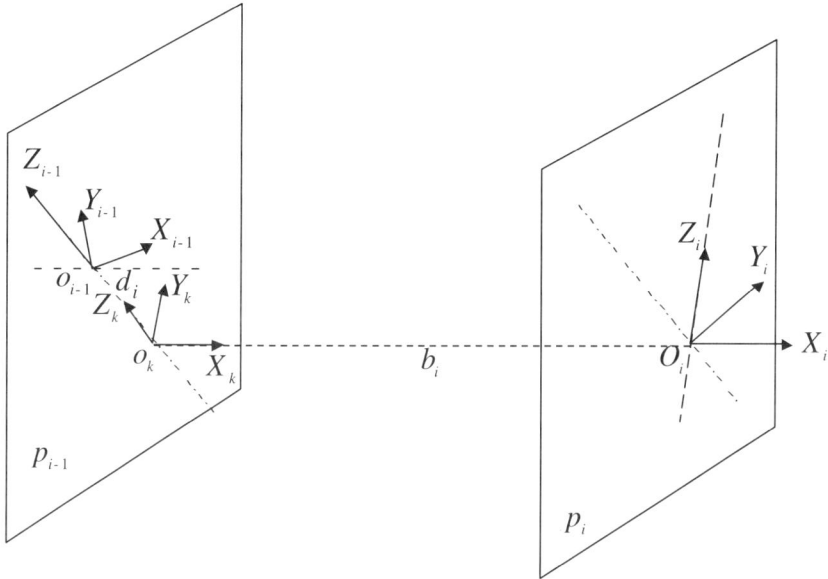

Fig. 3.17   Illustration of DH parameters, using two parallel planes.

Z_{i-1} axis. Refer to Fig. 3.17. Frame $i-1$ is placed in a general orientation which depends on link $i-1$.

Refer to Fig. 3.17 again. It is clear that the first DH parameter is the rotation angle β_i about the X_i axis which makes the Z_i axis parallel to the Z_{i-1} axis. This angle β_i indicates an elementary rotation of frame i about the X_i axis if we want to bring frame i back to frame $i-1$.

Obviously, the second DH parameter is the distance b_i between these two parallel planes. In fact, this distance b_i indicates the linear displacement of an elementary translation by frame i along the X_i axis if we want to bring frame i back to frame $i-1$.

By applying two successive elementary motions: the elementary rotation about the X_i axis followed by the elementary translation along the X_i axis, frame i will reach an intermediate configuration which is denoted by frame k in Fig. 3.17.

Now, it becomes clear that the third DH parameter is rotation angle θ_i about the Z_k axis which is aligned with the Z_{i-1} axis. This rotation aligns frame k parallel to frame $i-1$. If joint i of the simple open kinematic-chain is a revolute joint, then θ_i is the *joint variable* corresponding to link i (i.e. $q_i = \theta_i$).

Finally, it is necessary to make a final elementary translation along the Z_{i-1} axis in order to move the origin of frame k to coincide with the origin of frame $i - 1$. The linear displacement d_i of this elementary translation is the fourth DH parameter. It is clear that this fourth parameter d_i will be the *joint variable* if joint i is a prismatic joint (i.e. $q_i = d_i$).

In summary, DH representation only requires a set of four parameters to fully determine the relative posture between two consecutive frames. These four parameters are:

(1) β_i: rotation angle about the X_i axis;
(2) b_i: linear displacement along the X_i axis;
(3) θ_i: rotation angle about the Z_{i-1} axis, which is the *joint variable* if joint i is a revolute joint;
(4) d_i: linear displacement along the Z_{i-1} axis, which is the *joint variable* if joint i is a prismatic joint.

Example 3.13   In Fig. 3.9, the right arm and right leg are simple open kinematic-chains with four planar links (including the common base link). The frame assignments and the geometrical parameters of the right arm and right leg, shown in Fig. 3.9, are generically illustrated by Fig. 3.18. Based on the frame assignments, the DH parameters of the simple open kinematic-chain shown in Fig. 3.18 are:

Link	β_i	b_i	θ_i	d_i
1	0	l_1	θ_1	0
2	0	l_2	θ_2	0
3	0	l_3	θ_3	0

◇◇◇◇◇◇◇◇◇◇◇◇◇◇◇◇◇◇

From the above discussions, we can see that the DH parameters are not the same if we alter the designation of the base link and end-effector link for the same open kinematic-chain. As a result, an open kinematic-chain requires two sets of DH parameters to fully describe its kinematic property, especially in the case of the humanoid robot.

3.3.1.3   *Motion Transformation Matrix of a Link*

If we know the DH parameters of link i, we can quickly and systematically establish the motion transformation matrix which describes the position and orientation of frame i with respect to frame $i - 1$. Let us denote $^{i-1}M_i$

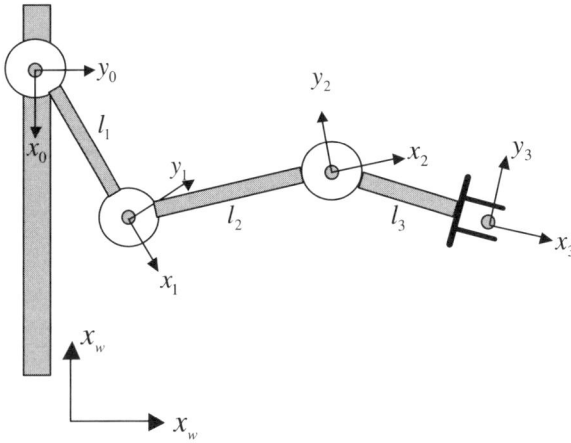

Fig. 3.18 Simple open kinematic-chain with four planar links (including the base link). Frame w is a world coordinate system.

the motion transformation matrix from frame i to frame $i - 1$. Based on this notation, we have to pay attention to the signs of the values of the DH parameters. Refer to Fig. 3.17. An angle takes a positive value if the rotation is clockwise about the axis of rotation, and a linear displacement takes a positive value if the translation is towards the negative axis of translation. For the example shown in Fig. 3.17, where θ_i is the joint variable, (β_i, d_i) take negative values and b_i has a positive value.

Refer to Fig. 3.17 again. It is easy to write the homogeneous motion-transformation matrix ${}^k M_i$, which describes the position and orientation of frame i with respect to frame k as follows:

$$
{}^k M_i = \begin{pmatrix} 1 & 0 & 0 & b_i \\ 0 & \cos(\beta_i) & -\sin(\beta_i) & 0 \\ 0 & \sin(\beta_i) & \cos(\beta_i) & 0 \\ 0 & 0 & 0 & 1 \end{pmatrix}. \tag{3.1}
$$

Similarly, the position and orientation of frame k, with respect to frame $i - 1$, is fully determined by the following homogeneous motion-transformation matrix:

$$
{}^{i-1} M_k = \begin{pmatrix} \cos(\theta_i) & -\sin(\theta_i) & 0 & 0 \\ \sin(\theta_i) & \cos(\theta_i) & 0 & 0 \\ 0 & 0 & 1 & d_i \\ 0 & 0 & 0 & 1 \end{pmatrix}. \tag{3.2}
$$

The combination of Eq. 3.1 and Eq. 3.2 yields

$$^{i-1}M_i = \begin{pmatrix} C\theta_i & -C\beta_i \bullet S\theta_i & S\beta_i \bullet S\theta_i & b_i \bullet C\theta_i \\ S\theta_i & C\beta_i \bullet C\theta_i & -S\beta_i \bullet C\theta_i & b_i \bullet S\theta_i \\ 0 & S\beta_i & C\beta_i & d_i \\ 0 & 0 & 0 & 1 \end{pmatrix} \qquad (3.3)$$

where

$$S\beta_i = \sin(\beta_i), \qquad C\beta_i = \cos(\beta_i),$$

$$S\theta_i = \sin(\theta_i), \qquad C\theta_i = \sin(\theta_i).$$

Eq. 3.3 is called the *link matrix* in robotics, and summarizes both the position and orientation of frame i with respect to frame $i-1$. The position of frame i, with respect to frame $i-1$, is

$$^{i-1}O_i = \begin{pmatrix} b_i \bullet C\theta_i \\ b_i \bullet S\theta_i \\ d_i \end{pmatrix}. \qquad (3.4)$$

And the orientation of frame i, with respect to frame $i-1$, is

$$^{i-1}R_i = \begin{pmatrix} C\theta_i & -C\beta_i \bullet S\theta_i & S\beta_i \bullet S\theta_i \\ S\theta_i & C\beta_i \bullet C\theta_i & -S\beta_i \bullet C\theta_i \\ 0 & S\beta_i & C\beta_i \end{pmatrix}. \qquad (3.5)$$

From Eq. 3.4, we can see that the position of frame i depends on (θ_i, d_i). This means that the position of frame i always undergoes changes, regardless of whether it is a revolute joint (meaning θ_i is the joint variable) or prismatic joint (meaning d_i is the joint variable). However, this is not the case for the orientation of frame i which will only undergo changes if the joint is revolute (i.e. θ_i is the joint variable).

3.3.1.4   *Linear/Circular Velocity of a Link*

Eq. 3.3 describes the configuration (or posture) of frame i with respect to frame $i-1$. Another important property of robot kinematics is the differential relationship between the joint variable and the configuration of frame i. This is called *motion kinematics* or *differential kinematics*.

If joint i is a prismatic joint, we have

$$q_i = d_i.$$

By differentiating Eq. 3.4 with respect to time, we can obtain the linear velocity vector at the origin of frame i with respect to frame $i-1$. That is,

$$^{i-1}\dot{O}_i = \frac{d\{^{i-1}O_i\}}{dt} = \begin{pmatrix} 0 \\ 0 \\ \frac{dq_i}{dt} \end{pmatrix} = \begin{pmatrix} 0 \\ 0 \\ 1 \end{pmatrix} \bullet \frac{dq_i}{dt}. \tag{3.6}$$

This linear velocity vector is in the direction of the Z_{i-1} axis (i.e. $^{i-1}\vec{k}_{i-1} = (0,0,1)^t$). The above equation can be rewritten, as follows:

$$^{i-1}\dot{O}_i = \, ^{i-1}\vec{k}_{i-1} \bullet \dot{q}_i. \tag{3.7}$$

Obviously, a prismatic joint will not cause any changes in the orientation of frame i. If we denote $^{i-1}w_i$ the angular velocity vector of frame i with respect to frame $i-1$, we have $^{i-1}w_i = \vec{0}$ if joint i is a prismatic joint.

On the other hand, the joint variable for a revolute joint i is

$$q_i = \theta_i.$$

Let $^{i-1}w_i$ be the angular velocity vector of frame i. Due to the instantaneous change of joint variable q_i about the Z_{i-1} axis, the circular velocity vector of any point P in frame i with respect to frame $i-1$ will be

$$^{i-1}\dot{P} = \frac{d\{^{i-1}P\}}{dt} = (^{i-1}w_i) \times (^{i-1}P) \tag{3.8}$$

where

$$^{i-1}w_i = \, ^{i-1}\vec{k}_{i-1} \bullet \dot{q}_i. \tag{3.9}$$

If we choose point P to be the origin of frame i, we have

$$^{i-1}\dot{O}_i = (^{i-1}w_i) \times (^{i-1}O_i) \tag{3.10}$$

or

$$^{i-1}\dot{O}_i = \{(^{i-1}\vec{k}_{i-1}) \times (^{i-1}O_i)\} \bullet \dot{q}_i. \tag{3.11}$$

In summary, the velocity vector at the origin of frame i, with respect to frame $i-1$, is

$$^{i-1}\dot{O}_i = \begin{cases} ^{i-1}\vec{k}_{i-1} \bullet \dot{q}_i & \text{if prismatic joint} \\ \\ \{(^{i-1}\vec{k}_i) \times (^{i-1}O_i)\} \bullet \dot{q}_i & \text{if revolute joint.} \end{cases} \tag{3.12}$$

3.3.1.5   *Angular Velocity of a Link*

From the above discussions, it is clear that there is no angular velocity if joint i is prismatic. That is,

$$^{i-1}\omega_i = \vec{0}. \tag{3.13}$$

If joint i is revolute, the angular velocity of frame i with respect to frame $i-1$ will be (see Eq. 3.9)

$$^{i-1}\omega_i = {}^{i-1}\vec{k}_{i-1} \bullet \dot{q}_i. \tag{3.14}$$

In summary, the angular velocity vector of frame i with respect to frame $i-1$ is

$$^{i-1}\omega_i = \begin{cases} \vec{0} & \text{if prismatic joint} \\ {}^{i-1}\vec{k}_{i-1} \bullet \dot{q}_i & \text{if revolute joint.} \end{cases} \tag{3.15}$$

3.3.2   *Forward Kinematics of Open Kinematic-Chains*

Let us consider a simple open kinematic-chain having $n+1$ links as shown in Fig 3.19. Link 0 is the base link and link n is the end-effector link. If we denote q_i the joint variable corresponding to link i, then joint variable vector q of the simple open kinematic-chain will be

$$q = (q_1, q_2, ..., q_n)^t. \tag{3.16}$$

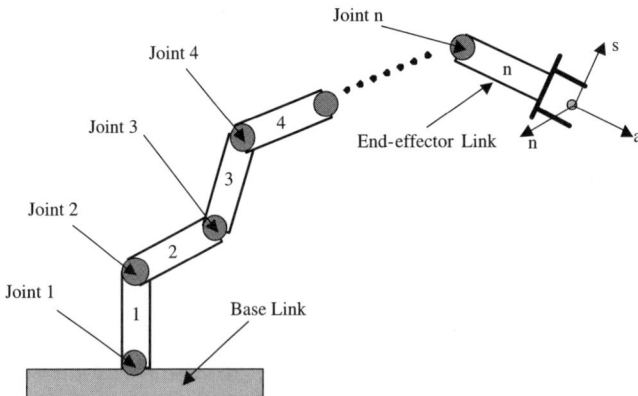

Fig. 3.19   A simple open kinematic-chain with $n+1$ links.

Here, the objective of the study of forward kinematics is to determine the configuration (i.e. position and orientation) and velocity of the end-effector's frame (i.e. frame n) if the joint variable vector and its velocity are known.

3.3.2.1 *Determination of the End-effector's Posture*

If we denote $^{i-1}M_i(q_i)$ the homogeneous motion-transformation of frame i with respect to frame $i-1$, then the homogeneous motion-transformation of frame n (the end-effector link's frame) with respect to frame 0 (the base link's frame) will be as follows:

$$^{0}M_n(q) = \{^{0}M_1(q_1)\} \bullet \{^{1}M_2(q_2)\} \bullet \dots \bullet \{^{n-1}M_n(q_n)\} \qquad (3.17)$$

or

$$^{0}M_n(q) = \begin{pmatrix} ^{0}R_n(q) & ^{0}T_n(q) \\ \vec{0} & 1 \end{pmatrix} = \prod_{i=1}^{n}\{^{i-1}M_i(q_i)\} \qquad (3.18)$$

(NOTE: $\vec{0} = (0,0,0)$).

The physical meaning of the above equation is that frame n will coincide with frame 0 after a series of successive motions, which pass through frames $n-1$, $n-2$, ..., and 1. The orientation of frame n, with respect to frame 0, is represented by rotation matrix $^{0}R_n(q)$, which is a function of the joint variable vector q. And, the position of frame n's origin, with respect to frame 0, is represented by translation vector $^{0}T_n(q)$, which is also a function of the joint variable vector q.

We have joint variable vector q, the position and orientation of frame n, with respect to frame 0, can be directly computed from Eq. 3.18. There-fore, $^{0}M_n(q)$ describes forward kinematics mapping from joint space to task space.

Example 3.14 Refer to Fig. 3.9. The right arm is treated as a simple open kinematic-chain with the upper body as its base link. The lengths of the upper link (Link 1), lower link (Link 2), and hand link (Link 3) of the right arm are $20cm$, $20cm$ and $8cm$ respectively. The origin of the base link's frame (frame 0) is at $(0, 80cm)$ in the world frame. If the input of the joint angles are as follows:

$$\begin{cases} q_1 = 37.5^0 \\ q_2 = 98.5^0 \\ q_3 = -36.0^0, \end{cases}$$

then the origin of the hand link's frame, with respect to the world frame, will be at

$$\begin{cases} x = 33.9469cm \\ y = 79.9089cm. \end{cases}$$

◇◇◇◇◇◇◇◇◇◇◇◇◇◇◇◇◇◇

Example 3.15   Refer to Fig. 3.9 again. The right leg is also treated as a simple open kinematic-chain with the upper body as its base link. The lengths of the upper link (Link 1), lower link (Link 2), and foot link (Link 3) of the right leg are 20cm, 20cm and 15cm respectively. The origin of the base link's frame (frame 0) is at $(0, 40cm)$ in the world frame. If the input of the joint angles are as follows:

$$\begin{cases} q_1 = 72.0^0 \\ q_2 = -90.0^0 \\ q_3 = 103.5^0, \end{cases}$$

then the origin of the foot link's frame, with respect to the world frame, is at

$$\begin{cases} x = 27.7946cm \\ y = 13.6216cm. \end{cases}$$

◇◇◇◇◇◇◇◇◇◇◇◇◇◇◇◇◇◇

3.3.2.2   *Determination of the End-effector's Velocity*

In a simple open kinematic-chain, the constraints imposed by the joints are independent of each other. In other words, joint 1 will independently move the assembly consisting of links 1, 2, ..., until n; and joint 2 will independently move the sub-assembly consisting of links 2, 3, until n etc. Now, the question is: What will the linear and angular velocities of frame n be, with respect to frame 0, when the velocities $\dot{q}$ of the joint variables q are given?

For simplicity's sake, we use the term *linear* to mean the combined effect of linear and circular velocities which act at the origin of a frame. Refer to Fig. 3.19. As the joints independently act on their corresponding sub-assemblies of links, the linear velocity vector (the combined effect of linear and circular velocities) at the origin of frame n will be the sum of results obtained from all the joints.

Now, let us examine the contribution of joint i to the linear and angular velocities of frame n:

We consider that the sub-assembly from links i until n forms a single link supporting frame n. By applying Eq. 3.12, the linear velocity vector caused by joint i at the origin of frame n, with respect to frame $i - 1$, will be

$$^{i-1}\dot{O}_n(\dot{q}_i) = \begin{cases} ^{i-1}\vec{k}_{i-1} \bullet \dot{q}_i & \text{if prismatic joint} \\ \{(^{i-1}\vec{k}_{i-1}) \times (^{i-1}O_n)\} \bullet \dot{q}_i & \text{if revolute joint} \end{cases} \quad (3.19)$$

where $^{i-1}O_n$ is the position vector from the origin of frame n to the origin of frame $i - 1$.

By multiplying rotation matrix $^0R_{i-1}$ to both sides of Eq. 3.19, we obtain the linear velocity vector at the origin of frame n, with respect to frame 0. That is,

$$^0\dot{O}_n(\dot{q}_i) = \begin{cases} ^0\vec{k}_{i-1} \bullet \dot{q}_i & \text{if prismatic joint} \\ \{(^0\vec{k}_{i-1}) \times (^0R_{i-1} \bullet ^{i-1}O_n)\} \bullet \dot{q}_i & \text{if revolute joint.} \end{cases} \quad (3.20)$$

As $^{i-1}O_n$ is the position vector from the origin of frame n to the origin of frame $i - 1$, we have

$$^0R_{i-1} \bullet ^{i-1}O_n = {}^0O_n - {}^0O_{i-1} \quad (3.21)$$

with

$$^0O_n = {}^0M_n \bullet \begin{pmatrix} 0 \\ 0 \\ 0 \\ 1 \end{pmatrix} \qquad ^0O_{i-1} = {}^0M_{i-1} \bullet \begin{pmatrix} 0 \\ 0 \\ 0 \\ 1 \end{pmatrix}. \quad (3.22)$$

If we substitute Eq. 3.21 into Eq. 3.20, and define

$$J_{oi} = \begin{cases} ^0\vec{k}_{i-1} & \text{if prismatic joint} \\ \{(^0\vec{k}_{i-1}) \times (^0O_n - {}^0O_{i-1})\} & \text{if revolute joint,} \end{cases} \quad (3.23)$$

we have

$$^0\dot{O}_n(\dot{q}_i) = J_{oi} \bullet \dot{q}_i. \quad (3.24)$$

Now, let us compute the sum of the results obtained from all the joints. The final result of the linear velocity at the origin of frame n, with respect to frame 0, will be

$$^{0}\dot{O}_n(\dot{q}) = \sum_{i=1}^{n}\{^{0}\dot{O}_n(\dot{q}_i)\} = \sum_{i=1}^{n}\{J_{oi} \bullet \dot{q}_i\} \tag{3.25}$$

or in matrix form

$$^{0}\dot{O}_n(\dot{q}) = J_o \bullet \dot{q} \tag{3.26}$$

with

$$\begin{cases} J_o = (J_{o1}, J_{o2}, ..., J_{on}) \\ \dot{q} = (\dot{q}_1, \dot{q}_2, ..., \dot{q}_n)^t. \end{cases} \tag{3.27}$$

Eq. 3.26 describes the mapping of the joint variables' velocities to the linear velocity at the origin of frame n.

Similarly, by applying Eq. 3.15, the angular velocity vector of frame n caused by joint i, with respect to frame $i-1$, will be

$$^{i-1}\omega_n(\dot{q}_i) = \begin{cases} \vec{0} & \text{if prismatic joint} \\ ^{i-1}\vec{k}_{i-1} \bullet \dot{q}_i & \text{if revolute joint.} \end{cases} \tag{3.28}$$

If we multiply rotation matrix $^{0}R_{i-1}$ to both sides of Eq. 3.28, and define

$$J_{\omega i} = \begin{cases} \vec{0} & \text{if prismatic joint} \\ ^{0}\vec{k}_{i-1} & \text{if revolute joint,} \end{cases} \tag{3.29}$$

then Eq. 3.28 can be rewritten as follows:

$$^{0}\omega_n(\dot{q}_i) = J_{\omega i} \bullet \dot{q}_i. \tag{3.30}$$

Now, if we compute the sum of results obtained from all the joints, the final result of the angular velocity vector of frame n, with respect to frame 0, will be

$$^{0}\omega_n(\dot{q}) = \sum_{i=1}^{n}\{^{0}\omega_n(\dot{q}_i)\} = \sum_{i=1}^{n}\{J_{\omega i} \bullet \dot{q}_i\} \tag{3.31}$$

or in matrix form

$$^{0}\omega_n(\dot{q}) = J_\omega \bullet \dot{q} \tag{3.32}$$

with

$$\begin{cases} J_\omega = (J_{\omega 1}, J_{\omega 2}, ..., J_{\omega n}) \\ \dot{q} = (\dot{q}_1, \dot{q}_2, ..., \dot{q}_n)^t. \end{cases} \tag{3.33}$$

Eq. 3.32 describes the mapping of the joint variables' velocities to the angular velocity vector of frame n.

If we denote

$$\dot{P} = \begin{pmatrix} {}^0\dot{O}_n \\ {}^0\omega_n \end{pmatrix} \tag{3.34}$$

the velocity vector of frame n, with respect to frame 0, and further define

$$J(q) = \begin{pmatrix} J_o \\ J_\omega \end{pmatrix}, \tag{3.35}$$

the combination of Eq. 3.26 and Eq. 3.32 will yield

$$\dot{P} = J(q) \bullet \dot{q}. \tag{3.36}$$

Eq. 3.36 describes the mapping from the joint variables' velocities to the velocity of frame n. This mapping is fully conditioned by matrix $J(q)$, commonly called the *Jacobian matrix*. If the dimension of $\dot{P}$ is m ($m \leq 6$), $J(q)$ is a $m \times n$ matrix. It is clear that the inverse mapping is many-to-one if $n > m$. It is worth noting that the Jacobian matrix is not a constant matrix and it depends on the actual values of the joint variables.

Example 3.16   Refer to Fig. 3.9. We name the upper link, the lower link and the hand link of the right arm as Links 1, 2 and 3 respectively. The velocity of frame 3, with respect to frame 0 (the upper body), will be

$$\dot{P} = J(q) \bullet \begin{pmatrix} \dot{\theta}_1 \\ \dot{\theta}_2 \\ \dot{\theta}_3 \end{pmatrix} = J(q) \bullet \dot{q}$$

with

$$J(q) = \begin{pmatrix} ({}^0\vec{k}_0) \times ({}^0O_3 - {}^0O_0) & ({}^0\vec{k}_1) \times ({}^0O_3 - {}^0O_1) & ({}^0\vec{k}_2) \times ({}^0O_3 - {}^0O_2) \\ {}^0\vec{k}_0 & {}^0\vec{k}_1 & {}^0\vec{k}_2 \end{pmatrix}$$

and

$$ {}^0O_3 = {}^0M_3 \bullet \begin{pmatrix} 0 \\ 0 \\ 0 \\ 1 \end{pmatrix} \qquad {}^0O_2 = {}^0M_2 \bullet \begin{pmatrix} 0 \\ 0 \\ 0 \\ 1 \end{pmatrix} \qquad {}^0O_1 = {}^0M_1 \bullet \begin{pmatrix} 0 \\ 0 \\ 0 \\ 1 \end{pmatrix}. $$

◇◇◇◇◇◇◇◇◇◇◇◇◇◇◇◇◇◇◇◇

3.3.3   *Inverse Kinematics of Open Kinematic-Chains*

The robot mechanism's output motion is usually at the end-effector link of a simple open kinematic-chain. If we know the joint variables and their velocities, it is simple to compute the robot mechanism's output motion because of the simple formulation of forward kinematics. Then, the question is: Why is there a need to study inverse kinematics?

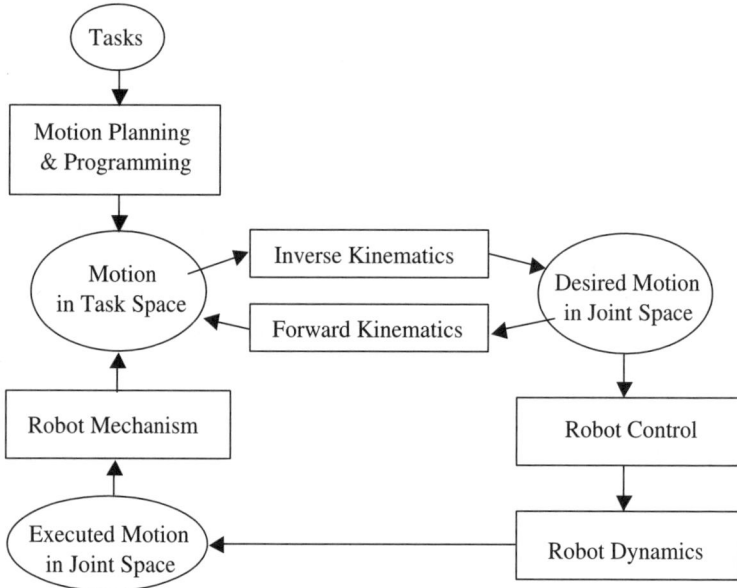

Fig. 3.20   The motion-related data flow inside a robot.

Let us examine the motion-related data flow inside a robot. From the illustration shown in Fig. 3.20, it is clear that the desired motion is specified in task space. The motion specification in task space needs to be mapped into joint space, as the direct input to robot control is the desired motion in joint space.

The purpose of robot control is to apply energy into the robot's mechanism which will alter the dynamic behavior of the robot. The direct outcome of robot control is the executed motion in joint space, which is the robot mechanism's input motion. Through the robot's mechanism, the executed motion in joint space is physically manifested by the robot mechanism's output motion in task space. Therefore, the robot mechanism's inverse kinematics is crucial, it determines the ability of a robot to fulfill a task through the execution of motion in joint space. Accordingly, the objective of inverse kinematics is to compute the desired motions of the joint variables for a given motion specification of the end-effector's frame in task space.

3.3.3.1 *Determination of Joint Displacements*

Refer to Fig. 3.19. The input for the determination of joint displacements is the posture or configuration (position and orientation) of the end-effector's frame with respect to the base frame. That is,

$$^0M_n = \begin{pmatrix} a_x & s_x & n_x & t_x \\ a_y & s_y & n_y & t_y \\ a_z & s_z & n_z & t_z \\ 0 & 0 & 0 & 1 \end{pmatrix} \tag{3.37}$$

where $\vec{a} = (a_x, a_y, a_z)^t$, $\vec{s} = (s_x, s_y, s_z)^t$ and $\vec{n} = (n_x, n_y, n_z)^t$ are called the *approaching vector*, the *sliding vector* and the *normal vector* of the end-effector's frame, as shown in Fig. 3.19. For consistency of notation with the planar open kinematic-chain, we consider that $\vec{a}$, $\vec{s}$ and $\vec{n}$ coincide respectively with the X, Y and Z axes of frame n.

Now, what we want to determine is the set of joint angles q, which satisfies the following equation:

$$^0M_n = \{^0M_1(q_1)\} \bullet \{^1M_2(q_2)\} \bullet \ldots \bullet \{^{n-1}M_n(q_n)\}. \tag{3.38}$$

Eq. 3.38 describes the equality of two matrices. It implies that their corresponding elements must be equal to each other. In general, it is unlikely for us to derive a closed form solution from Eq. 3.38.

Example 3.17   Refer to Fig. 3.9 and Fig. 3.18. Frame 3 is the hand's frame for the right arm (or the foot's frame, in the case of the right leg). Assume that, for a given task at a time-instant, the configuration of frame 3 needs to be

$$^0M_3 = \begin{pmatrix} ^0\vec{i_3} & ^0\vec{j_3} & ^0\vec{k_3} & ^0O_3 \\ 0 & 0 & 0 & 1 \end{pmatrix} = \begin{pmatrix} C\beta & -S\beta & 0 & x_3 \\ S\beta & C\beta & 0 & y_3 \\ 0 & 0 & 1 & 0 \\ 0 & 0 & 0 & 1 \end{pmatrix}$$

where $C\beta = \cos(\beta)$, $S\beta = \sin(\beta)$ and $(x_3, y_3, 0)^t$ is the origin of frame 3 with respect to frame 0. Now, the question is: What should the joint angles

$(\theta_1, \theta_2, \theta_3)$ be?

Refer to Example 3.14 and Eq. 3.3. The three link matrices of the right arm are as follows:

$$^0M_1 = \begin{pmatrix} C\theta_1 & -S\theta_1 & 0 & l_1C\theta_1 \\ S\theta_1 & C\theta_1 & 0 & l_1S\theta_1 \\ 0 & 0 & 1 & 0 \\ 0 & 0 & 0 & 1 \end{pmatrix},$$

$$^1M_2 = \begin{pmatrix} C\theta_2 & -S\theta_2 & 0 & l_2C\theta_2 \\ S\theta_2 & C\theta_2 & 0 & l_2S\theta_2 \\ 0 & 0 & 1 & 0 \\ 0 & 0 & 0 & 1 \end{pmatrix},$$

$$^2M_3 = \begin{pmatrix} C\theta_3 & -S\theta_3 & 0 & l_3C\theta_3 \\ S\theta_3 & C\theta_3 & 0 & l_3S\theta_3 \\ 0 & 0 & 1 & 0 \\ 0 & 0 & 0 & 1 \end{pmatrix}.$$

(NOTE: $C\theta = \cos(\theta)$ and $S\theta = \sin(\theta)$).

Substituting $^0M_1$ and $^1M_2$ into the equation

$$^0M_2 = \{^0M_1\} \bullet \{^1M_2\}$$

yields

$$^0M_2 = \begin{pmatrix} C(\theta_1 + \theta_2) & -S(\theta_1 + \theta_2) & 0 & l_2C(\theta_1 + \theta_2) + l_1C\theta_1 \\ S(\theta_1 + \theta_2) & C(\theta_1 + \theta_2) & 0 & l_2S(\theta_1 + \theta_2) + l_1S\theta_1 \\ 0 & 0 & 1 & 0 \\ 0 & 0 & 0 & 1 \end{pmatrix}.$$

According to the frame assignment shown in Fig. 3.18, frame 2's origin, with respect to frame 0, can be expressed in terms of frame 3's origin with an offset of l_3 along the X axis. That is,

$$^0O_2 = {}^0O_3 - l_3 \bullet \{^0\vec{i_3}\}.$$

If we denote $^0O_2 = (x_2, y_2, 0)^t$, we can easily establish the following equation:

$$\begin{cases} l_2C(\theta_1 + \theta_2) + l_1C\theta_1 = x_2 \\ l_2S(\theta_1 + \theta_2) + l_1S\theta_1 = y_2 \end{cases} \tag{3.39}$$

with

$$\begin{cases} x_2 = x_3 - l_3 C\beta \\ y_2 = y_3 - l_3 S\beta. \end{cases}$$

From Eq. 3.39, the computation of $x_2^2 + y_2^2$ yields

$$2l_1 l_2 C\theta_2 = x_2^2 + y_2^2 - l_1^2 - l_2^2.$$

Hence, the solution for the unknown θ_2 is

$$\theta_2 = \pm \arccos\left(\frac{x_2^2 + y_2^2 - l_1^2 - l_2^2}{2l_1 l_2}\right).$$

Two possible solutions for θ_2 mean that the mapping from task space to joint space is usually a one-to-many mapping (or conversely, the mapping from joint space to task space is usually a many-to-one mapping). Knowing θ_2, we can now rewrite Eq. 3.39 in a matrix form in terms of the unknown θ_1 as follows:

$$\begin{pmatrix} l_1 + l_2 C\theta_2 & -l_2 S\theta_2 \\ l_2 S\theta_2 & l_1 + l_2 C\theta_2 \end{pmatrix} \bullet \begin{pmatrix} C\theta_1 \\ S\theta_1 \end{pmatrix} = \begin{pmatrix} x_2 \\ y_2 \end{pmatrix}$$

or

$$\begin{pmatrix} C\theta_1 \\ S\theta_1 \end{pmatrix} = \frac{1}{k} \begin{pmatrix} l_1 + l_2 C\theta_2 & l_2 S\theta_2 \\ -l_2 S\theta_2 & l_1 + l_2 C\theta_2 \end{pmatrix} \begin{pmatrix} x_2 \\ y_2 \end{pmatrix}$$

with

$$k = (l_1 + l_2 C\theta_2)^2 + (l_2 S\theta_2)^2 = l_1^2 + l_2^2 + 2l_1 l_2 C\theta_2.$$

If we know $C\theta_1$ and $S\theta_1$, the solution for the unknown θ_1 will be

$$\theta_1 = \arctan\left(\frac{S\theta_1}{C\theta_1}\right).$$

Analytically, from the equality of the corresponding elements at location (1, 1) of the matrices in the following equation:

$$^0M_3 = \{^0M_1\} \bullet \{^1M_2\} \bullet \{^2M_3\},$$

we can easily obtain the solution for the unknown θ_3 as follows:

$$\theta_3 = \beta - \theta_1 - \theta_2.$$

◇◇◇◇◇◇◇◇◇◇◇◇◇◇◇◇◇◇

If we examine the solution for θ_2 and define

$$\gamma = \pi \pm \theta_2$$

we have

$$\cos(\gamma) = \cos(\pi \pm \theta_2) = \frac{l_1^2 + l_2^2 - x_2^2 - y_2^2}{2l_1 l_2}.$$

The above expression is precisely the proof of the cosine theorem applied to the triangle formed by the origins of frames 0, 1 and 2. And angle γ is the inner angle between the two edges corresponding to link 1 and link 2.

Example 3.18   Refer to Fig. 3.21. The posture of the world frame, with respect to the base frame (frame 0) of the right arm, is described by

$$^0M_w = \begin{pmatrix} 0 & -1 & 0 & 80 \\ 1 & 0 & 0 & 0 \\ 0 & 0 & 1 & 0 \\ 0 & 0 & 0 & 1 \end{pmatrix}.$$

Now, we want to move frame 3's origin to

$$\begin{cases} x_w = 29cm \\ y_w = 86cm \end{cases}$$

where the coordinates (x_w, y_w) are expressed in the world frame. The coordinates of this point, expressed in frame 0, will be

$$(x_3, y_3, 0, 1)^t = {}^0M_w \bullet (x_w, y_w, 0, 1)^t$$

or

$$\begin{cases} x_3 = 80 - y_w = -6cm \\ y_3 = x_w = 29cm. \end{cases}$$

By applying the inverse-kinematic solution discussed in the previous example, and choosing

$$\begin{cases} \beta = \arctan\left(\frac{y_3}{x_3}\right) \\ \theta_2 = \arccos\left(\frac{x_2^2 + y_2^2 - l_1^2 - l_2^2}{2l_1 l_2}\right), \end{cases}$$

the computed joint angles of the right arm (as shown in Fig. 3.21) are

$$\begin{cases} \theta_1 = 44.3972^0 \\ \theta_2 = 114.5844^0 \\ \theta_3 = -57.2922^0. \end{cases}$$

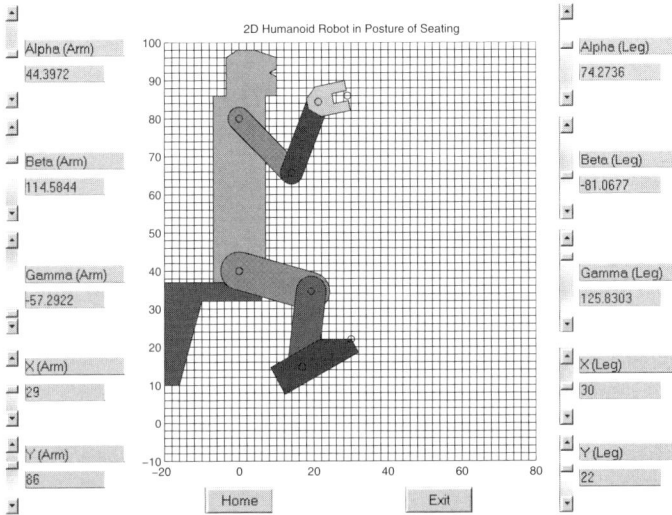

Fig. 3.21   Examples of an inverse-kinematic solution for the right arm and right leg.

Similarly, the posture of the world frame, with respect to the base frame (frame 0) of the right leg, is described by

$$
{}^0M_w = \begin{pmatrix} 0 & -1 & 0 & 40 \\ 1 & 0 & 0 & 0 \\ 0 & 0 & 1 & 0 \\ 0 & 0 & 0 & 1 \end{pmatrix}.
$$

For the posture of the right leg shown in Fig. 3.21, the origin of frame 3 of the right leg is

$$
\begin{cases} x_w = 30cm \\ y_w = 22cm \end{cases}
$$

expressed in the world frame. Its coordinates, with respect to the base frame (frame 0) are

$$
\begin{cases} x_3 = 40 - y_w = 18cm \\ y_3 = x_w = 30cm. \end{cases}
$$

If we choose

$$
\begin{cases} \beta = \arctan\left(\frac{y_3}{x_3}\right) + 60 \\ \theta_2 = -\arccos\left(\frac{x_2^2 + y_2^2 - l_1^2 - l_2^2}{2 l_1 l_2}\right), \end{cases}
$$

the computed joint angles of the right leg, as shown in Fig. 3.21, will be

$$\begin{cases} \theta_1 = 74.2736^0 \\ \theta_2 = -81.0677^0 \\ \theta_3 = 125.8303^0. \end{cases}$$

◇◇◇◇◇◇◇◇◇◇◇◇◇◇◇◇◇◇◇

For a simple open kinematic-chain with $n+1$ links, it is clearly difficult in practice to obtain a closed-form solution for determining the joint variables. But, it is possible to derive a general numerical solution from knowledge of the velocities of the joint variables.

Let us denote $\dot{q}(t)$ the joint velocity vector. If $\dot{q}(t)$ follows a trajectory from an initial configuration (i.e. in joint space) to a desired final configuration within time interval $[t_i, t_f]$, then the joint variables at time instant t_f will be

$$q(t_f) = q(t_i) + \int_{t_i}^{t_f} \dot{q}(t)dt. \tag{3.40}$$

If we perform the integration in a discrete time domain with the sampling step of $\triangle t$, Eq. 3.40 can numerically be computed as follows:

$$q(t_f) = q(t_i) + \sum_{k=0}^{N-1} \dot{q}(t_i + k \bullet \triangle t) \bullet \triangle t \tag{3.41}$$

with $N = \frac{t_f - t_i}{\triangle t}$. If we denote $t_k = t_i + k \bullet \triangle t$, Eq. 3.41 can also be written in the following recursive form:

$$q(t_{k+1}) = q(t_k) + \dot{q}(t_k) \bullet \triangle t, \quad \forall k \in [0, N-1]. \tag{3.42}$$

In the next section, we will discuss in detail the numerical solution for the computation of inverse kinematics.

3.3.3.2 *Determination of Joint Velocities*

The dynamics of task execution is more appropriately illustrated by the behavior of a trajectory following, as shown in Fig. 3.22. In general, all motions intrinsically involve velocity. For a given task, the desired motion for the fulfillment of the task is normally described in the form of a trajectory which is a spatial curve with a time constraint. Thus, it is crucial to know how to determine the joint variables' velocities $\dot{q}$ if the end-effector frame's velocity vector $\dot{P}$ is given as input (NOTE: P is a vector which denotes the posture or configuration of a frame). We will study how to plan the corresponding trajectory or spatial path for a given task in Chapter 9.

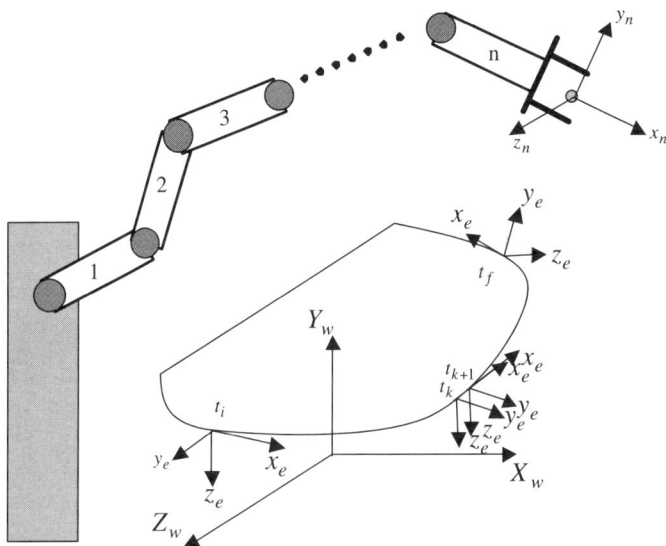

Fig. 3.22   Trajectory following by the end-effector's frame in task space.

A given task's planned spatial curve describes both the position and orientation of the tool's frame denoted by frame e, as shown in Fig. 3.22. Let us assume that frame n has already been taken into account the offset between the frame of the last link in a simple open kinematic-chain and the frame assigned to a tool. In this way, the end-effector's frame is considered to coincide with the tool's frame. Thus, robotic task execution is accomplished by the end-effector's frame which follows the planned trajectory represented by the time evolution of frame e.

Recall Eq. 3.36. The relationship between joint velocity vector $\dot{q}$ and velocity vector $\dot{P}$ of the end-effector's frame is as follows:

$$\dot{P} = J(q) \bullet \dot{q} \tag{3.43}$$

with

$$\dot{P} = \begin{pmatrix} {}^{0}\dot{O}_n \\ {}^{0}\omega_n \end{pmatrix}. \tag{3.44}$$

If $\dot{P}$ is a $m \times 1$ vector in a three-dimensional Cartesian space, the maximum dimension of $\dot{P}$ will be six (i.e. $m = 6$). The first three elements of $\dot{P}$ are the end-effector frame's linear velocity vector. The last three elements of $\dot{P}$ are the end-effector frame's angular velocity vector. If a simple open kinematic-chain has $n + 1$ links (inclusive of the base link), the dimension of the Jacobian matrix J will be $m \times n$.

From Eq. 3.43, it is easy to see that the unknown $\dot{q}$ is over-constrained if $m > n$. In this case, there may not be any feasible solution. But there is one approximate solution exists based on an optimal estimation. In general, if $m \geq n$, we can derive the solution for joint velocity vector $\dot{q}$ as follows:

- Step 1: Multiplication of the transpose of J to both sides of Eq. 3.44 yields

$$J^t \bullet \dot{P} = (J^t J) \bullet \dot{q}. \tag{3.45}$$

- Step 2: By multiplying the inverse of $(J^t J)$ to both sides of Eq. 3.45, we obtain

$$\dot{q} = J^\dagger \bullet \dot{P} \tag{3.46}$$

with

$$J^\dagger = (J^t J)^{-1} \bullet J^t.$$

Matrix $J^\dagger$ is commonly called the *pseudo-inverse* of the Jacobian matrix J. It is clear that the pseudo-inverse does not exist if matrix $J^t J$ is not of full rank. In this case, the simple open kinematic-chain is said to be in a singular posture. Therefore, when determining the joint velocity vector, it is necessary to pay attention to whether or not the simple open kinematic-chain is in a singular posture.

Refer to Eq. 3.43 again. Joint velocity vector $\dot{q}$ is under-constrained if $m < n$. In other words, the simple open kinematic-chain has more degrees of freedom than necessary to configure the end-effector's frame to a desired posture. In this situation, the simple open kinematic-chain is said to be kinematically redundant. From a mathematical point of view, there is no unique solution for $\dot{q}$ if there is a *kinematic redundancy* (i.e. $m < n$). In fact, the number of feasible solutions is infinite. This is a very advantageous situation in robotics because the redundancy allows us to impose constraints. In this way, optical solutions can be obtained by optimization.

For a given posture of the end-effector's frame, it is useful to find a solution among the infinite number of possible solutions that requires the least amount of energy. The minimum-energy solution is the one which guarantees the shortest path in joint space. In other words, it is the solution which minimizes the following cost function:

$$f(\dot{q}) = \frac{1}{2}(\dot{q}^t \bullet \dot{q}). \qquad (3.47)$$

The minimum-energy solution for $\dot{q}$ can be derived by using *Lagrangian multipliers* as follows:

- Step 1: Considering the primary constraint expressed by Eq. 3.43, a new cost function using Lagrangian Multipliers is constructed as follows:

$$f(\dot{q}, \lambda) = \frac{1}{2}(\dot{q}^t \bullet \dot{q}) + \lambda^t \bullet (\dot{P} - J \bullet \dot{q}) \qquad (3.48)$$

where λ is an $m \times 1$ vector whose elements are called Lagrangian Multipliers.

- Step 2: The solution for $\dot{q}$ that we are interested in must satisfy the following necessary conditions:

$$\begin{cases} \frac{\partial f}{\partial \dot{q}} = 0 \\ \\ \frac{\partial f}{\partial \lambda} = 0. \end{cases} \qquad (3.49)$$

- Step 3: Substituting Eq. 3.48 into Eq. 3.49 yields

$$\begin{cases} \dot{q} - J^t \bullet \lambda = 0 \\ \dot{P} - J \bullet \dot{q} = 0. \end{cases} \qquad (3.50)$$

- Step 4: From the first equation in Eq. 3.50, we have

$$\dot{q} = J^t \bullet \lambda. \qquad (3.51)$$

Substituting it into the second equation in Eq. 3.50 yields

$$\dot{P} = (J \bullet J^t) \bullet \lambda$$

or

$$\lambda = (J \bullet J^t)^{-1} \bullet \dot{P}.$$

- Step 5: Finally, substituting λ into Eq. 3.51 provides the solution for $\dot{q}$ as follows:

$$\dot{q} = J^\dagger \bullet \dot{P} \qquad (3.52)$$

with

$$J^\dagger = J^t \bullet (J \bullet J^t)^{-1}$$

where $J^\dagger$ is also called the pseudo-inverse of J.

Another useful constraint in robotics is called the *internal-motion* of a kinematically-redundant open kinematic-chain. The physical meaning of internal motion can mathematically be described as the difference between any two feasible solutions of $\dot{q}$ for a given input of $\dot{P}$. If we denote $\dot{q}^1$ as feasible solution 1 and $\dot{q}^2$ as feasible solution 2, the corresponding internal motion vector will be

$$\dot{q}^i = \dot{q}^2 - \dot{q}^1.$$

If $\dot{q}^n$ is the minimum-energy solution of $\dot{q}$, the sum of

$$\dot{q}^n + \dot{q}^i$$

is also a feasible solution of $\dot{q}$ which satisfies Eq. 3.43.

Since there are an infinite number of feasible solutions to $\dot{q}$ for a redundant open kinematic-chain, we have full freedom to impose an internal motion. However, it has to be clear that not any given velocity vector of $\dot{q}$ can be treated as an internal motion vector. Fortunately, there is an analytical method for us to map any given velocity vector of $\dot{q}$ to its corresponding internal motion vector. This powerful method can be described as follows:

For any given velocity vector $\dot{q}^a$ in joint (velocity) space, it is obvious that the sum of $\dot{q}^n + \dot{q}^a$ is generally not a solution for $\dot{q}$ satisfying Eq. 3.43. What we can hope to obtain is a feasible solution which is as close to $\dot{q}^a$ as possible. This requirement can easily be translated into a cost function as follows:

$$f(\dot{q}) = \frac{1}{2}\{(\dot{q} - \dot{q}^a)^t \bullet (\dot{q} - \dot{q}^a)\}. \qquad (3.53)$$

By applying *Lagrangian multipliers*, the solution satisfying Eq. 3.43 and Eq. 3.53 can be derived as follows:

- Step 1: Considering the primary constraint expressed by Eq. 3.43, a new cost function using Lagrangian Multipliers is constructed as follows:

$$f(\dot{q}, \lambda) = \frac{1}{2}\{(\dot{q} - \dot{q}^a)^t \bullet (\dot{q} - \dot{q}^a)\} + \lambda^t \bullet (\dot{P} - J \bullet \dot{q}) \qquad (3.54)$$

where λ is an $m \times 1$ vector, the elements of which are called Lagrangian Multipliers.

- Step 2: The solution for $\dot{q}$ that we're interested in must satisfy the following conditions:

$$\begin{cases} \frac{\partial f}{\partial \dot{q}} = 0 \\ \\ \frac{\partial f}{\partial \lambda} = 0. \end{cases} \tag{3.55}$$

- Step 3: Substituting Eq. 3.54 into Eq. 3.55 yields

$$\begin{cases} \dot{q} - \dot{q}^a - J^t \bullet \lambda = 0 \\ \dot{P} - J \bullet \dot{q} = 0. \end{cases} \tag{3.56}$$

- Step 4: From the first equation in Eq. 3.56, we have

$$\dot{q} = \dot{q}^a + J^t \bullet \lambda. \tag{3.57}$$

Substituting it into the second equation in Eq. 3.56 yields

$$\dot{P} = J \bullet \dot{q}^a + (J \bullet J^t) \bullet \lambda$$

or

$$\lambda = (J \bullet J^t)^{-1} \bullet (\dot{P} - J \bullet \dot{q}^a).$$

- Step 5: Finally, substituting λ into Eq. 3.57 gives the solution for $\dot{q}$, as follows:

$$\dot{q} = J^\dagger \bullet \dot{P} + [I - J^\dagger \bullet J] \bullet \dot{q}^a \tag{3.58}$$

with

$$J^\dagger = J^t \bullet (J \bullet J^t)^{-1}$$

where $J^\dagger$ is again the pseudo-inverse of J, and I is an identity matrix.

Now, assume that the end-effector frame stands still. Since $\dot{P} = \vec{0}$, Eq. 3.58 becomes

$$\dot{q} = [I - J^\dagger \bullet J] \bullet \dot{q}^a. \tag{3.59}$$

Eq. 3.59 physically means that the mapping of any given velocity vector in joint (velocity) space into its corresponding internal motion vector will not cause any change to the end-effector's frame (i.e. $J \bullet \dot{q} = 0$). If we denote $\dot{q}^i$ the internal motion vector corresponding to $\dot{q}^a$, we have

$$\dot{q}^i = [I - J^\dagger \bullet J] \bullet \dot{q}^a. \tag{3.60}$$

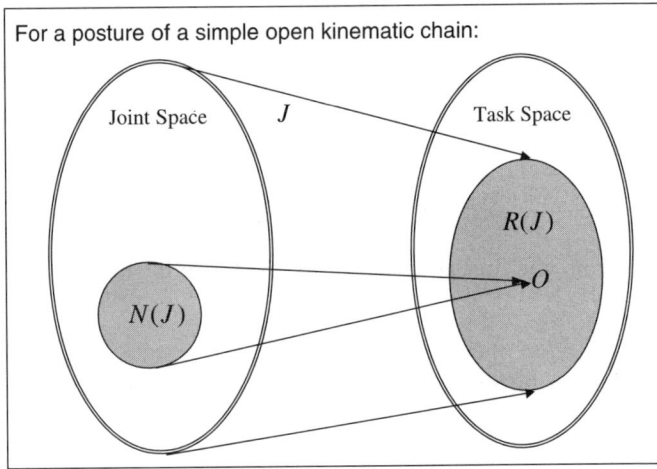

Fig. 3.23   Mapping from joint space to task space using the Jacobian J.

As illustrated in Fig. 3.23, all feasible velocity vectors $\dot{P}$ obtained from $\dot{q}$ by Eq. 3.43 are called the *range* of J, and denoted by $R(J)$. The range of J is the reachable subspace of task (velocity) space from the mapping using Jacobian Matrix J. On the other hand, all the internal motion vectors in joint (velocity) space form a subspace which is called the *Null space* of J. It is denoted by $N(J)$. In terms of the null space of J, Eq. 3.60 means the mapping from any given velocity vector $\dot{q}^a$ into the null space. And, Jacobian Matrix J maps the null space into the zero-velocity vector in task (velocity) space.

How to advantageously explore kinematic redundancy depends on the applications. The general guideline for choosing an internal motion constraint is simple. One common method is to align the internal motion vector in the direction that makes a simple open kinematic-chain either approach or move away from a particular configuration, such as: collision with obstacles, singularities, or joint limits. If we construct an objective function $g(q)$ which measures the extent to which a simple open kinematic-chain is reaching a particular configuration, the internal motion vector can

be chosen as

$$\dot{q}^a = \pm c \frac{\partial g(q)}{\partial q},$$

where c is a coefficient which determines the magnitude of the internal motion vector. If we choose the "+" sign in the above equation, this means that the simple open kinematic-chain should approach a desired configuration; Otherwise, it means it should move away from a particular configuration.

Example 3.19 Fig. 3.24 shows the results of internal motions applied to a simple open kinematic-chain with 20 degrees of freedom. Notice the end-effector frame stands still after successive applications of internal motions on one, two, and four joint variables.

(a)   (b)

(c)   (d)

Fig. 3.24 Illustration of the effects of internal motion on a simple open kinematic-chain with 20 degrees of freedom: a) the initial posture, b) effect due to the internal motion of one joint variable, c) effect due to the internal motion of two joint variables, and d) effect due to the internal motion of four joint variables.

◇◇◇◇◇◇◇◇◇◇◇◇◇◇◇◇◇◇◇◇◇◇

In summary, the joint velocity vector can be computed analytically from Eq. 3.46 or Eq. 3.52 for any given velocity vector of the end-effector's frame. These solutions are crucial for velocity control in joint space. These solutions also serve as the computational basis for determining the joint displacements if the end-effector frame's posture or configuration is given as input.

3.3.3.3   *Numerical Solution for Joint Displacements*

Refer to Fig. 3.22. Assume that the initial posture or configuration of the end-effector's frame is at time instant t_i, denoted by ${}^0M_n(t_i)$. In order to determine joint displacements, the input will be the final posture or configuration of the end-effector's frame at time instant t_f, which is denoted by ${}^0M_n(t_f)$. What we want to compute are the joint angles at time instant t_f (i.e. $q(t_f)$).

The basic idea underlying the numerical solution for determining $q(t_f)$ is to compute the series of incremental joint displacements $\dot{q}(t_k)$ (see Eq. 3.42). These displacements correspond to a series of incremental motion transformations ${}^eM_e(t_k, t_{k+1})$ of the end-effector's frame within a series of equal time intervals $[t_k, t_{k+1}]$. The origin of the end-effector's frame will travel along a spatial curve which connects the initial posture to the final posture.

Refer to Fig. 3.22 again. ${}^eM_e(t_k, t_{k+1})$ indicates the posture or configuration of the end-effector's frame at t_{k+1} with respect to itself at t_k. The symbol "e" highlights the fact that the incremental motion transformations are obtained from any chosen spatial curve regardless of the geometrical property of the simple open kinematic-chain under consideration. If we know a desired series of incremental motion transformations along a chosen spatial curve, the desired final posture of the end-effector's frame will be

$$ {}^0M_e(t_f) = {}^0M_e(t_i) \bullet \{ \prod_{k=0}^{N-1} [{}^eM_e(t_k, t_{k+1})] \} \qquad (3.61) $$

with

$$ {}^eM_e(t_k, t_{k+1}) = \{ {}^0M_e(t_k) \}^{-1} \bullet \{ {}^0M_e(t_{k+1}) \} \qquad (3.62) $$

and

$$ N = \frac{t_f - t_i}{\triangle t}. $$

(NOTE: $\triangle t$ is the sampling step in time. In practice, we can choose (t_i, t_f)

to be $(0, 1)$. Then $\triangle t$ is automatically computed once we decide the number N of iterations).

In a recursive form, Eq. 3.61 can equivalently be rewritten as follows:

$$^0M_e(t_{k+1}) = \{\,^0M_e(t_k)\} \bullet \{^eM_e(t_k), t_{k+1}), \quad \forall k \in [0, N-1]. \quad (3.63)$$

In fact, Eq. 3.63 is a discrete representation of a given spatial curve because one can choose $^eM_e(t_k, t_{k+1})$ to be a constant matrix if the spatial curve is linearly interpolated with a series of piecewise segments.

Now, let $\dot{P}_e(t_k)$ be the velocity vector of the end-effector's frame which executes the incremental motion transformation $^eM_e(t_k, t_{k+1})$ within time interval $[t_k, t_{k+1}]$. Substituting Eq. 3.58 into Eq. 3.42 yields

$$q(t_{k+1}) = q(t_k) + \{J^\dagger \bullet \dot{P}_e(t_k) + [I - J^\dagger \bullet J] \bullet \dot{q}^a\} \bullet \triangle t, \quad \forall k \in [0, N-1]. \quad (3.64)$$

If $^eM_e(t_k, t_{k+1})$ is expressed as follows:

$$^eM_e(t_k, t_{k+1}) = \begin{pmatrix} ^eR_e(t_k, t_{k+1}) & ^eT_e(t_k, t_{k+1}) \\ \vec{0} & 1 \end{pmatrix} \quad (3.65)$$

and we denote $(\triangle\theta,\ ^e\vec{r}_e)$ the equivalent rotation angle and equivalent rotation axis of $^eR_e(t_k, t_{k+1})$, the velocity vector $\dot{P}_e(t_k)$ of the end-effector's frame at t_k will be

$$\dot{P}_e(t_k) = \begin{pmatrix} ^0\dot{O}_n \\ ^0\omega_n \end{pmatrix} \quad (3.66)$$

with

$$\begin{cases} ^0\dot{O}_n = \frac{1}{\triangle t} \bullet \{^0R_n(t_k)\} \bullet \{^eT_e(t_k, t_{k+1})\} \\ \\ ^0\omega_n = \frac{\triangle\theta}{\triangle t} \bullet \{^0R_n(t_k)\} \bullet \{^e\vec{r}_e\} \end{cases}$$

where $^0R_n(t_k)$ is the rotation matrix which represents the orientation of the end-effector's frame at t_k with respect to frame 0 (base frame).

Eq. 3.64 is the recursive algorithm for the computation of the joint variables if the end-effector frame's final posture is given as input. The accuracy of the computation depends on the scale of the time interval $\triangle t$. In order to avoid an accumulation of numerical errors, a good strategy is to replace $^0M_e(t_k)$ in Eq. 3.62 with the computed $^0M_n(t_k)$ from the solution of forward kinematics. That is,

$$^0M_n(t_k) = {}^0M_n(q(t_k)) \quad (3.67)$$

where $q(t_k)$ is the actually computed joint angle vector from Eq. 3.64. This leads to the use of

$$^eM_e(t_k, t_{k+1}) = \{^0M_n(t_k)\}^{-1} \bullet \{^0M_e(t_{k+1})\} \tag{3.68}$$

to replace Eq. 3.65 for the computation of $\dot{P}_e(t_k)$.

Example 3.20   Fig. 3.25 is an example of the use of a numerical solution to solve inverse kinematics. The arm manipulator has 12 degrees of freedom. At time t_i, the robot is at its initial posture and the end-effector frame's final posture at time t_f is given as input. When applying the numerical solution to iteratively solve the inverse kinematics, the end-effector's frame reaches its desired final posture after 50 iterations (i.e. $N = 50$).

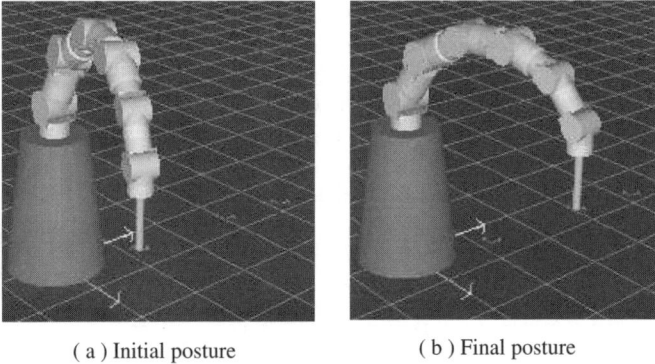

(a) Initial posture   (b) Final posture

Fig. 3.25   Example of the use of a numerical solution to solve inverse kinematics: a) robot at its initial posture and b) the final posture reached upon application of the numerical solution.

◇◇◇◇◇◇◇◇◇◇◇◇◇◇◇◇◇◇◇

3.3.3.4   *Effect of Singularity*

The determination of joint velocities and the numerical solution for estimating joint displacements all depend on the computation of the inverse of $(J \bullet J^t)$. When one or more eigenvalues of $(J \bullet J^t)$ approach zero, there is no inverse in the strict mathematical sense. In fact, when one eigenvalue is zero, it also means that $(J \bullet J^t)$ loses one rank. When $(J \bullet J^t)$ loses one rank, the physical meaning is that the end-effector's frame of a simple open kinematic-chain loses one degree of freedom or one mobility. When there is a loss of any mobility or degree of freedom, the simple open kinematic-chain is said to be in a singular posture. Therefore, it is important to know the effect of singularity on inverse kinematics.

Example 3.21   Fig. 3.26 shows a simulation of a humanoid robot in a standing position. In this position, it is clear that the end-effector's frame does not have any degree of freedom or mobility in the Y direction (i.e. vertical axis). Normally, a frame in a 2-D plane has three degrees of freedom: two for translation and one for rotation. The rank of $(J \bullet J^t)$ is 3 if the simple open kinematic-chain is not in a singular posture. As for the initial posture at t_i, as shown in Fig. 3.26, the rank of $(J \bullet J^t)$ is 2. Now, we choose three different final postures for the end-effector's frame to reach:

- It is easy for the robot to quit the singular posture when approaching final posture 1 because it does not require too much effort of movement in the vertical direction. The end-effector's frame can closely follow the chosen spatial path which connects the initial posture to final posture 1.
- It is a bit difficult for the robot to quit the singular posture when approaching final posture 2 because more effort is needed to move in the vertical direction. As a result, the singularity causes error to the end-effector frame's posture.
- It is even more difficult for the robot to quit the singular posture when approaching final posture 3 because a larger amount of motion in the vertical direction is required. As a result, the end-effector frame's posture is subject to a large amount of error due to the same singularity.

◇◇◇◇◇◇◇◇◇◇◇◇◇◇◇◇◇◇◇

Example 3.22   Due to disturbance, numerical error and singularity, it is necessary to compensate for the accumulated error in the numerical solution with the help of Eq. 3.68. Fig. 3.27 shows the benefit of using error compensation in the numerical solution for joint displacements. We can see that the end-effector's frame reaches precisely the desired final posture, despite the length of the path and the singularity of the initial posture of the simple open kinematic-chain.

If we imagine that the desired path is a real trajectory, this example is equivalent to the case when the end-effector's frame follows a pre-defined trajectory. The robot has been shifted away from its initial posture in order to have a better view of the first three iterations. We can see that the end-effector's frame deviates a lot from the desired trajectory at iterations 2

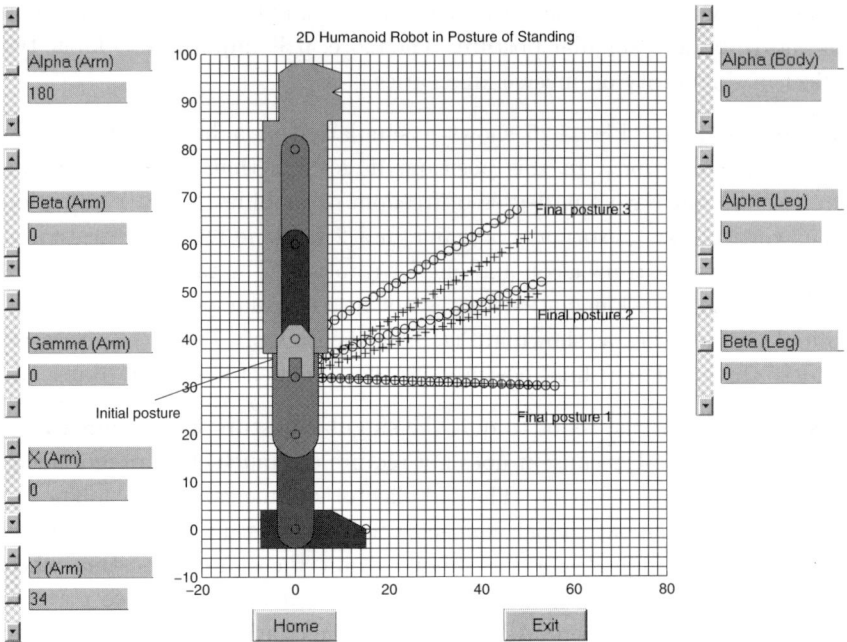

Fig. 3.26   Effect of singularity on the numerical solution of inverse kinematics without error compensation: "+" indicates the desired path of the end-effector's frame, and "o" indicates the actually computed path of the end-effector's frame.

and 3. This implies that the combined effect of error compensation and singularity will be disastrous for the scenario of the trajectory following.

◇◇◇◇◇◇◇◇◇◇◇◇◇◇◇◇◇◇◇

Based on the above two examples, we can make the following two statements:

- Error due to singularity can effectively be compensated by Eq. 3.68 in the numerical solution for computing joint displacements.
- Error compensation should be turned off when a simple open kinematic-chain is approaching singularity.

3.3.4 *Discrete Kinematic Mapping*

From the above discussions, we can highlight the following points regarding the conventional solutions of inverse kinematics:

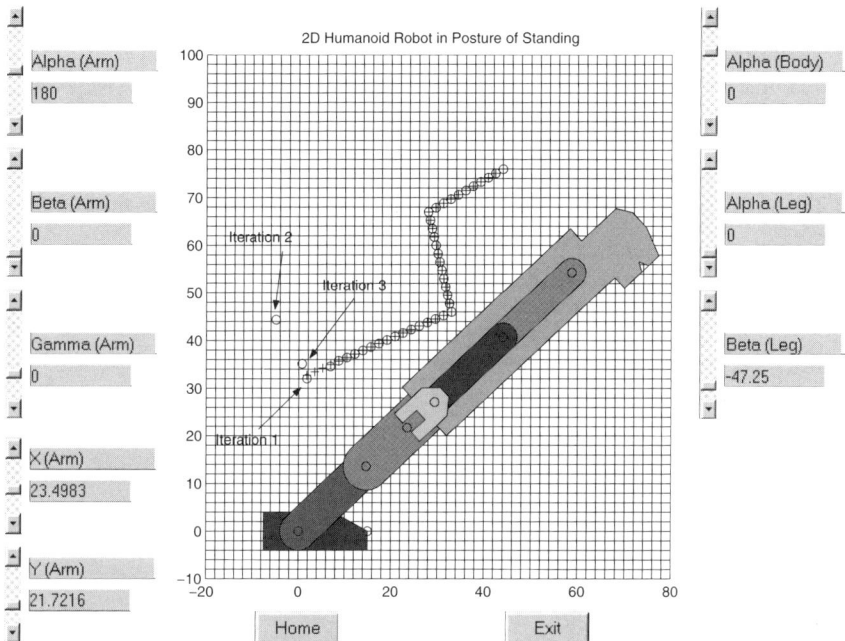

Fig. 3.27 The effect of singularity on error compensation used in the numerical solution of inverse kinematics: "+" indicates the desired path of the end-effector's frame, and "o" indicates the computed path of the end-effector's frame. The robot has been shifted away from its initial posture in order to give a better view of the first three iterations.

(1) In general, it is difficult to obtain a closed form solution for inverse kinematics.

(2) In general, mapping from task space to joint space is a one-to-many mapping. In practice, it is advantageous to explore this kind of kinematic redundancy.

(3) The range of joint variable q_i is specified by the *joint limits*, denoted by $(q_{i,min}, q_{i,max})$. The solutions of inverse kinematics usually do not explicitly take into account the limits of the joint variables. In practice, one must pay attention to the joint limits when programming a real or virtual robot.

(4) If we apply the solution for joint velocities to obtain a general numerical solution for recursively computing joint variables, it is necessary to be careful about singularity due to the inverse of $(J \bullet J^t)$.

(5) The numerical solution for computing joint displacements will not work properly if the initial configuration of the simple open kinematic-chain

is in singularity (i.e. the rank of $(J \bullet J^t)$ at t_i is less than the admissible number of degrees of freedom at the end-effector's frame).

(6) The numerical solution is computationally expensive as N is usually not a small number.

Clearly, there is room to improve the efficiency of the inverse-kinematic solutions. One interesting idea is to employ *discrete kinematic-mapping*.

3.3.4.1   *Discrete Forward Kinematic-Mapping*

For a simple open kinematic-chain with $n + 1$ links (inclusive of the base link), the corresponding joint space can be divided into a set of regular partitions which form a discrete joint space. If the range of joint variable q_i is $(q_{i,min}, q_{i,max})$, the discrete representation of this variable in joint space will be

$$q_i(k_i) = q_{i,min} + k_i \bullet \triangle q_i, \quad k_i \in [0, N_i - 1] \qquad (3.69)$$

with

$$N_i = \frac{q_{i,max} - q_{i,min}}{\triangle q_i}.$$

We denote $K = (k_1, k_2, ..., k_n)$ the index vector. When we have an index vector K, the corresponding posture of the end-effector's frame can be pre-calculated using the solution of forward kinematics (see Eq. 3.17), and stored if needed. All the pre-stored postures of the end-effector's frame can be arranged in the form of a reference table which is indexed by index vector K. This table describes *discrete mapping* from joint space to task space.

Since the granularity, in terms of $\{\triangle q_i, \ i = 1, 2, ..., n\}$, of the subdivision of joint space can incrementally be adjusted and fine-tuned during the developmental process of a robot, the idea of discrete kinematic-mapping naturally serves as a sound basis for the study of a developmental principle for a robot to acquire motion skills and kinematic modelling through experience, learning and real-time interaction with its environment.

The study of a developmental principle in robotics is emerging as a new research topic which directs the way towards the realization of an intelligent and autonomous robot, such as a humanoid robot. Most importantly, this principle appears to be an interesting approach to the study of a time-varying robot mechanism, the kinematic modelling of which is not available in advance.

3.3.4.2 *Discrete Inverse Kinematic-Mapping*

Similarly, we can divide task space into a set of regular partitions which are commonly called *voxels* in computer graphics. If the end-effector's effective workspace is within the volume of

$$\{(X_{min}, X_{max}),\ (Y_{min}, Y_{max}),\ (Z_{min}, Z_{max})\}$$

task space, in the form of voxels, can be represented by

$$
\begin{cases}
X(i_x) &= X_{min} + i_x \bullet \triangle X, \quad i_x \in [0, N_x - 1] \\[2mm]
Y(i_y) &= Y_{min} + i_y \bullet \triangle Y, \quad i_y \in [0, N_y - 1] \\[2mm]
Z(i_z) &= Z_{min} + i_z \bullet \triangle Z, \quad i_z \in [0, N_z - 1]
\end{cases}
\tag{3.70}
$$

with

$$
\begin{cases}
N_x = \frac{X_{max} - X_{min}}{\triangle X} \\[2mm]
N_y = \frac{Y_{max} - Y_{min}}{\triangle Y} \\[2mm]
N_z = \frac{Z_{max} - Z_{min}}{\triangle Z}.
\end{cases}
$$

Again, the granularity, in terms of $(\triangle X, \triangle Y, \triangle Z)$, of the subdivision of task space can also be incrementally adjusted and fine-tuned during the developmental process of robot.

Let $I = (i_x, i_y, i_z)$ be the index vector. Given an index vector I, the corresponding joint variable vector q can be pre-calculated by a conventional inverse-kinematics solution. For example, one can use the numerical solution. All the pre-calculated joint variables can be arranged in a reference table indexed by index vector I. Clearly, this table describes *discrete mapping* from task space to joint space.

Example 3.23 Fig. 3.28 shows the process of constructing discrete inverse kinematic-mapping by using the numerical solution. The discrete locations in task space, having the corresponding set of joint angles, are marked by the diamond symbol. In this example, we have set: $\triangle x = 10cm$ and $\triangle y = 10cm$.

◇◇◇◇◇◇◇◇◇◇◇◇◇◇◇◇◇◇◇◇

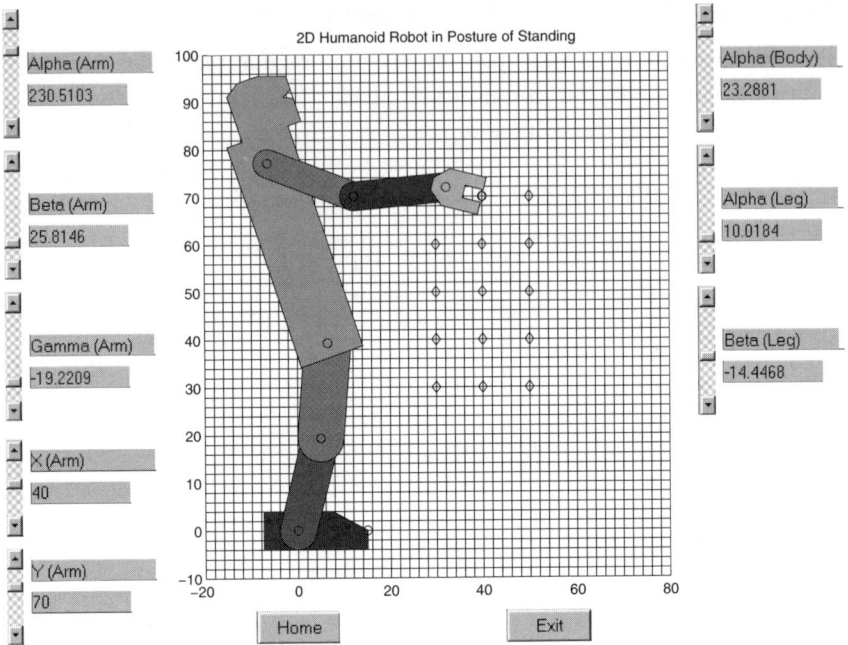

Fig. 3.28 Construction of the discrete inverse kinematic-mapping by using the numerical solution. The diamond symbol indicates the discrete locations in task space.

In the spirit of the developmental principle, if the precise kinematic modelling of a robot is not available, the discrete inverse kinematic-mapping can be established through a learning process based on real-time interaction between the robot and its environment,

3.3.4.3 *Bidirectional Discrete Kinematic-Mapping*

If there is no concern about the size of memory, it is theoretically possible to pre-calculate the mapping between index vectors I and K from the forward kinematic solution in Eq. 3.17. The results are not exact, but at a certain level of accuracy, depending on the levels of granularity. This mapping can be pre-stored into a database. If there is any change in the level of granularity of discrete task space or joint space, it is sufficient to recalculate the mapping and update the database.

With the mapping between index vectors I and K at a certain level of granularity, inverse kinematics solutions are known in advance and a basic computer can effortlessly retrieve a solution from the database. This is similar to the effortless way that human beings solve inverse kinematics. In addition, discrete kinematic-mapping automatically eliminates concern about joint limit and singularity because all these have been implicitly taken into account by the computation.

3.3.4.4   *Application of Discrete Kinematic-Mapping*

The numerical solution for determining joint displacements is very powerful but computationally demanding. In addition, it is necessary to pay attention to the effect of singularity. Therefore, it is useful to attempt to further improve the performance of the numerical solution.

In fact, the performance of the numerical solution can greatly be enhanced with the inclusion of discrete inverse kinematic-mapping. The idea is very simple. Instead of using a numerical solution to interpolate a chosen spatial curve which connects the end-effector frame's initial posture $^0M_n(t_i)$ to its final posture $^0M_n(t_f)$, we apply discrete inverse kinematic-mapping first to instantly obtain the joint angles which bring the end-effector's frame to an intermediate posture $^0M_n(t_j)$, which is close to the final posture $^0M_n(t_f)$. The difference depends on the granularity of discrete task space. Subsequently, we apply the numerical solution to compute the joint displacements which bring the end-effector's frame from the intermediate posture $^0M_n(t_j)$ to the final posture $^0M_n(t_f)$.

Example 3.24   Fig. 3.29 shows an example of the use of the numerical solution to determine joint displacements without any performance enhancement. The initial posture of the end-effector's frame is singular. Its final posture is at $(46cm, 54cm)$. After 26 iterations, the end-effector's frame reaches its final posture.

◇◇◇◇◇◇◇◇◇◇◇◇◇◇◇◇◇◇◇

Example 3.25   For the same example as shown in Fig. 3.29, we now explore the knowledge of discrete inverse kinematic-mapping. Instead of starting from the singular initial posture, the end-effector's frame is instantly moved over to an intermediate posture at $(50cm, 60cm)$, as shown in Fig. 3.30. From this intermediate posture, the end-effector's frame reaches its final posture within 5 iterations. In other words, the computational load with performance enhancement is three times less than the computational load required by the numerical method alone.

◇◇◇◇◇◇◇◇◇◇◇◇◇◇◇◇◇◇◇

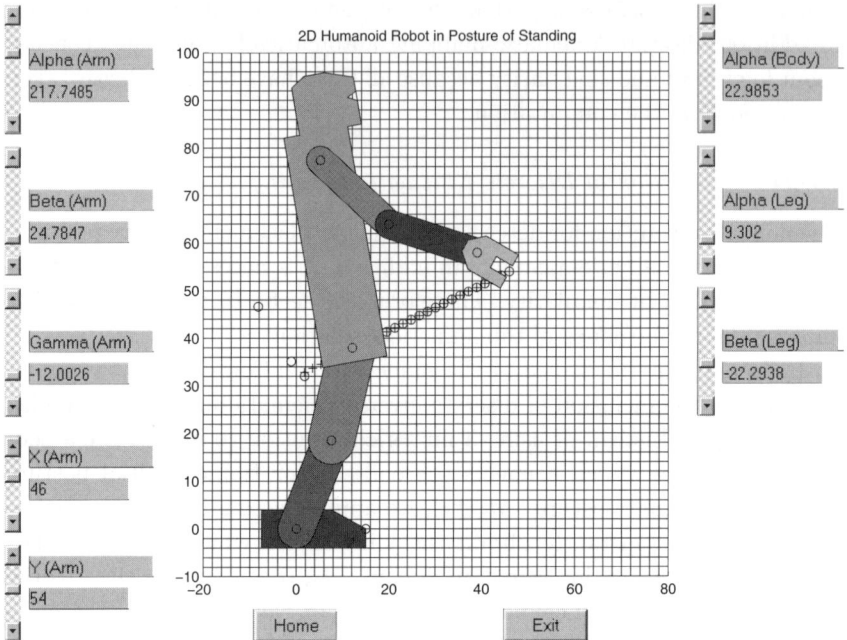

Fig. 3.29   Numerical solution for determining joint displacements without any enhancement.

3.4   Summary

A robot is able to perform motions because of its underlying mechanism. In this chapter, we started with the study of a robot's mechanism. We learned the concepts of link, joint, and open kinematic-chain. In order to facilitate the kinematic analysis of a humanoid robot, we discussed the concept of *simple open kinematics-chain.*

Subsequently, we studied a systematic way of representing a mechanism with a set of kinematic parameters, known as the DH representation. These parameters are sufficient to describe the relationship between the input and output motions governing the behavior of a robot's mechanism. For the sake of clarity, we discussed a new illustration of the DH representation. It makes use of two parallel planes to clearly explain the four kinematic parameters which uniquely describe the motion transformation between two consecutive links' frames.

Then, after the study of the kinematics of a single link, we covered in greater detail, the kinematic formulation of a simple open kinematic-chain.

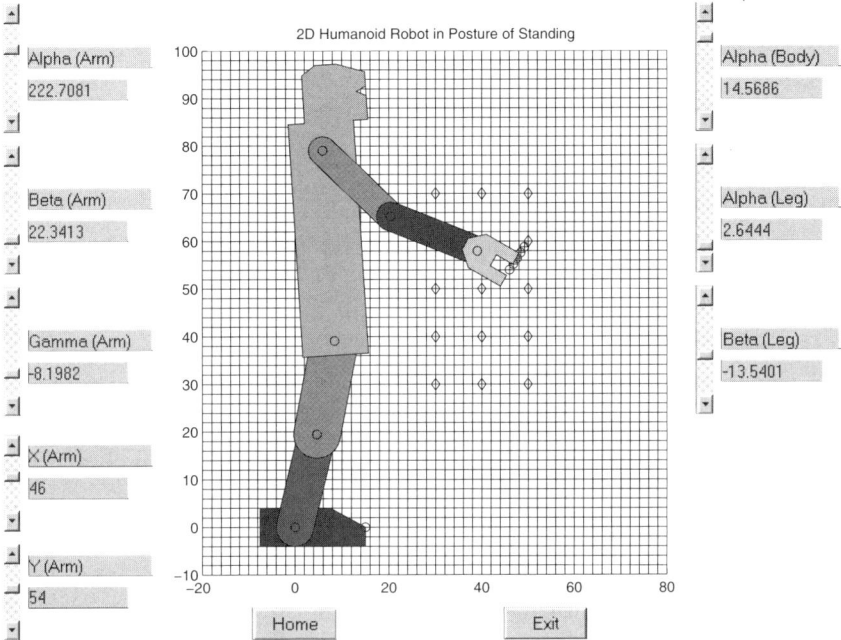

Fig. 3.30 Numerical solution for determining joint displacement with performance enhancement by discrete inverse kinematic-mapping.

We learned that there are two related problems under the topic of robot kinematics: a) forward kinematics and b) inverse kinematics.

We know that the objective of forward kinematics is to determine the output motion (in task space) of a robot's mechanism from the knowledge of its input motion (in joint space). As input to a robot system is a task, the most critical issue is how to determine the robot mechanism's input motions in joint space if the desired motion in task space is given. This is known as the inverse-kinematics problem. Subsequently, we studied different methods for solving inverse kinematics including *discrete kinematic mapping*. This method effectively complements the conventional numerical inverse-kinematics solution.

Now that we know the relationship between the input and output motions of a robot's mechanism, the next questions are: How do we physically produce the input motions to a robot's mechanism? And, how do we relate the input motions to the forces and torques applied to the joints? We will study the answers to these questions in the next chapter.

3.5 Exercises

(1) What is the purpose of studying a robot's mechanism?

(2) What is a simple open kinematic-chain?

(3) Illustrate the possible number of simple open kinematic-chains inside a humanoid robot's mechanism.

(4) What is the acceleration vector of a single link?

(5) Under what situation will Jacobian Matrix $J(q)$ be independent of joint variables q?

(6) Find the Jacobian matrix for the simple open kinematic-chain shown in Fig. 3.10.

(7) Comment on the applicability of the following algorithm to compute the joint angles of the planar open kinematic-chain shown in Fig. 3.9:

- Step 1: Multiply $^1M_0(q_1)$ to both sides of Eq. 3.38 to obtain:

$$^1M_0(q_1) \bullet \{^0M_n\} = \{^1M_2(q_2)\} \bullet ... \bullet \{^{n-1}M_n(q_n)\}.$$

- Step 2: The above equation describes 12 equalities between the corresponding elements of the two matrices of both sides. Among them, choose the simple one(s) to solve for q_1.

- Step 3: Multiply $^2M_1(q_2)$ to both sides of the above equation to obtain:

$$^2M_1(q_2) \bullet \{^1M_0(q_1)\} \bullet \{^0M_n\} = \{^2M_3(q_3)\} \bullet ... \bullet \{^{n-1}M_n(q_n)\}$$

and solve for q_2.

- Step 4: Repeat until q_n is solved.

(8) Verify the results of the inverse-kinematic solution applied to the right arm and right leg, as shown in Fig. 3.21.

(9) Explain why an arbitrary velocity vector, in joint space, cannot be treated as an internal-motion vector to a kinematically-redundant open kinematic-chain, the end-effector of which stands still in task space.

(10) How do you determine internal-motion vector $\dot{q}^a$ for a kinematically-redundant open kinematic-chain which is required to avoid the joint limits?

(11) How do you determine internal-motion vector $\dot{q}^a$ for a kinematically-redundant open kinematic-chain which is required to avoid the singularity caused by the inverse of $(J \bullet J^t)$?

(12) Look at Fig. 3.27. Discuss the reason for the large deviation of the end-effector's frame at iteration 2 from its desired trajectory.

3.6 Bibliography

(1) Asada, H. and J. J. E. Slotline (1986). *Robot Analysis and Control*, John Wiley and Sons.

(2) Denavit, J. and R. S. Hartenberg (1955). A Kinematic Notation for Lower-Pair Mechanisms Based on Matrices, *Journal of Applied Mechanics*, **22**, 215.

(3) Liegeois, A. (1977). Automatic supervisory control of the configuration and behavior of multi-body mechanisms, *IEEE Transaction on Systems, Man and Cybernetics*, **7**, 12.

(4) McKerrow, P. J. (1991). *Introduction to Robotics*, Addison-Wesley.

(5) Murray, R. M., Z. X. Li and S. S. Sastry (1994). *A Mathematical Introduction to Robotic Manipulator*, CRC Press.

(6) Norton, R. L. (1999). *Design of Machinery*, McGraw-Hill.

(7) Schilling, R. J. (1990). *Fundamentals of Robotics: Analysis and Control*, Prentice-Hall.

(8) Whitney, D. E. (1969). Resolved Motion Rate Control of Manipulators and Human Prostheses, *IEEE Transaction on Man-Machine Systems*, **10**, 2.

(9) (2002). *Proceedings of the IEEE Second International Conference on Development and Learning*, MIT, June 12 - 15.

Chapter 4

Electromechanical System of Robots

4.1 Introduction

The pure mechanical aspect of a robot is its underlying mechanism and structure. The study of the relationship between the input and output motions of a robot's mechanism is fully covered by kinematic analysis. From a systems point of view, it is logical to study the robot mechanism together with its kinematic analysis.

On the other hand, a robot is a machine which executes motion, as the result of energy consumption. Thus, it is indispensable to study the relationship between motion and the force/torque applied to a robot's mechanism. Naturally, it is also logical to study the actuation elements together with a robot's dynamic and static analysis.

In this chapter, we consider the case in which a robot's mechanical energy is obtained from electrical energy, because of the use of electric actuators. Accordingly, we will first study the concept of energy and the fundamentals underlying energy conversion from the electrical domain to the mechanical domain (i.e. electric actuators). Then, we will present, in detail, the mathematical principles for the establishment of *equations of motion*. A discussion about robot statics will come first before we move on to studying robot dynamics.

4.2 Origin of a Rigid Body's Motion

We all know that energy cannot be created nor destroyed. It can only be transformed from one form into another. For example, the internal combustion engine converts chemical energy into mechanical energy, in order to drive a car, while a portion of the chemical energy is wasted in the form of thermal energy. Another example is the rechargeable battery, which converts energy from the electrical domain to the electrochemical domain and vice-versa. In fact, any substance is an embodiment of energy. This is concisely described by the Einstein's famous equation $E = mc^2$.

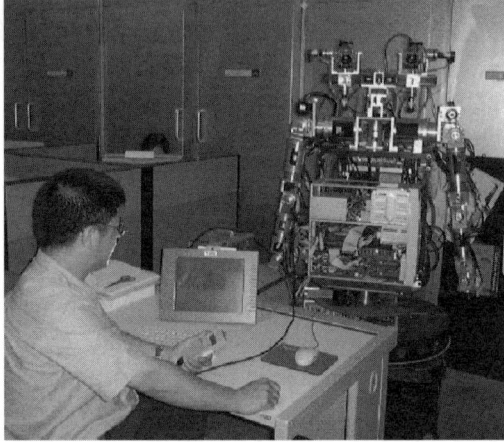

Fig. 4.1   A prototype of a human-like robot (HARO).

Now, the question is whether a robot's mechanism can move on its own without any extra element added to it. Obviously, the answer is negative. Fig. 4.1 shows a prototype of a human-like robot (called HARO-1). We can see that there are many components and devices in addition to the robot mechanism. Clearly, a robot's mechanism must be complemented with the extra actuation elements. In order to better understand this issue in robotics, it is important to study the origin of a rigid body's motion.

4.2.1   *Energy Conservation in a System*

Energy cannot be created nor destroyed. It can only be converted from one domain into another. This principle of energy conservation is also applicable to a system. As any substance is an embodiment of energy, a system can be treated as an energy storage device. Thus, at a time instant, a system has its own energy state, which corresponds to the amount of energy contained in the system. For convenience, we call the energy stored inside a system the *internal energy*. Then, the principle of energy conservation can be stated as: The change of internal energy (denoted by $\triangle E_{int}$) inside a system is equal to the sum of the energy added to the system (denoted by E_{in}), the energy removed from the system (denoted by E_{out}) and the energy dissipated from the system (denoted by E_{dis}). In other words, the following equality holds:

$$\triangle E_{int} = E_{in} - E_{out} - E_{dis}. \tag{4.1}$$

4.2.2 *Forces*

In robotics, an important concern is mechanical energy, and how to relate mechanical energy to motions. Mechanical energy is normally manifested in the form of *force or torque*. The nature of force is well described by Newton's second law. Before we briefly introduce this law, it is useful to examine the particle and its linear momentum. Formally, a particle can be defined as follows:

Definition 4.1 A particle is a body without a physical dimension. The entire mass of the body is concentrated at a single point.

If we denote m the mass of a particle and $v(t)$ its linear velocity vector with respect to a reference frame, then the linear momentum of the particle is defined as follows:

Definition 4.2 The linear momentum of a particle is the product of its mass and its linear velocity, that is,

$$p(t) = m \bullet v(t). \tag{4.2}$$

In fact, the linear momentum $p(t)$ of a particle characterizes the ability of the particle to maintain its body at a constant linear velocity. And, the mass of a particle describes the resistance to the change in its linear velocity. In other words, the mass measures the capacity of a body to store kinetic energy.

Now, Newton's first law simply states that a particle retains its linear momentum if no (external) forces act on the particle. Alternatively, the linear momentum $p(t)$ will change if there is an external force acting on it. The way in which this change is related to the acting forces is precisely described by Newton's second law. Newton's second law states that the rate of change of a particle's linear momentum is equal to the sum of all the (external) forces acting on it. If we denote $f_i(t)$ the individual force i acting on a particle and assume that there are n external forces (see Fig. 4.2), then we have

$$F(t) = \sum_{i=1}^{n} f_i(t) = \frac{dp(t)}{dt}. \tag{4.3}$$

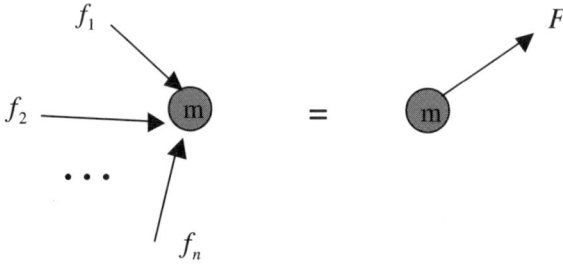

Fig. 4.2   External forces acting on a particle and the total sum of forces.

If a particle conserves its mass (i.e. m is a constant), substituting Eq. 4.2 into Eq. 4.3 yields

$$F(t) = \frac{dp(t)}{dt} = m \bullet \frac{dv(t)}{dt} = m \bullet a(t) \tag{4.4}$$

where $a(t) = \frac{dv(t)}{dt}$ is the *acceleration vector* of a particle.

As a rigid body (i.e. a body without any internal deformation) can be mathematically treated as a particle, Eq. 4.4 indicates that the motion of a rigid body originates from the forces acting on it. The formal definition of force is as follows:

Definition 4.3   A force is the effect of one body acting upon another body.

In the universe, there are two types of forces: a) contact force and b) field force. In the mechanical domain, a typical example of contact force is the elastic force generated by a spring. And a typical example of field force is gravitational force. When body A exerts a force upon body B, body A will receive a *reaction force* from body B. This phenomenon is precisely described by Newton's third law. This law states that when two particles mutually exert acting and reacting forces upon each other, these two forces are equal in magnitude and opposite in direction. If we denote $f_{i,j}$ the force exerted by particle i upon particle j, the reaction force exerted by particle

j upon particle i will be denoted by $f_{j,i}$. According to Newton's third law, we have

$$f_{i,j} = -f_{j,i}. \qquad (4.5)$$

4.2.3 *Torques*

If a force acts on a particle which is constrained to follow a circular path, as shown in Fig. 4.3, the energy state of the particle is conveniently described by angular momentum. Its definition is as follows:

Definition 4.4 The angular momentum of a particle moving along a circular path is equal to its mass times its angular velocity, that is,

$$H = m \bullet \vec{\omega} = m \bullet [\vec{r} \times v(t)] = \vec{r} \times [m \bullet v(t)] \qquad (4.6)$$

where $\vec{r}$ is the position vector of the particle, $v(t)$ its linear velocity vector, and $\vec{\omega}$ its angular velocity vector.

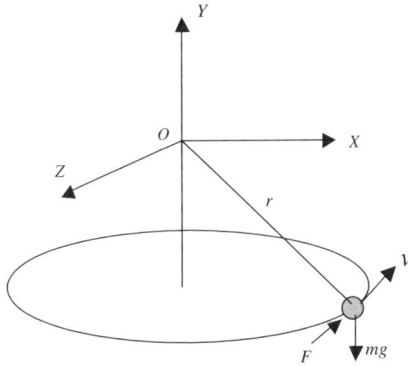

Fig. 4.3   A particle moving along a circular path.

Now, let us examine the variation of angular momentum. Differentiating Eq. 4.6 with respect to time gives

$$\frac{dH}{dt} = \frac{d\vec{r}}{dt} \times [m \bullet v(t)] + \vec{r} \times \left(\frac{d}{dt}[m \bullet v(t)] \right). \qquad (4.7)$$

Since the derivative of position vector $\frac{d\vec{r}}{dt}$ is in the direction of linear velocity vector $v(t)$, we have

$$\frac{d\vec{r}}{dt} \times [m \bullet v(t)] = 0. \tag{4.8}$$

Substituting Eq. 4.4 and Eq. 4.8 into Eq. 4.7 yields

$$\frac{dH}{dt} = \vec{r} \times F(t) \tag{4.9}$$

where $F(t)$ is a force vector expressed with respect to a reference frame.

If we define $\tau = \vec{r} \times F(t)$, Eq. 4.9 becomes

$$\frac{dH}{dt} = \tau = \vec{r} \times F(t). \tag{4.10}$$

By definition, vector τ is called the *moment of force*, or *torque* for short. In fact, Eq. 4.10 indicates that the angular momentum of a particle will remain unchanged if there is no external torque acting on particle (i.e. $\tau = 0$). In other words, the rate of change of a particle's angular momentum is equal to the torque acting on the particle. Physically, quantity τ depends on force F because the torque is equal to, by definition, the cross-product of a particle's position vector and the force vector acting on it.

We know that any acting force will have a corresponding reacting force. Similarly, any acting torque exerted by body A upon body B will also have a corresponding reacting torque exerted by body B on body A. For simplicity's sake, we use $\tau_{i,j}$ to denote the torque exerted by body i on body j. For convenience of description, the coupling between an acting force (or torque) and a reacting force (or torque) leads us to define two new terms, called a *dynamic pair* and a *dynamic chain*.

4.2.4   *Dynamic Pairs and Chains*

We use the term *dynamic pair* to define any pair of rigid bodies which exerts force or torque upon each other. Inside a dynamic pair, the body delivering the force, or torque, is called the *acting body*, while the one receiving the force or torque is logically called the *reacting body*. In practice, the acting body of a dynamic pair can exert multiple forces upon the reacting body. In robotics, we only consider the case where there is just one force or torque delivered by the acting body. Since a force can be a contact force or a field force, the dynamic pair's acting body will deliver either a contact force or a field force on the reacting body, as illustrated in Fig. 4.4.

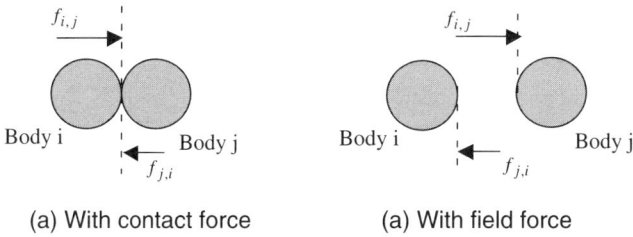

Fig. 4.4   Illustration of a dynamic pair: a) the acting body delivering a contact force, and b) the acting body delivering a field force.

The definition of a dynamic chain is similar to that of a kinematic chain. A dynamic chain is a set of rigid bodies which are arranged in a series, and mutually exert forces or torques on each other within a consecutive pair. Thus, a formal definition of dynamic chain can be stated as follows:

Definition 4.5   A dynamic chain is a set of rigid bodies arranged in a series which mutually exert forces, or torques, on each other within a consecutive pair.

A dynamic chain is said to be *open* if there is no direct coupling of force or torque between the first and last bodies in the chain. Otherwise, it is called a *closed dynamic chain.*

When dealing with robot dynamics, all forces in a dynamic chain are expressed by default with respect to the same reference frame (e.g. frame 0 of the base link in an open kinematic-chain). When we are not using the superscript, that implicitly indicates the reference frame.

Example 4.1   Fig. 4.5 shows an example of an open kinematic-chain with three links at rest. Due to gravitational force, these three links mutually exert acting and reacting forces and torques on each other. To study the effect of the forces and torques, we can break down the open kinematic-chain into a set of independent rigid bodies (i.e. links), which form an open dynamic chain as well. The interactions among the rigid bodies are governed by the acting and reacting forces/torques. In this example, we have

$$\begin{cases} f_{0,1} = -f_{1,0} \\ f_{1,2} = -f_{2,1} \end{cases}$$

and

$$\begin{cases} \tau_{0,1} = -\tau_{1,0} \\ \tau_{1,2} = -\tau_{2,1}. \end{cases}$$

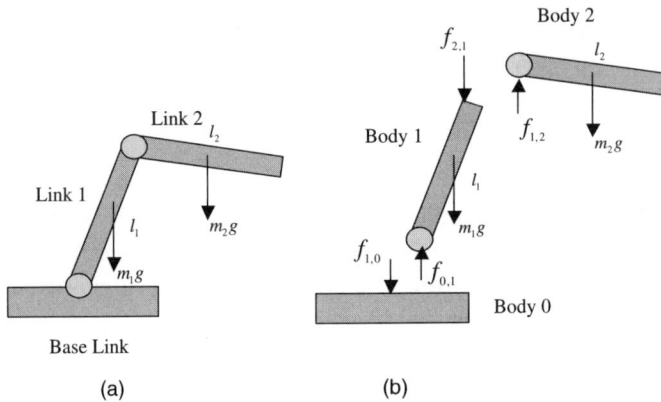

Fig. 4.5   Example of an open dynamic chain: a) an open kinematic-chain, and b) the corresponding open dynamic chain formed by the three rigid bodies with mutually acting and reacting forces/torques.

◇◇◇◇◇◇◇◇◇◇◇◇◇◇◇◇◇◇

4.2.5   *Incremental Works*

Motion originates from the force or torque of a body acting on another body. The rate of change of motion (i.e. linear or angular velocity) is proportional to the applied force or torque if the masses of the bodies inside a dynamic pair remain constant. Now, the question is: How is force, or torque, related to the physical quantity: energy?

Refer to Fig. 4.4. Assume that force $f_{i,j}$ acts on body j within time interval $[t_1, t_2]$, and body j undergoes a displacement along a linear or circular path. If we denote $d\vec{r}_j$ the differential displacement of body j caused by force $f_{i,j}$ (or its corresponding torque if the path is circular), by definition, the differential work done by body i exerting a force (or torque) on body j will be

$$dW = f_{i,j} \bullet d\vec{r}_j \tag{4.11}$$

where dW denotes the differential work done by force $f_{i,j}$ on body j.

The integration of Eq. 4.11 over time interval $[t_1, t_2]$ gives the expression of the incremental work done within the time interval, that is,

$$\triangle W = \int_{t_1}^{t_2} dW = \int_{t_1}^{t_2} f_{i,j} \bullet d\vec{r}_j. \tag{4.12}$$

Since $d\vec{r}_j = \frac{d\vec{r}_j}{dt} \bullet dt$ and $\frac{d\vec{r}_j}{dt} = v_j(t)$ (i.e. the linear velocity vector of body j), Eq. 4.12 can also be written as follows:

$$\triangle W = \int_{t_1}^{t_2} [f_{i,j} \bullet v_j(t)] \bullet dt. \tag{4.13}$$

By definition, the expression $f_{i,j} \bullet v_j(t)$ is called *power*, which characterizes the ability of body i to do work on body j.

4.2.6 *Potential Energy*

In Eq. 4.12, if force $f_{i,j}$ only depends on the position vector of body j and is independent of the time variable, Eq. 4.12 can be reformulated as follows:

$$\triangle W = -\triangle Q \tag{4.14}$$

with

$$\triangle Q = -\int_{\vec{r}_j(t_1)}^{\vec{r}_j(t_2)} f_{i,j}(\vec{r}_j) \bullet d\vec{r}_j. \tag{4.15}$$

In physics, work done by a force which only depends on a position vector is called *potential energy*. Eq. 4.15 expresses incremental potential energy done by force $f_{i,j}$ on body j during time interval $[t_1, t_2]$.

A typical example of potential energy is the work done by gravitational force. Assume that body i is the Earth. Then $f_{i,j}$ is the gravitational force acting on body j and is a constant force. In this case, Eq. 4.15 becomes

$$\triangle Q = -Q(t_2) + Q(t_1) \tag{4.16}$$

with

$$Q(t) = -f_{i,j}(\vec{r}_j(t)) \bullet \vec{r}_j(t). \tag{4.17}$$

Eq. 4.17 expresses potential energy done by gravitational force $f_{i,j}$. The negative sign inside the expression of potential energy indicates that the work done by the force is negative. And, energy is removed from body j if the displacement is in the direction of the force. For example, when an apple falls from a tree to the ground due to gravitational force, it loses potential energy. This lost potential energy is converted into kinetic energy.

4.2.7 *Kinetic Energy*

Refer to Eq. 4.12. Assume that the mass of body j is m_j and its linear velocity vector is $v_j(t)$. Force $f_{i,j}$ acting on body j will cause variation in the linear momentum. Then, we have

$$f_{i,j} = m_j \bullet \frac{dv_j(t)}{dt}. \tag{4.18}$$

Substituting Eq. 4.18 into Eq. 4.12 yields

$$\triangle W = \int_{t_1}^{t_2} m_j \bullet v_j(t) \bullet dv_j \tag{4.19}$$

or

$$\triangle W = \frac{1}{2} m_j [v_j^t(t_2) \bullet v_j(t_2)] - \frac{1}{2} m_j [v_j^t(t_1) \bullet v_j(t_1)]. \tag{4.20}$$

If we define

$$K(t) = \frac{1}{2} m_j [v_j^t(t) \bullet v_j(t)], \tag{4.21}$$

Eq. 4.20 becomes

$$\triangle W = K(t_2) - K(t_1).$$

By definition, Eq. 4.21 is an expression of kinetic energy. It describes the capacity to produce motion. When the displacement vector or motion of body j is in the direction of force $f_{i,j}$, its kinetic energy increases. For example, when we accelerate a car, its kinetic energy increases.

4.2.8 *Origin of Motions*

Now, we are able to precisely answer the question of what the origin of motion is. In the discussions above, we mentioned that a rigid body's motion originates from force or its corresponding torque applied to the body. At this point, it is clear that this answer is not completely correct. A simple example to prove this point is to look at the case when a person applies force to a building. Obviously, under normal circumstances, a person pushing a building will not cause any displacement. In other words, there will be no motion even if there is an applied force.

From the expression describing work done by an applied force or its corresponding torque, it becomes clear that the origin of motions is *energy* or *work* added to or removed from a body. In other words, the motion of any mechanical system is created by the addition or removal of mechanical energy to or from the system. Since a robot is a mechanical system, one may ask these questions:

- How do we add energy to or remove it from a robot's mechanism in order to produce the desired motions?
- How do we describe the relationship between the motions of a robot's mechanism and the forces/torques applied to it?

4.3   Actuation Elements

The purpose of a robot's mechanism is to shape the output motion, which is a function of the input motions of the robot's mechanism. In general, a robot's mechanism may include many kinematic chains.

As the motion originates from the addition or removal of energy, a robot's mechanism cannot produce any motion on its own. It is necessary to add extra elements to a robot's mechanism so that the motion can be generated in a controllable manner.

We know that a robot's mechanism can be treated as a set of kinematic pairs. Each kinematic pair determines the type of motion (prismatic or revolute) between its two links. A simple way to create controllable motion for a kinematic pair is to couple a dynamic pair to it, as the two bodies in a dynamic pair can exert force or torque upon each other.

This philosophy is concisely illustrated in Fig. 4.6. Conceptually, the mechanical system of a robot (or any machine) can be formally defined as a combination of kinematic pairs coupled with their corresponding dynamic pairs. This definition is helpful for those lacking a mechanical engineering background. It is also helpful in better understanding the design principle of the device called *actuator* or simply *force/torque generator*. Conceptually, an actuator or force/torque generator is the realization of a dynamic pair which consists of two bodies and their acting/reacting forces, or torques.

Example 4.2   For a land vehicle powered by an internal combustion engine, the kinematic pair is formed by the body of the vehicle and the wheels (treated as part of the ground). The corresponding dynamic pair is formed by the pistons and the engine block of the internal combustion engine. It

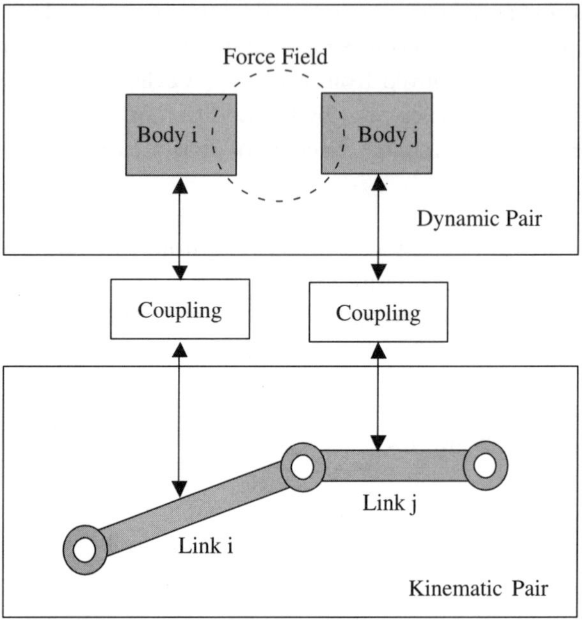

Fig. 4.6   The coupling of a dynamic pair with a kinematic pair in a robot's mechanical system.

goes without saying that a vehicle will stand still if there is no coupling between the kinematic pair and the dynamic pair.

◇◇◇◇◇◇◇◇◇◇◇◇◇◇◇◇◇◇

4.3.1   *Force and Torque Generators*

As mentioned earlier, there are two types of force: a) contact force and b) field force. In order to power a robot or any machine, a force (or torque) generator must satisfy the following two conditions:

(1) The generated force or torque must be controllable (i.e. its magnitude and direction are functions of certain controllable variables).
(2) The generated force or torque must be repeatable (i.e. one can produce a periodic and/or continuous force or torque as output).

A common method for the realization of a dynamic pair governed by a contact force or torque is illustrated in Fig. 4.7. As we can see, if pressure p_1 in the left chamber is not equal to pressure p_2 in the right chamber of body

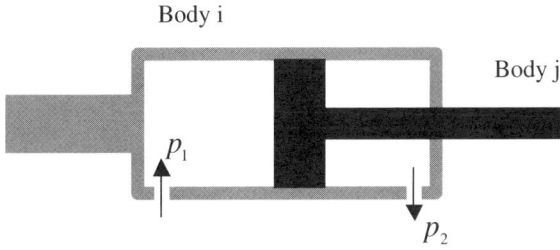

Fig. 4.7   A principle for the realization of a dynamic pair governed by a contact force.

i, body i will exert a force on body j through the medium of compressed gas (e.g. air), or liquid (e.g. oil). The generated force is both controllable and repeatable because it depends on quantity $(p_1 - p_2)$. The design of the *pneumatic actuators*, as well as *hydraulic actuators*, follows this layout of the dynamic pair.

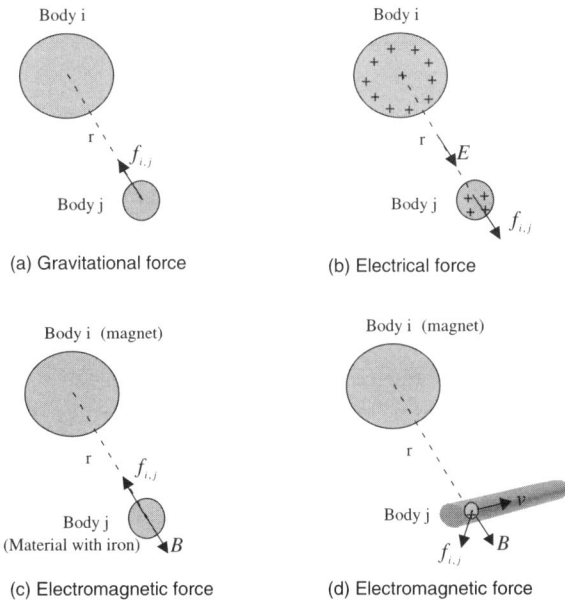

Fig. 4.8   Possible principles for the realization of a dynamic pair governed by a field force.

As for the realization of a dynamic pair governed by a field force, we have the following choices to consider (see Fig. 4.8):

Gravitational Field Forces

According to Newton's Law of Gravity, two bodies in space (i and j) will attract each other. The magnitude of the attraction is proportional to the product of the two bodies' masses and inversely proportional to the squared distance between them. When we have two bodies i and j separated by distance r, if their masses are m_i and m_j respectively, the gravitational force $f_{i,j}$ acting on body j by body i will be

$$f_{i,j} = G \frac{m_i \bullet m_j}{r^2} \tag{4.22}$$

where $G = 6.673 \times 10^{-11} \ m^3/(kg \bullet s^2)$ (m stands for meter, kg for kilogram, and s for second).

Now, assume that body i is the Earth. m_i will be about $5.976 \times 10^{24} \ kg$. If we consider the gravitational force field on the surface of the Earth, r will be about $6.378 \times 10^6 \ m$. The substitution of values of m_i and r into Eq. 4.22 yields

$$f_{i,j} = m_j \bullet g \tag{4.23}$$

where $g = 9.803 \ m/s^2$. g can be interpreted as the gravitational force per unit mass on the surface of the Earth. It characterizes the density of the gravitational force of the Earth at its surface.

From Eq. 4.23, it is clear that the density of the gravitational force on the surface of the Earth is almost constant, and not controllable. As a result, we cannot make use of this type of field force to realize a dynamic pair. Another way to explain this conclusion is the fact that a gravitational force is a conservative force. In other words, the conversion between force and potential energy is reversible. If we treat the pairing of the Earth and the moon as a dynamic pair, their interaction, or relative motion, is governed by a gravitational field force.

Electric Field Forces

Similarly, according to Coulomb's law, two electrically charged bodies, i and j, will exert *electric force* upon each other. This force is proportional to their electric charges (measured by C and called Coulomb) and inversely proportional to the distance between them. At an atomic level, an atom is composed of electron(s), proton(s) and neutron(s). The protons and neutrons of an atom form the *nucleus*.

For an electrically neutral atom, the number of electrons is equal to the number of protons. If some electrons are removed from an atom, the atom is said to be positively charged. Conversely, if more electrons are added to an atom, it is said to be negatively charged.

When we have two electrically charged bodies i and j, if their charges are q_i and q_j respectively, the electric force $f_{i,j}$ acting upon body j by body i will be

$$f_{i,j} = k\frac{|g_i| \bullet |q_j|}{r^2} \tag{4.24}$$

where $k = 8.988 \times 10^9 \ N \cdot m^2/C^2$ (N stands for Newton, m for meter and C for Coulomb). If we define

$$E(r) = k\frac{|g_i|}{r^2}, \tag{4.25}$$

Eq. 4.24 can be rewritten as

$$f_{i,j} = |q_j| \bullet E(r).$$

In fact, $E(r)$ describes the density of an electric field at distance r, created by body i having electric charge g_i. The electric force is controllable as it is possible to manipulate the electric charge q_j and the density $E(r)$ of the electric field. In theory, it is possible to design a device which forms a dynamic pair based on the principle of electric field force. However, this is not a practical solution because the high density of an electric field poses a safety challenge. Any incidental discharge may be dangerous and undesirable.

Electromagnetic Field Forces

We are all familiar with the phenomenon that two magnetic bars in proximity will attract or repel each other. A magnetic field will always exert an attractive force upon an object containing ferrous materials. And the direction of the force will point towards the source of the magnetic field. Strictly speaking, the interaction between a magnet and a body containing ferrous materials is not controllable because one cannot alter the direction of the interacting force between them. But, if we can design the two bodies in a cylindrical shape and choose an electromagnet to be the acting body, it is possible to alter the direction of the force acting on a body containing ferrous materials. This is precisely the working principle underlying electric motors.

There are three types of magnets: a) permanent magnets found in nature, b) cylindrical coils and c) electromagnets (i.e. coils wound around a ferrous core). A magnet will always produce a magnetic field B around it. If an electric charge travels inside a current conductor, this charge will receive a force proportional to the density of the magnetic field and proportional to the travelling speed of the charge. Moreover, the direction of the force acting upon an electric charge is perpendicular to both the velocity vector of the charge and the direction of the magnetic field.

Assume that body j is placed inside body i's magnetic field. If body j contains an electric charge q travelling inside it with velocity vector $\vec{v}$, force $f_{i,j}$ acting on body j by body i will be

$$f_{i,j} = q \bullet (\vec{v} \times B). \tag{4.26}$$

The net flow per unit time of electric charges at a conductor's cross-section area is called (electric) *current*, which is measured in Amperes (or A for short). If body j is a straight current conductor with a constant current I which flows inside it, the electric charges contained inside an infinitesimal length dl will be

$$dq = I \bullet dl.$$

By applying Eq. 4.26, force $df_{i,j}$ acting on the infinitesimal length by body i will be

$$df_{i,j} = I \bullet (\vec{v} \times B) \bullet dl$$

where $\vec{v}$ is a unit vector indicating the direction of current inside body j.

The integration of the above equation over the entire length of body j yields the total force acting upon body j by body i due to the magnetic field, that is,

$$f_{i,j} = I \bullet L \bullet (\vec{v} \times B). \tag{4.27}$$

From Eq. 4.27, we can see that the electromagnetic force between two bodies is controllable and repeatable. Most importantly, it can be easily manipulated by varying quantity I and/or B. Because of these desirable features, the working principle of all the electric motors is based on the use of electromagnetic force. Nowadays, electric motors are widely used in robotics for the realization of dynamic pairs in a robot's mechanical system.

4.3.1.1 *Working Principle of Electric Motors*

It is easy to control an interacting force caused by an electromagnetic field. As a result, devices which physically implement this principle of interaction are very popular in industry and are commonly called *electric motors*. Depending on the types of motion, motors are classified in different ways. It is called a *rotary motor* if the relative motion between the two bodies in a dynamic pair is rotational about a fixed axis. And, it is called a *linear motor* if the relative motion is along a straight line (i.e. linear trajectory). Here, we will discuss the basic working principle underlying electric rotary motors.

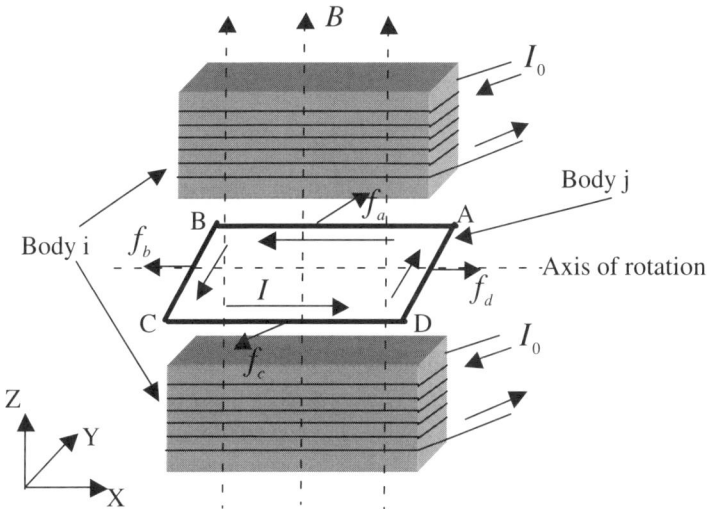

Fig. 4.9 Illustration of the basic working principle underlying electric rotary motors.

Refer to Fig. 4.9. Body i in the dynamic pair is an electromagnet. The density B of the generated magnetic field is a function of current I_0, which is a controllable variable. Body j in the dynamic pair is a rectangular closed-loop $ABCD$ which is constrained by an axis of rotation. In this example, body j is a rectangular coil having a single loop. However, a coil normally has multiple loops.

Assume that current I flows inside the coil. From Eq. 4.27, we know that the four edges of the coil will receive an electromagnetic force. Let us denote

- (f_a, f_b, f_c, f_d): the electromagnetic forces induced on edges AB, BC, CD and DA of the coil respectively;
- (l_a, l_b): the lengths of edges AB and BC of the coil respectively;
- $(\vec{v}_a, \vec{v}_b, \vec{v}_c, \vec{v}_d)$: the unit vectors indicating the directions of current I along edges AB, BC, CD and DA of the coil respectively.

By applying Eq. 4.27 to edges AB, BC, CD and DA, we will obtain

$$\begin{cases} f_a = I \bullet l_a \bullet (\vec{v}_a \times B) \\[2mm] f_b = I \bullet l_b \bullet (\vec{v}_b \times B) \\[2mm] f_c = I \bullet l_a \bullet (\vec{v}_c \times B) \\[2mm] f_d = I \bullet l_b \bullet (\vec{v}_d \times B). \end{cases} \qquad (4.28)$$

Then, the total forces received by body j in the X and Y directions will be

$$\begin{cases} f_x = f_b + f_d \\ f_y = f_a + f_c. \end{cases} \qquad (4.29)$$

Since $\vec{v}_a = -\vec{v}_c$ and $\vec{v}_b = -\vec{v}_d$, we have $f_a = -f_c$ and $f_b = -f_d$. This leads to the conclusion that $f_x = 0$, and $f_y = 0$. This means that the net force exerted on body j by body i is zero. Thus, body j will not undergo any linear displacement.

Now, assume that the plane containing the single-loop coil is not parallel to the XY plane. If the angle between the plane containing the coil and the XY plane is θ, from Fig. 4.9, it is clear that the net torque exerted on body j by body i is

$$\tau = |f_a| \bullet \frac{l_b}{2} \bullet sin(\theta) + |f_c| \bullet \frac{l_b}{2} \bullet sin(\theta) = |f_a| \bullet l_b \bullet sin(\theta). \qquad (4.30)$$

By substituting the value of f_a in Eq. 4.28 into Eq. 4.30, we obtain the final expression for the torque exerted upon body j by body i. That is,

$$\tau = A \bullet I \bullet |B| \bullet sin(\theta) \qquad (4.31)$$

with $A = l_a \bullet l_b$ (i.e. the rectangular area of the coil).

From Eq. 4.31, we can make the following observations:

- If the electromagnetic field is constant, the net torque exerted upon body j is a function of current I and angle θ.

- If body j is composed of a large number of rectangular loops, which are uniformly and symmetrically arranged about the common axis of rotation, the total net torque received by body j in a constant magnetic field is almost a function of the single variable I, and is expressed as

$$\tau = k_t \bullet I \qquad (4.32)$$

where k_t is called the electric motor's *torque constant*.
- If body j is an object containing ferrous materials, the torque exerted upon body j is solely a function of the magnetic density B, which is controllable by current I_0.

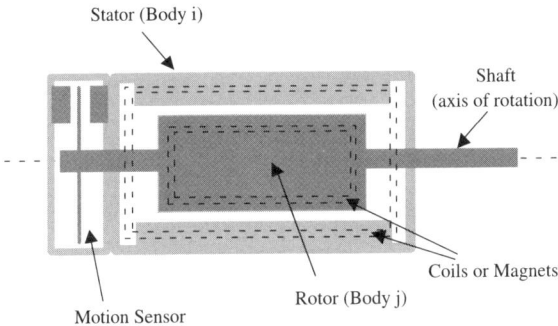

Fig. 4.10   A sectional view of the conceptual design for an electric rotary motor.

Eq. 4.31 and Eq. 4.32 mathematically describe the basic working principle underlying all electric rotary motors. All design solutions for electric motors invariably follow this principle. If we use terminology from electrical engineering, body i is called the *stator* and body j is called the *rotor*.

Fig 4.10 shows a sectional view of the conceptual design of an electric rotary motor. The controllable element inside an electric motor is the electromagnetic field which is produced by a set of coils wound on either the stator or rotor. In addition, a motion sensor is an indispensable element required by motion control, which we will discuss in more detail in the next chapter. Externally, all electric motors look similar. Fig. 4.11 shows one image of an electric motor.

In the following sections, we will discuss conceptual design solutions for electric stepper motors, DC brush-type motors, and DC brush-less motors.

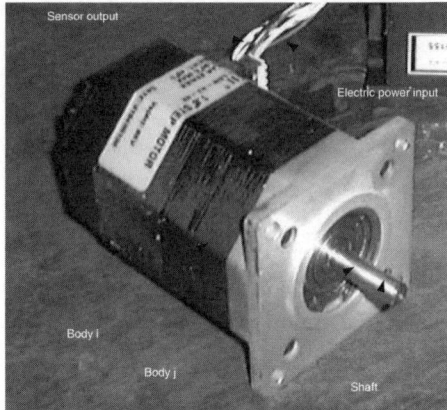

Fig. 4.11   An external view of an electric motor.

4.3.1.2   *Electric Stepper Motors*

Conceptually, an electric stepper motor is a torque generator which can produce step-wise motions between the two bodies in a dynamic pair (i.e. stator and rotor).

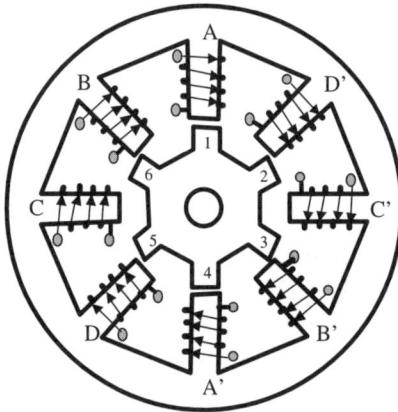

Fig. 4.12   A cross-sectional view of a stepper motor.

Fig. 4.12 shows a cross-sectional view of a conceptual design solution for a stepper motor. In this example, the rotor (body j) is a toothed cylinder which can be either a permanent magnet or an object containing ferrous materials. And, the stator (body i) is a toothed cylindrical tube. The coils wound around the teeth of the stator are grouped into four phases: AA', BB', CC' and DD'. If we sequentially energize these phases one at a time, we will create a rotating magnetic field around the shaft of the motor. The angle between two consecutive teeth is called the *pitch*. If the pitch of the rotor is different from the pitch of the stator, the rotating magnetic field will attract the rotor to rotate at the same speed as the magnetic field but in an opposite direction. This is clearly illustrated by Fig. 4.13.

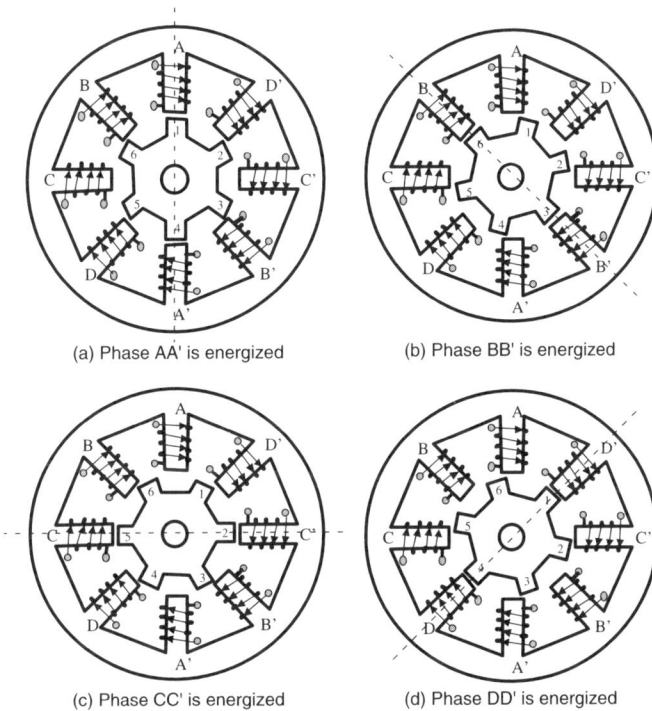

(a) Phase AA' is energized   (b) Phase BB' is energized

(c) Phase CC' is energized   (d) Phase DD' is energized

Fig. 4.13   Illustration of the working principle of a stepper motor.

When phase AA' is energized at a time instant, teeth 1 and 4 of the rotor will align with the magnetic flux generated by phase AA''s coils. At the next time instant, we switch off the electrical power of phase AA' and turn on the electrical power of phase BB', this will make teeth 6 and 3 of the rotor align with the magnetic flux of phase BB'. When the magnetic field of the stator rotates in a counterclockwise direction and a stepwise manner, the rotor will rotate in a clockwise direction and a stepwise manner as well.

If the numbers of teeth on the stator and rotor are N_s and N_r respectively, the step angle $\triangle\theta$ for each move will be the difference of their pitches, that is,

$$\triangle\theta = \frac{360^0}{N_r} - \frac{360^0}{N_s} = 360^0 \bullet \frac{N_s - N_r}{N_s \bullet N_r}. \tag{4.33}$$

How to control the direction of a stepper motor's rotation will be discussed in the next chapter, as this issue is closely related to a robot's control system.

4.3.1.3 *Brush-type DC Motors*

An electric stepper motor rotates in a stepwise manner. Obviously, it cannot run at a high speed. As we know, the product of the torque and angular velocity describes the mechanical power, which an electric motor can deliver. Thus, slow velocity means low mechanical power output. One way to overcome this drawback is to have the stator hold a pair of permanent magnets and the rotor carry a large number of rectangular coils. This design results in an electric motor known as a *brush-type DC motor*.

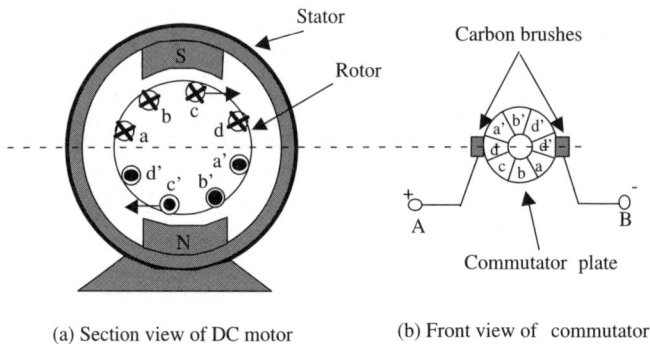

(a) Section view of DC motor   (b) Front view of commutator

Fig. 4.14   Illustration of the working principle behind the brush-type DC motors.

Fig 4.14 shows the conceptual design solution for a brush-type DC motor. A single, long wire is wound around the surface of the rotor according to a special winding diagram so that multiple rectangular coils are formed. In this example, there are four rectangular coils: aa', bb', cc', and dd'. Each coil has two terminals which are connected to an electrical power supply.

In fact, a special mechanical device called *commutator* groups all the terminals of the coils into a disk known as a *commutator plate*, as shown in Fig 4.14b. Two carbon brushes are placed symmetrically against the commutator plate for the purpose of supplying electrical power to the coils. The commutator plate rotates together with the rotor, but the two brushes remain stationary. Thus, there is friction between the carbon brushes and the commutator plate when the rotor is rotating.

In this example, at any time-instant, the terminals on the lower half of the commutator plate (below the dotted line in Fig 4.14) are electrically connected to the electric power supply's terminal A and the terminals on the upper half of the commutator plate (above the dotted line in Fig 4.14) are electrically connected to the electric power supply's terminal B. If terminal A has a higher voltage than terminal B (as shown in this example), edges a, b, c, and d will receive the current which goes inward (i.e. into the paper, as marked by the crosses), and edges a', b', c', and d' will receive the current which goes outward (i.e. out of the paper, as marked by the dots).

Since the direction of the magnetic field generated by permanent magnets is towards the upper side of the stator, according to Eq. 4.27, the direction of force induced on edges a, b, c and d is towards the right-hand side of the rotor, and the direction of force induced on edges a', b', c' and d' is towards the left-hand side of the rotor. As a result, the net torque induced on the rotor will make it rotate in a clockwise direction. If we make the voltage at terminal B higher than the voltage at terminal A, the rotor will rotate in a counterclockwise direction. And, it is obvious that the rotor will continuously rotate as long as the voltage at terminal A is not equal to the voltage at terminal B.

How to control the direction and velocity of a brush-type DC motor's rotation will be discussed in the next chapter as these issues are closely related to a robot's control system.

4.3.1.4 *Brush-less DC Motors*

There are several drawbacks to the design solution of a brush-type DC motor. The most notable one is the risk of generating electric sparks due to the friction between the commutator plate and the carbon brushes when the rotor is rotating at high speeds. Certain environments or workplaces may not tolerate this type of risk. A second drawback is the high inertia of the rotor if a large amount of coils have to be wound on the rotor in order to deliver high torque. A common solution to these drawbacks is to adopt a design, which places the coils on the stator and uses a cylindrical permanent magnet as the rotor. The result of this design is the *brush-less DC motor.*

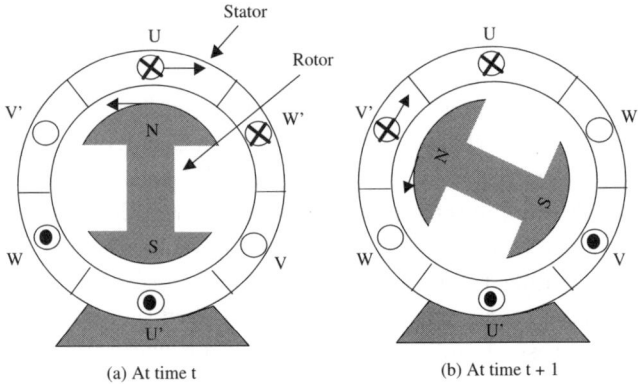

(a) At time t   (b) At time t + 1

Fig. 4.15   Illustration of the working principle of brush-less DC motors.

Fig. 4.15 shows a cross-sectional view of a conceptual design solution for the brush-less DC motor. The rotor inside a brush-less DC motor is a permanent magnet in a cylindrical shape. The magnetic polarity is shown in Fig. 4.15 (the axis of the north and south poles is perpendicular to the axis of the rotation of the rotor). There are three independent coils, each having multiple loops, which are symmetrically wound around the cylindrical inner surface of the stator. We call these coils UU', VV' and WW'. The directions of the currents, which go inside these coils, can be independently controlled in order to form a certain type of sequence.

At a time instant, only two coils are energized. Assume that the directions of the currents going inside the two energized coils, at time instant t, are as shown in Fig. 4.15a. At this time instant, the magnetic force induced on edge U points towards the right-hand side. This force will have a corresponding reacting force on the north pole of the rotor. Since the stator is stationary, the rotor will rotate in a counterclockwise direction as a result of the torque produced by the reacting force. Similarly, there will be a reacting force at the south pole of the rotor due to the induced magnetic force at edge U'. This reacting force will make the rotor rotate in a counterclockwise direction as well.

In order to keep the rotor continuously rotating, it is necessary to turn "on" the current which goes inside coil VV' at an appropriate next time-instant. For maximum efficiency, the current in coil VV' should be turned "on" at time-instant $t + 1$ when the north pole reaches a position underneath edge V. In this way, the electromagnetic field will keep the rotor continuously rotating with an induced torque at its maximum value. The closer a current conductor is to a magnetic pole, the larger the induced force will be.

In order to gain maximum efficiency, it is clear that the sequence of energizing the coils must be synchronized with the angular velocity of the rotor. Because of this, it is easy to understand that the electronic drive circuit for the control of a brush-less DC motor will be quite complicated.

We will discuss methods of controlling the magnitude of electrical power and the directions of the currents supplied to a brush-less DC motor in the next chapter, as these topics are closely related to a robot's control system.

4.3.2 *Force and Torque Amplifiers*

Conceptually, an electric motor is a force or torque generator which forms a dynamic pair for the purpose of creating motion. Since force and velocity are closely related to each other, we need to examine their relationship.

Refer to Eq. 4.13. Force acting on a body times the linear velocity of the body describes *power* in the mechanical domain. In the electrical domain, power is the voltage times the current supplied to an electrical load (e.g. a resistor). An electric motor is an electromechanical device which takes electrical power as input and produces mechanical power as output. If there is no dissipation of power, the output power must be equal to the input power.

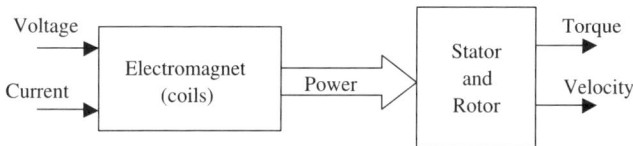

Fig. 4.16   Power conversion in an electric motor.

Fig .4.16 illustrates how an electric motor converts power from the electrical domain to the mechanical domain. If the input power in the electrical domain applied to an electric motor is a constant p, torque τ times angular velocity ω is a constant as well, that is,

$$\tau \bullet \omega = p. \tag{4.34}$$

It is easy to see that the higher the velocity is, the lower the effective torque will be. If an electric motor does not have any inertial load, its tendency will be to output a very high angular velocity (thousands of rotations per minute). As a result, the torque which an electric motor can deliver is very low. This is an undesirable feature because the static frictional force between two bodies at rest is usually very large.

In order to trigger the relative motion between two bodies at rest, one needs to apply strong force or torque to overcome the static frictional force. Therefore, it is necessary to alter the characteristic of an electric motor so that it can deliver the torque at a reasonable range of magnitude. From Eq. 4.34, we can see that the only way to increase torque is to reduce velocity if the power is to remain a constant. A device which allows us to reduce velocity without loss of power is commonly known as a *speed reducer*. With reference to force or torque, a speed reducer can also be called a *force or torque amplifier*. That is an indispensable element inside an actuation system.

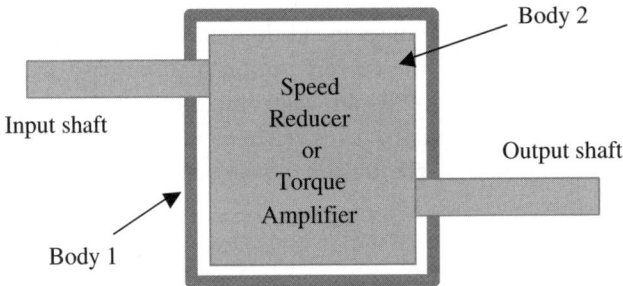

Fig. 4.17 Illustration of the input-output relationship in a speed reducer or torque amplifier for rotary motion.

Fig. 4.17 shows the relationship between the input and output in a speed reducer, or torque amplifier. To reduce speed or amplify torque related to a rotary motion, we connect this rotational motion to the *input shaft* of a speed reducer, or torque amplifier. The output motion from a speed reducer, or torque amplifier, will also be a rotational motion which is transmitted through the *output shaft*. If the velocity of the input motion is ω_0 and the velocity of the output motion is ω, the important parameter of the speed reducer is the *reduction ratio* k_r, that is,

$$k_r = \frac{\omega_0}{\omega}. \tag{4.35}$$

If the torque of the input motion is τ_0, then the amplified torque of the output motion will be

$$\tau = k_r \bullet \tau_0. \tag{4.36}$$

In fact, a speed reducer or torque amplifier is a coupling device because it is composed of two independent bodies, as shown in Fig. 4.17. By coupling device, we mean a device which can be physically connected to another pair of bodies without losing any degree of freedom. For example, we can interface a dynamic pair (e.g. a motor) with a kinematic pair (e.g. a pair of links) through a coupling device such as a speed reducer or torque amplifier. In this way, the degree of freedom of the kinematic pair is preserved.

In the following sections, we will discuss conceptually some examples of speed reducers or torque amplifiers.

4.3.2.1 *Gear Mechanisms*

A simple solution for speed reduction, or torque amplification, is to use a gear mechanism as shown in Fig. 4.18.

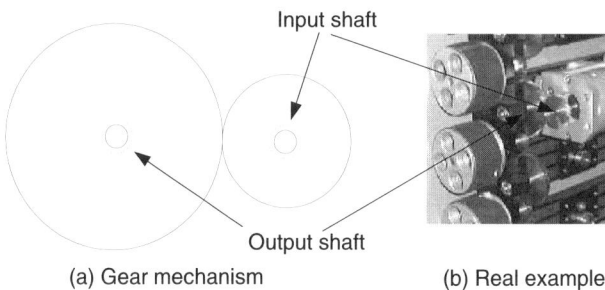

(a) Gear mechanism   (b) Real example

Fig. 4.18 A gear mechanism for speed reduction or torque amplification.

The input motion is supplied to the shaft of the gear having a smaller diameter, and the output motion is transmitted from the shaft of the gear having a larger diameter. If the numbers of teeth of the larger and smaller gears are N_o and N_i respectively, the ratio of speed reduction will be

$$k_r = \frac{N_o}{N_i}$$

k_r is also the ratio of amplification for the output torque.

However, there are three notable drawbacks with regard to the gear mechanism:

- First of all, the teeth of the two gears are in direct contact. After long-term use, there will be the problem of wear-and-tear. In addition to imprecision of machining and assembly, the undesirable problem of backslash (i.e. hysteresis) becomes inevitable.
- Secondly, the distance between the input shaft and the output shaft depends on the diameters of the two gears. More gears must be added if one wants to alter this distance.
- Thirdly, the ratio of reduction is not very high (normally less than 100).

4.3.2.2 *Pulley-and-Timing Belt Assemblies*

In order to overcome the first two drawbacks of a gear mechanism, a common solution is to use the *pulley-and-timing belt assembly* as shown in Fig. 4.19.

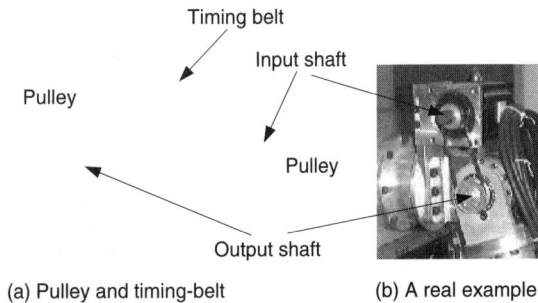

(a) Pulley and timing-belt   (b) A real example

Fig. 4.19   Pulley-and-timing belt assembly for speed reduction or torque amplification.

In a pulley-and-timing belt assembly, the timing belt is a flex spline with teeth. The whole timing belt is normally made of soft material (e.g. rubber). The direct contact of the two pulleys is therefore replaced by the soft contact between the timing belt and the two pulleys. As a result, the effect of wear and backslash is not as serious as that of a gear mechanism.

Since the length of the timing belt is selectable, it is easy to alter the distance between the input shaft and the output shaft. Just as with a gear mechanism, the speed reduction is equal to the ratio between the numbers of teeth of the two pulleys. If the number of teeth on the larger pulley is N_o and the number of teeth on the smaller pulley is N_i, the ratio of speed reduction (or the ratio of torque amplification) will be

$$k_r = \frac{N_o}{N_i}.$$

Again, this ratio is not very high (typically less than 100).

4.3.2.3 *Harmonic-Drive Devices*

The best device for speed reduction, or torque amplification, is the *harmonic drive*, as shown in Fig. 4.20.

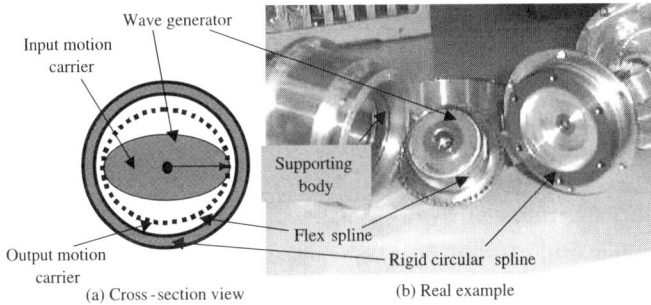

Fig. 4.20   Harmonic drive for speed reduction or torque amplification.

Refer to Fig. 4.20. A harmonic drive is composed of: a) an elliptical wave generator, b) a flex spline with teeth, and c) a rigid circular spline with teeth, and d) the supporting body. The elliptical wave generator is the input motion carrier. Its axis of rotation is the input shaft. And, the rigid circular spline with teeth is the output motion carrier, if the flex spline is fixed onto the supporting body. Alternatively, the flex spline is the output motion carrier, if the circular spline is fixed onto the supporting body. The axis of rotational motion is the output shaft.

Assume that the output motion carrier is the flex spline. The role of the wave generator is to keep the flex spline in contact with the circular spline at the two ends along the major axis of the elliptical wave generator. It is easy to see that there will be zero displacement between the flex spline and the circular spline if they have the same numbers of teeth. Let N_f be the number of teeth on the flex spline and N_c the number of teeth on the circular spline. If $N_f = N_c + 1$, it is easy to see that the circular spline will shift one tooth after one full rotation of the wave generator. Similarly, if $N_f = N_c + 2$, the circular spline will shift two teeth after one full rotation of the wave generator. In order to make the flex spline shift N_f teeth, the wave generator has to make $N_f/2$ rounds of full rotation. If $N_f = N_c + 2$, the ratio of speed reduction or torque amplification will be

$$k_r = N_f/2.$$

Interestingly, k_r is linearly proportional to N_f. This explains why the ratio of speed reduction of a harmonic drive can be very high (easily over several hundred). Since the contact forces between the flex spline and the circular spline are along the major axis of the wave generator (perpendicular to the tangential direction of each pair of teeth in contact), there is no backslash effect, as long as there is no slippage between a pair of teeth in contact.

The only problem with a harmonic drive is that the input and output shafts are coaxial. If this is not desirable, a good solution is to combine the use of the harmonic drive (for achieving better ratio and no backslash) with the pulley-and-timing belt assembly (for achieving the offset between input and output shafts).

4.4   Formation of a Robot's Electromechanical System

We now know that a robot's mechanism cannot move on its own without the addition of actuation elements, such as motors and torque amplifiers. In other words, the kinematic pairs of a robot's mechanism must be coupled with a set of dynamic pairs in order to produce controllable motions. This is because all motions must originate from a source of energy. The role of a dynamic pair (e.g. motor) is to convert energy from one domain (i.e. electrical) into the mechanical domain.

As it is necessary to couple kinematic pairs of a robot's mechanism with a set of dynamic pairs, the design of a robot not only embraces mechanism design but also deals with machine design (or mechatronic design of a machine).

4.4.1 *One-to-One Couplings*

A robot's mechanism can be treated as the combination of a set of kinematic pairs. A common way to form the coupling between the dynamic and kinematic pairs is to independently connect a dynamic pair to a kinematic pair. In this way, an entity called *kineto-dynamic pair* is formed. Thus, a robot can also be treated as the combination of a set of kineto-dynamic pairs.

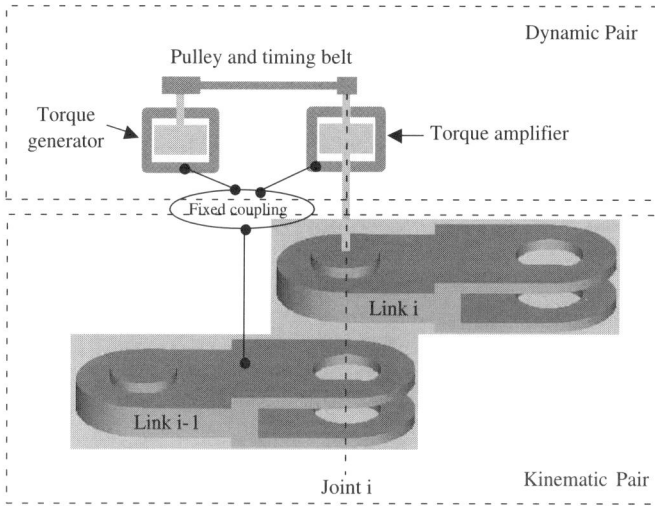

Fig. 4.21   Kineto-dynamic pair inside a robot.

Fig. 4.21 illustrates the coupling of a dynamic pair with a kinematic pair. Assume that the kinematic pair contains link $i - 1$ and link i. The output shaft of the torque amplifier coincides with the axis of rotation of joint i. And the output motion of the motor (i.e. torque generator) is coupled with the input motion of the torque amplifier through a pulley and timing-belt assembly. On the other hand, both the stator of the motor and the supporting body of the torque amplifier are fixed onto link $i - 1$.

In this illustration, the motion of link i comes from the torque applied to joint i. If we denote τ_i the torque received by joint i, τ_{mi} the output torque of the motor and k_{ri} the total ratio of torque amplification (including the one due to the pulley-and-timing belt assembly), we have

$$\tau_i = k_{ri} \bullet \tau_{mi}. \tag{4.37}$$

Similarly, if we denote ω_i the angular velocity vector of link i and ω_{mi} the angular velocity vector of the motor, we will have

$$\omega_i = \frac{1}{k_{ri}} \bullet \omega_{mi}. \tag{4.38}$$

As shown in Fig. 4.21, each kinematic pair is coupled to its corresponding dynamic pair. We call this type of kineto-dynamic coupling the *one-to-one coupling*.

Example 4.3   Fig. 4.22 shows an experimental set-up for research on robotic hand-eye coordination. An educational robot is used to serve as the arm manipulator. The kineto-dynamic pairs of this robot are designed by applying the one-to-one coupling scheme. In this example, the robot has five degrees of freedom and five kineto-dynamic pairs. All the commercially available industrial or educational arm manipulators make use of the one-to-one coupling scheme to form the kineto-dynamic pairs.

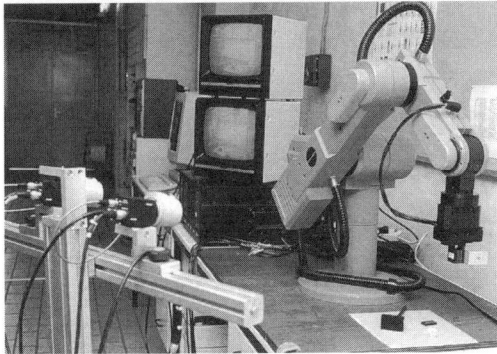

Fig. 4.22   An educational arm manipulator in an experimental set-up for research on robotic hand-eye coordination. The robot has five kineto-dynamic pairs, each of which is formed by applying the one-to-one coupling scheme.

◇◇◇◇◇◇◇◇◇◇◇◇◇◇◇◇◇◇

One notable advantage of the one-to-one coupling scheme is the fast motion response at each joint. As a result, it is easy to implement velocity feedback control which is necessary for applications involving the behavior of a *trajectory following*.

However, there are several drawbacks associated with this one-to-one coupling scheme. One serious drawback is the weight of the overall robot system because an electric motor is a heavy device compared with other materials and components. As a consequence, the effective payload of a robot is largely compromised due to the heavy weights of the electric motors, which must move together with the kinematic pairs. A second drawback is the cost. In general, an electric motor, together with its torque amplifier and power amplifier, is an expensive device. In fact, a major portion of the hardware cost of a robot is spent on electric motors, torque amplifiers, and power amplifiers.

4.4.2 *One-to-Many Couplings*

In order to overcome the drawbacks of the one-to-one coupling scheme, an alternative solution is the one-to-many coupling scheme. The philosophy of a one-to-many coupling scheme is to make use of a single dynamic pair (i.e. one motor) and to couple it with all, or a subset of independent kinematic pairs inside a robot's mechanism. In this way, the number of motors inside a robot is largely reduced. Moreover, the total cost and weight of a robot is greatly reduced as well. When there is no motor inside an open kinematic chain, it is possible to miniaturize a robot to suit medical needs such as a robotic arm for minimally-invasive surgery.

Refer to Fig. 4.21. The output torque, after the coupling between the motor and torque amplifier, is applied to joint i of the kinematic pair consisting of link $i - 1$ and link i. In the mechanical domain, link i is called an *inertial load* of the motor. If we attempt to use a single motor to independently drive multiple inertial loads, it is easy to consider how the motion from the motor is split and distributed to multiple inertial loads. A common scheme to achieve this objective is illustrated in Fig. 4.23. The additional new elements are: a) the motion splitters and b) the motion distributors.

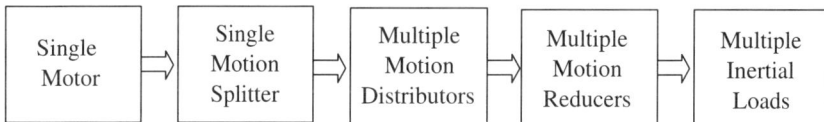

Fig. 4.23   Illustration of one-to-many coupling scheme.

4.4.2.1   *Motion Distributors*

In order to implement the one-to-many coupling scheme, we must address the following issues:

- How do we split the motion of the motor into multiple motions, visible at the multiple shafts supporting the multiple inertial loads?
- How do we handle the conflicting requirement on the directions of motions at these multiple shafts?

Let us examine the second issue first. Since we are using a single motor, the direction of the motion delivered by this motor is unique at one time instant (i.e. either clockwise or counterclockwise). However, the requirement on the motion directions at the multiple shafts is not fixed and is time-varying. This poses a problem. A simple solution is to build a new device called a *bi-directional clutch*. All the available clutches on the market are omnidirectional, meaning the direction of the output motion is exactly the same as the direction of the input motion. For a bi-directional clutch, the direction of the output motion is independent of the direction of the input motion and can be switched to any one of the two possible directions: clockwise or counterclockwise.

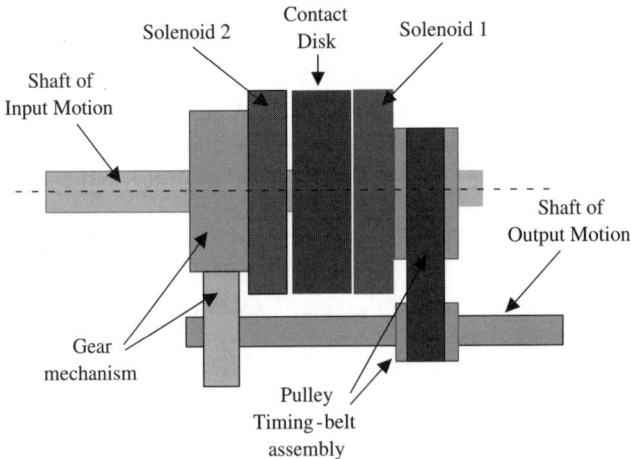

Fig. 4.24   Working principle of a bi-directional clutch.

The working principle of a bi-directional clutch is shown in Fig. 4.24. The contact disk rotates with the input shaft. When solenoid 1 is switched on, the contact disk, made of ferrous materials, is in contact with the pulley-and-timing belt assembly on the output shaft. In this way, the input motion is transmitted to the output shaft, while the direction of the output motion is the same as the direction of the input motion. Alternatively, when solenoid 2 is switched on, the contact disk is in contact with the gear mechanism on the output shaft and the input motion is inverted and transmitted to the output shaft. The inversion of the input motion is due to the effect of the gear mechanism (i.e. a pair of gears will invert the direction of the input motion).

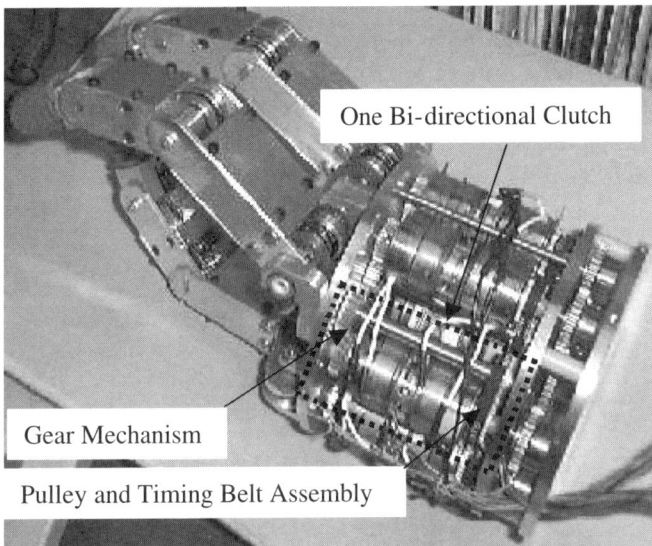

One Bi-directional Clutch

Gear Mechanism

Pulley and Timing Belt Assembly

Fig. 4.25   Single-motor-driven multiple-fingered hand which makes use of bi-directional clutches for motion distribution.

Fig. 4.25 shows a prototype of single-motor-driven multiple-fingered hand. Due to the use of bi-directional clutches, the whole hand (including the motor coaxially located at the middle of the wrist) is very compact. In this example, the total number of independent degrees of freedom for the three fingers is seven, and the multiple speed reducers are composed of worm gears and cables.

The drawback of the clutch-based motion distribution is that the mode of motion transmission is binary (i.e. either "on" or "off"). It does not produce smooth motion. Therefore, it is not suitable for velocity control, which requires the ability to smoothly vary the transmitted motion (i.e. to regulate the amount of energy released to the corresponding inertial load). One way to overcome this drawback is to introduce a device called a *continuous variable transmission* (or CVT), found in some automobiles. How to miniaturize a CVT device is still a research issue.

4.4.2.2 Parallel Splitter of Motion

The purpose of a motion splitter is to duplicate the input motion into multiple output motions. The input rotary-motion to a motion splitter comes from the output motion of a single motor. There are two ways to duplicate the input rotary motion. The first method consists of using a planar-gear mechanism to split the input motion into multiple output motions at multiple output shafts which are positioned in parallel (i.e. their axes of rotation are parallel). A device implementing this method is called a *parallel splitter of motion*.

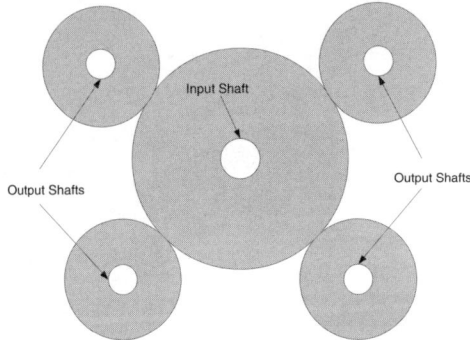

Fig. 4.26   Parallel splitter of input motion using a planar gear mechanism.

Fig. 4.26 shows an example of a planar-gear mechanism which splits the input rotary-motion into a set of four output rotary-motions at four output shafts. Each output motion of the splitter is coupled to a motion distributor, as was discussed above. Subsequently, the motion is further coupled to the inertial load through a speed reducer and a cable (or a pulley-and-timing belt mechanism).

Fig. 4.27 shows the application of the parallel splitter of motion to implement the single-motor-driven multiple-fingered hand. In this example, the motion of the single motor is duplicated into seven output motions at the output shafts of the planar-gear mechanism.

Fig. 4.27   Parallel splitter of motion inside a single-motor-driven multiple-fingered hand.

The advantage of a parallel motion splitter is its simplicity. However, a notable drawback is that a long cable or timing belt has to be used if the inertial load is located far away from the motion splitter. Therefore, the parallel splitter of motion is suitable for the implementation of a single-motor-driven multiple-fingered hand. However, it is not advantageous for the implementation of an open kinematic-chain, like an arm manipulator.

4.4.2.3   *Serial Splitter of Motion*

A humanoid robot normally has four limbs (two legs and two arms) which are open kinematic-chains. If one adopts the scheme of one-to-many coupling for the kineto-dynamic pairs in a limb, it is better to employ a device called a *serial splitter of motion*. A serial splitter of motion is a mechanism which duplicates the input rotary motion into multiple output motions at multiple output shafts aligned in a series.

In practice, there are many different methods of implementing a serial splitter of motion. However, the simplest solution is the one which makes use of the bevel-gear mechanism.

Fig 4.28 conceptually illustrates a serial splitter of motion having n consecutive units. Each unit has two output shafts aligned in a series. The first one is the shaft for twist rotation, and the second one is the shaft for pivotal rotation. The twist and pivotal axes of a unit are perpendicular to

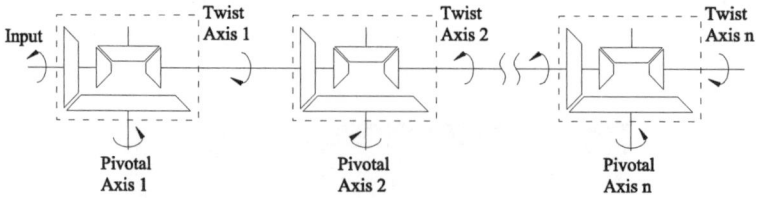

Fig. 4.28   Serial splitter of input motion using a series of bevel-gear mechanisms.

each other. Therefore, an open kinematic-chain can be easily formed by putting together a set of these units in a series.

Example 4.4   An example is shown in Fig. 4.29. In this example, we display two units. Each unit has two degrees of freedom: a) a twist rotation and b) a pivotal rotation.

(a) Real example

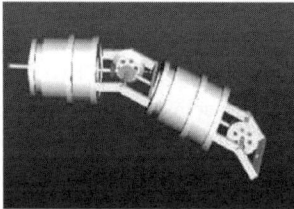

(b) Pivotal rotation   (c) Twist rotation

Fig. 4.29   Real example of an open kinematic-chain having a serial splitter of motion.

◇◇◇◇◇◇◇◇◇◇◇◇◇◇◇◇◇

It is advantageous to employ the scheme of one-to-many coupling to design the mechanisms of a humanoid robot, because the number of motors will largely be reduced. For example, a single motor may be sufficient to supply the necessary amount of energy to all the degrees of freedom inside one limb, including the multiple-fingered hand or foot. Fig. 4.30 shows a prototype of a humanoid robot. In this prototype, each of the two arm systems has six independent degrees of freedom. Each arm is driven by a single motor located inside the shoulder.

Fig. 4.30 A prototype of humanoid robot where each of the two arm systems is driven by a single motor.

4.4.2.4 *Torque at Inertial Loads*

Under the one-to-many coupling scheme, the output torque τ_0 and the angular velocity ω_0 of the motor are the same for all the inertial loads. Refer to Fig. 4.22. Let us denote k_s the ratio of speed reduction from the motion splitter, k_{di} the ratio of speed reduction from the motion distributor connected to inertial load i (link i in an open kinematic chain), and k_{ri} the ratio of speed reduction from the speed reducer coupled to inertial load i.

Then, the torque τ_i received by inertial load i will be

$$\tau_i = \begin{cases} k_s \bullet k_{di} \bullet k_{ri} \bullet \tau_0 & \text{if the clutch is ``on'';} \\ 0 & \text{otherwise.} \end{cases} \qquad (4.39)$$

It is clear that the torque applied to an inertial load is not linear. The only way to obtain a linearly-variable output torque to an inertial load is to introduce a CVT device which has a ratio of speed reduction that is a real number and is controllable.

4.4.3   *Open Kineto-Dynamic Chains*

A robot's mechanism is used to shape the output motion as a result of the input motions. In Chapter 3, we learned that the relationship between the input and output motions of a robot's mechanism is fully described by kinematics. We also know that a robot's mechanism cannot move on its own because all the mechanical motions must originate from the work done by the forces and/or torques acting at the joints of a robot's mechanism.

In order to apply forces or torques to a robot's mechanism, there must be a coupling between the kinematic pairs of a robot's mechanism and a set of dynamic pairs (or one dynamic pair if adopting the one-to-many coupling scheme). Accordingly, the kineto-dynamic pairs of a robot form the *electromechanical system*.

For kinematic analysis of a robot's mechanism, we employ the concept of an *open kinematic-chain* to make the description clear. In a similar way, it is helpful to define the concept of *open kineto-dynamic chain* to ease the study of robot statics and dynamics. A formal definition of *open kineto-dynamic chain* can be stated as follows:

Definition 4.6   An open kineto-dynamic chain is the entity formed by the coupling of a simple open kinematic-chain with a single or a set of dynamic pair(s).

Based on the concept of kineto-dynamic chain, the definition of robot statics can be easily defined as follows:

Definition 4.7   Robot statics is the study of forces or torques acting on a kineto-dynamic chain which keep the kineto-dynamic chain at rest.

Similarly, the definition of robot dynamics can be defined as follows:

Definition 4.8   Robot dynamics is the study of forces or torques acting on a kineto-dynamic chain which drive the kineto-dynamic pairs to produce the

Table 4.1   Parameters of links.

Link	Weight	Length	Center of Gravity (CG)	Offset of CG
1	m_1	l_1	r_{c1}	l_{c1}
2	m_2	l_2	r_{c2}	l_{c2}

desired input motions to a robot's mechanism, so as to obtain the desired output motion of the robot's mechanism.

In other words, robot dynamics is the study of the relationship between the forces (or torques) applied to a set of kineto-dynamic pairs, and the motions of the corresponding simple open kinematic-chain.

For the study of robot statics and dynamics, all physical quantities must be expressed in a common reference frame. By default, we consider that all the parameters and variables are expressed in the base frame of the simple open kinematic-chain. We do not explicitly indicate the common reference frame with any superscript, whenever it is not necessary to do so.

4.5   Robot Statics

Fig. 4.31 illustrates the problem of robot statics. The humanoid robot is standing still, and its task is to hold the wheel which has a rotation axis that is fixed at point A. If the robot releases its hold, the wheel will rotate in a clockwise direction due to the inertial load hanging onto the wheel. By holding the wheel, the robot receives a force and torque at the end-effector of the arm. Let us consider the case in which the robot's arm consists of two rigid links in the XY plane of the base frame. We assume that the parameters of the two links are as shown in Table 4.1, where the offset of the center of gravity (CG) of a link is measured from the link's proximal node.

In this example, the robot's body and link 1 of the arm forms the first kineto-dynamic pair, and links 1 and 2 of the arm form the second kineto-dynamic pair. Assume that the two joints are revolute joints. And, we denote r_e the acting point of the end-effector of the robot arm upon the wheel. Now, the question is: What should torques τ_1 and τ_2 applied to the two revolute joints be in order to keep the wheel at rest, if the end-effector receives force f_e from the wheel ?

In general, $f_e = (f_x, f_y, f_z)^t$, and $\tau_e = (\tau_x, \tau_y, \tau_z)^t$. In the following sections, we will discover the answer.

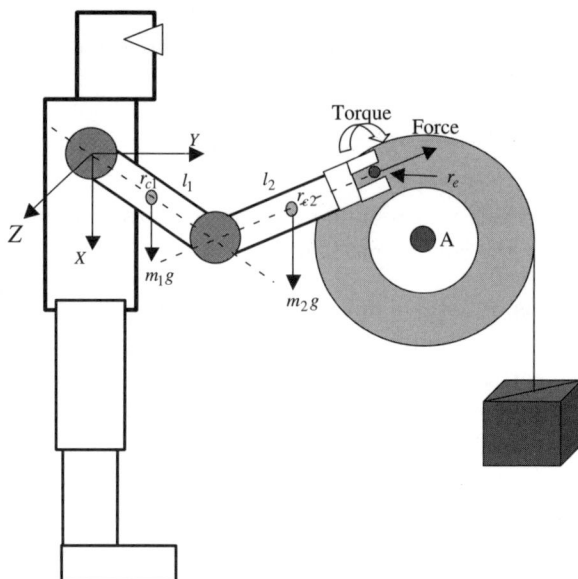

Fig. 4.31   Illustration of the problem of robot statics.

4.5.1 *Dynamic System of Particles*

The inertial load of the wheel can be treated as a force or torque generator. When the end-effector holds the wheel, it forms a dynamic pair and also, a special kinematic pair. The kinematic pair is special because the end-effector has more than one degree of freedom. In general, it has six degrees of freedom. For simplicity's sake, we can conceptually treat this coupling as another kineto-dynamic pair.

Therefore, for the example shown in Fig. 4.31, there are three kineto-dynamic pairs which form an open kineto-dynamic chain. Since all the bodies under consideration are rigid bodies, we can mathematically treat these kineto-dynamic pairs as a dynamic system of particles (three particles, in this example) with their masses concentrated at their centers of gravity. For the third kineto-dynamic pair, the weight of the wheel is supported by its rotation axis. It has no effect on the static relationship between the robot's arm and the wheel. As a result, the mass of the third kineto-dynamic pair is treated as zero, and its center of gravity is considered at r_e.

Fig. 4.32 illustrates the mathematical abstraction of the open kineto-

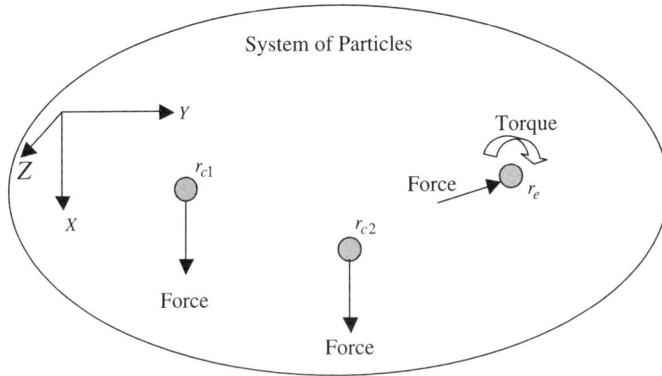

Fig. 4.32 Mathematical abstraction of an open kineto-dynamic chain by a dynamic system of particles.

dynamic chain by a dynamic system of particles. Since in this example, there are three kineto-dynamic pairs inside the open kineto-dynamic chain, the corresponding dynamic system has three particles.

In the dynamic system of particles, each particle has a mass and receives an externally exerted force (i.e. gravitational force). However, for the particle corresponding to the end-effector, it does not have any mass but may receive both externally-exerted force and torque.

Here, we only consider externally-exerted forces (or torques) to a system because all the internal forces (or torques) will not do any work to the system itself. Hence, we can ignore the effect of internal forces if all the bodies are rigid bodies (i.e. having no deformation).

4.5.2 *Generalized Coordinates and Forces*

In a three-dimensional space, the position of a particle is normally described by the X, Y and Z coordinates which are called the *physical coordinates* with respect to a Cartesian frame. For the study of certain problems (e.g. statics and dynamics), it is not helpful to directly use these coordinates. One idea, which is helpful, is to see whether there is an alternative set of coordinates which are easy to manipulate for mathematical formulation, and also permit the unique recovery of the physical coordinates (X, Y, Z). If a set of n new variables q_i $(i = 1, 2, ...n)$ is related to the physical coordinates

(X, Y, Z) in the following manner:

$$\begin{cases} X = f_x(q_1, q_2, ..., q_n) \\ Y = f_y(q_1, q_2, ..., q_n) \\ Z = f_z(q_1, q_2, ..., q_n), \end{cases} \quad (4.40)$$

and these new variables completely determine the positions of the particles in a dynamic system, these new coordinates are called *generalized coordinates*. The space defined by the generalized coordinates is called the *Configuration Space*.

In general, there are many sets of variables which satisfy the above requirement. This is an advantage because we have the freedom to choose whichever variable is easiest to manipulate. For example, in robotics, we choose the joint variables to be the generalized coordinates of a robot's dynamic system. However, one must be cautious and makes sure that the generalized coordinates can recover the physical coordinates without any ambiguity.

Example 4.5   Refer to Fig. 4.31 and Fig. 4.32. We choose the two joint variables θ_1 and θ_2 to be the generalized coordinates. For the position of particle 2 in Fig. 4.32, its physical coordinates will be

$$r_{c2} = \begin{pmatrix} x_{c2} \\ y_{c2} \end{pmatrix} = \begin{pmatrix} l_1 \bullet \cos(\theta_1) + l_{c2} \bullet \cos(\theta_1 + \theta_2) \\ l_1 \bullet \sin(\theta_1) + l_{c2} \bullet \sin(\theta_1 + \theta_2) \end{pmatrix}.$$

◇◇◇◇◇◇◇◇◇◇◇◇◇◇◇◇◇◇◇

For a generalized coordinate, we can imagine that there is a force associated to it. This force is called the *generalized force*. In robotics, if we choose joint variables to be the generalized coordinates, the generalized forces will be the forces or torques applied to the joints by the motors. The purpose of introducing the concept of generalized forces is to be able to compute the power which is equal to the generalized force times the displacement of the corresponding generalized coordinate.

4.5.3   *Constraints of a Dynamic System*

The mathematical abstraction of an open kineto-dynamic chain by a dynamic system of particles is valid if and only if the particles satisfy the constraints imposed upon the corresponding open kineto-dynamic chain. A constraint refers to the restriction imposed upon a pair of bodies which interacts with each other. For every constraint, there is a constraint equation and a constraint force or torque. Thus, a kineto-dynamic pair is a perfect constraint imposed upon a pair of bodies. And the constraint equation simply indicates the kinematic relationship among the generalized coordinates.

Example 4.6 Refer to Fig. 4.31 and Fig. 4.32. We choose the two joint variables θ_1 and θ_2 to be the generalized coordinates. Assume that the end-effector's position is r_e. Then, the generalized coordinates of the dynamic system of particles must satisfy the following constraint:

$$r_e = \begin{pmatrix} x_e \\ y_e \end{pmatrix} = \begin{pmatrix} l_1 \bullet \cos(\theta_1) + l_2 \bullet \cos(\theta_1 + \theta_2) \\ l_1 \bullet \sin(\theta_1) + l_2 \bullet \sin(\theta_1 + \theta_2) \end{pmatrix}.$$

◇◇◇◇◇◇◇◇◇◇◇◇◇◇◇◇◇◇◇

For a system with n generalized coordinates q_i $(i = 1, 2, ..., n)$, the constraints can be described in the following form:

$$f(q_1, q_2, ..., q_n, t) = 0. \tag{4.41}$$

One can omit the time variable if the constraint is independent of time. If Eq. 4.41 is differentiable, we have

$$\dot{f} = \frac{\partial f}{\partial q_1}\dot{q}_1 + \frac{\partial f}{\partial q_2}\dot{q}_2 + ... + \frac{\partial f}{\partial q_n}\dot{q}_n + \frac{\partial f}{\partial t} = 0. \tag{4.42}$$

A constraint expressed in the form of Eq. 4.41 is called a *configuration constraint*. A constraint expressed in the form of Eq. 4.42 is called a *velocity constraint*. If a constraint can be expressed in both the forms of Eq. 4.41 and Eq. 4.42, it is called a *holonomic constraint*. Otherwise, it is called a *non-holonomic constraint*. In robotics, all open kineto-dynamic chains impose holonomic constraints.

Example 4.7 Refer to Fig. 4.31 and Fig. 4.32. We choose the two joint variables θ_1 and θ_2 to be the generalized coordinates. Let us consider the position vectors of all the particles in the system. Then, the generalized coordinates of this dynamic system must satisfy the following constraints:

(1) For position vector r_{c1} of particle 1, we have

$$r_{c1} = \begin{pmatrix} x_{c1} \\ y_{c1} \end{pmatrix} = \begin{pmatrix} l_{c1} \bullet \cos(\theta_1) \\ l_{c1} \bullet \sin(\theta_1) \end{pmatrix}$$

or

$$r_{c1} = r_{c1}(\theta_1, \theta_2).$$

(2) For position vector r_{c2} of particle 2, we have

$$r_{c2} = \begin{pmatrix} x_{c2} \\ y_{c2} \end{pmatrix} = \begin{pmatrix} l_1 \bullet \cos(\theta_1) + l_{c2} \bullet \cos(\theta_1 + \theta_2) \\ l_1 \bullet \sin(\theta_1) + l_{c2} \bullet \sin(\theta_1 + \theta_2) \end{pmatrix}$$

or

$$r_{c2} = r_{c2}(\theta_1, \theta_2).$$

(3) For position vector r_e of particle 3, we have

$$r_e = \begin{pmatrix} x_e \\ y_e \end{pmatrix} = \begin{pmatrix} l_1 \bullet \cos(\theta_1) + l_2 \bullet \cos(\theta_1 + \theta_2) \\ l_1 \bullet \sin(\theta_1) + l_2 \bullet \sin(\theta_1 + \theta_2) \end{pmatrix}$$

or

$$r_e = r_e(\theta_1, \theta_2).$$

◇◇◇◇◇◇◇◇◇◇◇◇◇◇◇◇◇◇◇◇

4.5.4   *Virtual Displacements*

In the above example, we can see that a position vector of a particle in a dynamic system can be expressed as a function of the generalized coordinates. Assume that a dynamic system of particles is fully described with a set of n generalized coordinates q_i $(i = 1, 2, ..., n)$. Then, position vector r of a particle can be written as

$$r = r(q_1, q_2, ..., q_n). \tag{4.43}$$

Differentiating Eq. 4.43 with respect to time gives

$$dr = \frac{\partial r}{\partial q_1} dq_1 + \frac{\partial r}{\partial q_2} dq_2 + ... + \frac{\partial r}{\partial q_n} dq_n = \sum_{i=1}^{n} \left\{ \frac{\partial r}{\partial q_i} dq_i \right\}. \tag{4.44}$$

dr is the infinitesimal displacement within time interval dt. This displacement satisfies the velocity constraint in the form of Eq. 4.42. For the study of robot statics, the system is at rest and there is no real displacement. If we examine the tendency of particles' displacements, time interval dt can be ignored. In this way, the infinitesimal displacement dr is called the *virtual displacement*. In other words, it is not a real displacement. In dynamics, the virtual displacement corresponding to the differential dr is denoted by δr. If we denote δq_i the virtual displacement of the generalized coordinate q_i, Eq. 4.44 becomes

$$\delta r = \sum_{i=1}^{n} \left\{ \frac{\partial r}{\partial q_i} \bullet \delta q_i \right\}. \qquad (4.45)$$

Based on the above discussions, it is clear that a virtual displacement is an infinitesimal displacement. These virtual displacements are obtained by differentiation. As a result, all virtual displacements satisfy the velocity constraint of a dynamic system in the form of Eq. 4.42.

4.5.5 *Virtual Works*

Refer to Fig. 4.32. Assume that all the particles receive an externally-exerted force, and the particle corresponding to the end-effector may also receive an externally-exerted torque. Here, we assume that the end-effector receives an externally-exerted force only.

We denote (f_{c1}, f_{c2}) the external (e.g. gravitational) forces exerted on particles 1 and 2 respectively, and f_e the external force exerted on particle 3. By definition, the work done by these forces to the system will be

$$\delta W = f_{c1} \bullet \delta r_{c1} + f_{c2} \bullet \delta r_{c2} + f_e \bullet \delta r_e. \qquad (4.46)$$

Since the work done is not real because the displacements are not real, we call it *virtual work*.

4.5.6 *Principle of Virtual Work for Statics*

It is easy to understand at this point that a static system (i.e. a system at rest) will remain at rest if the virtual work done by the externally-exerted forces is zero. This is because the motion can only originate from work done to a system.

Refer to Fig. 4.32. Forces (f_{c1}, f_{c2}, f_e) are physically real. If there are no other external forces, the virtual work expressed in Eq. 4.46 will not be zero. In this case, real work will be done to the system and it will make the robot's arm and the wheel undertake motion.

If we want the robot to hold the wheel at rest, there must be some other external forces applied to the robot's arm in order to make the total virtual work be zero. (NOTE: A work can be negative if energy is removed from a system). In this example, the generalized coordinates of the system are (θ_1, θ_2). If we denote (τ_1, τ_2) as the generalized forces associated with these two generalized coordinates, the virtual work done by the generalized forces will be

$$\delta W_\tau = \tau_1 \bullet \delta\theta_1 + \tau_2 \bullet \delta\theta_2. \tag{4.47}$$

Now, we can state the principle of virtual work for statics as follows:

Principle of Virtual Work: For a static system to remain at rest, the total virtual work done by all external forces to the system must be equal to zero, i.e. $\delta W - \delta W_\tau = 0$.

4.5.7 *Statics Against Self-Inertial Loads*

If the goal is to keep the system at rest (i.e. statics), the purpose of the generalized forces will be to cancel out virtual work done by external forces acting upon the system. Refer to the example shown in Fig. 4.31 and Fig. 4.32. We first examine the generalized forces required to cancel out virtual work done by gravitational forces acting upon links 1 and 2.

In Example 4.7, the centers of gravity of links 1 and 2 can be expressed as functions of the generalized coordinates. That is,

$$\begin{cases} r_{c1} = r_{c1}(\theta_1, \theta_2) \\ r_{c2} = r_{c2}(\theta_1, \theta_2). \end{cases} \tag{4.48}$$

Differentiating Eq. 4.48 yields

$$\begin{cases} \delta r_{c1} = \frac{\partial r_{c1}}{\partial \theta_1} \bullet \delta\theta_1 + \frac{\partial r_{c1}}{\partial \theta_2} \bullet \delta\theta_2 \\ \\ \delta r_{c2} = \frac{\partial r_{c2}}{\partial \theta_1} \bullet \delta\theta_1 + \frac{\partial r_{c2}}{\partial \theta_2} \bullet \delta\theta_2. \end{cases} \tag{4.49}$$

These are expressions for the virtual displacements of the two centers of gravity.

By applying the principle of virtual work for statics, we have

$$\tau_1 \bullet \delta\theta_1 + \tau_2 \bullet \delta\theta_2 = f_{c1} \bullet \delta r_{c1} + f_{c2} \bullet \delta r_{c2}. \tag{4.50}$$

Substituting Eq. 4.49 into Eq. 4.50 yields the following two equalities:

$$\begin{cases} \tau_1 = f_{c1} \bullet \frac{\partial r_{c1}}{\partial \theta_1} + f_{c2} \bullet \frac{\partial r_{c2}}{\partial \theta_1} \\ \\ \tau_2 = f_{c1} \bullet \frac{\partial r_{c1}}{\partial \theta_2} + f_{c2} \bullet \frac{\partial r_{c2}}{\partial \theta_2}. \end{cases} \tag{4.51}$$

In general, in order to keep an open kineto-dynamic chain with n links at rest, the generalized force at joint i will be

$$\tau_i = \sum_{j=1}^{n} \left\{ f_{cj} \bullet \frac{\partial r_{cj}}{\partial q_i} \right\},$$ (4.52)

where $q_i = \theta_i$ if joint i is a revolute joint, and f_{cj} is the external (gravitational) force acting on link j at its center of gravity r_{cj}. Some textbooks treat Eq. 4.52 as the definition of generalized forces. Here, we treat it as the solution for determining generalized forces.

The work done by a gravitational force is called *potential energy*. An alternative way of deriving the expression in Eq. 4.52 is to make use of the concept of potential energy. For the example shown in Fig. 4.31, the potential energy of the robot's arm is

$$V = -f_{c1} \bullet r_{c1} - f_{c2} \bullet r_{c2}.$$ (4.53)

Since the centers of gravity of the arm's two links are the functions of the generalized coordinates, the potential energy can be implicitly expressed as follows:

$$V = V(\theta_1, \theta_2).$$

This means that the potential energy only depends on the positions which are the functions of the generalized coordinates.

The differentiation of the above equation yields virtual work done by gravitational forces f_{c1} and f_{c2}. That is,

$$\delta W_v = -\delta V = -\frac{\partial V}{\partial \theta_1} \bullet \delta \theta_1 - \frac{\partial V}{\partial \theta_2} \bullet \delta \theta_2.$$ (4.54)

(NOTE: The negative sign at expression $\delta W = -\delta V$ means that potential energy is removed from a body if the displacement of the body is in the direction of the gravitational force).

By applying the principle of virtual work (i.e. $\delta W_\tau = \delta W_v$), we obtain

$$\begin{cases} \tau_1 = -\frac{\partial V}{\partial \theta_1} \\ \tau_2 = -\frac{\partial V}{\partial \theta_2}. \end{cases}$$ (4.55)

Eq. 4.55 is equivalent to Eq. 4.51. In general, in order to maintain an open kineto-dynamic chain with n links at rest, the generalized force at joint i will be

$$\tau_i = -\frac{\partial V}{\partial q_i}. \tag{4.56}$$

In fact, Eq. 4.56 is equivalent to Eq. 4.52.

4.5.8   *Statics Against Inertial Loads at End-effector*

Refer to the example shown in Fig. 4.31 and Fig. 4.32 again. We will now examine what should be the generalized forces required to cancel out virtual work done by external force and torque acting upon the end-effector's frame in order to keep the robot's arm at rest.

From the study of robot kinematics, we know that velocity vector $\dot{P}_e$ of the end-effector's frame in a simple open kinematic-chain is composed of two vectors: a) the linear velocity vector $\dot{O}_e$ of the end-effector's frame and b) its angular velocity vector ω_e. That is,

$$\dot{P}_e = \begin{pmatrix} \dot{O}_e \\ \omega_e \end{pmatrix}.$$

(NOTE: Symbol P_e stands for the *posture* of the end-effector's frame).

If the Jacobian matrix of the robot's arm is J, we have

$$\dot{P}_e = J \bullet \dot{q} \tag{4.57}$$

where $\dot{q} = (\dot{\theta}_1, \dot{\theta}_2)^t$.

From Eq. 4.57 and by definition, the virtual displacement of the end-effector's frame will be

$$\delta P_e = J \bullet \delta q. \tag{4.58}$$

If we denote $\xi_e = (f_e, \tau_e)^t$ the vector of the external force and torque exerted upon the end-effector's frame, the virtual work done to the system (i.e. robot) at rest is as follows:

$$\delta W = \xi_e^t \bullet \delta P_e. \tag{4.59}$$

(NOTE: In general, $\xi_e = (f_x, f_y, f_z, \tau_x, \tau_y, \tau_z)^t$).

In order to cancel out virtual work done to the system at rest, we must apply generalized forces (τ_1, τ_2) to the system at rest. We know that virtual work done by generalized forces is

$$\delta W_\tau = \tau_1 \bullet \theta_1 + \tau_2 \bullet \theta_2 = \tau^t \bullet \delta q \tag{4.60}$$

where $\tau = (\tau_1, \tau_2)^t$.

By applying the principle of virtual work (i.e. $\delta W_\tau = \delta W$), we obtain

$$\tau^t = \xi_e^t \bullet J. \tag{4.61}$$

If we compute the transpose at both sides of Eq. 4.61, this equation will become

$$\tau = J^t \bullet \xi_e. \tag{4.62}$$

Eq. 4.62 describes the static relationship between external force/torque exerted upon the end-effector's frame, and generalized forces required on the open kineto-dynamic chain in order to keep the whole system at rest. This static relationship is very useful for the study of compliance or interactive control because ξ_e encapsulates the interacting force and/or torque between the robot's hand and its environment.

In general, Eq. 4.62 is also valid for any open kineto-dynamic chain with n rigid links. Here, we consider the case of revolute joints. If joint i is a prismatic joint, the corresponding generalized coordinate will be linear displacement d_i and the associated generalized force will be force f_i itself.

4.6   Robot Dynamics

The purpose of robot dynamics is to determine the generalized forces required to not only overcome the self-inertial loads (due to the weights of the links inside the robot) but also to produce the desired input motions to the robot's mechanism.

Fig 4.33 illustrates the problem of robot dynamics. In this example, we assume that the robot's body stands still. The robot's arm is holding a tool, which can be a laser welding gun or any other industrial tool. For the robot to fulfill a task (e.g. arc welding), the robot needs to move the tool to follow a predefined trajectory. It is clear that the robot's arm has to execute motions in order to carry the tool along that trajectory. Therefore, a certain amount of work (i.e. energy) must be done to the robot system. The work done to the robot will not only overcome the self-inertial load of the arm but also generate the desired motions to the joints of the arm so that the tool will follow the trajectory.

Now, the question is: What should the generalized forces applied to the joints of the robot arm be in order to produce the desired velocity and

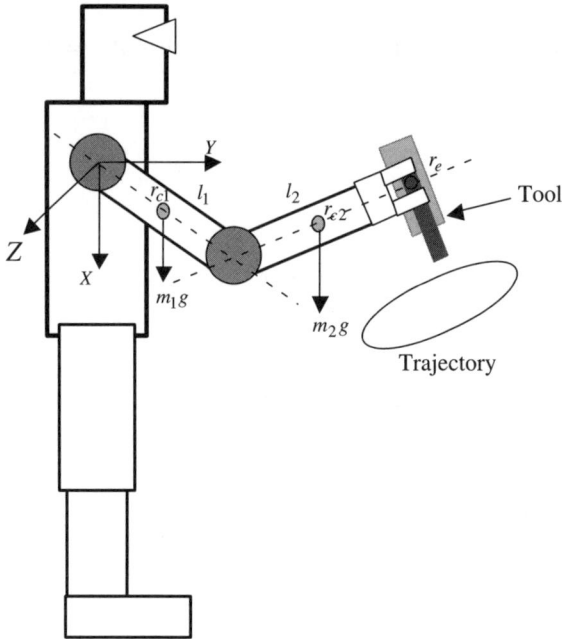

Fig. 4.33   Illustration of the problem of robot dynamics.

acceleration at the generalized coordinates (joint variables), which will subsequently produce the desired velocity and acceleration at the end-effector's
frame (forward kinematics of the robot's mechanism)?

4.6.1   *Dynamic System of Rigid Bodies*

Refer to Fig 4.33. The robot's arm is an open kineto-dynamic chain consisting of two rigid links. For the study of robot dynamics, an open kineto-
dynamic chain can be conveniently treated as a dynamic system of rigid
bodies. These bodies satisfy the constraints imposed by the kinematics of
the robot's mechanism, and also the constraints imposed by Newton's third
law (the relationship between a pair of acting and reacting forces).

Fig. 4.34 shows the mathematical abstraction of an open kineto-dynamic
chain by a dynamic system of rigid bodies. Link 0 is the supporting body of
link 1, while link 1 is the supporting body of link 2. In this example, links
1 and 2 are represented respectively by a mass, a center of gravity, a sum
of all externally-exerted forces upon the body and a sum of all externally-

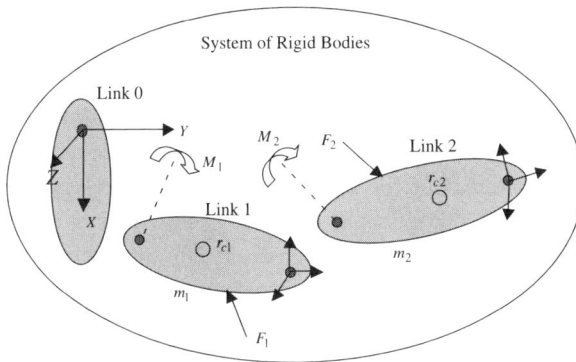

Fig. 4.34 Mathematical abstraction of an open kineto-dynamic chain by a dynamic system of rigid bodies.

Table 4.2 Parameters of links.

Link	Mass	Center of gravity	Summed force	Summed torque
1	m_1	r_{c1}	F_1	M_1
2	m_2	r_{c2}	F_2	M_2

exerted torques upon the body. Table 4.2 summarizes the parameters of links 1 and 2 in Fig. 4.34.

4.6.2 *Dynamics of a Rigid Body*

Before we develop the solution for robot dynamics, let us examine the solution of dynamics for a single rigid body.

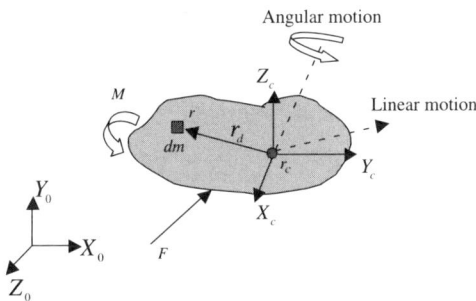

Fig. 4.35 Dynamics of a single rigid body.

Fig. 4.35 shows a single rigid body in a reference frame. r_c is the position vector indicating the center of gravity of the body; r is the position vector of differential mass element dm on the body; r_d is the displacement vector from the center of gravity to differential mass element dm. By default, all these vectors, r_c, r and r_d, are expressed in the common reference frame (i.e. frame 0). Assume that the body is subject to some external forces and torques. The sum of the external forces is denoted by F and the sum of the external torques is denoted by M. Now, the questions are: What will the relationship be between the motion parameters of a rigid body and the sum of (external) forces/torques exerted upon the body? And, how is the motion of a rigid body related to its energy state (i.e. potential and kinetic energy)?

4.6.2.1 Motion Equations of a Rigid Body

By definition, the mass of the body is

$$m = \int_{body} dm \tag{4.63}$$

where dm is a differential mass element. If the mass density of rigid body is $\rho(r)$, then dm is expressed as

$$dm = \rho(r) \bullet dx\ dy\ dz.$$

If we know the mass of the rigid body, then the position vector indicating the center of gravity of the body is

$$r_c = \frac{1}{m} \int_{body} r \bullet dm. \tag{4.64}$$

Since $r = r_c + r_d$, Eq. 4.64 becomes

$$r_c = \frac{1}{m} \int_{body} r_c \bullet dm + \frac{1}{m} \int_{body} r_d \bullet dm = r_c + \frac{1}{m} \int_{body} r_d \bullet dm.$$

Therefore, the following result is proven:

$$\int_{body} r_d \bullet dm = 0. \tag{4.65}$$

This equation illustrates that the sum of displacement vector r_d weighted by differential mass element dm is equal to zero.

Assume that the motion parameters of the body at the center of gravity are as follows:

v_c : linear velocity vector;
ω : angular velocity vector;
a_c : acceleration vector;
α : angular acceleration vector.

From the relationship $r = r_c + r_d$, the linear velocity vector at position r of differential mass element dm will be

$$v = \frac{dr}{dt} = \frac{dr_c}{dt} + \frac{dr_d}{dt}.$$

Since $\frac{dr_c}{dt} = v_c$ and $\frac{dr_d}{dt} = \omega \times r_d$, we have

$$v = v_c + \omega \times r_d. \tag{4.66}$$

Similarly, the differentiation of Eq. 4.66 yields the expression for the acceleration vector at position r, that is,

$$a = \frac{dv}{dt} = \frac{dv_c}{dt} + \frac{d}{dt}\{\omega \times r_d\}$$

or

$$a = a_c + \omega \times (\omega \times r_d). \tag{4.67}$$

By definition, the linear momentum at differential mass element dm is $v \bullet dm$. Then, the linear momentum of the rigid body will be

$$p = \int_{body} v \bullet dm = \int_{body} (v_c + \omega \times r_c) \bullet dm$$

or

$$p = \int_{body} v_c \bullet dm + \int_{body} (\omega \times r_d) \bullet dm.$$

Since $\int_{body} (\omega \times r_d) \bullet dm = \omega \times (\int_{body} r_d \bullet dm)$ and $\int_{body} r_d \bullet dm = 0$ (i.e. Eq 4.65), the linear momentum of the rigid body becomes

$$p = \int_{body} v_c \bullet dm = v_c \bullet m.$$

By definition, the variation rate of the linear momentum of a body is equal to the sum of external forces exerted upon the body. As a result, the equation of motion governing linear motion will be

$$F = \frac{dp}{dt} = m \bullet \frac{dv_c}{dt}$$

or

$$F = m \bullet a_c. \tag{4.68}$$

Now, we consider the general case in which the motion of the rigid body is composed of linear motion at its center of gravity, and angular motion about an axis at its center of gravity. If the linear momentum at differential mass element dm is $v \bullet dm$, by definition, the angular momentum of differential mass element dm, with respect to the center of gravity of the body, is $r_d \times (v \bullet dm)$. Then, the angular momentum of the rigid body, with respect to its center of gravity, will be

$$H = \int_{body} (r_d \times v) \bullet dm.$$

Substituting Eq. 4.66 into the above equation yields

$$H = \int_{body} (r_d \times r_c) \bullet dm + \int_{body} (r_d \times (\omega \times r_d)) \bullet dm. \tag{4.69}$$

If $S(\cdot)$ denotes the skew-symmetric matrix of a vector, the cross-product of any two vectors a and b can be expressed as

$$a \times b = S(a) \bullet b$$

or

$$a \times b = (-b) \times a = S^t(b) \bullet a.$$

By using the property of the skew-symmetric matrix, we have

$$\int_{body} (r_d \times r_c) \bullet dm = S^t(r_c) \bullet \int_{body} r_d \bullet dm = 0$$

(i.e. Eq. 4.65), and

$$r_d \times (\omega \times r_d) = S(r_d) \bullet S^t(r_d) \bullet \omega.$$

As a result, Eq. 4.69 can be simplified to:

$$H = \int_{body} S(r_d) \bullet S^t(r_d) \bullet \omega \bullet dm.$$

By applying the following property of a skew-symmetric matrix:

$$S(\cdot) \bullet S^t(\cdot) = S^t(\cdot) \bullet S(\cdot),$$

the angular moment H can also be written as

$$H = \int_{body} S^t(r_d) \bullet S(r_d) \bullet \omega \bullet dm.$$

If we define

$$I = \int_{body} S^t(r_d) \bullet S(r_d) \bullet dm, \qquad (4.70)$$

the final expression for the angular momentum of the rigid body becomes

$$H = I \bullet \omega. \qquad (4.71)$$

In fact, matrix I is the *inertial tensor* or *mass moments of inertia* (or simply *inertial matrix*). From Eq. 4.70, it is clear that the inertial matrix depends on the choice of reference frame because vector r_d is expressed with respect to the reference frame chosen.

By definition, the variation rate of the angular momentum of a body is equal to the sum of external torques exerted upon the body. Finally, the equation of motion, governing the angular motion, will be

$$M = \frac{dH}{dt} = I \bullet \frac{d\omega}{dt} + \frac{dI}{dt}\omega$$

or

$$M = I \bullet \alpha + \frac{dI}{dt} \bullet \omega. \qquad (4.72)$$

(NOTE: I is not a constant matrix).

In Eq. 4.72, inertial matrix I is computed with respect to frame 0. When the rigid body changes its position or orientation, this matrix has to be recalculated. Computationally, it is not desirable to keep calculating an inertial matrix based on Eq. 4.70. An alternative way is to compute the inertial matrix of a rigid body with respect to a fixed frame assigned to the rigid body itself, and then to calculate inertial matrix I with respect to frame 0.

In practice, it is convenient to choose a frame assigned to the center of gravity of a rigid body. Refer to Fig. 4.35. Frame c is attached to the rigid body at the center of gravity. Let $^c r_d$ denote the displacement vector of

r_d, expressed in frame c. By definition, inertial matrix ^cI of the rigid body, with respect to frame c, is

$$^cI = \int_{body} S^t(^cr_d) \bullet S(^cr_d) \bullet dm. \qquad (4.73)$$

If rotation matrix R_c (or more precisely ^0R_c) describes the orientation of frame c with respect to frame 0, then vector r_d expressed in frame 0 is related to the same vector ^cr_d expressed in frame c as follows:

$$r_d = R_c \bullet^c r_d. \qquad (4.74)$$

Substituting Eq. 4.74 into Eq. 4.70 yields

$$I = \int_{body} S^t(R_c \bullet^c r_d) \bullet S(R_c \bullet^c r_d) \bullet dm. \qquad (4.75)$$

Since $S(R_c \bullet^c r_d) = R_c \bullet S(^cr_d) \bullet R_c^t$ (see Chapter 2), Eq. 4.75 can be simplified as

$$I = R_c \bullet \left\{ \int_{body} S^t(^cr_d) \bullet S(^cr_d) \bullet dm \right\} \bullet R_c^t$$

or

$$I = R_c \bullet \{^cI\} \bullet R_c^t. \qquad (4.76)$$

Eq. 4.76 is a computationally efficient solution to calculate the inertial matrix of a rigid body with respect to a reference frame because inertial matrix ^cI of the same body, with respect to frame c assigned to the body, is a constant (and symmetric) matrix. Eq. 4.76 also allows us to evaluate the derivative $\frac{dI}{dt}$ in Eq. 4.72.

4.6.2.2 *Potential Energy of a Rigid Body*

Now, let us examine the relationship between the position of a rigid body and its potential energy.

Assume that the density vector of gravitational force on the surface of the Earth is g. g is a vector pointing towards the center of the Earth, and is expressed in the common reference frame. If the Z axis of the reference frame is pointing upward and perpendicular to the ground, $g = (0, 0, -9.803)^t$. If we consider differential mass element dm, its differential potential energy dV_m will be

$$dV_m = -g^t \bullet r \bullet dm = -g^t \bullet (r_c + r_d) \bullet dm.$$

As a result, the total potential energy of the body is

$$V = \int_{body} dV_m = -g^t \bullet \int_{body} (r_c + r_d) \bullet dm.$$

Since $\int_{body} r_d \bullet dm = 0$ and $\int_{body} r_c \bullet dm = r_c \bullet m$, the potential energy of a rigid body will be

$$V = -m \bullet g^t \bullet r_c. \tag{4.77}$$

4.6.2.3 *Kinetic Energy of a Rigid Body*

When a rigid body is in motion, it stores kinetic energy. In other words, if we want to move a rigid body which is initially at rest, we must apply energy to it. In robotics, the kinetic energy of a rigid body is very important in the understanding of robot dynamics.

Consider differential mass element dm at position r on the rigid body. If v is the velocity vector of this differential mass element, the kinetic energy of this differential mass element will be

$$dK_m = \frac{1}{2} v^t \bullet v \bullet dm.$$

Since $v = v_c + w \times r_d$ (i.e. Eq. 4.66), we have

$$dK_m = \frac{1}{2} (v_c + w \times r_d)^t \bullet (v_c + w \times r_d) \bullet dm.$$

The integration over the entire body of the above equation yields the total kinetic energy of the body, that is,

$$K = \int_{body} dK_m = \frac{1}{2} \int_{body} (v_c + w \times r_d)^t \bullet (v_c + w \times r_d) \bullet dm$$

or

$$K = \frac{1}{2} \int_{body} \left\{ v_c^t v_c + v_c^t (w \times r_d) + (w \times r_d)^t v_c + (w \times r_d)^t (w \times r_d) \right\} \bullet dm. \tag{4.78}$$

By applying the property $\int_{body} r_d dm = 0$, one can easily prove

$$\int_{body} \{ v_c^t (w \times r_d) + (w \times r_d)^t v_c \} \bullet dm = \int_{body} \{ v_c^t S(w) r_d + r_d^t S^t(w) v_c \} \bullet dm$$

$$= 0$$

where $S(w)$ is the skew-symmetric matrix of w.

If we use the property of the skew-symmetric matrix, we have

$$(\omega \times r_d)^t (\omega \times r_d) = [S^t(r_d)\omega]^t [S^t(r_d)\omega] = \omega^t [S(r_d)S^t(r_d)]\omega.$$

Therefore, Eq. 4.78 can be simplified as

$$K = \frac{1}{2} \int_{body} \left\{ v_c^t v_c + \omega^t [S(r_d)S^t(r_d)]\omega \right\} \bullet dm. \tag{4.79}$$

Substituting Eq. 4.63, Eq. 4.70 and Eq. 4.76 into Eq. 4.79 yields

$$K = \frac{1}{2}m \bullet v_c^t \bullet v_c + \frac{1}{2}\omega^t \bullet I \bullet \omega \tag{4.80}$$

or

$$K = \frac{1}{2}m \bullet v_c^t \bullet v_c + \frac{1}{2}\omega^t \bullet [R_c \bullet^c I \bullet R_c^t] \bullet \omega. \tag{4.81}$$

Eq. 4.81 describes the kinetic energy stored inside a rigid body in motion. The first term in the equation refers to kinetic energy from the linear motion, and the second term refers to kinetic energy from the angular motion. From this equation, it is clear that the motion of a rigid body is directly related to its kinetic energy. A simple way to produce the desired motion on a rigid body is to control the amount of energy added to, or removed from, the kinetic energy of the rigid body.

4.6.3   Newton-Euler Formula

If we know the solutions to a single rigid body's dynamics, it is easy to derive the equations of motion for an open kineto-dynamic chain. The first method is the *Newton-Euler formula*. The basic idea behind the Newton-Euler formula is to treat an open kineto-dynamic chain as a dynamic system of rigid bodies which satisfy: a) the kinematic constraint imposed upon the robot's mechanism, and b) the dynamic constraint imposed upon the kineto-dynamic couplings.

For the purpose of better illustrating kinematic and dynamic constraints, we redraw Fig. 4.34. The result is illustrated in Fig. 4.36. Now, let us consider an open kineto-dynamic chain with $n + 1$ links. For the example shown in Fig. 4.36, the robot's arm has three links ($n = 2$).

4.6.3.1   Kinematic Parameters of a Link

An open kineto-dynamic chain is the superset of the corresponding simple open kinematic-chain. Obviously, we must know the kinematic parameters

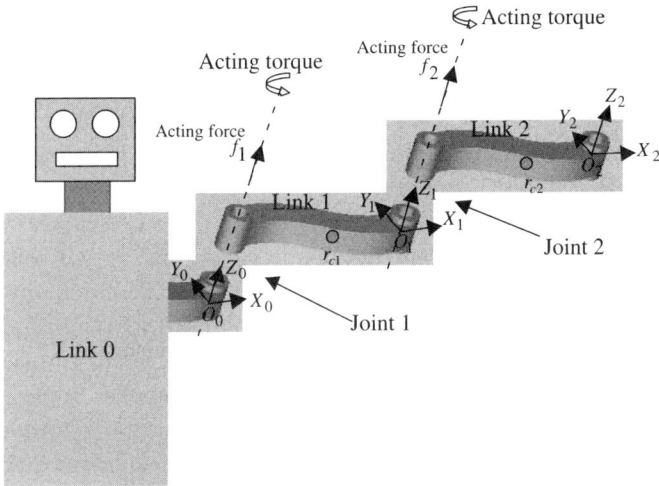

Fig. 4.36   Illustration for Newton-Euler formula.

of an open kineto-dynamic chain even if the purpose here is only to study the dynamics.

As shown in Fig. 4.36, each link is assigned a frame according to the DH representation (see Chapter 3). The frame assigned to a link is located at its distal node. For the study of dynamics, the center of gravity of a rigid body plays a useful role. Therefore, we must know where the center of a link's gravity is. In general, link i inside an open kineto-dynamic chain should have the following kinematic parameters:

- O_i: the origin of link i's frame;
- R_i: the rotation matrix describing the orientation of link i's frame;
- r_{ci}: the center of gravity of link i;
- v_i: the linear velocity vector of link i's frame;
- a_i: the linear acceleration vector of link i's frame;
- ω_i: the angular velocity vector of link i's frame which has Z_{i-1} as its axis of rotation;
- α_i: the angular acceleration of link i.

By default, all the kinematic parameters of a link are expressed with respect to a common reference frame. Usually, the base link's frame (i.e. frame 0) is chosen as the common reference frame.

4.6.3.2 *Dynamic Parameters of a Link*

Similarly, we must also know what the dynamic parameters related to each rigid body inside an open kineto-dynamic chain are. In general, the dynamic parameters of a rigid body include mass and inertial matrix. Due to the dynamic constraint imposed by kineto-dynamic couplings, the acting and reacting forces/torques are very important dynamic parameters as well.

Now, consider an open kineto-dynamic chain with $n + 1$ links where link i is supported by link $i - 1$ through joint i. The externally-exerted force or torque (except gravitational force) is applied to a link through the supporting joint. Therefore, it is logical to call the force or torque applied to a joint the *acting force or torque*. When joint i applies an acting force or torque to link i, joint i will also exert a reacting force or torque to link $i - 1$ according to Newton's third law. As a result, the dynamic parameters of link i inside an open kineto-dynamic chain will include

- m_i: the mass of link i;
- $^c I_i$: the inertial matrix of link i, calculated with respect to the center of the link's gravity;
- I_i: the inertial matrix of link i, calculated with respect to the common reference frame (frame 0);
- f_i: the acting force exerted by joint i on link i;
- $-f_{i+1}$: the reacting force exerted by joint $i + 1$ on link i;
- τ_i: the acting torque exerted by joint i on link i;
- $-\tau_{i+1}$: the reacting torque exerted by joint $i + 1$ on link i.
- g: the density vector of gravitational force on the surface of the Earth, expressed in the common reference frame (frame 0).

By default, all dynamic parameters are expressed with respect to the common reference frame (frame 0).

4.6.3.3 *Sum of Forces of a Link*

Before we derive equations of motion for each link, we need to know the sum of forces/torques exerted upon a link.

We denote F_i the sum of all exerted forces on link i. If we know the dynamic parameters of link i, the sum of forces F_i will be

$$F_i = f_i + (-f_{i+1}) + m_i \bullet g. \qquad (4.82)$$

g is a vector expressed in the common reference frame (frame 0). When studying a humanoid robot, this vector may not be a constant vector because the reference frame will have relative motion with respect to the Earth when a humanoid robot is in motion.

4.6.3.4 *Sum of Torques of a Link*

If joint i is a prismatic joint, it will not apply any torque to link i. Here, we consider the case of revolute joints and assume that all links can undergo angular motion.

When force is applied to a body, which can make rotation, this force will also induce torque on the body with respect to a reference point. Here, we consider the center of gravity of a body as a reference point. If we denote

- $r_{ci,i-1}$: the vector connecting the center of gravity r_{ci} of link i to origin O_{i-1} of frame $i-1$ ($r_{ci,i-1} = O_{i-1} - r_{ci}$), and
- $r_{ci,i}$: the vector connecting the center of gravity r_{ci} of link i to origin O_i of frame i ($r_{ci,i} = O_i - r_{ci}$),

the induced torques corresponding to acting force f_i and reacting $-f_{i+1}$ will be $r_{ci,i-1} \times f_i$ and $r_{ci,i} \times (-f_{i+1})$ respectively.

We further denote M_i the sum of all torques exerted on a link i. If we know the kinematic and dynamic parameters of link i, the sum of torques M_i will be

$$M_i = \tau_i + (-\tau_{i+1}) + r_{ci,i-1} \times f_i + (-r_{ci,i} \times f_{i+1}). \tag{4.83}$$

4.6.3.5 *Equation of Linear Motion of a Link*

By direct application of the result in Eq. 4.68, the equation of linear motion of link i is

$$f_i + (-f_{i+1}) + m_i \bullet g = m_i \bullet a_{ci} \tag{4.84}$$

where a_{ci} is the acceleration vector at the center of gravity of link i.

By differentiating the equation $r_{ci} = O_i - r_{ci,i}$, we obtain

$$v_{ci} = \frac{dr_{ci}}{dt} = \frac{dO_i}{dt} - \frac{d}{dt}r_{ci,i} = v_i - \omega_i \times r_{ci,i}.$$

Further differentiation of the above equation yields the expression for the computation of the linear acceleration vector at the center of gravity.

That is,

$$a_{ci} = \frac{dv_{ci}}{dt} = \frac{dv_i}{dt} - \frac{d\omega_i}{dt} \times r_{ci,i} - \omega_i \times \left(\frac{d}{dt} r_{ci,i} \right)$$

or

$$a_{ci} = a_i - \alpha_i \times r_{ci,i} - \omega_i \times (\omega_i \times r_{ci,i}). \qquad (4.85)$$

Finally, the expression of the equation of linear motion becomes

$$f_i + (-f_{i+1}) + m_i \bullet g = m_i \bullet \{a_i - \alpha_i \times r_{ci,i} - \omega_i \times (\omega_i \times r_{ci,i})\}. \qquad (4.86)$$

4.6.3.6   *Equation of Angular Motion of a Link*

Similarly, the direct application of the result in Eq. 4.72 yields the equation of angular motion for link i, that is,

$$\tau_i + (-\tau_{i+1}) + r_{ci,i-1} \times f_i + (-r_{ci,i} \times f_{i+1}) = I_i \bullet \alpha_i + \frac{dI_i}{dt} \bullet \omega_i. \qquad (4.87)$$

Since $I_i = R_i \bullet \{^c I_i\} \bullet R_i^t$ (i.e. Eq. 4.76), we have

$$\frac{dI_i}{dt} = \frac{dR_i}{dt} \bullet \{^c I_i\} \bullet R_i^t + R_i \bullet \{^c I_i\} \bullet \frac{dR_i^t}{dt}.$$

By applying the property $\frac{dR}{dt} = S(\omega) \bullet R$ where R is a rotation matrix describing the orientation of a frame, and ω is the angular velocity acting on the frame (see Chapter 2 for detail), the derivative of inertial matrix I_i becomes

$$\frac{dI_i}{dt} = S(\omega_i) \bullet R_i \bullet \{^c I_i\} \bullet R_i^t + R_i \bullet \{^c I_i\} \bullet R_i^t \bullet S^t(\omega_i)$$

or

$$\frac{dI_i}{dt} = S(\omega_i) \bullet I_i + I_i \bullet S^t(\omega_i).$$

As a result, the term $\frac{dI_i}{dt} \bullet \omega_i$ in Eq. 4.87 can be expressed as

$$\frac{dI_i}{dt} \bullet \omega_i = S(\omega_i) \bullet (I_i \bullet \omega_i) + I_i \bullet (S^t(\omega_i) \bullet \omega_i)$$

or

$$\frac{dI_i}{dt} \bullet \omega_i = \omega_i \times (I_i \bullet \omega_i) + I_i \bullet (S^t(\omega_i) \bullet \omega_i).$$

Since $S^t(\omega_i) \bullet \omega_i = \omega_i \times \omega_i = 0$, the derivative of inertial matrix I_i can be simplified as

$$\frac{dI_i}{dt} \bullet \omega_i = \omega_i \times (I_i \bullet \omega_i). \tag{4.88}$$

Substituting Eq. 4.88 into Eq. 4.87 yields

$$\tau_i + (-\tau_{i+1}) + r_{ci,i-1} \times f_i + (-r_{ci,i} \times f_{i+1}) = I_i \bullet \alpha_i + \omega_i \times (I_i \bullet \omega_i). \tag{4.89}$$

Eq. 4.89 is the final expression of the equation of angular motion.

4.6.3.7 *Recursive Algorithm for Forces/Torques*

A robot is used to perform actions and tasks through the execution of motions. The desired motions of the links inside a robot's mechanism are normally planned based on the input task or action that a robot must undertake. Therefore, given the kinematic parameters, and the planned motions of a robot, the required forces and torques can be computed by using the equations of linear and angular motion in a recursive manner.

In order to make this statement clear, let us write Eq. 4.86 and Eq. 4.89 in the following recursive form

$$\begin{cases} f_i = f_{i+1} + m_i \bullet (a_i - \alpha_i \times r_{ci,i} - \omega_i \times (\omega_i \times r_{ci,i})) - m_i \bullet g \\ \tau_i = \tau_{i+1} - r_{ci,i-1} \times f_i + r_{ci,i} \times f_{i+1} + I_i \bullet \alpha_i + \omega_i \times (I_i \bullet \omega_i), \end{cases} \tag{4.90}$$

$(\forall i = n, n - 1, ...1)$.

It is obvious that Eq. 4.90 describes a backward recursive algorithm for the computation of the required forces and torques at the joints in order to produce the desired motions to the links inside the robot's mechanism. It is a tedious process if we have to do the computation by hand. In practice, we will make use of a computer to do the complex computations.

4.6.3.8 *Recursive Algorithm for Velocities and Accelerations*

Strictly speaking, Eq. 4.90 does not explicitly relate the generalized forces (the forces or torques produced at the joints) to the generalized coordinates (the joint variables and their derivatives). This is because the motion parameters $(v_i, a_i, \omega_i, \alpha_i)$ are related to the frame assigned to link i and they are not the derivatives of the generalized coordinate q_i of joint i. As a result, we need to know how these motion parameters are related to the generalized coordinates and their derivatives.

Assume that q_i is the joint variable of joint i. According to DH representation, we have

$$q_i = \begin{cases} \theta_i & \text{if revolute joint;} \\ d_i & \text{if prismatic joint.} \end{cases} \qquad (4.91)$$

And, the axis (or direction) of motion at joint i is the Z_{i-1} axis (i.e. $\vec{k}_{i-1}$).

A joint can be either a revolute joint or a prismatic joint. Here, we imagine that each joint has two joint variables: θ and d. If joint i is a revolute joint, its two joint variables will be: θ_i and $d_i = 0$ (0 means that this variable is not in use). If joint i is a prismatic joint, its two joint variables are: $\theta_i = 0$ and d_i.

Let us first derive the recursive solution for the computation of the angular velocity vectors $\{\omega_i,\ i = 1, 2, ..., n\}$ and acceleration vectors $\{\alpha_i,\ i = 1, 2, ..., n\}$ of link frames. For joint i, we set $q_i = \theta_i$. Obviously, $q_i = 0$ if joint i is a prismatic joint. Refer to Fig. 4.36 or Chapter 3. The relative angular velocity vector of frame i with respect to frame $i-1$ will be

$$\dot{q}_i \bullet {}^{i-1}\vec{k}_{i-1}.$$

If R_{i-1} is the rotation matrix describing the orientation of frame $i-1$ with respect to the common reference frame (frame 0), the relationship between the angular velocity vector of frame i and the angular velocity of frame $i-1$ will be

$$\omega_i = \omega_{i-1} + R_{i-1} \bullet \{\dot{q}_i \bullet {}^{i-1}\vec{k}_{i-1}\}$$

or

$$\omega_i = \omega_{i-1} + \dot{q}_i \bullet \vec{k}_{i-1}. \qquad (4.92)$$

Differentiating Eq. 4.92 with respect to time gives

$$\alpha_i = \alpha_{i-1} + \ddot{q}_i \bullet \vec{k}_{i-1} + \dot{q}_i \bullet (\omega_{i-1} \times \vec{k}_{i-1}). \qquad (4.93)$$

(NOTE: $\frac{d}{dt}(\vec{k}_{i-1}) = \omega_{i-1} \times \vec{k}_{i-1}$.)

Eq. 4.92 and Eq. 4.93 are the forward recursive solutions for the computation of angular velocity vectors and angular acceleration vectors of link frames (i.e $i = 1, 2, ..., n$ if there are $n+1$ links in the open kineto-dynamic chain).

Now, we derive the recursive solutions for the computation of linear velocity vectors $\{v_i,\ i = 1, 2, ..., n\}$ and acceleration vectors $\{a_i,\ i =$

$1, 2, ..., n\}$ of link frames. In this case, we set $q_i = d_i$ for joint i. $q_i = 0$, if joint i is a revolute joint. Let $O_{i-1,i}$ denote the displacement vector from origin O_{i-1} of frame $i - 1$ to origin O_i of frame i. By default, all vectors are expressed with respect to the common reference frame. From a simple geometry, we know

$$O_i = O_{i-1} + O_{i-1,i}.$$

If rotation matrix R_i describes the orientation of frame i with respect to the common reference frame (frame 0), displacement vector $O_{i-1,i}$ can be expressed as

$$O_{i-1,i} = R_i \bullet {}^iO_{i-1,i}.$$

Therefore, origin O_i can be rewritten as

$$O_i = O_{i-1} + R_i \bullet {}^iO_{i-1,i}.$$

Differentiating the above equation with respect to time yields

$$v_i = v_{i-1} + \frac{dR_i}{dt} \bullet {}^iO_{i-1,i} + R_i \bullet \frac{d}{dt}\{{}^iO_{i-1,i}\}. \tag{4.94}$$

Since link i is a rigid body, we have

$$\frac{d}{dt}\{{}^iO_{i-1,i}\} = \dot{q}_i \bullet {}^i\vec{k}_{i-1}.$$

(NOTE: it will be zero if joint i is a revolute joint). Then, Eq. 4.94 becomes

$$v_i = v_{i-1} + S(\omega_i) \bullet R_i \bullet {}^iO_{i-1,i} + R_i \bullet \dot{q}_i \bullet {}^i\vec{k}_{i-1}$$

or

$$v_i = v_{i-1} + \omega_i \times O_{i-1,i} + \dot{q}_i \bullet \vec{k}_{i-1}. \tag{4.95}$$

Because $v_i = v_{i-1} + \frac{d}{dt}\{O_{i-1,i}\}$, Eq. 4.95 also proves the following equality

$$\frac{d}{dt}\{O_{i-1,i}\} = \omega_i \times O_{i-1,i} + \dot{q}_i \bullet \vec{k}_{i-1}. \tag{4.96}$$

By differentiating Eq. 4.95, we obtain

$$a_i = a_{i-1} + \alpha_i \times O_{i-1,i} + \omega_i \times \frac{d}{dt}\{O_{i-1,i}\} + \ddot{q}_i \bullet \vec{k}_{i-1} + \dot{q}_i \bullet (\omega_{i-1} \times \vec{k}_{i-1}). \tag{4.97}$$

(NOTE: $\frac{d}{dt}(\vec{k}_{i-1}) = \omega_{i-1} \times \vec{k}_{i-1}$).

Eq. 4.95 and Eq. 4.97 are the forward recursive solutions for the computation of linear velocity vectors and acceleration vectors of link frames (i.e $i = 1, 2, ..., n$ if there are $n + 1$ links in the open kineto-dynamic chain).

4.6.4 *Euler-Lagrange Formula*

The equations of motion derived from the Newton-Euler formula is recursive in nature. It is computationally efficient and is fundamentally based on Newton's Third Law governing the relationship among acting and reacting forces in a dynamic system. In some cases, it may be desirable to have a closed form solution of robot dynamics. In fact, we can apply the Euler-Lagrange Formula to derive a compact description of motion equations in a matrix form.

4.6.4.1 *D'Alembert Principle*

The Euler-Lagrange Formula is derived from the D'Alembert Principle. The D'Alembert Principle governs the balance of forces acting on a system of particles in motion.

The basic idea behind the D'Alembert Principle is simple. For a particle having mass m and moving at velocity v, its linear momentum is $p = m \bullet v$. If F denotes the sum of external forces acting on the particle, we have $F = \dot{p}$ (i.e. Newton's second law). If we define $-\dot{p}$ the *inertial force* of a particle, then $F + (-\dot{p}) = 0$. Conceptually, this means that the sum of external forces acting upon a particle, plus its inertial force, is equal to zero.

Now, let us consider a system of n particles. Assume that particle i is represented by the following physical quantities:

- m_i: mass;
- r_i: position vector;
- v_i: linear velocity vector;
- F_i: sum of external forces acting upon the particle;
- F_i': sum of internal forces acting upon the particle;
- $\dot{p}_i$: rate of change in its linear momentum.

According to Newton's Third Law, every acting force will have a corresponding reacting force. The sum of the acting and reacting forces is equal to zero. Therefore, we have

$$\sum_{i=1}^{n} F_i' = 0.$$

Since $F_i + F_i' + (-\dot{p}_i) = 0$ for particle i, then we have

$$\sum_{i=1}^{n} (F_i + F_i' + (-\dot{p}_i)) = 0. \tag{4.98}$$

Eq. 4.98 describes the D'Alembert Principle for the force balance governing a system of particles. It states that the sum of all forces inside a dynamic system of particles is equal to zero. In other words, a dynamic system is always dynamically in balance.

As the sum of internal forces is equal to zero, Eq. 4.98 can be rewritten as

$$\sum_{i=1}^{n} F_i = \sum_{i=1}^{n} \dot{p}_i. \tag{4.99}$$

If δr_i denotes the virtual displacement of particle i at a time-instant, the multiplication of δr_i to both sides of Eq. 4.99 yields

$$\sum_{i=1}^{n} (F_i \bullet \delta r_i) = \sum_{i=1}^{n} (\dot{p}_i \bullet \delta r_i). \tag{4.100}$$

If we denote $\delta W_F = \sum_{i=1}^{n} (F_i \bullet \delta r_i)$ and $\delta W_k = \sum_{i=1}^{n} (\dot{p}_i \bullet \delta r_i)$, Eq. 4.100 becomes

$$\delta W_F = \delta W_k. \tag{4.101}$$

Eq. 4.101 illustrates the balance of virtual work done to a system in dynamic equilibrium. In other words, the D'Alembert Principle also states that virtual work done by external forces to a system is equal to virtual work done by inertial forces. (NOTE: A work can be negative if energy is removed from a system). The principle of virtual work for statics is a special exception to the D'Alembert Principle (i.e. $\delta W_k = 0$ for a system at rest).

4.6.4.2 *Lagrange Formula*

If the particles in a system undergo linear motion, we can derive the well-known *Lagrange Formula* from the D'Alembert Principle.

Virtual Displacements

Assume that a system of n particles is presented by n generalized coordinates $\{q_i, i = 1, 2, ...n\}$. Then, position vector r_i of particle i is a function of the generalized coordinates as follows:

$$r_i = r_i(q_1, q_2, ..., q_n).$$

And, the virtual displacement δr_i will be

$$\delta r_i = \sum_{j=1}^{n} \left(\frac{\partial r_i}{\partial q_j} \bullet \delta q_j \right). \qquad (4.102)$$

As position vector r_i is a function of the generalized coordinates, its corresponding velocity vector v_i will be a function of the generalized coordinates and their velocities, that is:

$$v_i = \dot{r}_i = \sum_{j=1}^{n} \left(\frac{\partial r_i}{\partial q_j} \bullet \dot{q}_j \right). \qquad (4.103)$$

From Eq. 4.103, it is easy to see that the computation of the partial derivative $\frac{\partial \dot{r}_i}{\partial \dot{q}_i}$ yields

$$\frac{\partial \dot{r}_i}{\partial \dot{q}_i} = \frac{\partial r_i}{\partial q_i}. \qquad (4.104)$$

As a result, the virtual displacement δr_i can also be expressed as

$$\delta r_i = \sum_{j=1}^{n} \left(\frac{\partial \dot{r}_i}{\partial \dot{q}_j} \bullet \delta q_j \right). \qquad (4.105)$$

Kinetic Energy

Given particle i, its kinetic energy is

$$\frac{1}{2} m_i \bullet v_i^t \bullet v_i.$$

Hence, the total kinetic energy stored by the system of n particles will be

$$K = \sum_{i=1}^{n} \left(\frac{1}{2} m_i \bullet v_i^t \bullet v_i \right). \qquad (4.106)$$

From Eq. 4.103, we know that velocity vector v_i is a function of the generalized coordinates and their velocities. Then, Eq. 4.106 allows us to compute the following partial derivatives

$$\begin{cases} \frac{\partial K}{\partial q_j} = \sum_{i=1}^{n} m_i \bullet v_i^t \bullet \frac{\partial v_i}{\partial q_j}, \quad \forall j \in [1, n] \\ \frac{\partial K}{\partial \dot{q}_j} = \sum_{i=1}^{n} m_i \bullet v_i^t \bullet \frac{\partial v_i}{\partial \dot{q}_j}, \quad \forall j \in [1, n]. \end{cases} \quad (4.107)$$

Potential Energy

For particle i, its potential energy is

$$-m_i \bullet g^t \bullet r_i.$$

And, the total potential energy stored by the system of n particles will be

$$V = \sum_{i=1}^{n} (-m_i \bullet g^t \bullet r_i). \quad (4.108)$$

(NOTE: g is a vector expressed in the common reference frame).

Since position vector r_i is a function of the generalized coordinates, the potential energy will also be a function of the generalized coordinates as follows:

$$V = V(q_1, q_2, ..., q_n).$$

And, virtual work done to the system by gravitational force will be

$$\delta W_v = -\delta V = \sum_{i=1}^{n} \left(-\frac{\partial V}{\partial q_i} \bullet \delta q_i \right). \quad (4.109)$$

The potential energy is independent to the velocities (i.e. motions) of the generalized coordinates. It is obvious that the following equation holds:

$$\frac{\partial V}{\partial \dot{q}_i} = 0, \quad \forall i \in [1, n]. \quad (4.110)$$

Virtual Work done by External Forces

We denote τ_i the generalized force associated to the generalized coordinate q_i. Virtual work done by generalized forces to the system of n particles will be

$$\delta W_\tau = \sum_{i=1}^{n} (\tau_i \bullet \delta q_i). \quad (4.111)$$

In general, the external forces exerted upon the system of n particles include: a) gravitational forces and b) generalized forces. As a result, virtual work done to the system by external forces is

$$\delta W_F = \delta W_v + \delta W_\tau$$

or

$$\delta W_F = \sum_{i=1}^{n} \left(-\frac{\partial V}{\partial q_i} + \tau_i \right) \bullet \delta q_i. \tag{4.112}$$

Virtual Work done by Inertial Forces

The linear momentum of particle i is $m_i \bullet v_i$, and its corresponding inertial force is $(-m_i \bullet \dot{v}_i)$. Therefore, virtual work done by inertial forces will be

$$\delta W_k = \sum_{i=1}^{n} (m_i \bullet \dot{v}_i \bullet \delta r_i). \tag{4.113}$$

(NOTE: There is no negative sign. This means that the system absorbs energy).

From

$$\frac{d}{dt} \{v_i \bullet \delta r_i\} = \dot{v}_i \bullet \delta r_i + v_i \bullet \delta \dot{r}_i,$$

we have

$$\dot{v}_i \bullet \delta r_i = \frac{d}{dt} \{v_i \bullet \delta r_i\} - v_i \bullet \delta \dot{r}_i.$$

Then, Eq. 4.113 can be rewritten as

$$\delta W_k = A - B \tag{4.114}$$

where

$$A = \sum_{i=1}^{n} \left(m_i \bullet \frac{d}{dt} \{v_i \bullet \delta r_i\} \right),$$

$$B = \sum_{i=1}^{n} (m_i \bullet v_i \bullet \delta \dot{r}_i).$$

By applying Eq. 4.105, the first term in Eq. 4.114 can be expressed as

$$A = \sum_{i=1}^{n} \left(m_i \bullet \frac{d}{dt} \left\{ v_i \bullet \sum_{j=1}^{n} \frac{\partial \dot{r}_i}{\partial \dot{q}_j} \delta q_j \right\} \right).$$

The sequence of summations is permutable. So, the above equation can be rewritten as

$$A = \sum_{j=1}^{n} \frac{d}{dt} \left\{ \sum_{i=1}^{n} m_i \bullet v_i \bullet \frac{\partial \dot{r}_i}{\partial \dot{q}_j} \right\} \bullet \delta q_j.$$

By applying the second equality in Eq. 4.107, we have

$$A = \sum_{j=1}^{n} \frac{d}{dt} \left(\frac{\partial K}{\partial \dot{q}_j} \right) \bullet \delta q_j$$

or

$$A = \sum_{i=1}^{n} \frac{d}{dt} \left(\frac{\partial K}{\partial \dot{q}_i} \right) \bullet \delta q_i. \tag{4.115}$$

(NOTE: $\frac{\partial \dot{r}_i}{\partial \dot{q}_j} = \frac{\partial v_i}{\partial \dot{q}_j}$ because $v_i = \dot{r}_i$).

Now, we examine the second term in Eq. 4.114. As velocity vector v_i is a function of the generalized coordinates, virtual displacements of the generalized coordinates will cause virtual displacement of v_i. Therefore, δv_i can be expressed as follows:

$$\delta v_i = \delta \dot{r}_i = \sum_{j=1}^{n} \left(\frac{\partial v_i}{\partial q_j} \bullet \delta q_j \right).$$

As a result, the second term in Eq. 4.114 becomes

$$B = \sum_{i=1}^{n} \left\{ m_i \bullet v_i \bullet \sum_{j=1}^{n} \left(\frac{\partial v_i}{\partial q_j} \bullet \delta q_j \right) \right\}$$

or

$$B = \sum_{j=1}^{n} \left\{ \sum_{i=1}^{n} \left(m_i \bullet v_i \bullet \frac{\partial v_i}{\partial q_j} \right) \bullet \delta q_j \right\}.$$

By applying the first equality in Eq. 4.107, we obtain

$$B = \sum_{j=1}^{n} \left(\frac{\partial K}{\partial q_j} \bullet \delta q_j \right)$$

or

$$B = \sum_{i=1}^{n} \left(\frac{\partial K}{\partial q_i} \bullet \delta q_i \right). \tag{4.116}$$

Finally, virtual work done by inertial forces can be compactly expressed as a function of kinetic energy, that is,

$$\delta W_k = A - B = \sum_{i=1}^{n} \left\{ \left[\frac{d}{dt} \left(\frac{\partial K}{\partial \dot{q}_i} \right) - \frac{\partial K}{\partial q_i} \right] \bullet \delta q_i \right\}. \qquad (4.117)$$

Expression of Lagrange Formula

Now, it is easy to obtain the expression for the well-known *Lagrange Formula*. In fact, substitution of Eq. 4.112 and Eq. 4.117 into Eq. 4.101 yields

$$\sum_{i=1}^{n} \left\{ \left(-\frac{\partial V}{\partial q_i} + \tau_i \right) \bullet \delta q_i \right\} = \sum_{i=1}^{n} \left\{ \left[\frac{d}{dt} \left(\frac{\partial K}{\partial \dot{q}_i} \right) - \frac{\partial K}{\partial q_i} \right] \bullet \delta q_i \right\}. \qquad (4.118)$$

It can also be equivalently written as

$$-\frac{\partial V}{\partial q_i} + \tau_i = \frac{d}{dt} \left(\frac{\partial K}{\partial \dot{q}_i} \right) - \frac{\partial K}{\partial q_i}, \quad \forall i \in [1, n]. \qquad (4.119)$$

Gravitational force is a conservative force. This means that potential energy does not depend on any velocity variable. Thus, we have

$$\frac{d}{dt} \left(\frac{\partial V}{\partial \dot{q}_i} \right) = 0.$$

If we define $L = K - V$, Eq. 4.119 can be compactly expressed as

$$\tau_i = \frac{d}{dt} \left(\frac{\partial L}{\partial \dot{q}_i} \right) - \frac{\partial L}{\partial q_i}, \quad \forall i \in [1, n]. \qquad (4.120)$$

Eq. 4.120 is the famous Lagrange Formula. This expression is also valid for a system of n rigid bodies, which satisfy a holonomic constraint such as the kinematic constraint imposed upon a robot's mechanism. This extension is called the *Euler-Lagrange Formula*. It is a very useful formula for the derivation of a closed-form solution to the motion equations in robotics.

4.6.4.3 *Equations of Motion*

Refer to Fig. 4.36. An open kineto-dynamic chain can be treated as a dynamic system of rigid bodies which satisfy: a) the kinematic constraint imposed by the robot's mechanism and b) the dynamic constraint imposed by the kineto-dynamic couplings (i.e. Newton's Third Law on the relationship between acting and reacting forces). If a robot's mechanism is in the form of a simple open kinematic-chain, the kinematic constraint will be a holonomic constraint. In this case, we can directly apply the

Euler-Lagrange Formula to derive the closed-form solution to the motion equations. These motion equations relate torques (or forces) to the input motions of the robot's mechanism.

Let us consider an open kineto-dynamic chain with $n+1$ links which are connected in a series by revolute joints. For the example shown in Fig. 4.36, the robot's arm has three links (i.e. $n = 2$). Given joint i, we represent it with two variables: a) the joint variable q_i (generalized coordinate), and b) the torque τ_i (generalized force). And, we denote $q = (q_1, q_2, ..., q_n)^t$ the joint variable vector, and $\tau = (\tau_1, \tau_2, ..., \tau_n)^t$ the joint torque vector. In addition, given link i, we first assign a new frame ci at its center of gravity, and define the following parameters for it:

- m_i: mass;
- r_{ci}: its center of gravity;
- v_{ci}: linear velocity vector at r_{ci};
- ω_i: angular velocity vector;
- $^{ci}I_{ci}$: inertial matrix calculated with respect to the center of gravity, in frame ci;
- R_{ci}: rotation matrix describing the orientation of frame ci with respect to the common reference frame (frame 0);
- J_{ci}: the Jacobian matrix of frame ci with respect to the common reference frame.

Frame at the Center of Gravity of a Link

In order to apply Eq. 4.81 to determine the kinetic energy of a link, we need to assign frame ci at the center of gravity r_{ci} of link i. One simple way to do this is to duplicate the link's frame i and translate it to location r_{ci}. If we treat frame ci as an end-effector frame, the Jacobian matrix J_{ci} describes the motion kinematics of frame ci with respect to frame 0 (the common reference frame), and can be computed from the kinematic solutions discussed in Chapter 3.

Now, if $\dot{P}_{ci}$ denotes the velocity vector of frame ci with respect to frame 0, we have

$$\dot{P}_{ci} = \begin{pmatrix} v_{ci} \\ \omega_i \end{pmatrix} = J_{ci} \bullet \dot{q}. \tag{4.121}$$

Normally, Jacobian matrix J is a $6 \times n$ matrix and is composed of two sub-matrices: a) a $3 \times n$ matrix J_o governing linear velocity, and b) a $3 \times n$

matrix J_ω governing angular velocity. If we denote

$$J_{ci} = \begin{pmatrix} J^o_{ci} \\ J^\omega_{ci} \end{pmatrix},$$

Eq. 4.121 can be rewritten as

$$\begin{cases} v_{ci} = J^o_{ci} \bullet \dot{q} \\ \omega_i = J^\omega_{ci} \bullet \dot{q}. \end{cases} \tag{4.122}$$

Kinetic Energy of an Open Kineto-Dynamic Chain

The direct application of Eq. 4.81 yields the expression for kinetic energy stored in link i, that is,

$$K_i = \frac{1}{2} m_i \bullet v^t_{ci} \bullet v_{ci} + \frac{1}{2} \omega^t_i \bullet R_{ci} \bullet \{^{ci}I_{ci}\} \bullet R^t_{ci} \bullet \omega_i.$$

Substituting Eq. 4.122 into the above equation gives

$$K_i = \frac{1}{2} m_i \bullet (J^o_{ci} \bullet \dot{q})^t \bullet (J^o_{ci} \bullet \dot{q}) + \frac{1}{2} (J^\omega_{ci} \bullet \dot{q})^t \bullet R_{ci} \bullet \{^{ci}I_{ci}\} \bullet R^t_{ci} \bullet (J^\omega_{ci} \bullet \dot{q})$$

or

$$K_i = \frac{1}{2} \dot{q}^t \bullet \{ m_i \bullet (J^o_{ci})^t \bullet J^o_{ci} + (J^\omega_{ci})^t \bullet R_{ci} \bullet \{^{ci}I_{ci}\} \bullet R^t_{ci} \bullet J^\omega_{ci} \} \bullet \dot{q}. \tag{4.123}$$

Then, the kinetic energy stored in an open kineto-dynamic chain will be

$$K = \sum_{i=1}^{n} K_i$$

or

$$K = \frac{1}{2} \dot{q}^t \bullet \left\{ \sum_{i=1}^{n} [m_i \bullet (J^o_{ci})^t \bullet J^o_{ci} + (J^\omega_{ci})^t \bullet R_{ci} \bullet \{^{ci}I_{ci}\} \bullet R^t_{ci} \bullet J^\omega_{ci}] \right\} \bullet \dot{q}.$$

If we define

$$B(q) = \sum_{i=1}^{n} [m_i \bullet (J^o_{ci})^t \bullet J^o_{ci} + (J^\omega_{ci})^t \bullet R_{ci} \bullet \{^{ci}I_{ci}\} \bullet R^t_{ci} \bullet J^\omega_{ci}], \tag{4.124}$$

the expression for the total kinetic energy K can be compactly written as

$$K = \frac{1}{2} \dot{q}^t \bullet B(q) \bullet \dot{q} \tag{4.125}$$

where $B(q)$ is a $n \times n$ matrix, which depends on the generalized coordinates q. In robotics, $B(q)$ is commonly called *inertial matrix* of a kineto-dynamic chain. In fact, we can represent $B(q)$ as follows:

$$\{b_{ij}(q), \quad \forall i \in [1, n], \forall j \in [1, n]\}.$$

Accordingly, Eq. 4.125 can also be expressed as

$$K = \frac{1}{2} \sum_{i=1}^{n} \sum_{j=1}^{n} b_{ij}(q) \bullet \dot{q}_i \bullet \dot{q}_j. \tag{4.126}$$

Potential Energy of an Open Kineto-Dynamic Chain

Let g be the density vector (or acceleration vector) of the gravitational force field on the surface of the Earth. It is expressed with respect to the common reference frame (frame 0). It is not a constant vector if the reference frame changes during the action or motion undertaken by a robot.

By definition, the potential energy stored in link i is

$$V_i = -m_i \bullet g^t \bullet r_{ci}.$$

As a result, the potential energy stored in an open kineto-dynamic chain will be

$$V = \sum_{i=1}^{n} V_i = -\sum_{i=1}^{n} m_i \bullet g^t \bullet r_{ci}. \tag{4.127}$$

Motion Equations of an Open Kineto-Dynamic Chain

If we know the expressions of potential and kinetic energies stored in an open kineto-dynamic chain, we are ready to apply Eq. 4.119 (or Eq. 4.120) to derive the motion equations. For easy reference, let us rewrite Eq. 4.119 as follows:

$$\frac{d}{dt}\left(\frac{\partial K}{\partial \dot{q}_i}\right) - \frac{\partial K}{\partial q_i} + \frac{\partial V}{\partial q_i} = \tau_i, \forall i \in [1, n]. \tag{4.128}$$

Now, we need to evaluate the expressions of the three terms on the left side of Eq. 4.128.

From Eq. 4.126, we have

$$\frac{\partial K}{\partial \dot{q}_i} = \sum_{j=1}^{n} b_{ij}(q) \bullet \dot{q}_j.$$

The computation of the first-order derivative of the above equation yields

$$\frac{d}{dt}\left(\frac{\partial K}{\partial \dot{q}_i}\right) = \sum_{j=1}^{n}[\dot{b}_{ij}(q) \bullet \dot{q}_j + b_{ij}(q) \bullet \ddot{q}_j]. \tag{4.129}$$

The derivative $\dot{b}_{ij}(q)$ in Eq. 4.129 can be computed as follows:

$$\dot{b}_{ij}(q) = \frac{\partial b_{ij}}{\partial q_1} \bullet \dot{q}_1 + \frac{\partial b_{ij}}{\partial q_2} \bullet \dot{q}_2 + ... + \frac{\partial b_{ij}}{\partial q_n} \bullet \dot{q}_n$$

or

$$\dot{b}_{ij}(q) = \sum_{k=1}^{n}\frac{\partial b_{ij}}{\partial q_k} \bullet \dot{q}_k.$$

Accordingly, Eq. 4.129 becomes

$$\frac{d}{dt}\left(\frac{\partial K}{\partial \dot{q}_i}\right) = \sum_{j=1}^{n}b_{ij}(q) \bullet \ddot{q}_j + \sum_{j=1}^{n}\sum_{k=1}^{n}\frac{\partial b_{ij}}{\partial q_k} \bullet \dot{q}_k \bullet \dot{q}_j. \tag{4.130}$$

Now, let us examine the second term in Eq. 4.128. For convenience, we change the indices of summation in Eq. 4.126 and express it as:

$$K = \frac{1}{2}\sum_{j=1}^{n}\sum_{k=1}^{n}b_{jk}(q) \bullet \dot{q}_k \bullet \dot{q}_j. \tag{4.131}$$

Since the partial derivative of $b_{jk}(q) \bullet \dot{q}_k \bullet \dot{q}_j$ with respect to variable q_i is $\frac{\partial b_{jk}}{\partial q_i} \bullet \dot{q}_k \bullet \dot{q}_j$, the partial derivative of Eq. 4.131 with respect to variable q_i yields

$$\frac{\partial K}{\partial q_i} = \frac{1}{2}\sum_{j=1}^{n}\sum_{k=1}^{n}\frac{\partial b_{jk}}{\partial q_i} \bullet \dot{q}_k \bullet \dot{q}_j. \tag{4.132}$$

This is the expression of the second term in Eq. 4.128.

Finally, we have to determine the expression of the third term in Eq. 4.128. For convenience, we also change the index of summation in Eq. 4.127 and express it as

$$V = -\sum_{j=1}^{n}m_j \bullet g^t \bullet r_{cj}.$$

The partial derivative of the above equation, with respect to variable q_i, yields

$$\frac{\partial V}{\partial q_i} = -\sum_{j=1}^{n} m_j \bullet g^t \bullet \frac{\partial r_{cj}}{\partial q_i}. \tag{4.133}$$

From Eq. 4.104, we have

$$\frac{\partial r_{cj}}{\partial q_i} = \frac{\partial \dot{r}_{cj}}{\partial \dot{q}_i} = \frac{\partial v_{cj}}{\partial \dot{q}_i}.$$

We know $v_{cj} = J_{cj}^o \bullet \dot{q}$ (i.e. Eq. 4.122). If we define

$$J_{cj}^o = \left(J_{cj}^o(1) \quad J_{cj}^o(2) \quad ... \quad J_{cj}^o(n) \right)_{3 \times n}$$

where $J_{cj}^o(i)$ denotes the *ith* column of matrix J_{cj}^o, we have

$$\frac{\partial v_{cj}}{\partial \dot{q}_i} = J_{cj}^o(i).$$

Accordingly, Eq. 4.133 becomes

$$\frac{\partial V}{\partial q_i} = -\sum_{j=1}^{n} m_j \bullet g^t \bullet J_{cj}^o(i). \tag{4.134}$$

Now, substituting Eq. 4.130, Eq. 4.132 and Eq. 4.134 into Eq. 4.128 yields

$$\sum_{j=1}^{n} b_{ij}(q) \bullet \ddot{q}_j + \sum_{j=1}^{n} \left\{ \sum_{k=1}^{n} \left[\frac{\partial b_{ij}}{\partial q_k} - \frac{1}{2} \frac{\partial b_{jk}}{\partial q_i} \right] \bullet \dot{q}_k \right\} \bullet \dot{q}_j - \sum_{j=1}^{n} m_j \bullet g^t \bullet J_{cj}^o(i) = \tau_i, \tag{4.135}$$

$\forall i \in [1, n]$.

If we define

$$\begin{cases} c_{ij}(q, \dot{q}) = \sum_{k=1}^{n} \left[\frac{\partial b_{ij}}{\partial q_k} - \frac{1}{2} \frac{\partial b_{jk}}{\partial q_i} \right] \bullet \dot{q}_k \\ g_i(q) = -\sum_{j=1}^{n} m_j \bullet g^t \bullet J_{cj}^o(i), \end{cases} \tag{4.136}$$

Eq. 4.135 can be rewritten as

$$\sum_{j=1}^{n} b_{ij}(q) \bullet \ddot{q}_j + \sum_{j=1}^{n} c_{ij}(q, \dot{q}) \bullet \dot{q}_j + g_i(q) = \tau_i, \ \forall i \in [1, n]$$

or in a compact matrix form as:

$$B(q) \bullet \ddot{q} + C(q, \dot{q}) \bullet \dot{q} + G(q) = \tau \tag{4.137}$$

where

$$
\begin{cases}
q = (q_1, q_2, ..., q_n)^t \\[2mm]
\tau = (\tau_1, \tau_2, ..., \tau_n)^t \\[2mm]
B(q) = \{b_{ij}(q), \forall i \in [1, n], \quad \forall j \in [1, n]\} \\[2mm]
C(q, \dot{q}) = \{c_{ij}(q, \dot{q}), \quad \forall i \in [1, n], \forall j \in [1, n]\} \\[2mm]
G(q) = \{g_i(q), \quad \forall i \in [1, n]\}.
\end{cases}
\tag{4.138}
$$

Eq. 4.137 describes the dynamic model of an open kineto-dynamic chain with $n + 1$ links. Matrix $B(q)$ is the *inertial matrix*. Matrix $C(q, \dot{q})$ accounts for centrifugal and Coriolis effects. It is a function of the generalized coordinates and their velocities. And, vector $G(q)$ accounts for the effect caused by gravitational force.

In practice, it is difficult to obtain an exact expression for the dynamic model of a robot's electromechanical system because the computation of inertial matrices poses a challenge. Nevertheless, knowledge of robot dynamics is important and useful for dynamic simulation and control system design.

4.7 Summary

In this chapter, we started with the study of the origin of a rigid body's motion. Then, after a presentation of the important concepts on force, work and energy, we discussed a new concept on dynamic pairs and dynamic chains. This motivated the study of various schemes of generating the force field underlying the design of force and torque generator. We learned the working principles behind electric motors, as well as the associated actuation elements, such as force or torque amplifier.

We studied, in detail, two ways of forming a robot's electromechanical system through the coupling of kinematic and dynamic pairs. The first scheme consisted of one-to-one coupling for each kinematic pair and its corresponding dynamic pair. Under this scheme, each degree of freedom in a robot's mechanism has one actuator and a set of associated actuation elements. The advantages of this scheme are the fast response of force or torque output and the efficiency in performing velocity and force/torque control. The notable drawbacks include the weight of the robot system and the cost of the robot's hardware. An alternative scheme is the one-to-many coupling. The notable advantages of this new scheme are the robot's compact size, light weight, and low cost.

From a kinematic point of view, a robot is a mechanism or pure mechanical system. But, from a dynamic point of view, a robot is also an electromechanical system. Mathematically, it is important to know: a) how to apply forces and torques in order to keep the robot at rest, and b) how to apply forces and torques to produce the desired motions from a robot's mechanism. We have discussed the issue and solution of robot statics. Subsequently, we have learned two methods for formulating *equations of motion*. The first method is the Newton-Euler Formula. This method is based on Newton's Second and Third Laws, and allows us to develop a recursive solution for the computation of required forces and torques which will produce required motions from a robot's mechanism. A second method is the Euler-Lagrange Formula which allows us to derive a compact expression for the equations of motion.

4.8   Exercises

(1) Explain the origin of motion.
(2) Give a conceptual definition of what the mechanical system underlying a robot (or any machine) is.
(3) Explain the physical meaning of potential energy and kinetic energy.
(4) What is a dynamic pair ? Apply the concept of a dynamic pair to propose a design solution for an electric linear motor.
(5) From Eq. 4.13, prove that the mechanical power of a rotary motor is equal to the product of its torque and angular velocity.
(6) Comment on the difference, advantages and drawbacks between the one-to-one coupling and the one-to-many coupling schemes of kineto-dynamic pairs.
(7) Prove the expression in Eq. 4.52.
(8) Search the website of US Patent Databases (it is a free service) to survey the various design solutions for CVT devices (CVT stands for Continuous Variable Transmission). Propose a new solution for one-to-many coupling between kinematic pairs and one dynamic pair which incorporates CVT devices for motion distribution.
(9) If $S(b)$ is a skew-symmetric matrix of three-dimensional vector b, prove

the following equality:

$$S(b) \bullet S^t(b) = S^t(b) \bullet S(b) = \begin{pmatrix} y^2 + z^2 & -xy & -xz \\ -xy & x^2 + z^2 & -yz \\ -xz & -yz & x^2 + y^2 \end{pmatrix}.$$

(10) In the study of differential kinematics of an open kinematic-chain, we have the relationship

$$v_e = \dot{P}_e = J \bullet \dot{q}$$

where v_e is the end-effector frame's velocity vector and $\dot{q}$ is a robot's joint velocity vector. At configuration $q(t)$, assume that the robot moves with a unit joint velocity vector (i.e. $q^t \bullet q = 1$). In other words, the joint velocity vector is on a spherical surface of unit radius. Explain what the spatial distribution of the end-effector frame's velocity v_e will be, and, also, its meaning.

(11) In the study of robot statics of an open kineto-dynamic chain, we have the relationship

$$\tau = J^t \bullet \xi_e$$

where τ is the joint torque vector required to balance the externally-exerted force/torque ξ_e acting upon the end-effector frame in order to keep a robot at rest. Now, assume that a robot is at rest and starts to exert a unit torque vector, i.e. $\tau^t \bullet \tau = 1$. Geometrically, the unit torque vector is located on a spherical surface of unit radius. Explain what will be the spatial distribution of the external force/torque exerted upon the end-effector's frame in order to keep the robot at rest. Also, explain the meaning of this distribution.

(12) Write the equations of motion of the robot arm shown in Fig. 4.36 according to the Newton-Euler formula.

(13) Write the equations of motion of the robot arm shown in Fig. 4.36, according to the Euler-Lagrange formula.

(14) Refer to Eq. 4.137. If we define a new matrix $N(q, \dot{q})$ as follows:

$$N(q, \dot{q}) = \dot{B}(q) - C(q, \dot{q}),$$

prove that matrix $N(q, \dot{q})$ is a skew-symmetrical matrix (i.e. $n_{ij} = -n_{ji}$, if $N(q, \dot{q})$ is represented by $\{n_{ij}, \ \forall i \in [1, n], \forall j \in [1, n]\}$).

(15) If N is a $n \times n$ skew-symmetrical matrix and q is a $n \times 1$ vector, prove the following result:

$$q^t \bullet N \bullet q = 0.$$

4.9   Bibliography

(1) Armstrong, M. W. (1979). Recursive Solution to the Equations of Motion of an n-Link Manipulator, *Proceedings of 5th World Congress on Theory of Machines and Mechanisms*, Montreal, Canada.
(2) Ryff, P. F., D. Platnick and J. A. Karnas (1987). *Electrical Machines and Transformers*, Prentice-Hall.
(3) Spong, M. W. and M. Vidyasagar (1989). *Robot Dynamics and Control*, John Willey & Sons.
(4) Stadler, W. (1995). *Analytical Robotics and Mechatronics*, McGraw-Hill.
(5) Uicker, J. J. (1967). Dynamic Force Analysis of Spatial Linkages, *ASME Journal of Applied Mechanics*, **34**.
(6) Vukobratovic, M. (1978). Dynamics of Active Articulated Mechanisms and Synthesis of Artificial Motion, *Mechanism and Machine Theory*, **13**.
(7) Young, H. D and R. A. Freedman (1996). *University Physics*, Addison-Wesley.

Chapter 5

Control System of Robots

5.1 Introduction

Control is a buzz-word employed almost everywhere. At the macroscopic level, the duty of a government is to control the social stability, economic growth, financial order, national security etc. On a large scale, traffic control of air, land and on-water vehicles are indispensable for the proper functioning of our modern industry and society. On the scale of enterprise, control is exerted at various levels from process control, equipment control, production control, and all the way up to supply-chain control and human-resource management.

The magic of control arises from the scheme of an automatic-feedback control loop which responds to a cost function of errors (and state-feedback if working in a state space). The ultimate goal of control is to minimize the cost function in an efficient and robust manner (i.e. resistance to noise, disturbance, and parameter variation). In this way, a desired output can be easily achieved even if the intrinsic dynamic model of a system under control is not accurately known.

A robot is a dynamic system, the primary job of which is to accomplish tasks or actions through the execution of motion. In the study of electrome-chanical aspect of the robot system, we know that it is quite difficult to obtain an exact dynamic model of a robot because it is almost impossible to theoretically compute the inertial matrices for all the links inside the robot. Consequently, it is useless to attempt to run a robot for motion execution without invoking the automatic-feedback control scheme.

In this chapter, we will discuss how to apply the automatic-feedback control scheme to control a robot's motion execution even if the dynamic model of the robot is not accurately known.

It is well-known that the core of a feedback control system consists of three basic building blocks: a) the system under control (the *plant*), b) the control elements (a set of elements which transform error signals into corresponding power signal to be applied to the system under control), and c) the sensing elements (a set of elements which measure actual output as feedback to enable the computation of error signals). Therefore, it is helpful to study all these related topics together. The emphasis of this chapter is on the application of the automatic-feedback control scheme for efficient motion execution by a robot. In-depth coverage on the theoretical aspects of an automatic control system is beyond the scope of a robotics textbook as there are many textbooks dealing with this vast body of knowledge in control engineering.

5.2   Automatic-Feedback Control System

So far, we have treated the robot as the combination of a mechanical system and an electromechanical system. The output motion from a robot's mechanical system is at the end-effector, and its input motions are at the joints of the underlying mechanism. As for a robot's electromechanical system, we know that input is the forces or torques applied to the joints of the robot's mechanism and output is the joints' motions. The relationship between input and output is known as the dynamic model of a robot, or more precisely, the dynamic model of a robot's electromechanical system.

5.2.1   *System Concept*

Using the robot as an example, it is easy to formally introduce the concept of system. In general, a system can be defined as follows:

Definition 5.1   A system is a combination of elements which act together and perform a certain task.

Normally, a system has input and produces output according to the characteristics of the system's dynamics. In the real world, however, a system is often subject to noise or disturbance. The internal parameters of a system may also undergo change as a function of time. If a system exhibits certain dynamic behaviors, (i.e. the system's output varies with time when being stimulated by the system's input or a disturbance), the internal state of the system can be described by a set of variables. These variables are called the *state variables*. Fig. 5.1 illustrates the concept of a system.

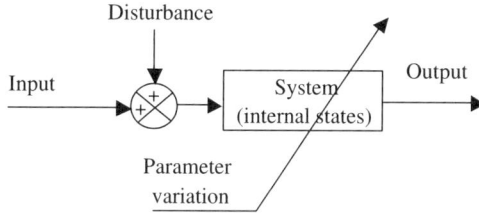

Fig. 5.1   Illustration of a system concept.

When a system at rest is stimulated by input or noise, its output starts to vary. After a certain period of time, it is expected that the system will reach a new state and be at rest again. If not, the system is considered to be *unstable* (i.e. the output oscillates or converges to infinity). For a stable system, the output when a system is at rest is called the *steady-state* response. And the output when a system is making the transition from its initial state into a new state is called the *transient response*. Formally, the steady-state and transient responses of a system can be defined as follows:

Definition 5.2   The steady-state response of a system is the input-output relationship when the system is at rest.

Definition 5.3   The transient response of a system is the input-output relationship when the system is evolving from an initial state into a new state.

To a great extent, the study of the design, analysis and control of a system aims at achieving the desired steady-state and transient responses. In fact, the steady-state response directly indicates a system's *absolute stability* (i.e. whether a system is stable or not). And the transient response describes a system's *relative stability* (i.e. how smooth a system is evolving to a new state).

5.2.2   *Closed-loop Control Scheme*

If there is no feedback verification or checking, a system then operates in an open-loop manner. In this mode, a system directly responds to the system's input, as shown in Fig. 5.1. Under the open-loop control scheme, the control action is the system's input itself. If we denote $y(t)$ the system's output, $u(t)$ the system's input, $d(t)$ the system's disturbance or noise, c the system's internal parameter vector and $x(t)$ the system's internal state vector (i.e. a vector of state variables), the steady-state response of a system can be analytically expressed as a function $f(.)$ of input, state vector, parameter vector and noise as follows:

$$y(t) = f(x(t), u(t), c, d(t)). \tag{5.1}$$

(NOTE: These notations are consistent with those used in control engineering textbooks).

From Eq. 5.1, we can express the variation of output as a function of the variations of input, state vector, parameter vector and disturbance as follows:

$$\triangle y(t) = \frac{\partial f}{\partial x} \bullet \triangle x(t) + \frac{\partial f}{\partial u} \bullet \triangle u(t) + \frac{\partial f}{\partial c} \bullet \triangle c + \frac{\partial f}{\partial d} \bullet \triangle d(t). \tag{5.2}$$

Now, it becomes clear that an open-loop system does not have the ability to compensate for the variations due to disturbance $d(t)$ and parameter variations $\triangle c$. On the other hand, the variation of output depends on the term $\frac{\partial f}{\partial u} \bullet \triangle u(t)$. This means that the control action by input $u(t)$ requires knowledge about the system's description in terms of function $f(.)$. In other words, the system's dynamic modelling must be known in advance. Any uncertainty about a system's dynamic modelling will result in an unpredictable system output. These two notable drawbacks to an open-loop system clearly demonstrate the inappropriateness of adopting an open-loop control scheme to operate a complex engineering system.

An alternative scheme is the *closed-loop control*. The basic philosophy behind the closed-loop control scheme is to build a system which responds to an error signal instead of an input signal. Under the closed-loop control scheme, the control objective is to minimize the error signal. And the control action to a system under control is determined by a function of the error signal.

Fig. 5.2 illustrates the closed-loop control scheme. The working principle behind a closed-loop control system is very simple. As long as the error signal is not equal to zero, the control signal will continuously act on the system to alter its output so that the error signal will converge to zero (assuming that the system is stable). When the error signal is defined to be the difference between desired output and the measurement of actual output, the actual output will be equal to the desired output if the error

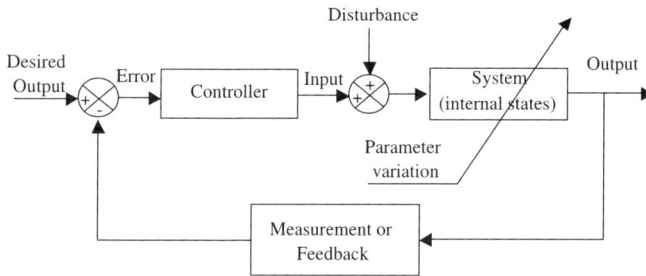

Fig. 5.2   Illustration of closed-loop control scheme.

signal vanishes. For a stable closed-loop control system, the control action of minimizing the error signal does not require any precise knowledge about the system's dynamics. Thus, a closed-loop control system is robust in the presence of model uncertainty, disturbance, and parameter variation.

In order to construct a closed-loop control system, one must consider the following issues:

- Specification of the Desired Output:
  A system is normally dedicated to achieving an outcome based on a predefined specification. Therefore, one must know how to specify the desired output that a system has to produce.
- Sensory Feedback:
  Before a control action can be determined, one must be clear about what error signal the system is to minimize. Normally, an error signal is simply the difference between desired output and actual output. Hence, one must know how to measure actual output by using a set of sensing elements.
- Design of Control Algorithms or Laws:
  The ultimate goal in studying the feedback control system is to design a control algorithm or law which will make the steady-state and transient responses of the system meet certain performance specifications (e.g. stable, fast, accurate etc). For example, the *PID control law* and its variations are the most popular control methods employed in industry. In general, the purpose of designing a control law is to find out a suitable control function which determines the appropriate control action (or state-feedback if working in state space) in order to meet the predefined control performance.
- Conversion of Control Signals to Power Signals:

A system under control is usually known as a *plant*. For example, a robot's electromechanical system is a plant which is controlled for the purpose of achieving desired motion output. The input to a plant is normally related to a certain form of energy. And the control signal is usually calculated by a computer and is in the form of electrical signal without any associated electrical power. As a result, it is necessary to convert a control signal to a corresponding power signal which is the actual input to a plant under control. This conversion is normally done by a set of control elements.

- Analysis of a Control System's Performance:
  Given a closed-loop control system, it is desirable to analyze its performance with respect to its steady-state and transient responses. Normally, we look at three important performance indicators: a) the absolute stability of the system (is the system stable or not?), b) the relative stability of the system in terms of response time (how fast does the system's output converge to the desired output?), and c) the relative stability of the system, in terms of accuracy (how small is the error between the desired output and the steady-state output?).

5.2.3 *System Dynamics*

By definition, all systems will exhibit certain types of dynamic behaviors caused by the interaction among the elements inside the systems. And a system's dynamics means the relationship between its input and output during the period of its transient response.

A system is said to be a *continuous-time system* if all variables inside the system are continuous-time variables. In the time domain, the dynamics of a continuous-time system can be conveniently described by a differential equation.

Example 5.1 Let us consider a motion controller implementing a PID control algorithm. The input to the controller is error signal $e(t)$ and the output from the controller is control signal $c(t)$. Thus, the dynamics underlying the PID control algorithm is described by

$$c(t) = k_p \bullet \left[e(t) + \frac{1}{T_i} \bullet \int_0^t e(t) \bullet dt + T_d \bullet \frac{de(t)}{dt} \right] \qquad (5.3)$$

where k_p is proportional control gain, k_p/T_i integral control gain and $k_p \bullet T_d$ derivative control gain.

PID = Proportional Integral Derivative

◇◇◇◇◇◇◇◇◇◇◇◇◇◇◇◇◇◇◇◇

As a result of the wide use of digital computers, almost all closed-loop control systems have both continuous-time variables and discrete-time variables. Thus, it is necessary to study discrete-time systems. A second reason for studying discrete-time systems is because a control system may involve the sampling process in the sensory-feedback channel. A typical example is the statistical process control in which output is sampled and measured in regular discrete time intervals. As a result, a statistical process control system is intrinsically a discrete-time system.

By definition, any system having discrete-time variables is called a *discrete-time system*. Therefore, it is important to study discrete-time systems. In the time domain, the dynamics of a discrete-time system is usually described by a difference equation.

Example 5.2   Assume that a PID control algorithm is implemented on a digital computer. We treat this digital computer as a discrete-time system. Input to the system is error signal $e(k)$ at discrete time-instant k, and output from the system is control signal $c(k)$ at discrete time-instant k. The difference equation relating output to input can be derived in the following way:

Differentiating Eq. 5.3 with respect to time gives

$$\frac{dc(t)}{dt} = k_p \bullet \left[\frac{de(t)}{dt} + \frac{1}{T_i} \bullet e(t) + T_d \bullet \frac{d^2 e(t)}{dt^2} \right]. \qquad (5.4)$$

By definition, derivative $\frac{dc(t)}{dt}$ at time-instant $t = kT$ is computed as follows:

$$\frac{dc(kT)}{dt} = \lim_{T \to 0} \frac{c(kT) - c((k-1)T)}{T}$$

or

$$\frac{dc(k)}{dt} = \lim_{T \to 0} \frac{c(k) - c(k-1)}{T}$$

where T is the time interval between two consecutive time-instants.

Now, at time-instant $t = kT$, we substitute derivative $\frac{dc(t)}{dt}$ with $\frac{c(k)-c(k-1)}{T}$, derivative $\frac{de(t)}{dt}$ with $\frac{e(k)-e(k-1)}{T}$, and second-order derivative $\frac{d^2 e(t)}{dt^2}$ with $\frac{e(k)-2e(k-1)+e(k-2)}{T^2}$. As a result, Eq. 5.4 becomes

$$c(k) = c(k-1) + a_0 \bullet e(k) + a_1 \bullet e(k-1) + a_2 \bullet e(k-2) \qquad (5.5)$$

with

$$\begin{cases} a_0 = k_p \bullet \left(1 + \frac{T}{T_i} + \frac{T_d}{T}\right) \\[2ex] a_1 = -k_p \bullet \left(1 + \frac{2T_d}{T}\right) \\[2ex] a_2 = k_p \bullet \frac{T_d}{T}. \end{cases}$$

Eq. 5.5 is the difference equation describing the dynamics of a discrete-time system which is also called a digital PID controller.

◇◇◇◇◇◇◇◇◇◇◇◇◇◇◇◇◇◇◇◇

5.2.4   *Transfer Functions*

It is not an easy task to solve a linear differential equation in the time domain. It becomes even more difficult if the equation in the time domain includes trigonometric and exponential functions. Fortunately, a practical and powerful mathematical tool exists to deal with the issue of solving linear differential equations. This tool is the well-known *Laplace Transform*. A formal definition of Laplace Transform can be stated as follows:

Definition 5.4   For continuous-time function $f(t)$ in the time domain such that $f(t) = 0$ if $t < 0$, Laplace Transform $F(s)$ of function $f(t)$ is

$$F(s) = \int_0^\infty f(t) \bullet e^{-st} \bullet dt$$

where s is a complex variable.

There are many notable advantages to using the Laplace Transform:

- A linear differential equation in the time domain becomes an algebraic equation in the complex s domain (or s-plane).
- The differentiation and integration operations in the time domain become algebraic operations in the complex s domain.
- The trigonometric and exponential functions in the time domain become algebraic functions in the complex s domain.
- The inverse of the Laplace Transform contains both transient and steady-state responses to the system.
- System performance, in terms of absolute and relative stabilities, can be graphically predicted in the s-plane.

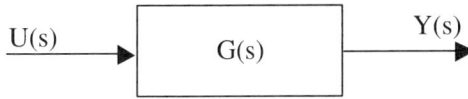

Fig. 5.3   Illustration of transfer functions.

For a linear continuous-time system, the ratio between the output's Laplace Transform and the input's Laplace Transform is called the *transfer function*. As shown in Fig. 5.3, if $Y(s)$ is the Laplace Transform of output $y(t)$ and $U(s)$ is the Laplace Transform of input $u(t)$, the system's transfer function will be

$$G(s) = \frac{Y(s)}{U(s)}. \qquad (5.6)$$

If the input to a system is unit-impulse function $\delta(t)$, its Laplace Transform $U(s)$ is 1. When $U(s) = 1$, $G(s) = Y(s)$. This means that the transfer function of a linear continuous-time system is equal to the Laplace Transform of output in response to the unit-impulse as input.

Example 5.3   The differential equation describing a PID control algorithm (or law) is expressed in Eq. 5.3. If we apply the Laplace Transform, the transfer function of a PID control algorithm is

$$\frac{C(s)}{E(s)} = k_p \bullet \left(1 + \frac{1}{T_i \bullet s} + T_d \bullet s \right) \qquad (5.7)$$

where $C(s)$ is the Laplace Transform of control signal $c(t)$, and $E(s)$ is the Laplace Transform of error signal $e(t)$.

◇◇◇◇◇◇◇◇◇◇◇◇◇◇◇◇◇◇◇◇

For a discrete-time system, signals are in the form of a series of discrete values. These discrete values are normally obtained by a sampling process over a continuous-time function. If $f(t)$ is a continuous-time function and is sampled at a series of time-instants t_k (i.e. $t_k = k \bullet T$ where T is the sampling period or interval), the corresponding discrete-time function $f_k(t)$ will be

$$f_k(t) = \sum_{k=0}^{\infty} f(t_k) \bullet \delta(t - t_k) \qquad (5.8)$$

where $\delta(.)$ is the unit impulse function and is expressed as follows:

$$\delta(t - t_k) = \begin{cases} 0 & \text{if } t \neq t_k \\ \lim_{\triangle T \to 0} \left(\frac{1}{\triangle T} \right) & \text{if } t = t_k. \end{cases}$$

(NOTE: If $t < 0$, $f_k(t) = 0$).

The extension of the Laplace Transform to the discrete-time domain is known as *Z-Transform*. By definition, the Z-Transform of discrete-time function $f_k(t)$ is

$$F(z) = \sum_{k=0}^{\infty} f(t_k) \bullet z^{-k} \tag{5.9}$$

where z is a complex variable and is related to the complex variable s in the following way:

$$z = e^{Ts}. \tag{5.10}$$

Eq. 5.9 can also be expressed as

$$F(z) = \sum_{k=0}^{\infty} f(k) \bullet z^{-k} \tag{5.11}$$

where k is the discrete-time index.

Since Z-Transform is an extension of the Laplace Transform and the complex variable z is analytically related to the complex variable s (i.e. Eq. 5.10), all properties of the Laplace Transform can be extended to the Z-Transform through the application of Eq. 5.10.

Similarly, for a discrete-time system, the ratio between the output's Z-Transform and the input's Z-Transform is called the *Transfer Function*.

Example 5.4   The difference equation describing a discrete PID control algorithm (or law) is expressed in Eq. 5.5. Multiplying term z^{-k} to both sides of Eq. 5.5 gives

$$c(k) \bullet z^{-k} = c(k-1) \bullet z^{-k} + a_0 \bullet e(k) \bullet z^{-k} + a_1 \bullet e(k-1) \bullet z^{-k} + a_2 \bullet e(k-2) \bullet z^{-k}. \tag{5.12}$$

By applying the following relations

$$\begin{cases} C(z) = \sum_{k=0}^{\infty} c(k) \bullet z^{-k} \\[2mm] \sum_{k=0}^{\infty} c(k-1) \bullet z^{-(k-1)} \bullet z^{-1} = C(z) \bullet z^{-1} \\[2mm] E(z) = \sum_{k=0}^{\infty} e(k) \bullet z^{-k} \\[2mm] \sum_{k=0}^{\infty} e(k-1) \bullet z^{-(k-1)} \bullet z^{-1} = E(z) \bullet z^{-1} \\[2mm] \sum_{k=0}^{\infty} e(k-2) \bullet z^{-(k-2)} \bullet z^{-2} = E(z) \bullet z^{-2}, \end{cases} \quad (5.13)$$

Eq. 5.12 becomes

$$C(z) = C(z) \bullet z^{-1} + a_0 \bullet E(z) + a_1 \bullet E(z) \bullet z^{-1} + a_2 \bullet E(z) \bullet z^{-2}.$$

As a result, the transfer function of the discrete PID control algorithm (or law) is

$$G(z) = \frac{C(z)}{E(z)} = \frac{a_0 + a_1 \bullet z^{-1} + a_2 \bullet z^{-2}}{1 - z^{-1}}. \quad (5.14)$$

◇◇◇◇◇◇◇◇◇◇◇◇◇◇◇◇◇◇◇◇

5.2.5 System Performance

Based on the transfer function, it is easy to analyze the performance of a closed-loop control system. Fig. 5.4 shows a simple closed-loop control system with a proportional control law. We denote $G_p(s)$ the transfer function of the plant, K the control gain of the proportional control law, and $H(s)$ the transfer function of the feedback sensor.

Tracking of the Desired Output

Assume that the Laplace Transform of the desired output is $R(s)$. If there is no disturbance (i.e. $D(s) = 0$) and no sensor noise (i.e. $N(s) = 0$), the transfer function of $Y(s)$ to $R(s)$ will be

$$\frac{Y(s)}{R(s)} = \frac{K \bullet G_p(s)}{1 + K \bullet G_p(s) \bullet H(s)}. \quad (5.15)$$

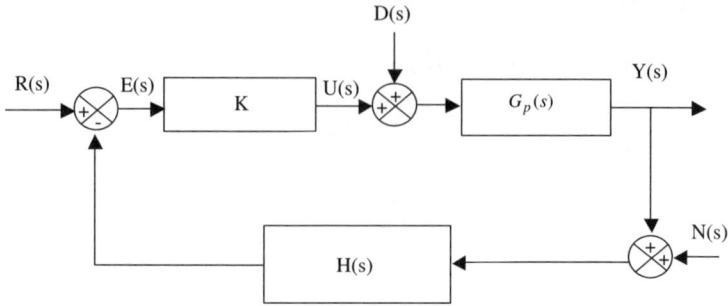

Fig. 5.4   Block diagram of a closed-loop control system with a proportional control law.

When $K \to \infty$, Eq. 5.15 becomes

$$\frac{Y(s)}{R(s)} \simeq \frac{1}{H(s)}.$$

Normally, the Laplace Transform of a feedback sensor is 1. It is clear that a large control gain K will result in a system which closely follows desired output $R(s)$ if the plant itself (i.e. $G_p(s)$) is intrinsically stable.

Example 5.5   Refer to Fig. 5.4. The plant's transfer function is

$$G_p(s) = \frac{1}{s^2 + s + 1}$$

Assume that the feedback sensor's transfer function is 1 (i.e. $H(s) = 1$), and input to the closed-loop control system is a unit step function (i.e. $R(s) = \frac{1}{s}$).

Now, we use the Simulink of MATLAB software to simulate this closed-loop control system and choose two values for the proportional control gain: 1 and 100. The two corresponding responses are shown in Fig. 5.5. When $K = 1$ (see the left sub-plot), the steady-state response of the closed-loop control system is 0.5. When the proportional control gain is increased to 100, the steady-state response is almost 1 (i.e. almost no error).

◇◇◇◇◇◇◇◇◇◇◇◇◇◇◇◇◇◇◇

Disturbance Cancellation

Assume that the system under control is at rest and there is no sensor noise. In this case, we only examine the output caused by disturbance $D(s)$. From

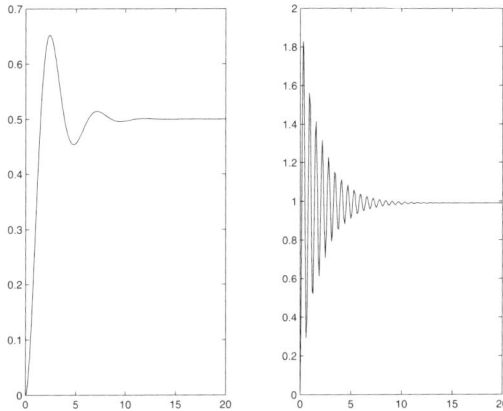

Fig. 5.5 Example of tracking the desired output with different proportional control gains: The left sub-plot is the result when $K = 1$ and the right sub-plot is the result when $K = 100$. The horizontal axis is the time axis with the unit in seconds.

Fig. 5.4, the transfer function of $Y(s)$ to $D(s)$ is

$$\frac{Y(s)}{D(s)} = \frac{G_p(s)}{1 + K \bullet G_p(s) \bullet H(s)}. \tag{5.16}$$

When $K \to \infty$, Eq. 5.16 becomes

$$\frac{Y(s)}{D(s)} \simeq 0.$$

This means that a large proportional control gain will reduce disturbance.

Example 5.6 Refer to Fig. 5.4. The plant's transfer function is

$$G_p(s) = \frac{1}{s^2 + 2 \bullet s + 3}$$

and the feedback sensor's transfer function is 1 (i.e. $H(s) = 1$). Assume that there is no input (i.e. $R(s) = 0$) and the system is subject to a periodic disturbance described by a sinusoidal function as follows:

$$d(t) = \sin(t)$$

or

$$D(s) = \frac{1}{s^2 + 1}.$$

Now, we use the Simulink of MATLAB software to simulate this closed-loop control system and choose two values for the proportional control gain: 1 and 100. The two corresponding responses are shown in Fig. 5.6. When $K = 1$ (see the left sub-plot), the amplitude of output is about 0.28. When the proportional gain is increased to 100, the output's amplitude is diminished to about 0.01.

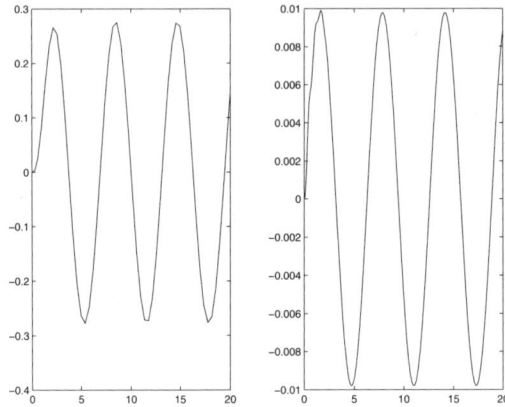

Fig. 5.6   Example of disturbance cancellation with different proportional control gains: The left sub-plot is the result when $K = 1$ and the right sub-plot is the result when $K = 100$. The horizontal axis is the time axis with the unit in seconds.

◇◇◇◇◇◇◇◇◇◇◇◇◇◇◇◇◇◇◇

Effect of Feedback-Sensor's Noise

Assume that the system under control is at rest and there is no disturbance. Now, let us examine what happens if the feedback sensor introduces noise (i.e. $N(s) \neq 0$).

From Fig. 5.4, the transfer function of $Y(s)$ to $N(s)$ is

$$\frac{Y(s)}{D(s)} = \frac{-K \bullet G_p(s) \bullet H(s)}{1 + K \bullet G_p(s) \bullet H(s)}. \tag{5.17}$$

When $K \to \infty$, Eq. 5.17 becomes

$$\frac{Y(s)}{N(s)} \simeq -1.$$

This means that 100% of feedback sensor's noise adds to the system's output if a large proportional control gain is chosen. In other words, the performance of a closed-loop control system depends on the quality of the feedback sensor.

5.2.6 Analysis of Absolute Stability

Refer to Fig. 5.5. The system is stable but the transient response is oscillatory. Ideally, the design of a control law should make a system not only stable but also have a timely and smooth transient-response. Since a closed-loop control system is a system that determines control action based on the error signal, the dynamics of error signal $e(t)$ not only influences the system's transient response but also its absolute stability.

Intuitively, if $e(t) \to \infty$ when $t \to \infty$, a closed-loop control system is definitely unstable. Thus, the primary concern in the analysis of a closed-loop control system is absolute stability. Without a guarantee of absolute stability, all discussions about timely and smooth transient response are useless.

In control engineering, there are many powerful mathematical tools for studying a system's stability. If the dynamic model of a system under control is known, it is easy to analyze the stability of the closed-loop control system. However, the big challenge is how to design a closed-loop control system when the dynamic model of a plant is not exactly known. In robotics, we normally face this situation. The exact dynamic model of a robot's electromechanical system is rarely known in advance.

5.2.6.1 Root-Locus Method

Let us consider a linear continuous-time system. If $G(s)$ is the transfer function of a closed-loop control system, the Laplace Transform of the output, in response to the unit impulse as input, will be

$$Y(s) = G(s). \tag{5.18}$$

Normally, $G(s)$ is in the form of the ratio of two polynomial functions in terms of the complex variable s. Assume that $G(s)$ can be expressed as

$$G(s) = \frac{A(s)}{B(s)} = \frac{K_0 \bullet (s + a_1)(s + a_2)...(s + a_m)}{(s + p_1)(s + p_2)...(s + p_n)} \tag{5.19}$$

where $m < n$ and K_0 is a gain coefficient. By definition, the roots of the numerator in Eq. 5.19 are called the *zeros* of a closed-loop control system, and the roots of the denominator in Eq. 5.19 are known as the *poles* of a closed-loop control system. The partial-fraction expansion of Eq. 5.19 yields

$$G(s) = \frac{b_1}{s + p_1} + \frac{b_2}{s + p_2} + ... + \frac{b_n}{s + p_n} \tag{5.20}$$

where

$$b_i = \left\{ (s + p_i) \bullet \frac{A(s)}{B(s)} \right\} \Big|_{s=-p_i}, \ \forall i \in [1, n]. \tag{5.21}$$

If we compute the inverse Laplace Transform of Eq. 5.20, the output of a closed-loop control system responding to the unit impulse input will be

$$y(t) = b_1 \bullet e^{-p_1 t} + b_2 \bullet e^{-p_2 t} + ... + b_n \bullet e^{-p_n t}. \tag{5.22}$$

It becomes clear that $e^{-p_i t} \rightarrow 0$ when $t \rightarrow \infty$ if p_i is positive. Otherwise, if p_i is negative, $e^{-p_i t} \rightarrow \infty$ when $t \rightarrow \infty$. When the impulse response of a system goes to infinity, it simply means that the system is not stable.

When p_i is positive, the root $s_i = -p_i$ is located at the left half-plane of the complex s domain. In other words, a closed-loop control system will be stable if and only if all the poles of the closed-loop control system's transfer function are located in the left half-plane of the complex s domain. This method of analyzing a system's stability is known as the *Root-Locus method.*

Example 5.7   The transfer function of a closed-loop control system is

$$G(s) = \frac{6 \bullet (s + 1)}{(s + 2)(s + 3)}.$$

The two poles of the transfer functions are: $s_1 = -2$ and $s_2 = -3$. Therefore, the system is stable because the two poles are in the left half-plane of the complex s domain.

Fig. 5.7 shows the unit-step response (i.e. $R(s) = \frac{1}{s}$) of this system.

◇◇◇◇◇◇◇◇◇◇◇◇◇◇◇◇◇◇

Example 5.8   Refer to the above example. Assume that the transfer function of the closed-loop control system becomes

$$G(s) = \frac{6 \bullet (s + 1)}{(s - 2)(s + 3)}.$$

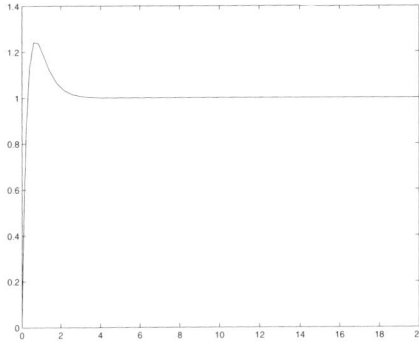

Fig. 5.7   Unit step response of a stable system. The horizontal axis is the time axis, with the unit in seconds.

Pole $s_1 = 2$ is now located at the right half-plane of the complex s domain. The system becomes unstable. Fig. 5.8 shows the unit step response (i.e. $R(s) = \frac{1}{s}$) of this system.

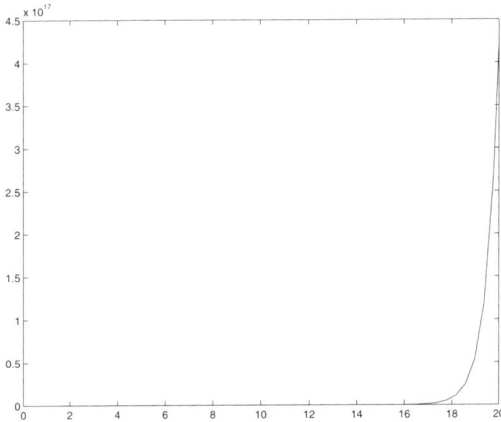

Fig. 5.8   Unit step response of an unstable system. The horizontal axis is the time axis, with the unit in seconds.

◇◇◇◇◇◇◇◇◇◇◇◇◇◇◇◇◇

5.2.6.2 *Lyapunov's Method*

In practice, we frequently face the situation in which the transfer function of a closed-loop control system is not exactly known for various reasons (e.g.

non-linearity, difficulty in obtaining an exact dynamic model, un-modelled dynamics etc). In this case, we can apply the Lyapunov's method to test the asymptotical stability of a closed-loop control system. The basic idea behind Lyapunov's method(s) can be explained as follows:

Lyapunov Function

When we have scalar function $V(x)$ which is a function of vector x, this function is called a *Lyapunov Function* if and only if the following conditions hold:

$$\begin{cases} V(x) > 0 & \text{if } x \neq 0 \text{ and for all } t > 0 \\ V(x) = 0 & \text{if } x = 0 \text{ and for all } t > 0 \end{cases} \quad (5.23)$$

and $V(x)$ has a continuous-time derivative (i.e. $\dot{V}(x)$ exists for all $t > 0$).

In practice, there are many scalar functions which are qualified to be Lyapunov Functions. This is an advantage. For example, if $x(t)$ is a continuous-time vector, $V(x) = \frac{1}{2}x^t(t) \bullet x(t)$ is a Lyapunov Function.

In fact, a Lyapunov Function is also known as a *control objective function* because the objective of control action is to make this function converge to zero.

Lyaponuv's Stability Theorem

Let $x(t)$ be the state vector of a system. If we are able to specify Lyapunov Function $V(x)$ and if its first-order derivative $\dot{V}(x)$ satisfies the following condition:

$$\dot{V}(x) < 0 \quad \forall t > 0, \quad (5.24)$$

the system is said to be *asymptotical stable* at the state when $x(t) = 0$ for all $t > 0$.

In fact, when $\dot{V}(x) < 0$, scalar function $V(x)$ is converging to zero. As $V(x)$ is a Lyapunov Function, vector x will also be converging to zero if $V(x)$ goes to zero. As a result, the system will reach a state of equilibrium, when $x(t) = 0$ for all $t > 0$.

Control System Design Using Lyapunov's Stability Theorem

Lyapunov's Stability Theorem does not help much with the stability analysis of an existing closed-loop control system but is useful in the design of an asymptotically stable system. In other words, we can make use of the

Lyapunov Stability Theorem to design a control law which will make the system under control asymptotically stable.

Refer to Fig. 5.2. A closed-loop control system is a system that responds to the error signal denoted by $e(t)$. The control law has a great influence over the dynamic response of a closed-loop control system. By definition, output from a control law is called the control signal (or action) denoted by $u(t)$. Because of the sensor feedback, error signal $e(t)$ is indirectly a function of control signal $u(t)$ as well. When designing a control algorithm or law, the objective becomes to find a Lyapunov Function $V(e(t))$ and a control law in which control signal $u(t)$ guarantees the following conditions:

$$\begin{cases} V(e(t)) > 0 & \text{for all } u(t) \text{ and } t > 0 \\ \dot{V}(e(t)) < 0 & \text{for all } u(t) \text{ and } t > 0. \end{cases} \tag{5.25}$$

If such a control law exists, $V(e(t))$ will shrink to zero with time. When $V(e(t))$ shrinks to zero, $e(t)$ will also converge to zero. Consequently, the system is asymptotically stable and will remain at rest, when $e(t) = 0$ for all $t > 0$. This is an expected and desirable behavior for a closed-loop control system.

Note that the conditions expressed in Eq. 5.25 are sufficient conditions, but not necessary for a system to be stable.

5.2.7 *Tuning of PID Control Algorithms*

In real life, there are many systems, the dynamics of which is not exactly known. However, many practical control systems (including robots) rely on the popular PID control law (or its variations) to achieve satisfactory control performance. Therefore, it is important to know how to empirically tune a PID control law.

Refer to Eq. 5.3. The tuning of a PID control law is to empirically find suitable values for proportional control gain k_p, integral time constant T_i and derivative time constant T_d so that the overall dynamic response of a closed-loop control system is empirically judged to be satisfactory. The process of adjusting the parameters of a control law is called *controller tuning*.

There are many empirical methods for tuning a PID control law. For example, Ziegler-Nichols methods are popular in control engineering. The first Ziegler-Nichols method requires recording the output of a plant (i.e. system under control) in response to a unit-step function. If the response looks like an S-shape curve, then this method is applicable. The second

Ziegler-Nichols method requires the closed-loop control system to exhibit a continuous oscillatory behavior when a large proportional control gain is chosen for a pure proportional control law. In practice, these two methods are not convenient for tuning the PID control laws inside a robot's control system.

An alternative method is to manually tune a PID control law in three simple steps. For the purpose of illustration, we choose a plant which the transfer function is

$$G_p = \frac{1}{(s+2)(s+3)}. \tag{5.26}$$

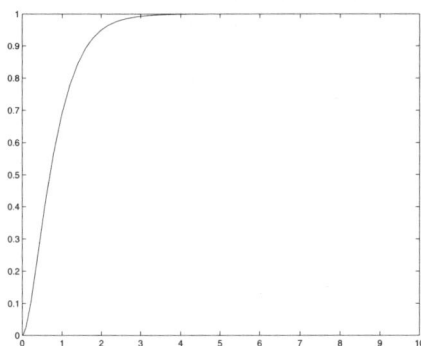

Fig. 5.9   Response of an open-loop system to a unit-step function with a proportional gain of 6 added to G_p.

Fig. 5.9 shows the response of the system, described by Eq. 5.26, to a unit-step function. For the sake of comparison with the closed-loop control system, a proportional gain of 6 is added to G_p in order to bring the steady-state response to the final value 1.

Now, let us use a PID control law to improve the dynamic response of plant G_p in Eq. 5.26. Fig. 5.10 shows a block diagram of the closed-loop control system with a PID control law. Tuning the parameters of the PID control law can be accomplished in the following steps:

Tuning of Proportional Gain k_p

The proportional control gain has an overall influence on the performance of a closed-loop control system. The first objective in tuning is to determine a proportional control gain which makes the closed-loop control system stable, reasonably smooth and accurate.

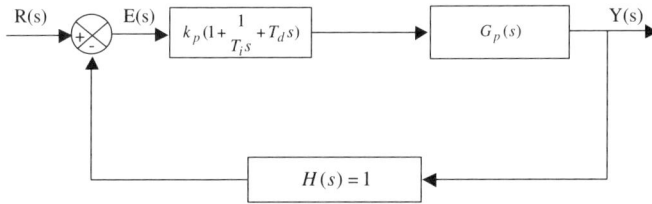

Fig. 5.10   Block diagram of the closed-loop control system with a PID control law.

At this step, we set $k_p = 0$, $T_i = \infty$ and $T_d = 0$. We gradually increase the proportional control gain until the unit-step response of the closed-loop control system reaches value 1 at a certain time-instant. This time-instant is known as the *rising time* of a closed-loop control system.

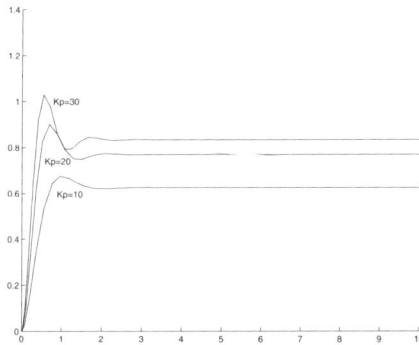

Fig. 5.11   Effect of proportional gain. The horizontal axis is the time axis with the unit in seconds.

Fig. 5.11 shows the results of tuning k_p for the example in Fig. 5.10. We can see that the proportional control gain at $k_p = 30$ permits the unit-step response to slightly overshoot value 1.

Tuning of Integral Time-Constant T_i

When the proportional gain is set as $k_p = 30$, the closed-loop control system is stable, but the steady-state error is quite large. It is natural to turn "on" the control action from the integral part of the PID control law. A control action proportional to the integral of the error signal in a PID control law allows to reduce accumulated errors over time. As a result, it will make the steady-state response converge to the desired output (i.e. 1). Thus, the second step of tuning is to adjust the integral time-constant T_i while turning "off" the derivative time-constant (i.e $T_d = 0$).

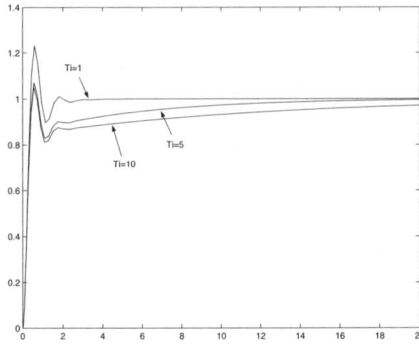

Fig. 5.12   Effect of the integral time-constant. The horizontal axis is the time axis with the unit in seconds.

Fig. 5.12 shows the results of tuning T_i for the example in Fig. 5.10. When the integral time-constant is set as $T_i = 1.0$, the unit-step response will quickly converge to its steady-state value. The time-instant when the actual output converges to the desired output is known as the *settling time*.

Tuning of Derivative Time-Constant T_d

As we can see in Fig. 5.12, the accuracy of the closed-loop control system is achieved with a large integral time-constant. However, the system lacks a good transient response (i.e. the transient response is oscillatory). In other words, the stiffness of the closed-loop control system will be reduced with a large integral time-constant. In order to compensate for the loss of stiffness, it is natural to turn "on" the derivative part of the PID control law. Thus, the last step in tuning is to adjust the derivative time-constant.

Fig. 5.13 shows the results of tuning T_d for the example in Fig. 5.10. When the derivative time-constant is set as $T_d = 0.1$, the unit-step response quickly converges to its steady-state value while the overshoot is reasonable and quite small.

Comparing Fig. 5.9 and Fig. 5.13, we can see that the settling time of the closed-loop control system is less than two seconds while for the open-loop system it is more than three seconds.

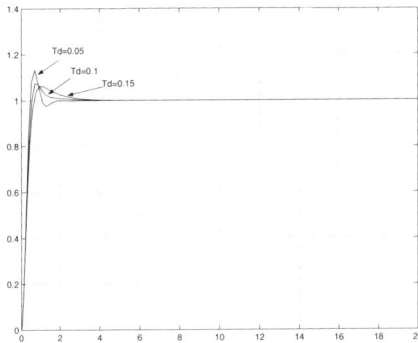

Fig. 5.13 Effect of the derivative time-constant. The horizontal axis is the time axis with the unit in seconds.

5.3 Control Elements

Our ultimate goal is for a robot to perform tasks by executing motions. A robot is a system in which the input is the tasks and the output is the executed motions. On the other hand, we understand that a closed-loop control system is superior to an open-loop system. As a result, the control of a robot's motion execution must rely on a closed-loop control scheme, in order to achieve a desired performance. However, the formation of a closed-loop control system must address the following issues:

- How do we appropriately define the error signal?
- How do we efficiently determine the control signal based on the error signal?
- How do we act, in a timely and energetic manner, in response to the control signal?

These issues are closely related to sensing, decision-making, and action-taking devices, all of which a closed-loop control system should have. These devices are also known as the *control elements*.

Since robot control implies a closed-loop control of *constrained or unconstrained* motions required for the performance of tasks or actions, the error signal of a robot's control system should be the difference between the desired motions and the actual motions. And, the purpose of the control algorithm(s) inside a robot's control system is to determine the control signal which will make the error signal converge to zero in order to produce the desired output.

Nowadays, almost all control algorithms are executed inside digital computers, also known as the *digital motion controllers*. Therefore, at this point, we can say that a robot is a combination of mechanical, electromechanical, and control systems, as shown in Fig. 5.14.

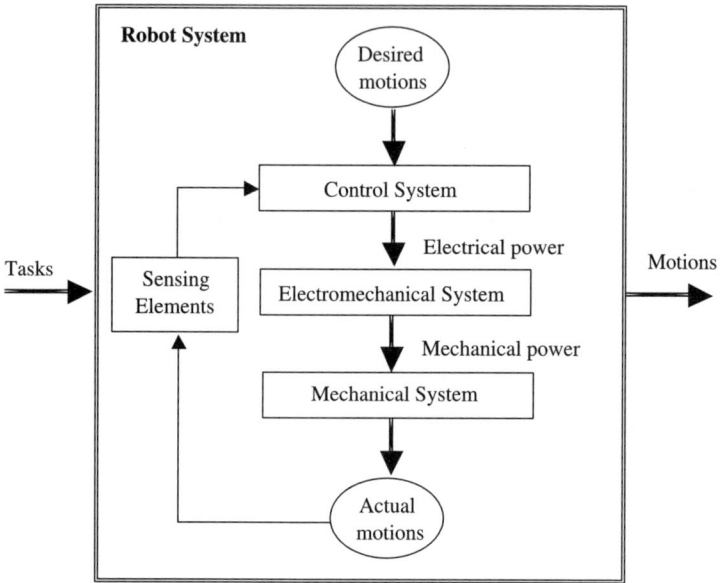

Fig. 5.14   A hierarchical view of a robot system, which includes the closed-loop control system for motion execution.

In Chapter 4, we studied the working principles of electric motors. The rotor and stator inside an electric motor constitute a dynamic pair in which the interaction is governed by electromagnetic force. From a control point of view, there are two concerns regarding the appropriate operation of an electric motor:

- How do we alter the direction of motion of an electric motor?
- How do we regulate the input of electrical energy to an electric motor?

Thus, a robot's control system is not simply a digital motion-controller and its electromechanical system. There must be other control elements handling the alteration of motion direction and regulating the input of electrical energy to the electric actuators inside a robot. For a motion control system, the device which alters motion direction is called the *power switch.* And the device which regulates the input of electrical power to an electric actuator is called the *power drive.* A product which incorporates both power switch and power drive is commonly referred to as the *power amplifier.*

Accordingly, a robot's control elements will necessarily include at least one digital motion controller, a power switch, and a power drive. Fig. 5.15 illustrates one set of control elements for the motion control loop governing one degree of freedom (one actuator) inside a robot.

We will study sensing elements in a separate section. However, here we simply highlight three basic units necessary for the construction of a power switch and power drive. These units are the wave generator, power generator and switch circuit, as shown in Fig. 5.15. The wave generator is responsible for producing control-logic signals or pulse waveforms (pulse-width modulated waveforms or PWM waveforms for short). Depending on the type of electric motor under control, a wave generator may include two sub-units: a) a logic wave generator and b) a PWM wave generator. On the other hand, the power generator can be as simple as a power source or supplier. We will study this in further detail in the later part of this section.

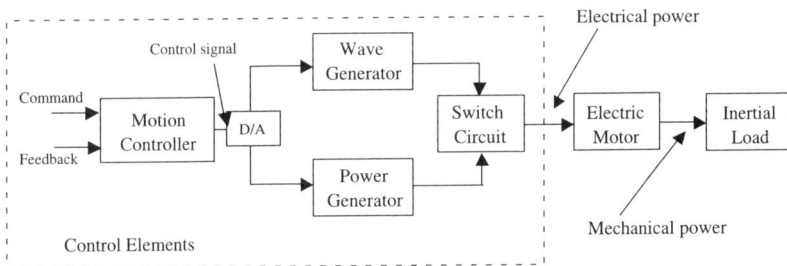

Fig. 5.15 Control elements necessary in a motion-control loop.

5.3.1 *Power Switches*

Here, we consider the case of rotary electric motors. According to the working principles behind the electric motor, interaction between the rotor and stator is caused by electromagnetic force. In order to change the direction of motion, we can alter the polarity or direction of the electromagnetic field. This is usually done in two ways:

- Change the direction of currents which flow into the coils inside an electric motor.
- Change the order of currents applied to the coils (alter the order in which we energize the coils).

A simple way to change the direction (or order) of currents is to use a switch circuit, also known as an *inverter*. The input to a switch circuit is a set of control-logic signals which determine the sequence of commutation of the logic states for switching ("on" or "off"). These control-logic signals are necessary for a switch circuit to function properly. A unit which generates a set of control-logic signals (i.e. rectangular electrical pulses) is called a logic-wave generator.

Thus, a power switch will consist of two units: a) a logic-wave generator and b) a switch circuit.

5.3.1.1   *Power Switches for Stepper Motors*

Refer to Fig. 4.13. The coils wound around the teeth of the stator are grouped into four phases: AA', BB', CC' and DD'. In order to make a stepwise motion, these phases are sequentially energized. This can be done with an appropriate power switch.

Switch Transistors

In electronics, there are two types of devices: a) passive devices (e.g. resistors, capacitors, transformers etc) and b) active devices (e.g. transistors, operational amplifiers etc). A typical active device is a transistor because it can amplify the power of an input electrical signal. (NOTE: The product of a signal's current and voltage is its electrical power). As shown in Fig. 5.16, a transistor has three terminals, namely: a) the base, b) the collector, and c) the emitter. The key characteristics of a *npn* transistor include:

(1) From the base to the emitter, it is similar to a diode. The voltage drop is about 0.6 (volts) or 0.6V (i.e. $V_B - V_E = 0.6V$).
(2) From the collector to the base, it is similar to an inverted diode. As a result, no current will flow from the collector to the base.
(3) If I_B is the current which flows to the base (from the base to the emitter), and I_C is the current which flows to the collector (from the collector to the emitter), then $I_C = \beta \bullet I_B$ when an input signal is applied to the base. The coefficient β is between 50 to 250.
(4) When no input signal is applied to the base, $I_B = 0$ and $I_C = 0$.

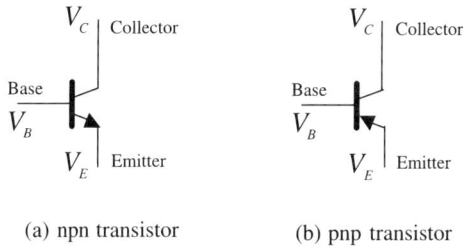

(a) npn transistor   (b) pnp transistor

Fig. 5.16   Symbolic representations of transistors.

As a result of the third and fourth characteristics, a transistor is a perfect electronic switch. When an input signal is applied to the base, we electronically connect the collector to the emitter because a large current can flow from the collector to the emitter (very low impedance between the collector and the emitter). Conversely, we electronically disconnect the collector from the emitter (i.e. $I_C = 0$) when there is no input signal at the base.

Switch Circuit

For the stepper motor, as illustrated in Fig. 4.12, the coils are divided into four phases. Thus, it is necessary for a switch circuit to handle the order of energizing these four phases. A conceptual solution is shown in Fig. 5.17.

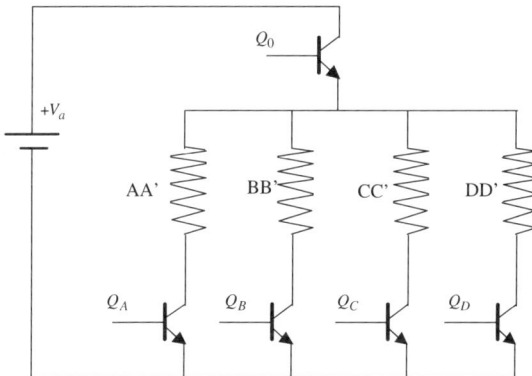

Fig. 5.17   Conceptual illustration of a switch circuit for a four-phase stepper motor.

Signal Q_0 serves as an enabling signal. When switch Q_A is "on" and the other switches are "off" (assuming that Q_0 is always "on"), phase AA' is energized. In order to control the direction of motion, it is necessary to supply a set of control-logic signals to Q_A, Q_B, Q_C and Q_D.

Logic Wave Generator

Refer to Fig. 5.15. The logic wave generator is responsible for generating the control-logic signals that control the order in which the coil phases are energized. In a digital system, the logic signal is in the form of rectangular pulse waveform. Usually, high voltage (+5V) corresponds to the logic "high" or "1", and low voltage (0V) corresponds to the logic "low" or "0".

For the example shown in Fig. 4.12, the coils are divided into four phases. Each phase needs one control-logic signal (i.e. one bit or one digital signal-line) to turn the current from a power supply, "on" or "off". If we combine all the bits of the control-logic signals together, we obtain a variable which is called the *state of the logic wave generator*. If we use "1" to represent the logic "high" and "0" the logic "low", logic state 1000 means that phase AA' is energized, logic state 0100 means that phase BB' is energized, logic state 0010 means that phase CC' is energized, and logic state 0001 means that phase DD' is energized.

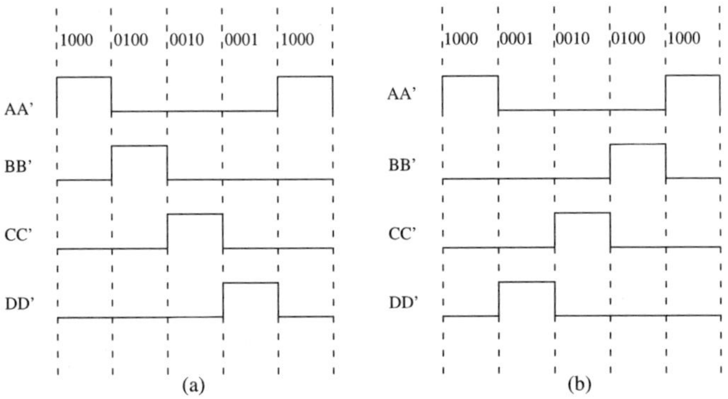

Fig. 5.18 Pulse waveforms of the control-logic signals from a wave generator for a four-phase stepper motor: a) spinning in a clockwise direction and b) spinning in a counterclockwise direction.

As shown in Fig. 5.18, the following sequential order of logic state commutation

$$1000 \rightarrow 0100 \rightarrow 0010 \rightarrow 0001 \rightarrow 1000 \rightarrow ...$$

will produce a clockwise rotation of the stepper motor shown in Fig. 4.12. And the following sequential order of logic state commutation

$$1000 \rightarrow 0001 \rightarrow 0010 \rightarrow 0100 \rightarrow 1000 \rightarrow \ldots$$

will produce a counterclockwise rotation of the stepper motor.

Since a stepper motor makes stepwise rotation, its speed obviously depends on the frequency (or speed) of commutating the logic states. Therefore, the speed control in a stepper motor is simply achieved by regulating the frequency of commutating the logic states of its wave generator.

5.3.1.2 *Power Switches for Brush-type DC Motors*

Refer to Fig. 4.14. Because of the mechanical commutator inside a brush-type DC motor, the direction of rotation depends only on the polarity of the direct voltage applied to the coils wound around the rotor of the motor. As a result, the power switch for a brush-type DC motor is very simple.

Switch Circuit

The switch circuit, which is commonly used to alter the polarity of the direct voltage applied to the coils of a brush-type DC motor, is the *H-bridge circuit*. Refer to Fig. 5.19. The purpose of the diode attached to each transistor is to allow the back electro-motive force (i.e. emf current) to die off quickly when the switch is turned "off".

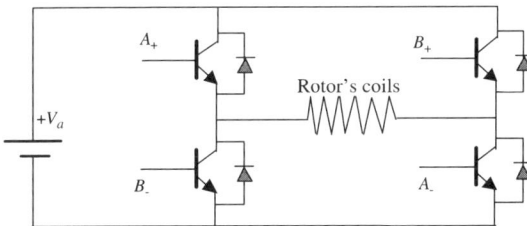

Fig. 5.19   Conceptual illustration of a switch circuit for a brush-type DC motor.

It is easy to see that when only switches A_+ and A_- are "on", the motor will rotate in one direction. Alternatively, if only switches B_+ and B_- are "on", the motor will change its direction of motion.

Logic-Wave Generator

Refer to Fig. 5.15. The purpose of the logic-wave generator (a sub-unit in the wave generator) is to produce the control-logic signals applied to the four electronic switches. Similarly, if we form a logic state variable by combining all the bits of the control-logic signals together (i.e. $A_+A_-B_+B_-$), the logic state of 1100 will make the brush-type DC motor rotate in one direction, and the commutation of this logic state to 0011 will alter the direction of motion. The pulse waveforms of the control-logic signals are shown in Fig. 5.20.

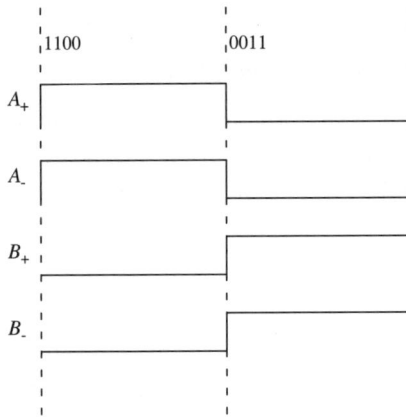

Fig. 5.20   Pulse waveforms of the control-logic signals from a wave generator for a brush-type DC motor.

5.3.1.3   *Power Switches for Brush-less DC Motors*

Refer to Fig. 4.15. A brush-less DC motor's coils are wound around the toothed structure of the stator. These coils are divided into three phases: UU', VV' and WW'. The direction of rotation can only be altered through the electronic commutation, both order and direction of the currents which flow independently into these three phases.

Power Switch

According to the working principle of a brush-less DC motor, at any time-instant, only two phases are energized. If we examine the power switch for a brush-type DC motor, we know that the electronic commutation of the current's direction inside a single coil requires four switches. If we independently control the commutation of the current for each of the three phases in a brush-less DC motor, it is necessary to have 12 switches. However, if we connect one end of each phase's coil to a common point, any pair of two energized coils will form a single circuit. In this way, the number of switches is reduced to 6, as shown in Fig. 5.21. (NOTE: This switch circuit needs both positive and negative voltage supplies).

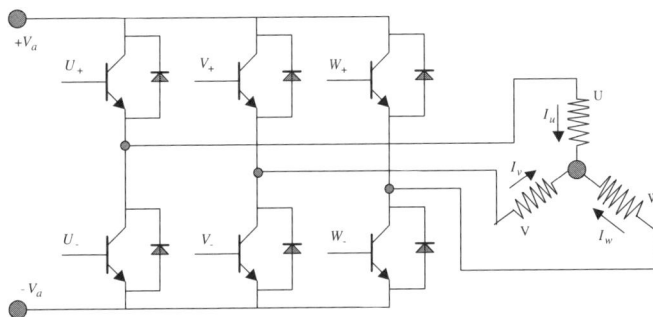

Fig. 5.21   Conceptual illustration of a switch circuit for a brush-less DC motor.

Refer to Fig. 5.21. At any time-instant, only one switch on the upper row and one switch on the lower row are turned "on". For example, if switches U_+ and V_- are turned "on", phases UU' and VV' are energized with $I_u > 0$ and $I_v < 0$. Alternatively, if switches V_+ and U_- are turned "on", phases UU' and VV' are also energized but $I_u < 0$ and $I_v > 0$.

Logic-Wave Generator

Similarly, the role of the logic-wave generator here is also to produce a set of control-logic signals. We can form a logic state variable by combining all the bits of the control-logic signals applied to switches U_+, V_+, W_+, U_-, V_- and W_-. Refer to Fig. 4.15 and Fig. 5.21. The following sequential order of the logic-state commutation

$$100001 \rightarrow 100010 \rightarrow 001010 \rightarrow 001100 \rightarrow 010100 \rightarrow 010001 \rightarrow 100001 \rightarrow \cdots$$

will produce a counterclockwise rotation. The corresponding pulse waveforms of the 6 control-logic signals are shown in Fig. 5.22.

Alternatively, the following sequential order of the logic-state commu-

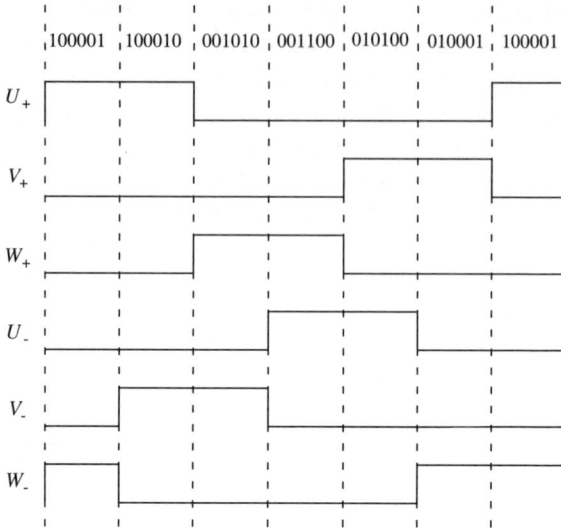

Fig. 5.22   Pulse waveforms from the wave generator for a brush-less DC motor spinning in a counterclockwise direction.

tation

$$100010 \rightarrow 100001 \rightarrow 010001 \rightarrow 010100 \rightarrow 001100 \rightarrow 001010 \rightarrow 100010 \rightarrow \ldots$$

will produce a clockwise rotation. And the corresponding pulse waveforms of the 6 control-logic signals are shown in Fig. 5.23.

It is important to note that for the proper operation of a brush-less DC motor, the frequency (or speed) of the logic state commutation must be synchronized with the speed of the rotor. This explains why the power amplifier for a brush-less DC motor requires velocity feedback from the rotor.

5.3.2   *Power Drives*

The primary concern of a control system is to design a control law for satisfactory control performance. However, almost all control laws are implemented on digital computers. And the control signals calculated by digital computers do not carry any energy after the digital-to-analog (D/A) signal conversion. These signals must be directly, or indirectly, amplified into the corresponding power signals in order to drive the motors inside a robot or any machine. As we mentioned before, a device which performs this direct

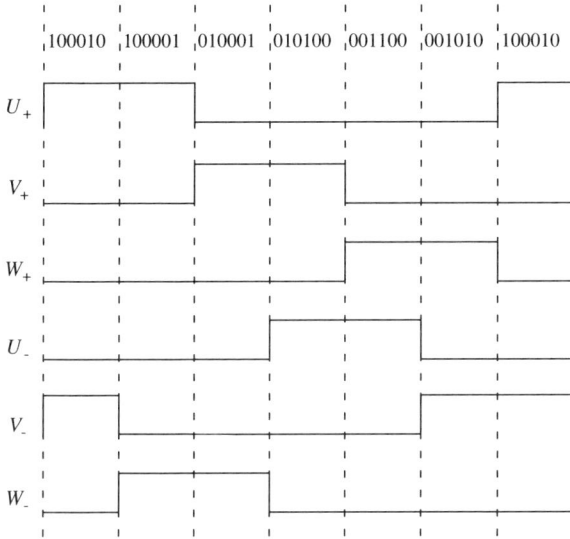

Fig. 5.23 Pulse waveforms from the wave generator for a brush-less DC motor spinning in a clockwise direction.

or indirect signal amplification is called a *power drive*.

In physics, we know that electrical power is equal to the product of voltage and current exerted on an electrical load (e.g. resistor, capacitor, inductor etc). If we denote V_a the voltage and I_a the current exerted on the coils inside an electric motor, the electrical energy applied to the motor within time interval $[t_s, t_f]$ will be

$$W_e = \int_{t_s}^{t_f} (V_a \bullet I_a) \bullet dt. \tag{5.27}$$

Clearly, we can manipulate either time interval $[t_s, t_f]$ or voltage V_a (or current I_a) to alter the amount of electrical energy applied to an electric motor.

Accordingly, there are two basic working principles for amplifying a control signal into a corresponding power signal:

- Direct Power Amplification:
  This method directly amplifies a control signal into a corresponding voltage signal, which is capable of supplying a sufficient amount of current, as well. Due to the low efficiency caused by heat dissipation, this method is suitable for low power output (low power motors).

- Indirect Power Amplification:
  For an electric stepper motor, the voltage applied to the coils directly comes from a power supply. As for DC motors, the widely-employed method of amplifying control signals is the *pulse-width modulation* or *pulse-width modulated* (PWM) method. This method is easy to operate (it is easy to generate a regular pulse waveform) and efficient (there's less heat dissipation). Therefore, it is commonly used to produce high power output (for high power motors).

5.3.2.1 *Linear Power Drives*

Linear power drive directly amplifies a control signal (input), and produces a voltage (output), which is linearly proportional to the control signal. Subsequently, the voltage is applied to a motor through a switch circuit. Fig. 5.24 shows a schematic illustration of a linear power drive together with power switch. A linear power drive is basically built on top of an active device called the *operational amplifier* or *op-amp* for short. Its role is to regulate the velocity of an electric motor.

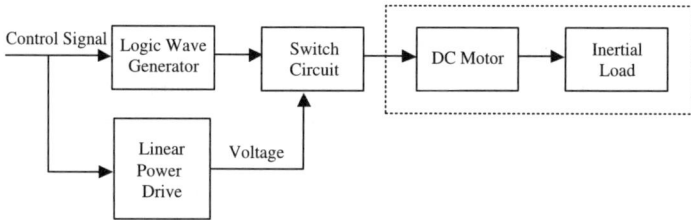

Fig. 5.24   Schematic illustration of a linear power drive together with a power switch.

As shown in Fig. 5.25, an op-amp is a device with two input terminals and one output terminal. The key characteristics of an op-amp include

- An op-amp's input impedance is very high while its output impedance is very low. As a result, no current flows into the two input terminals of an op-amp (i.e. the"+" terminal and the "-" terminal) .
- The output voltage is proportional to the difference between the two input voltages at the input terminals. The gain of an op-amp is very high (in the order of 10^5 to 10^6). As a result, an op-amp has to be operated with a negative feedback. In this way, the input voltages at the two input terminals can be treated as equal (the difference is practically zero).

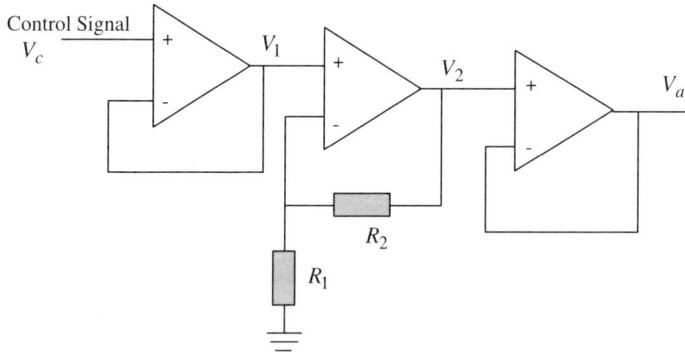

Fig. 5.25   Conceptual illustration of a noninverting linear power amplifier .

With unity negative feedback, an op-amp can serve as a perfect voltage follower. In other words, the output voltage is equal to the input voltage. This is useful to isolate electrical signals. Fig. 5.25 shows a conceptual solution for the design of a noninverting linear power amplifier. Since the current that flows through the resistor R_1 is

$$I_{R_1} = \frac{V_1}{R_1},$$

the voltage V_2 will be

$$V_2 = I_{R_1} \bullet (R_1 + R_2)$$

or

$$V_2 = \frac{V_1}{R_1} \bullet (R_1 + R_2).$$

(NOTE: The current at resistor R_2 is equal to the current at resistor R_1).
By applying the equalities: $V_1 = V_c$ and $V_a = V_2$, the relationship between input and output voltages will be

$$V_a = \left(1 + \frac{R_2}{R_1}\right) \bullet V_c. \qquad (5.28)$$

If I_a is the current from voltage V_a applied to the coils of a motor, the electrical energy released to the motor within time interval $[t_s, t_f]$ will be

$$W_e = \int_{t_s}^{t_f} \left(1 + \frac{R_2}{R_1}\right) \bullet V_c \bullet I_a \bullet dt \qquad (5.29)$$

It becomes clear that control signal V_c, computed from a control algorithm or law, directly regulates the electrical energy applied to an electric motor.

From Eq. 5.28, we can see that the signs of the input and output voltages of the power drive are the same. This is why the circuit, shown in Fig. 5.25, is called the *noninverting linear power drive*.

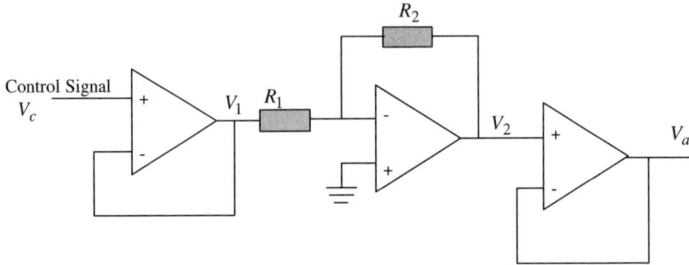

Fig. 5.26   Conceptual illustration of an inverting linear power-amplifier.

If an electric motor requires negative voltage to be applied to its coils, we can easily design an *inverting linear power drive*. Fig. 5.26 shows a conceptual solution for the design of an inverting linear power drive. The current that flows into resistor R_1 is

$$I_{R_1} = \frac{V_1}{R_1}.$$

Since the current at R_2 is equal to the current at R_1, voltage V_2 will be

$$V_2 = -R_2 \bullet I_{R_1}$$

or

$$V_2 = -R_2 \bullet \frac{V_1}{R_1}.$$

As $V_1 = V_c$ and $V_a = V_2$, the relationship between the input and output voltages becomes

$$V_a = -\frac{R_2}{R_1} \bullet V_c. \qquad (5.30)$$

Similarly, electrical energy applied to the electric motor within time

interval $[t_s, t_f]$ will be

$$W_e = \int_{t_s}^{t_f} \left(\frac{R_2}{R_1} \bullet V_c \bullet I_a \right) \bullet dt. \tag{5.31}$$

(NOTE: We can ignore the negative voltage sign).

5.3.2.2 *PWM Power Drives*

Refer to Eq. 5.27. Electrical energy applied to an electric motor depends on time interval $[t_s, t_f]$, voltage V_a, and current I_a. We know that all electronic components suffer from heat deterioration, which in turn, affects performance. As a result, the longer time interval $[t_s, t_f]$ is, the more severe the heat deterioration effect. Due to thermal inertia, an electronic component will not heat up if it operates for short periods of time. Therefore, it is good to control the electrical energy by regulating time interval $[t_s, t_f]$ while keeping term $V_a \bullet I_a$ at its maximum value.

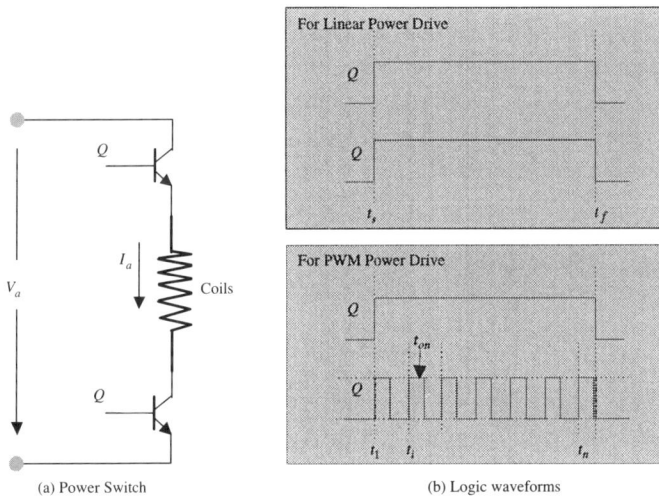

Fig. 5.27   Illustration of the difference between a linear power drive and a PWM power drive.

Here, we use a single phase to illustrate the difference between a linear power drive and a PWM power drive. As shown in Fig. 5.27, in a linear power drive, control-logic signals Q_+ and Q_- will be at the logic "high" for the whole period of time interval $[t_s, t_f]$. This means that when the two switches are turned "on" within time interval $[t_s, t_f]$, continuous current I_a

flows through these two switches. The control of the electrical energy is based on the control of input voltage V_a (or current I_a).

On the other hand, if we superimpose a periodic pulse waveform on control-logic signal Q_-, something which is easily done with an "AND" Boolean logic circuit, input voltage V_a (or current I_a) will only be applied to the coils when the combined logic signal, applied to switch Q_-, is at the logic "high". The duration of the logic high state at Q_- depends on width t_{on} of the positive pulse.

It is clear that we can fix input voltage V_a. Normally, we choose it to be equal to the voltage of the power supply. And we can regulate pulse-width t_{on}. Assume that the periodic pulse waveform has a fixed time period (cycle). If time interval $[t_s, t_f]$ contains n cycles of periodic pulses, the electrical energy applied to a motor will be

$$W_e = \sum_{i=1}^{n} \left\{ \int_{t_i}^{t_i+t_{on}} (V_a \bullet I_a) \bullet dt \right\}. \tag{5.32}$$

From Fig. 5.27 and Eq. 5.32, we can see that switches Q_+ and Q_- will carry current I_a only when the control-logic signal applied to Q_- is at the logic "high". Clearly, the PWM method has much less power dissipation caused by the heat deterioration of electronic components. Thus, a PWM power drive is more energy efficient than a linear power drive.

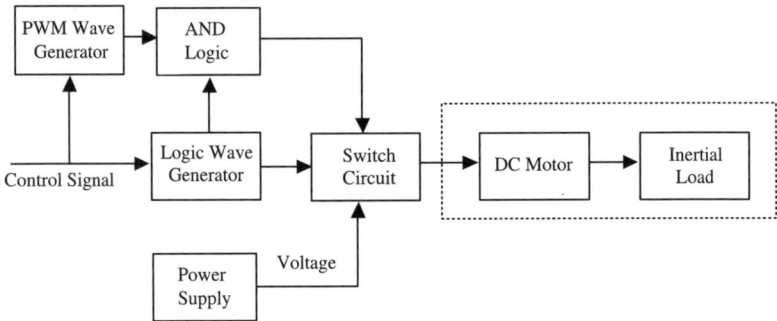

Fig. 5.28   Schematic illustration of a PWM power drive together with power switch.

Fig. 5.28 shows a schematic illustration of a PWM power drive together with a power switch. Since the control signal is not directly converted into any voltage or current signal, the PWM method falls into the category of indirect power amplification.

5.3.3 *Digital Motion Controllers*

Refer to Fig. 5.15. The most important control element inside a robot's motion control system is the digital motion controller. A digital motion-controller is a decision-making device and should be treated as part of a robot's information system.

In fact, a robot's information system, the details of which will be studied in the next chapter, is like a human brain. It primarily plays five important roles:

- Computation:
  A typical computational load is the execution of a control algorithm in order to determine the control signal from the error signal inside a robot's closed-loop control system.
- Decision-making:
  A typical example is the determination of a control action exerted by an actuator (i.e. an action device) upon a joint. A control action can be determined by a simple computation or a complicated decision-making process.
- Coordination:
  Since a robot has multiple degrees of freedom, one obvious coordination requirement is how to synchronize the executed motions at a robot's multiple joints.
- Communication:
  At the lowest level, a digital computer can interface with the outside world through its input/output subsystems (I/O, for short). For example, the interface between a digital motion controller and a sensing device requires either a serial or a parallel communication channel.
- Cognition:
  For high-level control or decision-making, it is necessary for a humanoid robot to have the ability to incrementally build its own internal representations of the real world. In this way, a humanoid robot will be able to interact with humans through the conceptual symbols of a natural language. The cognitive ability of an artificial life or system is, intrinsically, computation-dependent.

The neural system of a human brain is a highly parallel neural architecture with about a 100 billion neurons. Obviously, the computing system of a humanoid robot must also be a parallel or distributed system with multiple central-processing units (CPUs). Due to the complexity of a hu-

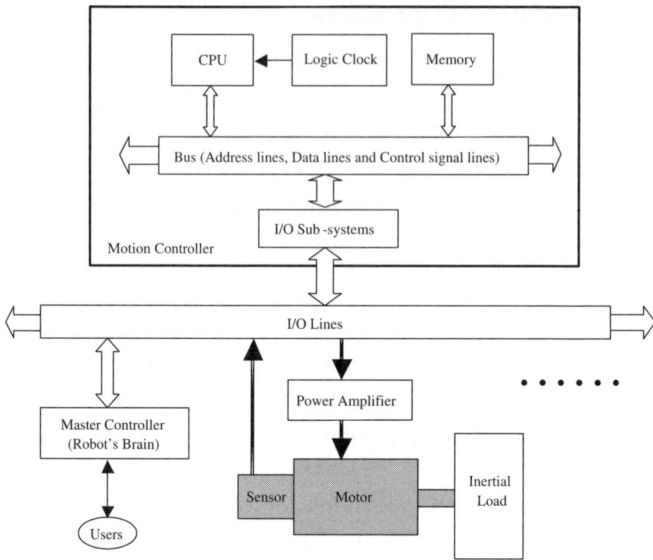

Fig. 5.29   Schematic illustration of a digital motion controller together with a robot's master controller and a motor.

manoid robot (a large number of degrees of freedom), it is impossible to use a single super-computer to handle all the computation, decision-making, coordination, communication, and cognition required. A cheap and affordable solution is to deploy an embedded and distributed computing system inside a humanoid robot. Fig. 5.29 illustrates a distributed computing system which includes a robot's master controller (at least one) and a digital motion controller.

The basic building blocks of a digital computer include: a) CPU (Central Processing Unit), b) memory (e.g. registers, RAM, ROM, EPROM), c) I/O subsystems and d) bus (address lines, data lines, and control signals/strobe lines). All computation-related tasks are executed by the CPU. All these tasks must be coded in the form of programs and stored inside the memory.

It is important to note that the digital nature of a computer is dictated by the cycle-by-cycle operation of the CPU. In fact, all the operations of the CPU are synchronized by the regular pulse waveform of a logic clock. The frequency of the logic clock primarily determines a digital computer's speed or computational power.

Consider the case of motion control of a single actuator shown in

Fig. 5.29. The digital motion controller will read the sensor's data through a I/O subsystem at discrete time-instants. Based on the error signal, which is normally the difference between the desired motion and the actual motion measured by the sensing device, a control signal is calculated by a control program executed on the digital motion controller. The control signal is then sent to the power amplifier through an I/O subsystem at discrete time-instants as well. The output of the power amplifier subsequently drives the actuator to accomplish the control action. If the closed-loop motion control system is stable, then the control action will make the actual motion of the actuator converge to the desired motion.

Refer to Eq. 5.5. If a PID control law is used to determine control signal V_c, the control program inside a digital motion controller simply implements the following recursive algorithm:

$$V_c(k) = V_c(k-1) + a_0 \bullet e(k) + a_1 \bullet e(k-1) + a_2 \bullet e(k-2) \qquad (5.33)$$

with

$$\begin{cases} a_0 = k_p \bullet \left(1 + \frac{T}{T_i} + \frac{T_d}{T}\right) \\[2mm] a_1 = -k_p \bullet \left(1 + \frac{2T_d}{T}\right) \\[2mm] a_2 = k_p \bullet \frac{T_d}{T} \end{cases}$$

where k_p is the proportional control gain of the PID control law, T_i the integral time-constant, and T_d the derivative time-constant. These parameters can be experimentally tuned if the system dynamics are not exactly known. Parameter T, in the above equations, refers to the sampling interval between two consecutive readings of the sensor's data. Alternatively, T also refers to the control action interval between two consecutive outputs of control signal V_c to the actuator.

5.4 Sensing Elements

A human sensory system can be divided into five distinct subsystems, namely: visual, auditory, kinesthetic, gustatory and olfactory. These sensory subsystems provide input to the brain, so it can not only make decisions but also build personalized internal representations of the external real world. In order to make a humanoid robot demonstrate human-like behaviors, its sensory system should be as complete and efficient as a human's. At the present, however, the robot's sensory system is far from comparable.

We hope, however, that the emergence of the humanoid robot and MEMS (i.e. Micro-Electro-Mechanical System) will inspire the development of smart sensors. One area for improvement is the optical encoder.

It is a well-known fact that almost all industrial robots use the incremental optical-encoders for position and velocity sensing. The primary reason behind this choice is that incremental optical encoders cost 1/10 the price of absolute optical-encoders. As a consequence, the robots using incremental optical-encoders must find the reference positions of the encoders before they can function properly. This is done by a process called *homing*.

The homing procedure involves unpredictable movement done by a robot. Thus, it is not at all desirable for a humanoid robot to do homing because this presents a potential danger outside the industrial environment. Society will not accept this potentially-dangerous behavior.

On the positive side, as the result of research on human computer interaction (HCI), speech recognition is a relatively mature technology. In addition, continuous efforts to develop a robust speech-recognition system will certainly benefit the auditory-sensory subsystem of future humanoid robots.

In this book, we will study the visual-sensory system of robots in a separate chapter. Here, we limit the discussion to the sensors which are closely related to a robot's motion execution. These sensors fall under the category of the kinesthetic sensory subsystem.

5.4.1 *Generic Sensing Principle*

For a general sensing instrument, the requirement of a sensor is that it somehow converts the variables of a system into corresponding signals suitable for display. But for a robot with automatic-feedback control using a digital computer, the requirement of a sensor is slightly different. The purpose of sensors is to convert the variables of a physical system under control into corresponding digital signals.

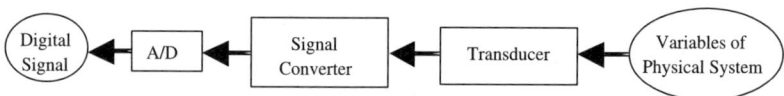

Fig. 5.30   Generic sensing procedure in an automatic feedback control system.

Fig. 5.30 shows the generic sensing procedure. In robotics, the variables under consideration may include: position, velocity, acceleration, force, torque, sonar, and lighting. These physical variables can be converted into electrical signals such as voltage, current, resistance, capacitance and inductance by using a device called a *transducer*. An ideal transducer is a device which converts a signal from one domain into another without removing any energy from the system under measurement. For example, the speed sensor of a motor should not consume any mechanical energy from the motor itself.

In physics, for the purpose of sensing, many working principles can be applied to the development of transducers. Depending on working principles, the sensors can be classified into the following typical categories:

- Electric Sensor:
  This refers to a sensor which converts a physical variable directly into a corresponding electrical signal. A typical example is the potentiometer, which converts a position variable into a corresponding resistance value. A strain gauge sensor for the measurement of force and torque also falls under this category.
- Electro-magnetic Sensor:
  This refers to a sensor which makes use of the magnetic field to do the signal conversion. A typical example is the *tacho-meter*. A tacho-meter makes use of a rotational conducting loop to pick up the current/voltage, which is proportional to the rotational velocity of the conducting loop in a magnetic field.
- Optical Sensor:
  This refers to a sensor which makes use of light as a medium to do the signal conversion. Optical encoders fall under this category.

As shown in Fig. 5.30, it is necessary to use a signal converter if the output from a transducer is not in the form of voltage. Therefore, the role of a signal converter is to convert the output of a transducer into corresponding voltage which will then be sent to an analog-to-digital (A/D) converter.

If we denote $y(t)$ the input to a sensor and $s(t)$ its output, the key features of a sensor typically include:

- Accuracy:
  This refers to the accurate relationship between the sensor's input and output. Ideally, we would like to have $s(t) = k_s \bullet y(t)$ with $k_s = 1$.

- Precision:
  This refers to the statistical distribution of output $s(t)$ when input $y(t)$ is set at a fixed value. The smaller the statistical distribution, the more precise it will be.
- Resolution or Sensitivity:
  This refers to the smallest amount of variation in input which will trigger variation in output from the sensor. If $s(t) = k_s \bullet y(t)$, we have $\frac{\triangle s(t)}{\triangle y(t)} = k_s$. The higher the value k_s, the better the resolution.
- Operating Range:
  This refers to the range within which input $y(t)$ is allowed to vary.
- Response Time:
  This refers to the dynamics of output $s(t)$ in response to input $y(t)$. Ideally, we would like to have a sensor which does nt exhibit any transient response. In other words, $s(t) = k_s \bullet y(t)$ with k_s being a constant gain.
- Reliability:
  Reliability is equal to the inverse of the failure rate of a sensor. Obviously, the smaller the failure rate, the higher the reliability.

5.4.2   *Safety Sensors*

In the future, humanoid robots will not only operate in industry but also co-exist with humans in society. Consequently, safety is a crucial issue as society will never accept a mechanized creature that is unpredictably dangerous. Ideally, a humanoid robot should not only obey humans but also be friendly towards them.

Here, it is worthy to cite Asimov's laws published in 1950:

- Law 1: A robot should not injure a human being.
- Law 2: A robot should obey human commands as long as these commands are not in conflict with Law 1.
- Law 3: A robot should protect itself as long as such protection is not in conflict with Law 1.

In order to ensure people's safety, all of a robot's actions should only be enabled if they are judged to be safe to human beings and their environment. Thus, a humanoid robot should have an emergency stop-sensor, a proximity sensor, and visual perception system. These sensors are necessary to ensure the protection of human beings, the environment, and the robot.

5.4.2.1 *Emergency Stop Sensors*

In the event of an emergency, humans should be able to interrupt the actions of a humanoid robot. This can be easily achieved by providing one or more emergency stop-buttons which serve as simple "on-off" sensors. The output signal (the "on" signal) from an emergency stop sensor should have the highest priority to halt all the power drives inside a humanoid robot. Fig. 5.31 shows a prototype of a humanoid robot with two emergency stop-sensors: one on the front part of the body and one at the back.

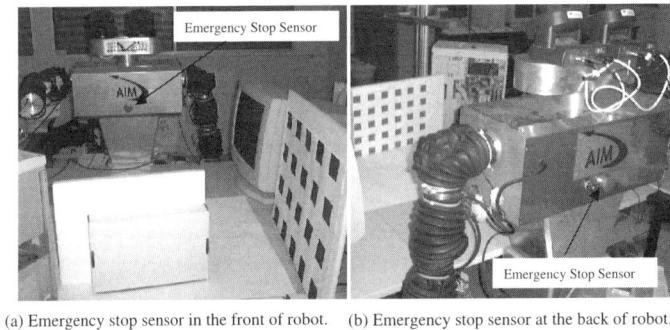

(a) Emergency stop sensor in the front of robot.   (b) Emergency stop sensor at the back of robot.

Fig. 5.31   Example of two emergency stop-sensors on a prototype of a humanoid robot.

5.4.2.2 *Proximity Sensing Systems*

An emergency-stop sensor is considered a passive safety sensor because it is manually activated by a human or other equipment. Alternatively, one may propose an active sensing system for safety. A typical example of an active safety sensor is the *proximity system* such as ultrasonic or infra-red sensors.

An ultrasonic sensor is a low-cost device which is able to measure the distance to an object within a typical range of 0.3m to 10m. Fig. 5.32a shows an example of ultrasonic sensor (i.e. the model Polaroid 6500) which is composed of two parts: a) the buzzer and b) the drive circuit.

An ultrasonic sensor is an active-sensing device because its buzzer not only transmits sonar waves but also receives the echoes. The working principle behind an ultrasonic sensor is based on the measurement of the flight time of the sonar wave because the speed of a sonar wave is constant (about 344m/s).

(a) Buzzer and the drive circuit.   (b) Single sensor with a micro -controller.

(c) A ring of sensors with a micro -controller.

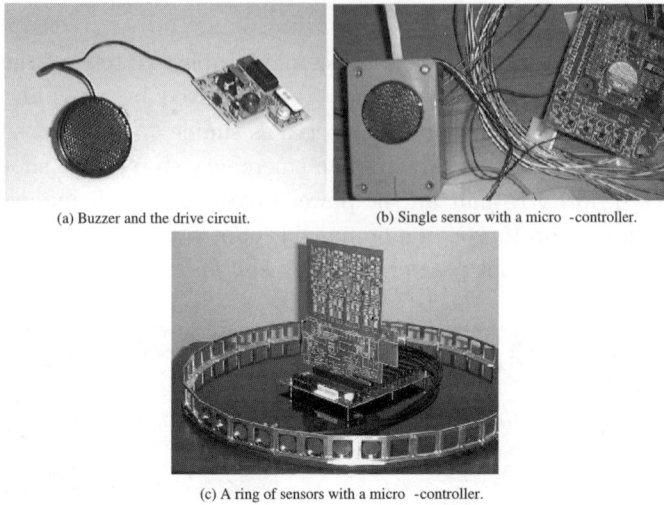

Fig. 5.32   Example of proximity systems based on ultrasonic sensor(s): a) the sensor's buzzer and drive circuit, b) a proximity system with a single sensor and c) a proximity system with a ring of sensors.

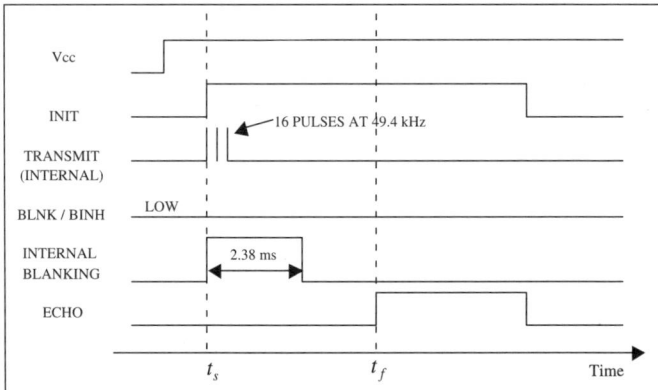

Fig. 5.33   Signal diagrams for the operation of an ultrasonic sensor.

For an ultrasonic sensor, there are two basic modes of operation: a) the single-echo mode and b) the multiple-echo mode. Fig. 5.33 shows the signal diagrams corresponding to the single-echo mode.

When an ultrasonic sensor (e.g. the model Polaroid 6500) is turned "on" by applying voltage to line V_{cc}, a minimum of 5ms must elapse before the

INIT signal is driven to the logic "high" in order to trigger the transmission of a sonar wave at time-instant t_s. When the INIT signal is driven to the logic "high", a series of 16 pulses at a frequency of 49.4kHz will excite the buzzer (the transducer), which generates and transmits a sonar wave. Since any mechanical device has a ringing effect due to vibration, an internal blanking signal must be driven to the logic "high" for about 2.38ms in order to avoid the detection of the buzzer's ringing.

After the blanking period, the buzzer is ready to serve as a receiver and detect an echo. Once an echo is detected at time-instant t_f, the echo signal will be instantly amplified by the drive circuit and the logic "high" signal will be issued to the ECHO signal line, which is an output line of the ultrasonic sensor's drive circuit. By driving the INIT signal line to the logic "low" again, this will reset the ultrasonic sensor's drive circuit and the ECHO line will be driven to the logic "low". Then, the ultrasonic sensor is ready to start a new sensing cycle.

Since the sonar wave travels at a constant speed in air, the distance to a detected object is linearly proportional to the time of flight $t_f - t_s$. This time of flight can be easily measured by the programmable timer I/O system of a micro-controller (e.g. M68HC11).

Depending on the intended application, one can easily configure a proximity system with one ultrasonic sensor or a ring of ultrasonic sensors as shown in Fig. 5.32b and Fig. 5.32c.

5.4.2.3   *Detection of a Human's Presence*

There are two problems with the active-sensing technique. First of all, a sonar wave can be reflected in different directions. This will result in multiple echoes, which in turn, will cause false detection of the presence of objects. Secondly, the response time is proportional to the spatial coverage and range of the active sensor. The larger the spatial coverage and range, the slower the response time will be.

A human being primarily relies on the visual sensory system to gather information about his/her surroundings. Some notable advantages to a visual-sensory system include the rich information, high resolution, fast response (parallel and spontaneous), and lack of interference with environment (passive sensing). Just as a visual sensory/perception system is important for a human being, a visual sensory/perception system is crucial in order for a humanoid robot to accomplish activities autonomously. We will study the visual perception system of a humanoid robot in Chapter 8.

Here, we show an example of detecting a human's presence. The algorithm is based on a probabilistic RCE neural network. As shown in Fig. 5.34, the identification of pixels having the appearance of skin color is a simple process. It allows the robot to make a quick assessment of whether or not a human being is present. Obviously, a more elaborate approach is for the robot to be able to detect the face and hand gestures.

(a) Original (color) image.   (b) Blacked pixels of skin color.

Fig. 5.34   Detection of the presence of a human being through the identification of pixels having the appearance of skin color. (NOTE: The original image is in color).

5.4.3   *Motion Sensors*

A humanoid robot's mechanism is highly complex with many degrees of freedom. As each degree of freedom requires at least one motion sensor to provide sensory feedback, a humanoid robot will need a large number of motion sensors. And the performance of these motion sensors is crucial to the proper control of motions executed by a humanoid robot.

As we mentioned before, motion sensors inside an industrial robot are typically the incremental-encoder type. Consequently, an industrial robot must perform a homing procedure before it is ready to work. For a humanoid robot, however, this homing procedure is not desirable because it involves unpredictable movements. Obviously, the absolute-motion sensor is the most appropriate choice for a humanoid robot. With the absolute-motion senor, the homing procedure is unnecessary. However, for the purpose of comparison, let us study the working principles behind both absolute and incremental motion sensors.

Regardless of the type of motion sensor, one must be clear that the purpose of a motion sensor is to provide motion feedback by measuring the motion parameters at a joint inside a robot's mechanism. As a result, there are two concerns:

- Where do we place a motion sensor?
- How do we measure the motion parameters?

5.4.3.1 *Placement of Motion Sensors*

For the sake of clarity, let us consider the motion sensor for rotary motors. In Chapter 4, we studied the coupling of a dynamic pair with a kinematic pair. We know that a dynamic pair is normally coupled with a kinematic pair through a device called a *speed reducer* or *torque amplifier*. This means that the motion of a motor is not exactly equal to the motion of the corresponding joint driven by the motor. Now, the question is: Should a motion sensor be placed before the speed reducer or after the speed reducer?

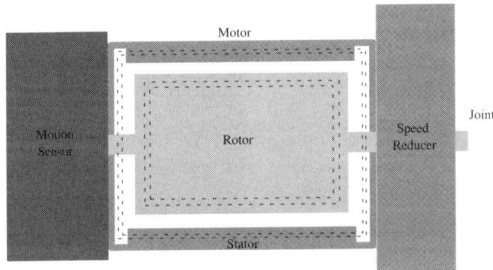

Fig. 5.35   Sectional view showing the placement of a motion sensor before the speed reducer.

The answer depends on the type of sensor and the application requirement. Ideally, we would prefer to place the motion sensor before the speed reducer, as shown in Fig. 5.35. In this way, we can simply measure the motion parameters of a motor. The output from a motion sensor is then divided by the reduction ratio of the speed reducer in order to obtain the motion parameters at the corresponding joint. Let us denote q_i the angular position of joint i, q_{mi} the angular position of corresponding motor i, and k_{ri} the reduction ratio of corresponding speed reducer i. If a motion sensor is placed before speed reducer i, the output from the sensor is q_{mi}. But, the required output for motion feedback is q_i, which can be easily determined as follows:

$$q_i = \frac{q_{mi}}{k_{ri}}. \tag{5.34}$$

The differentiation of Eq. 5.34 yields

$$dq_i = \frac{1}{k_{ri}} \bullet dq_{mi}.$$

This simply means that the error caused by the sensor's output is reduced by a factor of k_{ri}. Clearly, the advantage of placing a motion sensor before a speed reducer is to obtain accurate motion feedback even if the motion sensor itself is not absolutely precise. If the working principle behind a motion sensor requires the physical contact between one fixed element and one moving element, the placement of a motion sensor before the speed reducer is not suitable because angular velocity $\dot{q}_{mi}$ is k_{ri} times higher than angular velocity $\dot{q}_i$. The higher the velocity, the more severe the wear-and-tear due to contact friction.

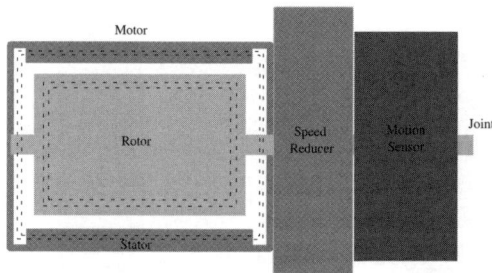

Fig. 5.36   Sectional view showing the placement of a motion sensor after the speed reducer.

Alternatively, one may choose to place a motion sensor after the speed reducer, as shown in Fig. 5.36. In this way, we can directly measure the motion parameters at a joint. In other words, the output from a motion sensor directly corresponds to the motion parameter(s) of a joint.

5.4.3.2   *Potentiometers*

All of us should be familiar with the potentiometer. When we adjust the frequency of a radio, we rotate a rotary disk. And this rotary disk drives a potentiometer to alter its output resistance value inside the internal circuitry of the radio. Physically, the potentiometer of a radio is a motion sensor because the input is the angular position of a rotary disk and the output is its resistance value.

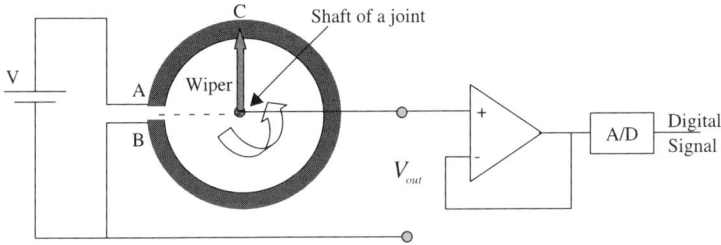

Fig. 5.37   Illustration of the sensing elements built on top of a rotary potentiometer.

Fig. 5.37 shows an example of a rotary potentiometer and its associated sensing elements (i.e. resistance-to-voltage signal converter and A/D converter). A potentiometer is composed of two elements: one fixed circular resistor and one movable wiper. As we discussed above, a potentiometer should be placed after a speed reducer because it is a contact-based sensing device. Therefore, the axis of a potentiometer should be coupled with the shaft of a joint. Assume that point B is the reference position of potentiometer i coupled with joint i. Let us denote R the resistance value of arc $\widehat{AB}$, r the resistance value of arc $\widehat{CB}$, and q_i the angular position of the wiper with respect to reference position B. If the circular resistor is uniform and the gap between points A and B is negligible, we have

$$\frac{r}{R} = \frac{\widehat{AB}}{\widehat{CB}} = \frac{q_i}{360^0}. \tag{5.35}$$

Assume that input voltage V is applied to the potentiometer. If the output voltage is picked up at the wiper, we have

$$\frac{V_{out}}{V} = \frac{r}{R} = \frac{q_i}{360^0}$$

or

$$V_{out} = \frac{V}{360^0} \bullet q_i. \tag{5.36}$$

Clearly, the measurement of output voltage V_{out} allows us to exactly determine angular position q_i.

5.4.3.3 *Absolute Optical Encoders*

A contact-based motion sensor has some obvious drawbacks: It has short lifetime due to the wear-and-tear. And, it has low accuracy of feedback output. As we mentioned above, the ideal place to put a motion sensor is somewhere before the speed reducer. However, this is only possible for non-contact motion sensors. Accordingly, it is desirable to develop non-contact motion sensors. Nowadays, one of the most widely-used non-contact motion sensors is called the *optical encoder.*

Working Principle Behind Optical Encoders

The basic working principle behind an optical encoder is the conversion of a motion into a series of light pulses, which are subsequently converted into a corresponding series of electrical pulses (digital logic signals).

In order to generate light pulses, it is necessary to have a light emitting device. One good light emitting device is the light-emitting diode, or LED, for short. A diode is a nonlinear electric component. It has two terminals: a) the anode and b) the cathode. The current can effortlessly flow from the anode to the cathode of a diode when the forward voltage drop is within the range of 0.5V to 0.8V. No current can flow in reverse from the cathode to the anode of a diode. A light-emitting diode continuously emits the light rays as long as there is a forward current which flows from the anode to the cathode. Under normal working conditions, current flowing through an LED is within the range of 5mA to 20mA. Therefore, it is necessary to have a current-limit resistor in order to protect an LED.

After generating a series of light pulses, it is necessary to convert the light pulses into electrical pulses. This is usually done with a photo-detection device such as a photo-transistor (or photo-diode). A photo-transistor is a special transistor which will produce a current from the collector to the emitter when the base is exposed to a light source (or excited by a light pulse). When combined with an op-amp circuit, an electrical pulse can be easily produced when there is a light pulse present at the base of a photo-transistor.

Fig. 5.38 illustrates the working principle behind an optical encoder. An optical encoder consists of: a) a light-emitting device with an optical lens to produce focused parallel light rays, b) a light-detecting device with a logic-signal converter and c) a moving slotted (metal) sheet which can be interfaced with a rotary motor. From Fig. 5.38, it is easy to see that the logic-signal output will be high (i.e. "1") if light from the LED can

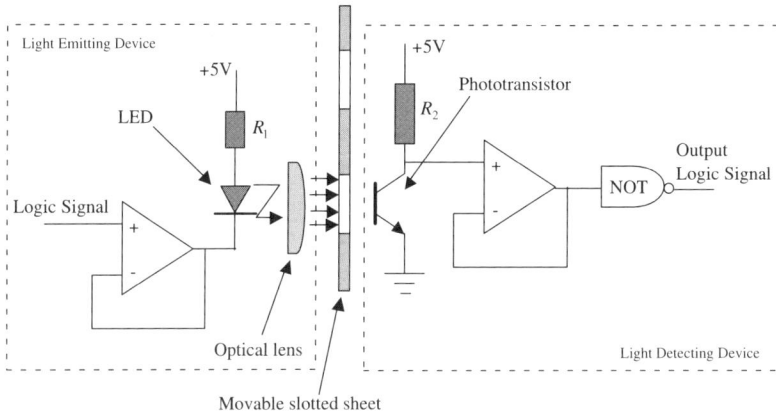

Fig. 5.38   Illustration of the working principle behind an optical encoder.

pass through a slot which is aligned with the photo-transistor. Otherwise, if light is blocked by the opaque (metal) sheet, the logic-signal output will be low ("0").

Working Principle Behind Absolute Optical Encoders

By definition, an absolute optical encoder is a device which is able to pin-point the absolute angular positions of a rotary motor. As output from an optical encoder are logic signals, an optical encoder is intrinsically a digital device.

For a digital system, one logic-signal line corresponds to one digital bit. And a digital bit only carries one piece of information (i.e. either "on" or "off"). If we put all the bits of an absolute optical encoder together to form a logic state variable, the number of logic states that an absolute optical encoder has will determine the number of absolute angular positions which can be measured.

For example, if we would like to measure a full rotation of 360^0 at an accuracy of 0.1^0, an absolute optical encoder has to have at least 3600 logic states as output. In other words, an absolute optical encoder should have at least 3600 commutating logic states. As one bit only allows the commutation between two logic states (i.e. "on" or "off"), 3600 logic states require at least 12 bits (NOTE: 11 bits correspond to 2048 logic states, and 12 bits will have 4096 logic states).

Therefore, one important design parameter for an absolute optical encoder is the number of commutating logic states. The specification of this parameter depends on where the absolute encoder is placed on the motor, for example:

- If the absolute optical encoder is placed before the speed reducer of a motor, the accuracy of an absolute optical encoder will be increased by k_r times, where k_r is the reduction ratio of the speed reducer.
- If the absolute optical encoder is placed after the speed reducer of a motor (i.e. to measure the angular position of a joint), the number of commutating logic states has to be specified directly according to the desired accuracy of the measured angle at a joint.

Once the number of commutating logic states has been determined, the next issue is how to engrave these commutating logic states onto the movable slotted (metal) sheet, which can be in the shape of either a disk (the default option) or a cylindrical tube. With the advent and maturity of MEMS technology, the option of engraving the commutating logic states onto a cylindrical tube sounds promising.

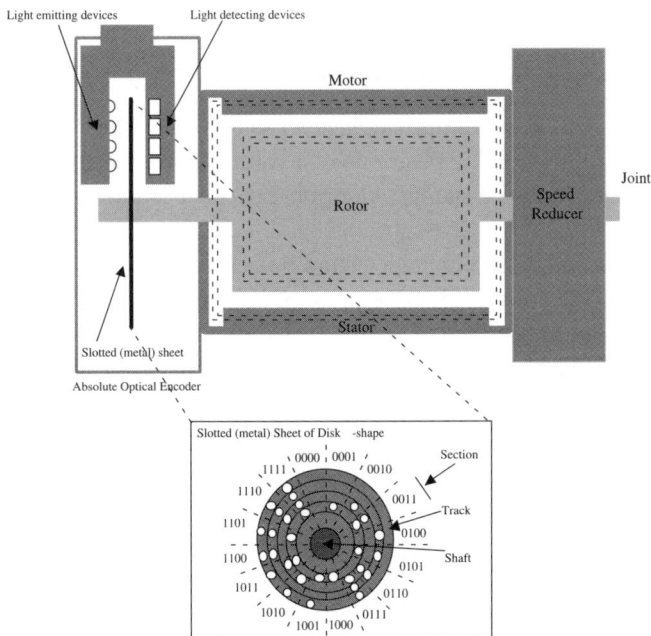

Fig. 5.39   Illustration of a conceptual design of an absolute optical encoder with 16 logic states being engraved on a slotted, disk-shape sheet of metal.

For the sake of clarity, we illustrate the case of designing an absolute optical encoder with 16 logic states (4 bits). Fig. 5.39 shows a conceptual design of an absolute optical encoder with 16 logic states being engraving on a slotted, disk-shape sheet of metal. The engraved disk of an absolute optical encoder is normally fixed on the shaft of a motor. Since there are four logic bits, it is necessary to have four pairs of light-emitting and light-detecting devices in order to read out the logic states.

From Fig. 5.39, we can see that the slotted, disk-shape sheet is divided into four tracks (corresponding to the number of logic bits in an absolute optical encoder) and 16 sections (corresponding to the number of logic states in an absolute encoder). When the disk rotates with the motor, the logic states will commutate either from 0000 to 1111, or from 1111 to 0000. As a result, an absolute optical encoder allows us to obtain not only the absolute angular positions but also the direction of motion.

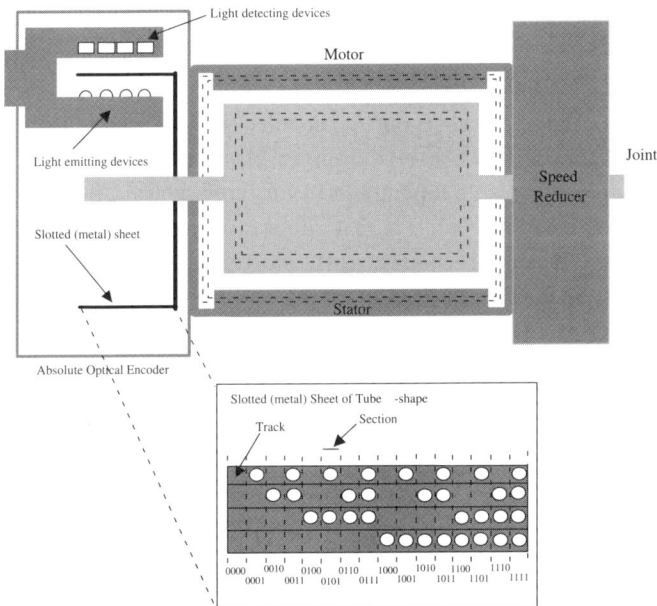

Fig. 5.40   Illustration of a conceptual design of an absolute optical encoder with 16 logic states being engraved on a slotted, tube-shape sheet of metal.

When tracks are engraved on a disk, their dimensions are not uniform. The inner track will have the smallest circumference. Therefore, the density of the sections (the number of sections per track) is limited by the inner track. The nonuniformity of the dimensions of the tracks also implies that the process of manufacturing slotted, disk-shape sheets of metal may be costly. However, this drawback can easily be overcome by engraving the logic states on a tube-shape sheet of metal, as shown in Fig. 5.40.

When a tube is extended onto a planar surface, it becomes a rectangular sheet. All the tracks and sections engraved on it will have a uniform dimension and shape. Hence, it poses no challenge to manufacture the slotted, tube-shape sheets of metal. Furthermore, it is easy to increase the number of tracks engraved on a tube-shape sheet for better accuracy.

In summary, an absolute optical encoder can either be placed before the speed reducer or after the speed reducer of a motor. The output from an absolute optical encoder is the value of a logic state which indicates the absolute angular position. If the rotated angle exceeds 360^0, it can be easily detected by counting the number of commutations from the smallest logic state value to the highest logic state value, or vice-versa.

5.4.3.4 *Incremental Optical Encoders*

Due to the concern about cost, one of the most widely-used optical encoders for industrial robots and other equipment like printers is the *incremental optical encoder*. A notable advantage to an incremental optical encoder is its compact size and low cost. Fig. 5.41 illustrates a dismantled incremental optical encoder. It has three parts: a) the slotted, disk-shape sheet of metal, b) the integrated portion of light-emitting and light-detecting devices, and c) the compact casing.

Working Principle Behind Incremental Optical Encoders

The idea behind the design solution of an incremental optical encoder is to reduce the number of tracks on the disk-shape metal sheet to one. Instead of engraving the logic states onto a disk-shape (metal) sheet, as in the case of an absolute optical encoder, one simply engraves a periodic slotted pattern on a single track, as shown in Fig. 5.42. Refer to Fig. 5.42. The slotted metal disk has the periodic slotted pattern with eight cycles engraved on a single track.

In principle, a single track of periodic slots only requires one pair of light-emitting and light-detecting devices. In this way, a series of electrical pulses will be generated when the slotted disk of an incremental optical encoder rotates with a motor. In order to measure angular displacements,

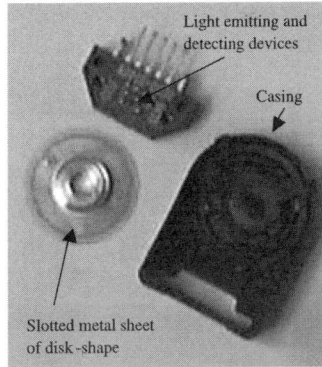

Fig. 5.41   Picture of a dismantled incremental optical encoder.

one simply counts the number of pulses which is linearly proportional to angular displacements.

For the example shown in Fig. 5.42, eight (8) pulses indicate an angular displacement of 360^0 with an accuracy of $\pm 45^0$ (i.e. $\frac{360^0}{8}$). Obviously, the larger the number of cycles of the slotted pattern, the more accurate an incremental optical encoder. If n_c denotes the number of cycles of the slotted pattern on the disk, the accuracy of an incremental optical encoder, denoted by $\triangle \theta$, will be

$$\triangle \theta = \frac{360^0}{n_c}. \tag{5.37}$$

A larger value of n_c indicates an increase in difficulty or cost for the manufacturing of the slotted disks. In practice, the accuracy of an incremental optical encoder is in the range of 1^0 to 0.1^0. Consequently, an incremental optical encoder must be placed before the speed reducer of a motor.

Because of its simple structure, the incremental optical encoder with a single track is only able to measure angular displacements without knowing the direction of motion. But, what we need for the motion feedback in a closed-loop control system is the measurement of an angular position or velocity in the absolute sense.

In order to determine the absolute angular position from the measurement of angular displacements, one must address these two issues:

- How do we register a reference position or home position?
- How do we determine the direction of motion?

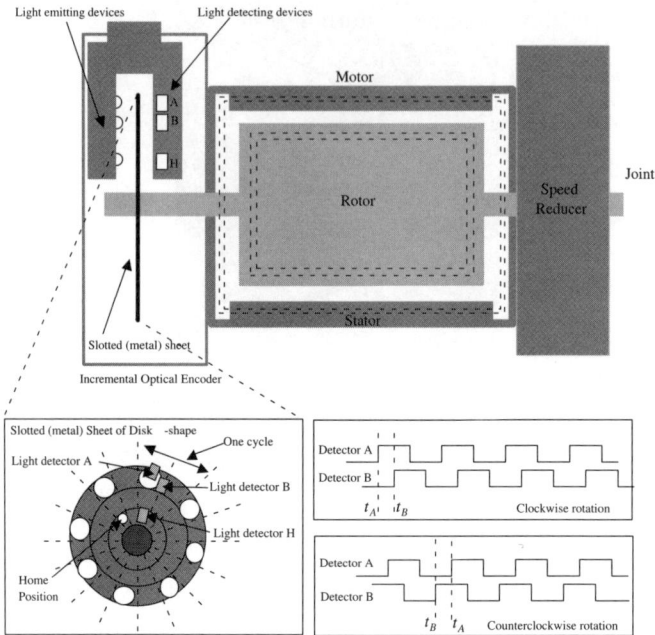

Fig. 5.42   Illustration of a conceptual design of an incremental optical encoder.

Determining the Home Position

The first question is easy to answer. A common solution is to create a reference slot on the disk. This reference slot indicates a reference, or home position, as shown in Fig. 5.42.

With an added reference slot, there must be a pair of light-emitting and light-detecting devices to locate this home position. In Fig. 5.42, detector H is to identify the home position. As it is necessary to identify a home position, a robot or any machine which uses the incremental optical encoders must undertake a homing procedure before it is ready for actions or operations.

A homing procedure for one motor normally involves two rotations at most. A motor will first rotate in one direction (either clockwise or counterclockwise) and its range of movement will be no more than 180^0. If no home position is detected in the first rotation, the motor will rotate in the opposite direction until a home position is found. As we mentioned earlier, this homing procedure is not desirable for a humanoid robot to perform because the movement of a robot during the homing is unpredictable.

Determining the Direction of Motion

The second issue is how to determine the direction of motion. This issue is slightly more complicated. A single light-detector (detector A) normally produces one series of pulses when the slotted disk of an incremental optical encoder rotates with a motor. Obviously, it is impossible to determine the direction of motion from a single source of pulses. In order to solve this issue, a common solution is to introduce a second light-detector (detector B).

Since the slotted pattern on a disk is composed of periodic cycles, the output from a light-detector will be a series of periodic pulses when the disk is rotating. If we use the term *phase* in signal processing, a cycle will correspond to a full phase of 360^0. If the position of detector B has no phase-shift with respect to the position of detector A, the logic signal from detector B will be identical to the logic signal from detector A. Similarly, if the position of detector B has a phase-shift of 180^0 with respect to the position of detector A, the logic signal from detector B will be the complement of the logic signal from detector A. In other words, the logic signals from these two detectors carry the same information (they are logically equal). However, if the phase-shift between the two detectors is not equal to 0^0 (or the multiple of 360^0) nor 180^0 (or the multiple of 180^0), the logic signals from these two detectors will not be logically equal. In practice, a common design solution is to choose a phase-shift of 90^0 between the two detectors, as shown in Fig. 5.42.

For the example shown in Fig. 5.42, detector B has a phase-shift (phase-lag) of 90^0 behind detector A in a clockwise direction. When the slotted disk rotates in a clockwise direction, the cycle of the waveform from detector A will have the phase-lead of 90^0 with respect to the cycle of the waveform from detector B. By default, a new cycle of a periodic logic signal starts with the rising edge which marks the transition from the logic "low" to the logic "high". For any given cycle of the slotted pattern on a rotating disk, let t_A denote the starting time of a generated electrical pulse from detector A, and t_B the starting time of a generated electrical pulse from detector B. A phase-lead of the waveform from detector A with respect to the waveform from detector B simply means

$$t_A < t_B. \tag{5.38}$$

Alternatively, if we choose detector B as a reference, detector A has a phase-lag of 90^0 behind detector B in a counterclockwise direction. Accordingly, if the disk rotates in a counterclockwise direction, the cycle of the waveform from detector B will have a phase-lead of 90^0 with respect to the cycle of the waveform from detector A. In other words, we have

$$t_B < t_A. \tag{5.39}$$

Clearly, the direction of motion is easily determined by measuring t_A and t_B.

In summary, an incremental optical encoder is normally placed before the speed reducer of a motor. This is because the accuracy of an incremental optical encoder itself is not very high due to the limitation on the number of cycles of the slotted pattern. The output from an incremental optical encoder normally includes four signal lines: one for the pulse waveform from detector A, one for the pulse waveform from detector B, one for the pulse waveform from detector H, and one for the common ground. A digital motion controller must interpret the received logic signals from these four signal lines in order to determine rotated angle or velocity, and direction of motion etc. Finally, an incremental optical encoder will impose a homing procedure on all robots or other equipment like printers which use it for motion feedback.

5.4.4   *Force/Torque Sensors*

So far, we studied ways to alter the direction of motion, to regulate the electrical energy applied to a motor for the purpose of varying its velocity, and to measure the output motion of a motor either before the speed reducer or after the speed reducer for the purpose of motion feedback. Practically, we are ready to close the feedback-control loop and concentrate on the design of control algorithms. However, in order for the robot to perform certain tasks, the motion executed by a limb (e.g. arm-hand or leg-foot) may involve direct contact between its end-effector and the environment or workpiece.

For example, when a humanoid robot writes on paper with a pen, its hand will experience contact force between the paper and the pen. In other words, for motion control, it may be necessary to consider the force or torque feedback as well in order to produce the desired interaction force or torque between a robot and its environment.

5.4.4.1 *Indirect Measurement*

In Chapter 4, we studied robot statics. When a robot's limb is at rest, external force/torque ξ_e exerted on its end-effector's frame is related to torque vector τ applied to the actuators of the robot's limb in the following way:

$$\tau = J^t \bullet \xi_e$$

where J is the Jacobian matrix of the robot's limb.

Jacobian matrix J is easy to evaluate if the angular positions of the joints are known. Now, one may question whether it is possible to determine external force/torque ξ_e exerted on the end-effector's frame of a limb from measured torque vector τ at the limb's actuators. The answer to this question depends on the following cases:

- If the robot is operating in outer space where there is no gravitational force, it is possible to directly determine ξ_e from measured torque vector τ.
- However, if the robot is operating on Earth, it is not easy to determine ξ_e based on the measurement of τ because a portion of τ is spent to balance gravitational forces acting on the links of a limb.

5.4.4.2 *Direct Measurement*

As most robots operate on Earth where there is gravitational force, it is necessary to develop a sensor which is able to measure the force/torque externally exerted on the end-effector (hand or foot) of a robot's limb.

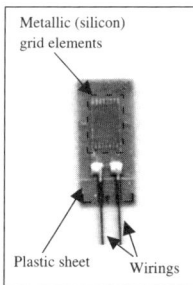

(a) Enlarged view of strain gauge.   (b) Open-up view of force/torque sensor.

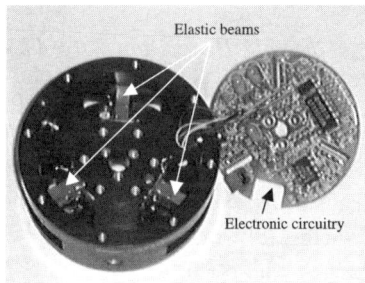

Fig. 5.43   Example of a force/torque sensor: a) the enlarged view of a strain gauge and b) the internal view of a force/torque sensor.

In robotics, a widely-used method of measuring forces and/or torques is with a strain gauge. Fig. 5.43 shows an example of a force/torque sensor. It is also known as a *wrist sensor* because a force/torque sensor is normally mounted between the end-effector of a robot's arm and its hand (or a tool).

As shown in Fig. 5.43, a strain gauge is a metallic (e.g. silicon) grid element of foil which can come in different shapes and sizes. A strain gauge is normally attached to a plastic sheet. In order to measure an applied force or torque, an element called an *elastic beam* is used. When an applied force or torque is coupled with an elastic beam, the beam deforms in response to the stress caused by the applied force or torque. This deformation can be detected by placing one or two strain gauges onto one or two surfaces of the beam.

When a beam deforms in response to the stress caused by an applied force or torque, the strain gauges placed on it will also deform. Thus, the electrical resistances of the metallic grid elements in the strain gauges will change in a specific way, either increasing or decreasing. As a result, a strain gauge behaves like a variable resistor, the electrical resistance of which changes as a function of its deformation. Accordingly, in order to design a force/torque sensor based on strain gauges, one must address these issues:

- How do we couple the applied forces and/or torques to a structure of elastic beams?
- How do we convert (very small) variations in electrical resistances of the strain gauges to variations in corresponding voltage signals?
- How do we map the voltage signals to the measurement of the applied forces and/or torques?

Structure of Elastic Beams for Force/Torque Sensors

A sensor should not interfere with the proper operation of a system under measurement. This rule implies that a force/torque sensor should not cause any extra motion to a robot system. In other words, a force/torque sensor should structurally behave like a rigid body as long as the applied force & torque is within its working range.

In robotics, the structure of a force/torque sensor is commonly based on an elastic beam assembly. As shown in Fig. 5.44, a force/torque sensor's structure consists of three parts: a) rigid body 1 (e.g. a tube), b) rigid body 2 (e.g. a tube) and c) a set of coupling elastic beams. Rigid body 1 is coupled to rigid body 2 through the supporting beams. When body

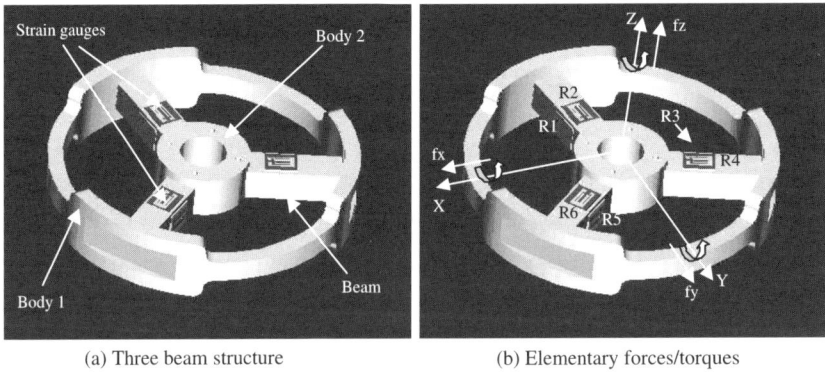

<table>
<tr><td>(a) Three beam structure</td><td>(b) Elementary forces/torques</td></tr>
</table>

Fig. 5.44 Example of a three-beam structure for strain-gauge based force & torque sensors.

2 is subject to the external forces and torques, the built-up stress between bodies 1 and 2 will cause the beams to deform. By placing some strain gauges on the surfaces of the beams, these deformations will be converted into the variations in electrical resistances of the strain gauges.

In a three-dimensional space, force/torque vector ξ_e acting on the end-effector of a robot's limb will have six elements: a) three for the elementary forces in X, Y, and Z directions, and b) three for the elementary torques about the X, Y and Z axes. If a beam is in a shape of a cube, it is able to capture the deformations in two directions if its two ends are fixed onto two rigid bodies respectively. As a result, it is necessary to have a minimum number of three beams to produce six values of output, which can almost be linearly mapped into a corresponding measurement of external force/torque vector ξ_e with six elements.

Fig. 5.44 illustrates an example of a three-beam structure for a force/torque sensor. Each beam has two strain gauges placed on its two adjacent surfaces. In total, there will be six strain gauges. Since a strain gauge behaves like a variable resistor, we will have six variable resistors denoted as R_1, R_2, R_3, R_4, R_5 and R_6.

In practice, body 1 (the outer ring or tube) of a force/torque sensor is attached to the end-effector's link of a robot's limb while body 2 (the inner ring or tube) is attached to a robot's hand, foot or tool. For the sake of convenience, we can have the end-effector's frame at the center of a force/torque sensor's beam structure. In general, when an external force & torque acts on body 2, force/torque vector ξ_e experienced by the end-effector's frame will be

$$\xi_e = (f_x, f_y, f_z, \tau_x, \tau_y, \tau_z)^t \qquad (5.40)$$

where $(f_x, f_y, f_z)^t$ is the force vector and $(\tau_x, \tau_y, \tau_z)^t$ the torque vector. This force/torque vector ξ_e will cause variations in electrical resistances of the strain gauges. That is: $\triangle R_1$, $\triangle R_2$, $\triangle R_3$, $\triangle R_4$, $\triangle R_5$ and $\triangle R_6$.

Resistance to Voltage Conversion

In electronics, a common way to convert the variation of electrical resistance to the corresponding variation of voltage is to use the *Wheatstone bridge* circuit, as shown in Fig. 5.45a.

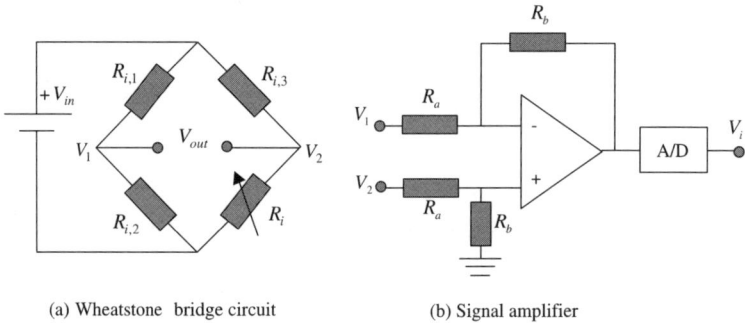

(a) Wheatstone bridge circuit   (b) Signal amplifier

Fig. 5.45   a) Wheatstone bridge circuit for measuring resistance variation and b) voltage signal amplifier.

The input voltage to the Wheatstone bridge circuit is V_{in} and the output voltage is V_{out}. And, the output voltage is equal to $V_1 - V_2$. A Wheatstone bridge circuit has two branches. Each branch has two resistors. Assume that we convert the variation of resistor R_i $(i = 1, 2, ..., 6)$ to a corresponding variation of the voltage signal. By adding another three high-quality resistors $R_{i,1}$, $R_{i,2}$ and $R_{i,3}$, we will form a Wheatstone bridge circuit, as shown in Fig. 5.45a. Ideally, these four resistors should have similar thermal characteristics in order to balance out noise caused by thermal effect.

From Fig. 5.45a, voltage V_1 is

$$V_1 = \frac{R_{i,2}}{R_{i,1} + R_{i,2}} \bullet V_{in}. \qquad (5.41)$$

Similarly, voltage V_2 will be

$$V_2 = \frac{R_i}{R_i + R_{i,3}} \bullet V_{in}. \tag{5.42}$$

As $V_{out} = V_1 - V_2$, we have

$$V_{out} = \left(\frac{R_{i,2}}{R_{i,1} + R_{i,2}} - \frac{R_i}{R_i + R_{i,3}} \right) \bullet V_{in}. \tag{5.43}$$

Assume that the electrical resistances of resistors $R_{i,1}$, $R_{i,2}$ and $R_{i,3}$ are constant. Differentiating Eq. 5.43 with respect to time yields

$$dV_{out} = \frac{-R_{i,3} \bullet V_{in}}{(R_i + R_{i,3})^2} \bullet dR_i. \tag{5.44}$$

Consequently, the relationship between the variation of the electrical resistance of strain gauge i and the corresponding variation of the voltage signal is

$$\triangle V_{out} = \frac{-R_{i,3} \bullet V_{in}}{(R_i + R_{i,3})^2} \bullet \triangle R_i. \tag{5.45}$$

A beam's deformation is very small, so the corresponding variation $\triangle R_i$ is very small as well. As a result, voltage signal $\triangle V_{out}$ needs to be amplified to an appropriate level before it can be further converted into a corresponding digital signal. (NOTE: V_{in} is usually within +5V).

A typical circuit for voltage amplification is shown in Fig. 5.45b. The input voltages are V_1 and V_2. And the output voltage (i.e. V_i) is

$$V_i = \frac{R_b}{R_a} \bullet (V_2 - V_1) = -\frac{R_b}{R_a} \bullet V_{out}. \tag{5.46}$$

By differentiating Eq. 5.46 with respect to time, we can obtain

$$\triangle V_i = -\frac{R_b}{R_a} \bullet \triangle V_{out}. \tag{5.47}$$

Finally, substituting Eq. 5.45 into Eq. 5.47 yields

$$\triangle V_i = \frac{R_b}{R_a} \bullet \frac{R_{i,3} \bullet V_{in}}{(R_i + R_{i,3})^2} \bullet \triangle R_i. \tag{5.48}$$

Eq. 5.48 indicates that electrical resistance's variation $\triangle R_i$ is linearly converted to the corresponding voltage signal $\triangle V_i$ (i.e. in digital form after the A/D converter). And, $\triangle V_i$ is the output from strain gauge i ($i = 1, 2, ..., 6$).

Voltage to Force/Torque Mapping

For the force/torque sensor, shown in Fig. 5.43 and Fig. 5.44, the input to the sensor is the actual force/toruqe vector

$$\xi_e = (f_x, f_y, f_z, \tau_x, \tau_y, \tau_z)^t.$$

And, the direct output from the sensor is voltage vector

$$\triangle V = (\triangle V_1, \triangle V_2, \triangle V_3, \triangle V_4, \triangle V_5, \triangle V_6)^t.$$

What we would like to obtain is an estimation of the actual force/torque vector. Let ξ_e^* denote the estimated force/toruqe vector. Now, the question is how to compute ξ_e^* based on the knowledge of $\triangle V$.

In theory, it is difficult to establish an exact analytical relationship between ξ_e and $\triangle V$. In practice, however, we know that $\triangle V$ has an approximate linear relationship with ξ_e. This explains why the practical solution is to use a constant mapping matrix to convert measurement $\triangle V$ to an estimated force/torque vector ξ_e^*.

Let $W = \{w_{ij}, \forall i \in [1,6], \forall j \in [1,6]\}$ be the constant mapping matrix. Then, we have

$$\xi_e^* = W \bullet \triangle V. \tag{5.49}$$

W is a 6×6 matrix with 36 elements. When we have a pair of known ξ_e and its corresponding $\triangle V$, Eq. 5.49 provides six equations for the 36 elements in matrix W. If there are six pairs of known ξ_e and its corresponding $\triangle V$, we will have 36 equations which allow us to uniquely determine the 36 elements. If more than six pairs of known ξ_e and corresponding $\triangle V$ are available, matrix W can be estimated by a least square method. This procedure is known as the *calibration*. The purpose of this calibration is to determine matrix W which describes the voltage to force/torque mapping.

5.4.5   *Tactile Sensors*

When a humanoid robot operates, it is inevitable that it will touch objects in the vicinity. For example, when standing, its two feet are in contact with the floor/ground. When manipulating objects with its hands, the robot's fingers are in contact with the objects. Therefore, tactile sensors are also the important feedback sensors.

By definition, a tactile sensor is a device which measures the physical quantities related to touch. A touch normally involves force and/or torque when one object (e.g. a robot's hand) holds another object (e.g. a tool). But it may also contain other information such as the rolling and slippage effects between two objects in contact. As a result, a tactile sensor is a kind of force/torque sensor. However, a tactile sensor does not require an accurate measurement of the force and/or torque caused by the contact.

Refer to Fig. 5.30. A tactile sensor is also built on top of a transducer and a signal converter. In robotics, there are many ways to design transducers for tactile sensors. Basically, a tactile sensor's transducer has three elements: a) a deformable substrate, b) an array of touch detectors and c) a soft-cover tissue (like skin). Depending on the intended application, a touch detector can be as simple as an "on-off" switch which indicates the logic state of in-touch or not-in-touch. Other common touch detectors are the capacitive transducer which converts contact into corresponding capacitive impedance, and the LVDT (Linear Variable Differential Transformer) which measures the linear displacement caused by contact.

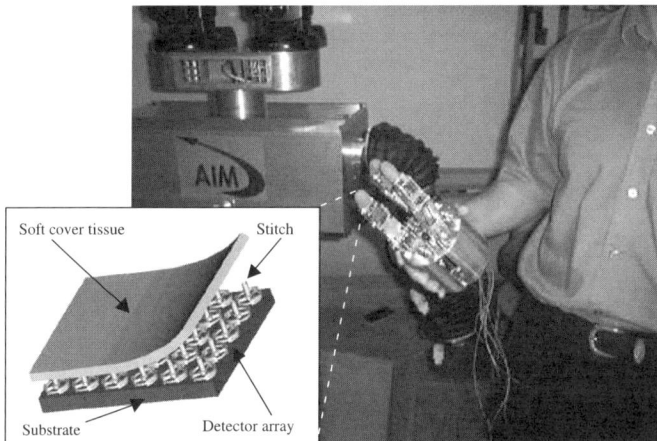

Fig. 5.46   Illustration of the composition of a tactile sensor.

In fact, an ideal transducer for a tactile sensor is a device consisting of an array of micro-force/torque sensors as shown in Fig. 5.46. Each micro-force/torque sensor is similar to the one illustrated in Fig. 5.44. And, it is able to measure external force/torque exerted on its stitch. In this way, we can obtain a complete measurement of contact. Of course, the challenge here is how to fabricate the tactile sensor based on the micro- or nano-detector.

5.5   Control System Design

If we know the sensor feedback, it is easy to close the loop of a motion control system. Then, the next important issue becomes how to design the control algorithms.

5.5.1   *Issues*

As the robot has multiple degrees of freedom, its motion controller must be able to handle multiple closed-loops of motion control. Despite the presence of multiple closed-loops inside a robot's controller, the basic structure of a closed-loop control system is still similar to the one shown in Fig. 5.47. In fact, a robot's system under control is its kineto-dynamic chain. Strictly speaking, the *inertial load* of a robot is the robot mechanism itself including the tool or payload, if any, attached to its end-effector.

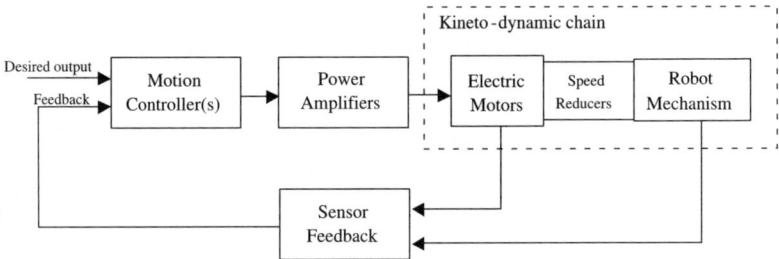

Fig. 5.47   Illustration of the closed-loop control system for a robot's motion execution.

As we discussed earlier, a closed-loop control system is a system which responds to an error signal. Let q denote the actual output of a closed-loop motion control system, and q^d its desired output. The error signal will be

$$e(t) = q^d(t) - q(t).$$

The minimization of the above equation is a typical example of the *control objective* in a closed-loop control system.

Now, the immediate challenge is how to design a control algorithm which will make $e(t)$ converge to zero after a certain period of time. In general, the design of a control algorithm is not a simple matter.

Although vast body of knowledge about control has been developed in control engineering, almost 90% of today's industrial control applications are still based on the popular PID control law. Accordingly, we will study some control schemes based on PID control law which are applicable to a robot's control system.

As the robot is a complex system, it is important to properly address the following issues relevant to the design of a robot's control system:

- How do we specify the desired output of a closed-loop control system in a robot?
- How do we obtain the exact dynamic model of a robot's system under control?
- What is the available sensory feedback information? And, how good is the sensory feedback?
- What is the control law which guarantees a satisfactory control performance in terms of transient and steady-state responses ?

5.5.2 *Planning of Desired Outputs*

In robotics, the specification of the desired output of a robot's control system is not easy. As we mentioned earlier, a task to be performed by a robot is normally specified in Cartesian space, also known as the task/configuration space. Thus, it is logical to specify the desired output of a robot's control system in task space. However, the output motion from a robot's mechanism depends on its input motion. And, a robot's motion control simply means the control of input motions at the joints of a robot's mechanism. Thus, there is strong reason to believe that the desired output of a robot's control system should be the desired values of some variables in joint space.

In fact, the desired output of a robot's control system can be specified in three spaces:

- Planning of the Desired Output in Joint Space:
  In Chapter 4, we studied the statics and dynamics of an open kineto-dynamic chain underlying a robot's mechanism. We know that the motion of a robot's mechanism originates from the energy which drives the dynamic pairs. Thus, the motion control inside a robot means the control of the input motions at the joints of a robot's mechanism. As a result, the desired output of a robot's control system should be the desired values of the motion variables at the joints. On the other hand, a robot is designed to perform tasks which are usually specified in Cartesian space (task space). Thus, it is inevitable to encounter the situation in which the specification in task space has to be translated

into the corresponding desired output in joint space.

- Planning of the Desired Output in Task Space:
  Since a task is normally specified in Cartesian space (task space), it appears logical to directly specify the desired output of a robot's control system in task space. For some motion-related variables (force/torque acting on an environment), it is easy to adopt the control scheme in task space. But for motion variables, like position, velocity and acceleration, it is not such a simple matter to do this. A hybrid-control scheme in both joint and task spaces may be a good solution.

- Planning of the Desired Output in Image Space:
  An image is a projection of a three-dimensional (3-D) space onto a two-dimensional (2-D) plane (image space). A control scheme in image space may be considered a special case of task space control. We can say that the majority of our daily activities rely on our visual perception system. We will study visual sensory and perception systems in Chapters 7 and 8. Here, we simply point out the fact that a human vision system is qualitative (not metric). Despite this, humans are skillful in performing visually-guided activities. Therefore, it is meaningful to develop engineering approaches to imitate human-like visually-guided motion execution.

5.5.3 *A Robot's System Dynamics*

Refer to Fig. 5.47. The plant of a robot's closed-loop control system is the kineto-dynamic chain. Strictly speaking, the dynamic model of a robot's system under control (the plant) should include the dynamics of a robot's mechanism, the dynamics of speed reducers and the dynamics of electric motors.

5.5.3.1 *Dynamics of a Robot's Mechanism*

In Chapter 4, we studied the dynamic model governing a robot's mechanism. This model is commonly called *robot dynamics*. More appropriately, it should be called the dynamics of a robot's mechanism. Let us consider an open kineto-dynamic chain having n kineto-dynamic pairs.

Refer to Eq. 4.137. The general form of the dynamic model of an open kinemto-dynamic chain with n kineto-dynamic pairs is

$$B(q) \bullet \ddot{q} + C(q, \dot{q}) \bullet \dot{q} + G(q) = \tau \qquad (5.50)$$

where:

- $q = (q_1, q_2, ..., q_n)^t$ is the vector of generalized coordinates and q_i is the angular displacement if joint i is a revolute joint. Otherwise, q_i is the linear displacement variable. A displacement with respect to a reference indicates the absolute position of a joint.
- $\tau = (\tau_1, \tau_2, ..., \tau_n)^t$ is the vector of generalized forces and τ_i is the torque exerted at joint i if it is a revolute joint. Otherwise, it refers to the force exerted at joint i.
- $B(q)$ is the inertial matrix of the open kineto-dynamic chain.
- $C(q, \dot{q})$ is the Coriolis matrix which accounts for the Coriolis and Centrifugal effects.
- $G(q)$ is equal to a derivative vector of the potential energy of the open kineto-dynamic chain. It accounts for the effect caused by gravitational force.

This dynamic model relates joint variables q to the joint torques (or forces for prismatic joints). In other words, it only describes the dynamics of the open kineto-dynamic chain underlying a mechanism. From our study of a robot's control elements, we know that the controllable variables for altering a robot's motions are the variables, either voltage or current, directly related to the electrical energy applied to the electric motors. As a result, Eq. 5.50 is not a complete dynamic model of a robot's system under control. We need to further understand: a) the dynamics of the speed reducers and b) the dynamics of the electric motors.

5.5.3.2 *Dynamics of Speed Reducers*

In Chapter 4, we studied the couplings between dynamic and kinematic pairs. The coupling devices are commonly called the *speed reducers* or *torque amplifiers*. (NOTE: The bearings involved in a coupling can be treated as part of a speed reducer).

For the sake of simplicity, let us consider the case where the motion sensors are placed before the speed reducers coupled with the electric motors inside a robot. With regard to an electric motor coupled to a kinematic pair, the joint variables will include angular (or linear) displacement and torque (or force) at the joint. Similarly, the variables at the motor's shaft will also include angular displacement and torque. Thus, let us denote

- $q = (q_1, q_2, ..., q_n)^t$ the vector of displacement variables at the joints,
- $\tau = (\tau_1, \tau_2, ..., \tau_n)^t$ the vector of torques (or forces) at the joints,

- $q_m = (q_{m1}, q_{m2}, ..., q_{mn})^t$ the vector of angular displacements at the motor shafts,
- $\tau_m = (\tau_{m1}, \tau_{m2}, ..., \tau_{mn})^t$ the vector of torques at the motor shafts.

If the reduction ratio of speed reducer i coupled with electric motor i is k_{ri}, then

$$\begin{cases} q_i = \frac{1}{k_{ri}} \bullet q_{mi} \\[2mm] \tau_i = k_{ri} \bullet \tau_{mi} - k_{vi} \bullet \dot{q}_i \end{cases} \qquad (5.51)$$

where k_{vi} is the viscous damping coefficient of speed reducer i (i.e. the viscous friction coefficient). Here, we have omitted the static frictional force of a speed reducer.

Accordingly, the relationship between (q, τ) and (q_m, τ_m) is

$$\begin{cases} q = (\frac{1}{k_{r1}} \bullet q_{m1}, \frac{1}{k_{r2}} \bullet q_{m2}, ..., \frac{1}{k_{rn}} \bullet q_{mn})^t \\[2mm] \tau = (k_{r1} \bullet \tau_{m1} - k_{v1} \bullet \dot{q}_1, k_{r2} \bullet \tau_{m2} - k_{v2} \bullet \dot{q}_2, ..., k_{rn} \bullet \tau_{mn} - k_{vn} \bullet \dot{q}_n)^t. \end{cases} \qquad (5.52)$$

If we define

$$K_r = diag(k_{r1}, k_{r2}, ..., k_{rn}) = \begin{pmatrix} k_{r1} & 0 & ... & 0 \\ 0 & k_{r2} & ... & 0 \\ \vdots & \vdots & \ddots & \vdots \\ 0 & 0 & ... & k_{rn} \end{pmatrix}$$

and

$$K_v = diag(k_{v1}, k_{v2}, ..., k_{vn}) = \begin{pmatrix} k_{v1} & 0 & ... & 0 \\ 0 & k_{v2} & ... & 0 \\ \vdots & \vdots & \ddots & \vdots \\ 0 & 0 & ... & k_{vn} \end{pmatrix},$$

Eq. 5.52 can be compactly written as

$$\begin{cases} q = K_r^{-1} \bullet q_m \\[2mm] \tau = K_r \bullet \tau_m - K_v \bullet \dot{q} \end{cases} \qquad (5.53)$$

where K_r^{-1} is the inverse of K_r. K_r can be called the *reduction ratio matrix*. This equation describes the dynamics of speed reducers.

Finally, substituting Eq. 5.53 into Eq. 5.50 yields

$$B_m(q) \bullet \ddot{q}_m + C_m(q, \dot{q}) \bullet \dot{q}_m + G_m(q) = \tau_m - K_{mv} \bullet \dot{q}_m \qquad (5.54)$$

with

$$\begin{cases} B_m(q) = K_r^{-1} \bullet B(q) \bullet K_r^{-1} \\[2mm] C_m(q, \dot{q}) = K_r^{-1} \bullet C(q, \dot{q}) \bullet K_r^{-1} \\[2mm] G_m(q) = K_r^{-1} \bullet G(q) \bullet K_r^{-1} \\[2mm] K_{mv} = K_r^{-1} \bullet K_v \bullet K_r^{-1}. \end{cases} \qquad (5.55)$$

Eq. 5.54 describes the dynamic model of an open kineto-dynamic chain in terms of the variables at the output shafts (before the speed reducers) of the electric motors.

5.5.3.3 *Dynamics of Electric Motors*

For an electric motor coupled with a kinematic pair, the relationship between the torque at its shaft and the voltage (or current) applied to the coils of its rotor or stator is not simple. Theoretically, if the motor does not dissipate any energy, electrical energy W_e applied to an electric motor should be equal to its mechanical energy W_m delivered at the output shaft.

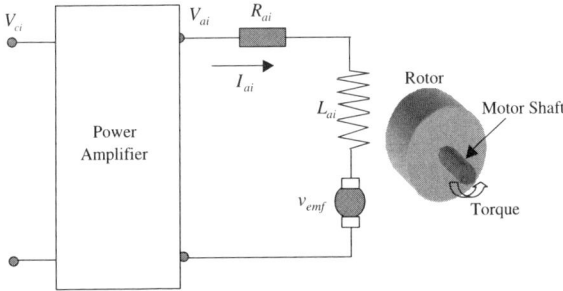

Fig. 5.48   Illustration of the equivalent electronic circuit of an electric motor.

In general, the equivalent electronic circuit describing an electric motor i looks like the one shown in Fig. 5.48. The parameters relevant to an electric motor i include

- V_{ai}: input voltage applied to its coil(s),

- I_{ai}: input current applied to its coil(s),
- R_{ai}: resistance of its coil(s),
- L_{ai}: inductance of its coil(s),
- v_{emf}: back electro-motive force (i.e. the induced voltage when a coil rotates inside a magnetic field),
- τ_{mi}: output torque at the motor's shaft,
- q_{mi}: angular position or displacement at the motor's shaft.

By definition, electrical energy applied to motor i within time interval $[t_s, t_f]$ is

$$W_e = \int_{t_s}^{t_f} V_{ai} \bullet I_{ai} \bullet dt.$$

And, the corresponding mechanical energy output from motor i will be

$$W_m = \int_{t_s}^{t_f} \tau_{mi} \bullet \dot{q}_{mi} \bullet dt.$$

By applying the principle of energy conservation, we have

$$\int_{t_s}^{t_f} V_{ai} \bullet I_{ai} \; dt = \int_{t_s}^{t_f} \tau_{mi} \bullet \dot{q}_{mi} \bullet dt. \tag{5.56}$$

Clearly, it is not easy to predict the exact relationship between the voltage (or current) applied to an electric motor and the torque at its shaft. However, if we consider brush-type DC motor i driven by a linear power amplifier, we can establish the following equation with regard to the equivalent electronic circuit as shown in Fig. 5.48:

$$V_{ai} = R_{ai} \bullet I_{ai} + L_{ai} \frac{dI_{ai}}{dt} + v_{emf} \tag{5.57}$$

with

$$v_{emf} = k_{ai} \bullet \dot{q}_{mi}$$

where k_{ai} is a constant.

In Chapter 4, we studied the relationship between current and torque in a brush-type DC motor (i.e. Eq. 4.32). If k_{ti} is the torque constant of brush-type DC motor i, we have

$$\tau_{mi} = k_{ti} \bullet I_{ai}.$$

By applying the above equation to Eq. 5.57 and omitting back electromotive force v_{emf}, we will obtain

$$V_{ai} = \frac{R_{ai}}{k_{ti}} \bullet \tau_{mi} + \frac{L_{ai}}{k_{ti}} \bullet \frac{d\tau_{mi}}{dt}. \tag{5.58}$$

Now, let us assume that the linear power amplifier for motor i is modelled by Eq. 5.28. Then, we have

$$V_{ai} = \frac{R_1 + R_2}{R_1} \bullet V_{ci}$$

where V_{ci} is the control voltage signal applied to electric motor i. This control voltage signal comes from a digital motion controller. The application of the above equation to Eq. 5.58 yields

$$V_{ci} = \frac{1}{k_{mi}} \bullet \left(\tau_{mi} + T_{mi} \bullet \frac{d\tau_{mi}}{dt} \right) \tag{5.59}$$

with

$$
\begin{cases}
k_{mi} = \frac{k_{ti} \bullet (R_1 + R_2)}{R_{ai} \bullet R_1} \\
T_{mi} = \frac{L_{ai}}{R_{ai}}.
\end{cases}
$$

In fact, T_{mi} is the *time-constant* of electric motor i and k_{mi} is its *voltage-to-torque gain*.

Eq. 5.59 is a simplified model which describes the dynamics of an electric motor. The combination of Eq. 5.54 and Eq. 5.59 fully describes the dynamics of a robot's system under control.

Example 5.9 Refer to Eq. 5.59. Let $k_{mi} = 1.2$ (N-m/volt) and $T_{mi} = 5$ (Ohms/Henrys). The (voltage-to-torque) transfer function of the motor will be

$$\frac{\tau_{mi}(s)}{V_{ci}(s)} = \frac{1.2}{1 + 5s}.$$

Fig. 5.49 shows the response of a motor (without inertial load) to a unit-step voltage signal. And, Fig. 5.50 shows the response of motor (without inertial load) to the input voltage signal of a unit-sine function (i.e. $\sin(t)$). We can see that the motor responds to the input voltage signal in a timely manner.

$$\diamond\diamond\diamond\diamond\diamond\diamond\diamond\diamond\diamond\diamond\diamond\diamond\diamond\diamond\diamond\diamond\diamond$$

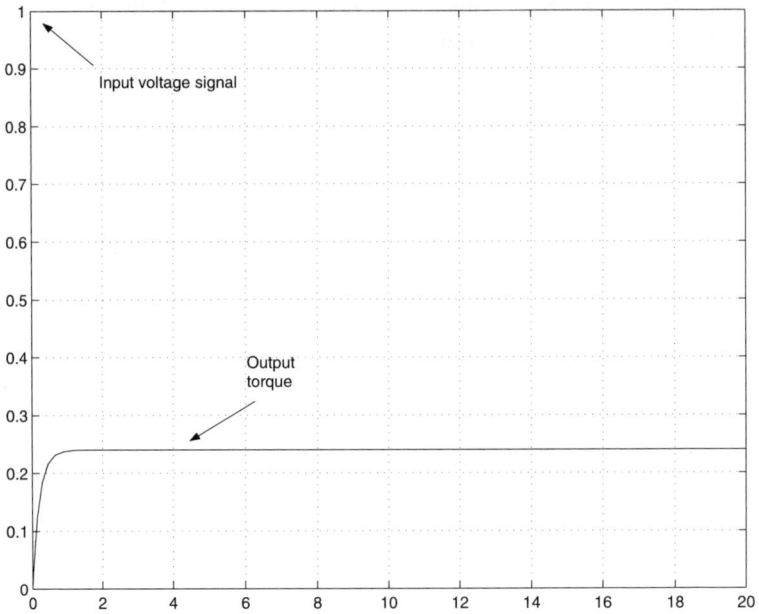

Fig. 5.49   Transient response to the input voltage signal of a unit step function. The horizontal axis is the time axis with the unit in seconds.

5.5.4   *Sensory Feedback*

Refer to Fig. 5.47. Sensory feedback is indispensable in a closed-loop control system. For motion control, the feedback measurements may include position, velocity, acceleration and force/torque. As we discussed earlier, a motion sensor can be placed either before or after a speed reducer. As for the force/toque sensor, it is always attached to the end-effector link.

The nature of the measured variables inside a closed-loop control system determines the nature of the desired output. For example, if the position variable is the only variable which is measurable, we can only specify the desired value for position. On the other hand, if the velocity variable is also measurable, both the desired values for position and velocity can be specified.

Since position, velocity, and acceleration are mathematically related to each other, one measurement may be sufficient, if the sensor's noise is negligible. For example, if a sensor's output $s(t)$ is the angular position at

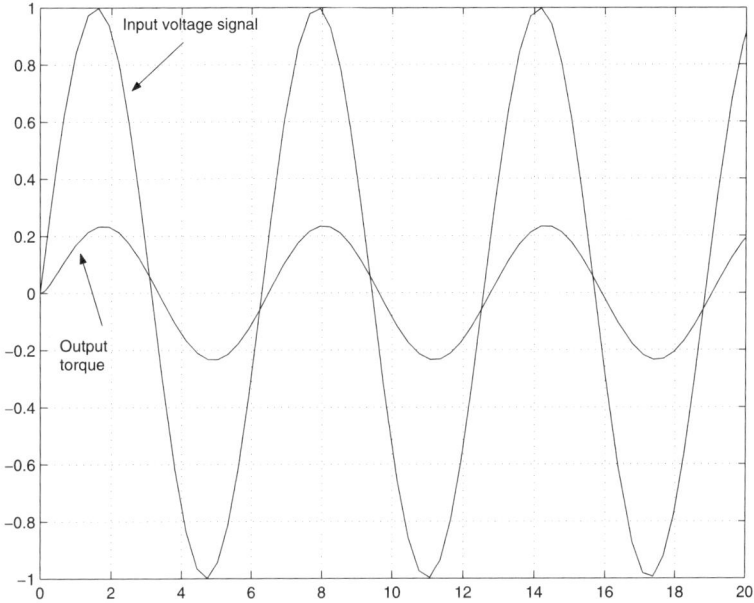

Fig. 5.50 Transient response to the input voltage signal of a unit sine function (i.e. $V_{ci}(t) = \sin(t)$. The horizontal axis is the time axis with the unit in seconds.

a motor's shaft, the corresponding angular velocity and acceleration will be

$$\begin{cases} v(t) = \frac{ds(t)}{dt} \\[2mm] a(t) = \frac{d^2 s(t)}{dt^2}. \end{cases}$$

Refer to Eq. 5.15. The transfer function of a feedback sensor will appear at the denominator of the closed-loop transfer function. This means that the dynamics of a feedback sensor will slow down the response of the closed-loop control system. Ideally, a feedback sensor should not exhibit any dynamic behavior (i.e. no transient response) and should behave like a unity gain channel.

Refer to Eq. 5.17. If the control gain is very large, almost 100% of the noise caused by a feedback sensor will appear in the output. This means that the output of a closed-loop control is very sensitive to the noise of feedback sensor. In order to achieve better control performance, the feedback sensors should not have any noise.

5.5.5 *Control Algorithms and Performances*

A closed-loop control system consists of three parts: a) the system under control (i.e. the plant), b) the controller (i.e. control law or algorithm), and c) the feedback sensor(s). Sometimes, a control system may just have a single input and a single output (SISO). In this case, it is called an *SISO system*. If a control system has multiple input and multiple output (MIMO), it is called a *MIMO system*.

As shown in Fig. 5.51, the transfer function of a robot's closed-loop control system can be generically written as

$$G(s) = \frac{G_c(s) \bullet G_p(s)}{1 + H(s) \bullet G_c(s) \bullet G_p(s)} \qquad (5.60)$$

where $G_p(s)$ is the transfer function of the system under control, $G_c(s)$ the transfer function of the control law or algorithm and $H(s)$ the transfer function of the sensing device.

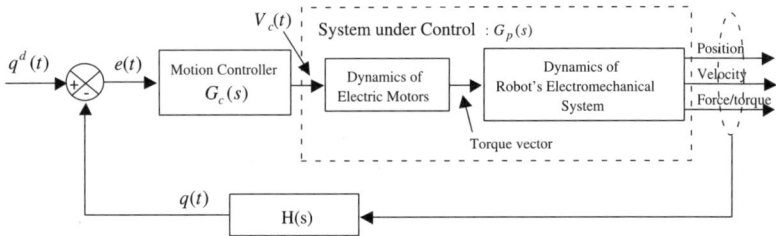

Fig. 5.51   Generic block diagram of a robot's closed-loop control system.

Clearly, the dynamics of a closed-loop control system not only depends on the dynamics of the system under control but also the dynamics of the control law or algorithm. By properly designing control law $G_c(s)$, we will be able to build a closed-loop control system with satisfactory performance in terms of transient and steady-state responses.

In control engineering, there are many available control laws. Some require knowledge about the exact dynamic model of a system under control and some do not. When a control objective is given, a good control law is judged by its achievable absolute and relative stabilities:

- Absolute Stability:
  The presence of the control law's dynamics $G_c(s)$ in Eq. 5.60 should not render the final system unstable. A convenient way to test a control

system's absolute stability is by the Lyapunov stability criterion. For example, we can define a Lyapunov Function as follows:

$$V(e(t)) = \frac{1}{2} \, e^t(t) \bullet e(t). \tag{5.61}$$

(NOTE: Assume that the error signal $e(t)$ is a vector).
A system under control with a given control law will be asymptotically stable as long as the following condition holds:

$$\dot{V}(e(t)) = e^t(t) \bullet \dot{e}(t) < 0, \quad \forall t > 0. \tag{5.62}$$

(NOTE: For a stable system, this is a sufficient, but not necessary condition).

- Relative Stability:
  Relative stability refers to a control system's transient response. However, the transient response depends on the dynamics of error signal $e(t)$. A good control law should minimize the following cost function:

$$L(t) = \int_0^t \left\{ e^t(t) \bullet P \bullet e(t) \right\} \bullet dt \tag{5.63}$$

where P is a constant positive-definite matrix.
If the control effort is directly measured by control signal V_c from a digital controller, a good control law should be able to achieve a desired outcome with minimum effort. In this case, the cost function to be minimized will be

$$L(t) = \int_0^t \left\{ e^t(t) \bullet P \bullet e(t) + V_c^t(t) \bullet Q \bullet V_c(t) \right\} \bullet dt \tag{5.64}$$

where Q is another constant positive-definite matrix.

5.5.5.1 *PID Control Laws*

Despite the large number of available control laws studied in control engineering, the most popular control law in industry is still the PID control law or algorithm. The main reasons for this include:

(1) It has a clearly-defined control objective as a PID control law's transfer function can be simply expressed as

$$G_c(s) = k_p \bullet \left(1 + \frac{1}{T_i \bullet s} + T_d \bullet s \right). \tag{5.65}$$

(2) It is able to respond quickly (i.e. due to the proportional law acting on the present signal).
(3) It is able to respond accurately and with a better steady-state (i.e. due to the integral law acting on past signals).
(4) It is able to respond stably to the predictive signal (i.e. the derivative law).
(5) It does not require knowledge of the dynamic model of a system under control.
(6) There are many methods of manually or automatically tuning proportional control gain k_p, integral time-constant T_i, and derivative time-constant T_d.
(7) It is easy to implement a PID control law on a micro-processor or micro-controller (see Eq. 5.5).

Even though the PID control law is widely-used in industry, there is no theoretical proof to guarantee the stability of a system under its control. However, this is not a critical drawback as a PID control law is usually implemented on a digital computer. It is not difficult to define a Lyapunov Function similar to Eq. 5.61 and automatically tune the parameters of a PID control law in order to guarantee the following conditions:

$$\begin{cases} V(e(t)) \longrightarrow 0 \\ \dot{V}(e(t)) \longrightarrow 0 \end{cases} \tag{5.66}$$

when $t > t_s$. Thus, t_s can be specified as the expected settling time of the system under control.

Based on experience, one may also come up with some empirical rules to support the automatic tuning of PID control laws. Table 5.1 shows one set of empirical rules.

Table 5.1   Empirical rules for the tuning of a PID control law

Parameters	L(t) in Eq. 5.63	Stability
k_p increases	decreases	decreases
T_i decreases	decreases	decreases
T_d increases	increases	increases

Example 5.10   Refer to Fig. 5.51. The plant's transfer function is

$$G_p(s) = \frac{1}{s^2 + 5\,s + 6}$$

and a PID control law (i.e. Eq. 5.65) is used.

Fig. 5.52 shows the responses of the closed-loop control system and the L function in Eq. 5.63 when k_p increases. We can see that the value of L decreases along with the stability (i.e. more oscillation).

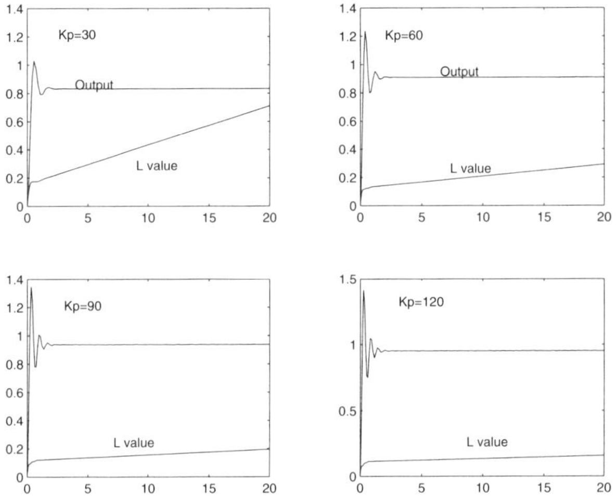

Fig. 5.52 Transient response when k_p=30, 60, 90 and 120 while $T_i = \infty$ and $T_d = 0$. The horizontal axis is the time axis with the unit in seconds.

Fig. 5.53 shows the responses of the closed-loop control system and the L function in Eq. 5.63 when T_i decreases. We can see that the value of L decreases along with the stability (i.e. more oscillation).

Fig. 5.54 shows the responses of the closed-loop control system and the L function in Eq. 5.63 when T_d increases. We can see that the value of L increases along with the stability (i.e. oscillation disappears).

◇◇◇◇◇◇◇◇◇◇◇◇◇◇◇◇◇◇◇

5.5.5.2 *Variable Structure Control*

Refer to Eq. 5.50. A robot is intrinsically a MIMO, nonlinear and time-varying system. The dynamics of a robot's mechanism is configuration dependent (i.e. depending on q and Jacobian matrix J). Thus, a robot is a typical example of the *variable-configuration system*. Clearly, a monolithic control law will not efficiently control a variable-configuration system. An alternative approach is the *variable-structure control methodology*. This

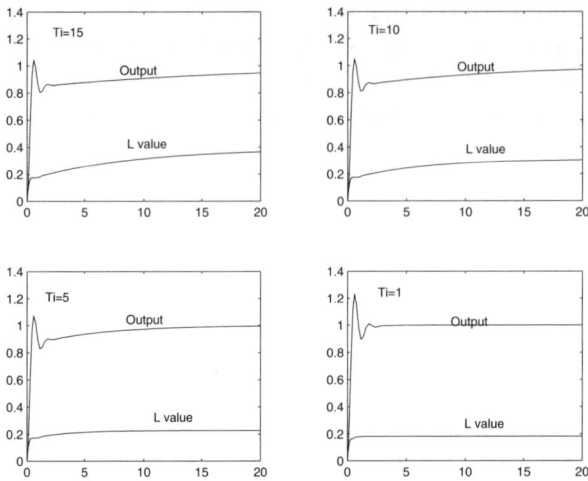

Fig. 5.53   Transient response when T_i=15, 10, 5 and 1 while $k_p = 30$ and $T_d = 0$. The horizontal axis is the time axis with the unit in seconds.

idea originated in Russia, and was proposed by Emelyanov and Barbashin in the early 1960s.

The philosophy behind the variable-structure control scheme is to use a set of feedback control laws to command a variable configuration system having unmodelled dynamics and uncertainty. The choice of a feedback control law for a specific type of system configurations is decided by a *switching function* $s(t)$. The switching function monitors the system's behaviors based on measurements related to the system's state variables. The well-known *sliding mode control* is a typical example of variable structure control. In a sliding mode control scheme, switching function $s(t)$ typically looks like

$$s(t) = S \bullet x(t) \tag{5.67}$$

where S is a $m \times n$ matrix (called a *switching function matrix*), and $x(t)$ is a $n \times 1$ state variable vector of the system under control.

We can extend the philosophy of variable-structure control to the case of multiple PID control scheme. Let $e(t)$ be the error vector of a closed-loop control system, the dynamics of which is unknown or unmodelled. We can construct the following switching function:

$$s(t) = \int_0^t \left\{ \alpha \left[e^t(t) \bullet e(t) \right] + (1 - \alpha) \left[\dot{e}^t(t) \bullet \dot{e}(t) \right] \right\} \bullet dt \tag{5.68}$$

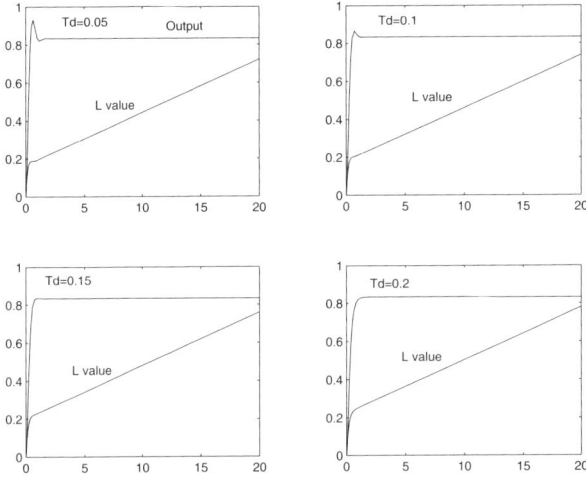

Fig. 5.54 Transient response when T_d=0.05, 0.1, 0.15 and 0.2 while $k_p = 30$ and $T_i = \infty$. The horizontal axis is the time axis with the unit in seconds.

where $0 \le \alpha \le 1$. α is a weighting coefficient which conditions the selection of a control law. We call Eq. 5.68 a *generalized switching function*.

By adjusting the coefficient α, we can balance the convergence to zero by both error signal $e(t)$ and its derivative. When choosing $\alpha = 1$, a control law which tends to quickly converge the error signal to zero will be selected. But this will be at the expense of causing unacceptable oscillation and overshoot. On the other hand, when choosing $\alpha = 0$, a control law which tends to minimize the derivative of the error signal will be selected. This will produce a very smooth output, but the response time may be unacceptably long.

Given a set of possible PID control laws predefined for a variable-configuration system, like a robot, a decision rule can be established based on the output of the generalized switching function $s(t)$. The decision rule will determine the appropriate control law when a variable-configuration system enters a new type of configurations. For example, a predefined PID control law which minimizes $s(t)$ is selected as the current control law.

The PID control laws, the generalized switching function $s(t)$ and the decision rules can be evaluated and fine-tuned through real-time interaction between a robot and its environment. Because of the nature of a robot which is programmed to perform repetitive tasks/actions, this interaction is intrinsically repetitive. Therefore, the idea of variable-structure control for a humanoid robot matches the spirit of developmental principles underlying the mental and physical development (e.g. decision-making and action-taking capabilities) of artificial systems or machines.

Example 5.11   Refer to Fig. 5.51. The plant's transfer function is

$$G_p(s) = \frac{1}{s^2 + a \bullet s + b}$$

and a set of four PID control laws are predefined as follows:

Control Law	k_p	T_i	T_d
PID1	4	3	0.25
PID2	8	3	0.25
PID3	12	3	0.25
PID4	16	3	0.25

In the generalized switching function (i.e. Eq. 5.68), we set $\alpha = 0.7$.

Now, let us have the plant's transfer function be as follows:

$$G_p(s) = \frac{1}{s^2 + 2 \bullet s + 1}$$

and choose a unit-step function as input.

Fig. 5.55 shows the responses of the closed-loop control system and the outputs of function $s(t)$ (i.e. Eq. 5.68) corresponding to the four predefined PID control laws. From Fig. 5.55, we can see that the selected control law, based on generalized switching function $s(t)$, will be PID1.

Then, let us modify the coefficients of the plant's transfer function to simulate the variation in the plant's dynamics. We set $a = 4$ and $b = 1$. The plant's transfer function becomes

$$G_p(s) = \frac{1}{s^2 + 4 \bullet s + 1}.$$

And the responses of the closed-loop control system and the outputs of the $s(t)$ function in Eq. 5.68 corresponding to the four predefined PID control laws are shown in Fig. 5.56. We can see that the selected control law, based on the output of the generalized switching function, will be PID2.

Let us further change the plant's transfer function to

$$G_p(s) = \frac{1}{s^2 + 4 \bullet s + 4}.$$

The responses of the closed-loop control system, and the outputs of the $s(t)$ function in Eq. 5.68 corresponding to the four predefined PID control laws,

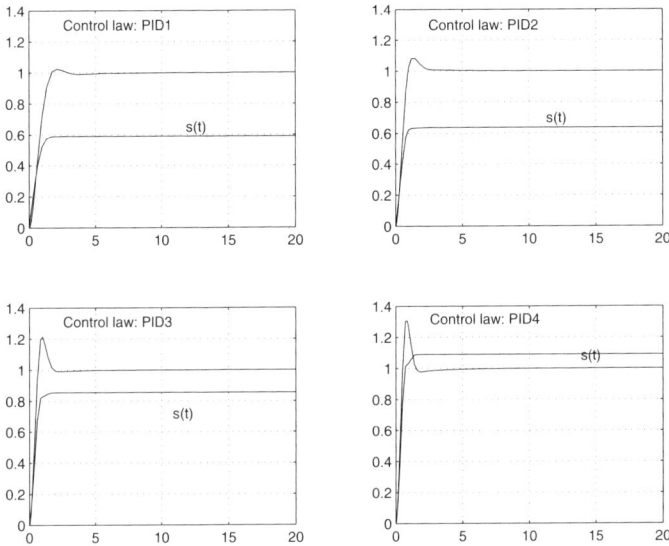

Fig. 5.55 Transient responses and the outputs of generalized switching function $s(t)$ corresponding to the four PID control laws when $a = 2$ and $b = 1$. The horizontal axis is the time axis with the unit in seconds.

are shown in Fig. 5.57. We can see that the selected control law, based on the output of the generalized switching function, will be PID3.

◇◇◇◇◇◇◇◇◇◇◇◇◇◇◇◇◇◇

5.5.6 *Joint-Space Control*

A robot is a machine which is skilled at executing motions. The motion performed by a robot is the output motion of its mechanism. And the output motion of a robot's mechanism depends on its input motion (i.e. the motions at the joints). Therefore, the control of the output motion of a robot's mechanism is equivalent to the control of its input motion. This explains why a robot's motion control often takes place at the joint level. Thus, the control scheme at the lowest level of a robot's motion control system is called *joint-space control*.

5.5.6.1 *Planning of Desired Outputs*

In a joint-space control scheme, the desired output should be desired joint variable vector q^d and its derivative $\dot{q}^d$. Sometimes, we can also specify

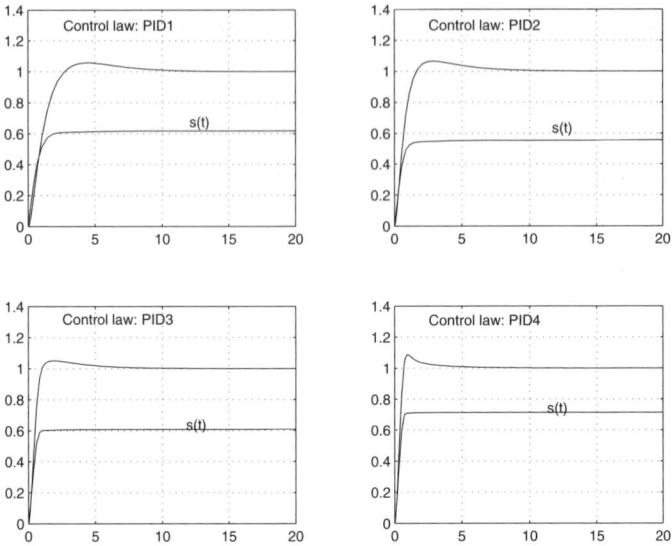

Fig. 5.56   Transient responses and the outputs of generalized switching function $s(t)$ corresponding to the four PID control laws when $a = 4$ and $b = 1$. The horizontal axis is the time axis with the unit in seconds.

desired acceleration vector $\ddot{q}^d$ at the joints.

Refer to Fig. 5.14. A robot's input is the tasks/actions and the output is the executed motions by the robot. Normally, a task/action is specified in Cartesian space (task space). As a result, in planning the desired output for a robot's control scheme in joint space, one must address two issues:

- How do we plan motion in terms of path and trajectory in task space?
- How do we transform planned motions in task space into the corresponding ones in joint space?

We will address the first issue in Chapter 9. The solution to the second issue is the robot's motion kinematics, studied in Chapter 3. Let J denote the Jacobian matrix of an open kineto-dynamic chain and $J^\dagger$ the pseudo-inverse of J. Then, the desired velocity vector (i.e. $\dot{q}^d$) of the joints is related to desired velocity vector $\dot{P}_e^d$ of the end-effector's frame as follows:

$$\dot{q}^d = J^\dagger \bullet \dot{P}_e^d. \tag{5.69}$$

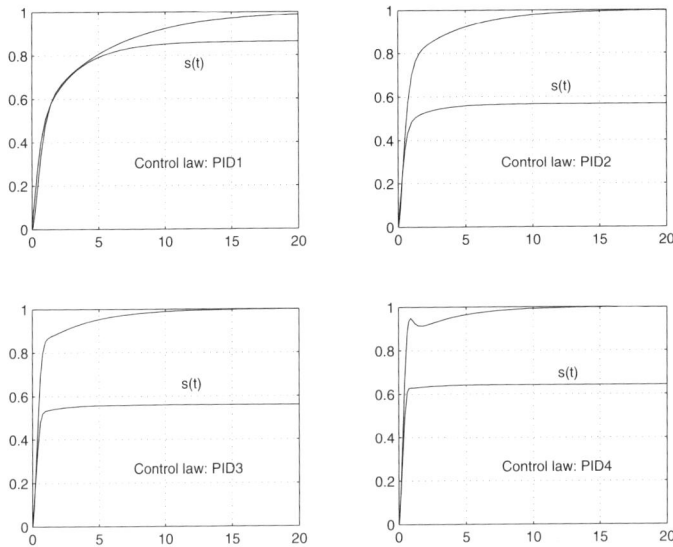

Fig. 5.57   Transient responses and the outputs of generalized switching function $s(t)$ corresponding to the four PID control laws when $a = 4$ and $b = 4$. The horizontal axis is the time axis with the unit in seconds.

5.5.6.2   *System Dynamics*

Refer to Eq. 5.54 and Eq. 5.59. The dynamics of a robot's system under control is

$$\begin{cases} B_m(q) \bullet \ddot{q}_m + [C_m(q, \dot{q}) + K_{mv}] \bullet \dot{q}_m + G_m(q) = \tau_m \\ V_c = K_m^{-1} \bullet (\tau_m + T_m \bullet \dot{\tau}_m) \end{cases} \tag{5.70}$$

with $K_m = diag\{k_{m1}, k_{m2}, ..., k_{mn}\}$ and $T_m = diag\{T_{m1}, T_{m2}, ..., T_{mn}\}$.

For the sake of simplicity, let us consider that electric motors inside a robot can be operated in the torque-controlled mode instead of the voltage-controlled mode. Thus, we can ignore the dynamics of electric motors and simply treat $V_c = \tau_m$. Then, Eq. 5.70 becomes

$$B_m(q) \bullet \ddot{q}_m + [C_m(q, \dot{q}) + K_{mv}] \bullet \dot{q}_m + G_m(q) = \tau_m. \tag{5.71}$$

In practice, this means that there is an unmodelled dynamics with regard to a robot's system under control.

5.5.6.3 *Centralized Joint Control Scheme*

The purpose of a control law is to determine control signal τ_m which will act on the system under control in order to achieve the desired control objective and performance.

Let the desired output be specified as the desired joint angular-position vector, denoted by q_m^d. Then, the error vector of the robot's closed-loop control system will be

$$e_m(t) = q_m^d - q_m(t) \qquad (5.72)$$

or, simply

$$e_m = q_m^d - q_m$$

where $q_m(t)$ is the measurement of the actual joint angle vector.

Subsequently, we define the following control objective function (Lyapunov Function):

$$V = V(\dot{q}_m, e_m) = \frac{1}{2}\,\dot{q}_m^t \bullet B_m(q) \bullet \dot{q}_m + \frac{1}{2}\,e_m^t \bullet K_1 \bullet e_m. \qquad (5.73)$$

Here, the control objective is to find a suitable control law which makes V converge to zero. The first term in Eq. 5.73 describes the kinetic energy of the robot. So, it is always a positive term. If we choose K_1 to be a positive-definite matrix, then we have

$$V(\dot{q}_m, e_m) > 0 \quad \forall \dot{q}_m \neq 0 \text{ and } e_m \neq 0.$$

And, $V(\dot{q}_m, e_m) = 0$ if and only if $\dot{q}_m = 0$ and $e_m = 0$. Differentiating Eq. 5.73 with respect to time yields

$$\dot{V} = \dot{q}_m^t \bullet B_m(q) \bullet \ddot{q}_m + \frac{1}{2}\,\dot{q}_m^t \bullet \dot{B}_m(q) \bullet \dot{q}_m + \dot{e}_m^t \bullet K_1 \bullet e_m. \qquad (5.74)$$

From Eq. 5.72, we have $\dot{e}_m(t) = -\dot{q}(t)_m$. By applying this result and substituting Eq.5.71 into Eq.5.74, we obtain

$$\dot{V} = \frac{1}{2}\,\dot{q}_m^t \bullet (\dot{B}_m - 2\,C_m) \bullet \dot{q}_m - \dot{q}_m^t \bullet K_{mv} \bullet \dot{q}_m + \dot{q}_m^t \bullet (\tau_m - G_m - K_1 \bullet e_m) \quad (5.75)$$

where $B_m = B_m(q)$, $C_m = C_m(q, \dot{q})$ and $G_m = G_m(q)$.

From the dynamic model of a robot's electromechanical system studied in Chapter 4, we can prove that matrix $\dot{B}_m - 2\,C_m$ is a skew-symmetric matrix. For skew-symmetric matrix N, we can also prove that

$$v^t \bullet N \bullet v = 0$$

for any vector v. This result implies that the first term in Eq. 5.75 is equal to zero. Then, Eq. 5.75 becomes

$$\dot{V} = -\dot{q}_m^t \bullet K_{mv} \bullet \dot{q}_m + \dot{q}_m^t \bullet (\tau_m - G_m - K_1 \bullet e_m). \tag{5.76}$$

K_{mv} is a diagonal matrix with the elements being the coefficients of the viscous frictions. Hence, K_{mv} is a positive-definite matrix. Consequently, the first term in Eq. 5.76 is negative for all $\dot{q}_m \neq 0$.

If we choose the control law to be

$$\tau_m = G_m + K_1 \bullet e_m, \tag{5.77}$$

then we can guarantee the following conditions

$$\begin{cases} \dot{V} < 0 & \forall \dot{q}_m \neq 0 \\ \dot{V} = 0 & \forall \dot{q}_m = 0. \end{cases} \tag{5.78}$$

And Eq. 5.76 will become

$$\dot{V} = -\dot{q}_m^t \bullet K_{mv} \bullet \dot{q}_m. \tag{5.79}$$

According to Lyapunov's stability criterion, the control law in Eq. 5.77 will make the closed-loop control system asymptotically stable because of the conditions guaranteed by Eq. 5.78.

If a robot's mechanical system is energy efficient, then the coefficients of the viscous frictions are quite small. This implies that the convergence to zero by the system under control is quite slow because the norm of $\dot{V}$ in Eq. 5.79 is small. A good way to improve response time is to increase the norm of $\dot{V}$. This can be achieved by introducing term $\dot{q}_m^t \bullet K_2 \bullet \dot{q}_m - \dot{q}_m^t \bullet K_2 \bullet \dot{q}_m$, which is equal to zero, into Eq. 5.76. In this way, we obtain

$$\dot{V} = -\dot{q}_m^t \bullet (K_{mv} + K_2) \bullet \dot{q}_m \tag{5.80}$$

if we choose the control law to be

$$\tau_m = G_m + K_1 \bullet e_m + K_2 \bullet \dot{e}_m \tag{5.81}$$

where K_2 is a positive-definite matrix and $\dot{e}_m = -\dot{q}_m$ (see Eq. 5.72).

From Eq. 5.81, we can see that the second term is the proportional control law of a PID control scheme and the third term is the derivative control law. Moreover, the first term in Eq. 5.77 is to compensate for the gravitational force acting on the robot. The control law expressed in Eq. 5.81 is known as the *PD control with gravity compensation*. If a robot is operating in outer space where there is no gravity, the control law in Eq. 5.81 is simply a PD control because $G_m = 0$.

5.5.6.4   *Independent Joint Control Scheme*

We know that matrix $B(q)$ in Eq. 5.50 is an $n \times n$ inertial matrix. It can be represented as follows:

$$B(q) = \{b_{ij}(q), \quad \forall i \in [1, n] \text{ and } \forall j \in [1, n]\}.$$

Now, we purposely split matrix $B(q)$ into the sum of two matrices, as shown below:

$$B(q) = B_1(q) + B_2(q) \tag{5.82}$$

with

$$\begin{cases} B_1(q) = \{b_{ii}(q), \quad \forall i \in [1, n]\} \\[2mm] B_2(q) = \{b_{ij}(q), \quad \forall i \neq j, \ \forall i \in [1, n] \text{ and } \forall j \in [1, n]\}. \end{cases}$$

From Eq. 5.55, we know that $B_m(q) = K_r^{-1} \bullet B(q) \bullet K_r^{-1}$. As a result, $B_m(q)$ can be expressed as

$$B_m(q) = B_{m1}(q) + B_{m2}(q) \tag{5.83}$$

with

$$\begin{cases} B_{m1}(q) = \left\{ \frac{b_{ii}(q)}{k_{ri}^2}, \quad \forall i \in [1, n] \right\} \\[3mm] B_{m2}(q) = \left\{ \frac{b_{ij}(q)}{k_{ri}^2}, \quad \forall i \neq j, \ \forall i \in [1, n] \text{ and } \forall j \in [1, n] \right\} \end{cases}$$

where $\{k_{ri}, i = 1, 2, ..., n\}$ are the diagonal elements of reduction-ratio matrix K_r.

Then, substituting Eq. 5.83 into Eq. 5.71 yields

$$B_{m1} \bullet \ddot{q}_m + N_m(q, \dot{q}) = \tau_m \tag{5.84}$$

with

$$N_m(q, \dot{q}) = B_{m2} \bullet \ddot{q}_m + [C_m(q, \dot{q}) + K_{mv}] \bullet \dot{q}_m + G_m(q).$$

As we can see from Eq. 5.55, the elements of matrices $B(q)$, $C(q, \dot{q})$, $G(q)$ and K_v are divided by the square of the corresponding elements in reduction-ratio matrix K_r. When the elements (i.e. the reduction ratios) in K_r are big (> 100), elements inside matrices $B_m(q)$, $C_m(q, \dot{q})$, $G_m(q)$ and K_{mv} will be negligible. Based on this, we can omit matrix $N_m(q, \dot{q})$ in Eq. 5.84. As a result, a robot's system dynamics is simplified to

$$B_{m1} \bullet \ddot{q}_m \simeq \tau_m. \tag{5.85}$$

Since $B_{m1}(q)$ is a diagonal matrix, Eq. 5.85 can be equivalently written as

$$\frac{b_{ii}(q)}{k_{ri}^2} \ddot{q}_{mi} = \tau_{mi} \quad \forall i \in [1, n]. \tag{5.86}$$

Eq. 5.86 indicates that the links of an open kineto-dynamic chain are decoupled. Thus, we can independently design a control law for the motion-control loop of each joint variable q_{mi} ($i = 1, 2, ..., n$). By default, we can use the PID control law for each control loop. Here, we propose an alternative solution.

Let us consider the motion-control loop for joint variable q_{mi} ($i = 1, 2, ..., n$). Assume that the desired output is q_{mi}^d. Then, the error signal will be

$$e_{mi}(t) = q_{mi}^d - q_{mi}(t) \tag{5.87}$$

or simply

$$e_{mi} = q_{mi}^d - q_{mi}$$

where $q_{mi}(t)$ is the measurement of the actual angular position of motor i.

Now, let us define the following Lyapunov Function

$$V = V(\dot{q}_{mi}, e_{mi}) = \frac{1}{2} (\dot{q}_{mi}^2 + e_{mi}^2). \tag{5.88}$$

Differentiating Eq. 5.88 with respect to time gives

$$\dot{V} = \dot{q}_{mi} \bullet \ddot{q}_{mi} + \dot{e}_{mi} \bullet e_{mi}. \tag{5.89}$$

From Eq. 5.87, we have $\dot{e}_{mi} = -\dot{q}_{mi}$. By applying this result and substituting Eq. 5.86 into Eq. 5.89, we obtain

$$\dot{V} = \dot{q}_{mi} \left(\frac{k_{ri}^2}{b_{ii}} \bullet \tau_{mi} - e_{mi} \right). \tag{5.90}$$

Now, we choose a proportional control law as follows:

$$\tau_{mi} = k_{pi} \bullet e_{mi}. \tag{5.91}$$

Then the question is: What should proportional gain k_{pi} be in order to make the closed-loop control system asymptotically stable?

Substituting Eq. 5.91 into Eq. 5.90 yields

$$V = \dot{q}_{mi} \bullet e_{mi} \left(\frac{k_{ri}^2}{b_{ii}} \bullet k_{pi} - 1 \right). \tag{5.92}$$

Let us define the switching function $s(.)$ as follows

$$s(\dot{q}_{mi}, e_{mi}) = \dot{q}_{mi} \bullet e_{mi} \tag{5.93}$$

And we schedule the proportional control gain according to the following decision rules:

$$k_{pi} = \begin{cases} \alpha \bullet \frac{b_{ii}}{k_{ri}^2} & \text{if } s(\dot{q}_{mi}, e_{mi}) < 0 \\[2mm] (1 - \beta) \bullet \frac{b_{ii}}{k_{ri}^2} & \text{if } s(\dot{q}_{mi}, e_{mi}) > 0 \end{cases} \tag{5.94}$$

where $\alpha > 1$ and $\beta > 0$.

As we can see from Eq. 5.92 and Eq. 5.94, for any set of admissible values of (α, β), the control law in Eq. 5.91 satisfies the following conditions:

$$\begin{cases} \dot{V} < 0 & \forall \dot{q}_{mi} \neq 0 \text{ and } \forall e_{mi} \neq 0 \\[2mm] \dot{V} = 0 & \forall e_{mi} = 0. \end{cases} \tag{5.95}$$

Clearly, Eq. 5.91, Eq. 5.93 and Eq. 5.94 describe a variable-structure control (similar to the sliding mode control) which guarantees the asymptotical stability of the closed-loop control system for joint variable q_{mi} $(i = 1, 2, ..., n)$.

One interesting advantage of this control scheme is that parameters (α, β) can be learned during a robot's real-time interaction with its environment. Therefore, this method is in line with the spirit of the developmental principles for the physical development of an artificial system, like a humanoid robot.

5.5.7   Task-Space Control

As we mentioned earlier, tasks or actions are specified in task space. Thus, it is natural to consider whether the motion control loop can be formed from the sensory feedback in task space.

5.5.7.1 *Planning of Desired Outputs*

Refer to Fig. 5.14. A robot's input is a task or action which is normally accomplished by a sequence of ordered motions. Let $\dot{P}_e = (\dot{x}, \dot{y}, \dot{z}, \omega_x, \omega_y, \omega_z)^t$ denote the angular velocity vector of the end-effector's frame in an open kineto-dynamic chain. By default, all vectors are referenced to the base frame (frame 0) of an open kineto-dynamic chain. The first three elements of $\dot{P}_e$ form the linear velocity vector of the end-effector's frame while the last three elements of $\dot{P}_e$ form its angular velocity vector. The linear and angular velocity vectors can be determined if the trajectory of motion in task space can be planned according to the specification of a task or action. If only path is planned, $\dot{P}_e$ is replaced by its differential variation $\triangle P_e$. More detail on motion planning will be studied in Chapter 9.

When a robot is not in direct contact with its environment while performing a task or action, its motion is called *unconstrained motion*. Otherwise, it is called *constrained motion*. When a robot performs an unconstrained motion, the desired motion in task space can be defined by position and/or velocity vectors: $(P_e^d, \dot{P}_e^d)$.

If a robot's motion is constrained by direct contact with its environment, it is necessary to define one extra desired output which will be desired force/torque vector ξ_e^d. In general, ξ_e^d consists of two vectors: a) desired force vector f_e^d and b) desired torque vector τ_e^d.

5.5.7.2 *Posture Control in Task Space*

Let P_e^d denote the desired posture of the end-effector's frame in an open kineto-dynamic chain. Then the error vector will be

$$e(t) = P_e^d - P_e(t) \tag{5.96}$$

or simply

$$e = P_e^d - P_e$$

where $P_e(t)$ is the measurement of the actual posture of the end-effector's frame. This can be obtained by the robot's vision system or from forward kinematics if joint angle vector q_m is measured as feedback.

Now, let us define a Lyaponuv function as follows:

$$V = V(\dot{q}_m, e_m) = \frac{1}{2} \left[\dot{q}_m^t \bullet B_m(q) \bullet \dot{q}_m + e^t \bullet K_1 \bullet e \right] \tag{5.97}$$

where K_1 is a positive definite matrix.

Differentiating Eq. 5.97 with respect to time gives

$$\dot{V} = \dot{q}_m^t \bullet B_m(q) \bullet \ddot{q}_m + \frac{1}{2} \dot{q}_m^t \bullet \dot{B}_m(q) \bullet \dot{q}_m + \dot{e}^t \bullet K_1 \bullet e. \qquad (5.98)$$

By applying the result $\dot{e} = -\dot{P}_e$ (see Eq. 5.96) and substituting Eq. 5.71 into Eq. 5.98, we obtain

$$\dot{V} = \frac{1}{2} \dot{q}_m^t \bullet (\dot{B}_m - 2\,C_m - 2\,K_{mv}) \bullet \dot{q}_m + \dot{q}_m^t [\tau_m - G_m] - \dot{P}_e^t \bullet K_1 \bullet e \quad (5.99)$$

where $B_m = B_m(q,\dot{q})$, $C_m = C_m(q,\dot{q})$ and $G_m(q) = G_m$.

Next, let J denote the Jacobian matrix of the open kineto-dynamic chain. From our study of robot kinematics, we know that

$$\dot{P}_e = J \bullet \dot{q} = J \bullet K_r^{-1} \bullet \dot{q}_m.$$

The transpose of the above equation is

$$\dot{P}_e^t = \dot{q}_m^t \bullet J^t \bullet K_r^{-1}. \qquad (5.100)$$

(NOTE: K_r is a reduction-ratio matrix which is a diagonal matrix).

By applying the result

$$\dot{q}_m^t \bullet (\dot{B}_m - 2\,C_m) \bullet \dot{q}_m = 0$$

and substituting Eq. 5.100 into Eq. 5.99, we obtain

$$\dot{V} = -\dot{q}_m^t \bullet K_{mv} \bullet \dot{q}_m + \dot{q}_m^t \left[\tau_m - G_m - J^t \bullet K_r^{-1} \bullet K_1 \bullet e\right]. \qquad (5.101)$$

Accordingly, we choose the control law to be

$$\tau_m = G_m + J^t \bullet K_r^{-1} \bullet K_1 \bullet e - K_2 \bullet \dot{q}_m \qquad (5.102)$$

where K_2 is a positive-definite matrix.

Then Eq. 5.101 becomes

$$\dot{V} = -\dot{q}_m^t \bullet [K_{mv} + K_2] \bullet \dot{q}_m. \qquad (5.103)$$

Clearly, the control law in Eq. 5.102 guarantees the following conditions:

$$\begin{cases} \dot{V} < 0 & \forall \dot{q}_m \neq 0 \\[2mm] \dot{V} = 0 & \forall \dot{q}_m = 0. \end{cases} \qquad (5.104)$$

This simply means that the system under control is asymptotically stable. However, if the robot is operating in outer space, then $G_m = 0$ and the control law in Eq. 5.102 is a PD control (i.e. having position and velocity feedback).

5.5.7.3 *Force Compliance Control*

Assume that the end-effector of an open kineto-dynamic chain has a force/torque sensor to measure the actual contact force/torque between the end-effector's frame and its environment. If desired force/torque vector ξ_e^d can be specified, the error signal will be

$$e(t) = \xi_e^d - \xi_e(t)$$

where ξ_e is the measurement of the actual contact force/torque vector.

From our study of robot statics, we know that torque vector τ of the joints required to compensate for or maintain the contact force/torque at the end-effector is

$$\tau = J^t \bullet \xi_e \qquad (5.105)$$

or

$$\tau_m = K_r^{-1} \bullet J^t \bullet \xi_e$$

where J is the Jacobian matrix of the open kineto-dynamic chain and K_r is the reduction-ratio matrix (a diagonal matrix). (NOTE: $\tau = K_r \bullet \tau_m$).

If the control objective is purely to maintain desired contact force and torque ξ_e^d, from Eq. 5.105, the control law for force-compliance control can simply be

$$\tau_m = K_r^{-1} \bullet J^t \bullet \xi_e^d + K_1 \bullet e(t) \qquad (5.106)$$

where K_1 is a proportional control gain matrix which is a positive-definite matrix. This is a proportional control law. Alternatively, one can also choose a PI or PID control scheme.

5.5.7.4 *Hybrid Force and Trajectory Control*

For tasks to be performed by a robot, we may expect the end-effector's frame in an open kineto-dynamic chain to follow a predefined trajectory and also to maintain a predefined contact force and torque with its environment. In this case, the torques of the actuators will not only contribute to the dynamic behaviors for trajectory tracking but also to the desired contact force and torque. Accordingly, the dynamic model of the robot's system under control will become

$$B_m \bullet \ddot{q}_m + [C_m + K_{mv}] \bullet \dot{q}_m + G_m + J_m^t \bullet \xi_e = \tau_m \qquad (5.107)$$

where $B_m = B_m(q)$, $C_m = C_m(q, \dot{q})$, $G_m = G_m(q)$ and $J_m = J \bullet K_r^{-1}$.

Assume that the desired output for hybrid force and trajectory control are: ξ_e^d, q_m^d, $\dot{q}_m^d$ and $\ddot{q}_m^d$. Now, we make use of the *inverse dynamics control* scheme to derive a suitable control law.

Let us denote

$$N_m = [C_m + K_{mv}] \bullet \dot{q}_m + G_m,$$

Then Eq. 5.107 becomes

$$B_m \bullet \ddot{q}_m + N_m + J_m^t \bullet \xi_e = \tau_m. \qquad (5.108)$$

Since B_m is the inertial-matrix which has the inverse, Eq. 5.108 can be equivalently written as

$$\tau_m = B_m \bullet y + N_m \qquad (5.109)$$

with

$$y = \ddot{q}_m + B_m^{-1} \bullet J_m^t \bullet \xi_e. \qquad (5.110)$$

Under the inverse-dynamics control scheme, we assume that the matrices in the dynamic model are known. In this case, one can simply suggest the following control law:

$$\tau_m = B_m \bullet y^d + N_m + K \bullet (y^d - y) \qquad (5.111)$$

with

$$y^d = \ddot{q}_m^d + B_m^{-1} \bullet J_m^t \bullet \xi_e^d$$

and K is a proportional control gain matrix.

However, this control law does not consider desired output q_m^d and $\dot{q}_m^d$. The idea underlying inverse-dynamics control is to linearize Eq. 5.110 around the desired output with

$$y = y^d + K_1 \bullet (q_m^d - q_m) + K_2 \bullet (\dot{q}_m^d - \dot{q}_m) + K_3 \bullet (\xi_e^d - \xi_e) \qquad (5.112)$$

where $y^d = \ddot{q}_m^d + B_m^{-1} \bullet J^t \bullet \xi_e^d$ (the desired output of y) and (K_1, K_2, K_3) are the proportional control gain matrices. These matrices should be positive-definite matrices. In this way, the control law under inverse-dynamics control scheme is

$$\begin{cases} \tau_m = B_m \bullet y + N_m \\ y = y^d + K_1 \bullet (q_m^d - q_m) + K_2 \bullet (\dot{q}_m^d - \dot{q}_m) + K_3 \bullet (\xi_e^d - \xi_e). \end{cases} \quad (5.113)$$

Clearly, the system under control will reach dynamic equilibrium when $q_m = q_m^d$, $\dot{q}_m = \dot{q}_m^d$ and $\xi_e = \xi_e^d$.

5.5.7.5 *Impedance Control*

Let us examine, in the previous example, the dynamics of error due to linearizing with Eq. 5.112. In fact, error caused by linearizing is the difference between Eq. 5.112 and Eq. 5.110. That is:

$$(\ddot{q}^d - \ddot{q}_m) + K_1 \bullet (q_m^d - q_m) + K_2 \bullet (\dot{q}_m^d - \dot{q}_m) + (B_m^{-1} \bullet J_m^t + K_3) \bullet (\xi_e^d - \xi_e) = 0 \quad (5.114)$$

or simply

$$\ddot{e}_q + K_2 \bullet \dot{e}_q + K_1 \bullet e_q + (B_m^{-1} \bullet J_m^t + K_3) \bullet e_\xi = 0$$

where $e_q = q_m^d - q_m$ and $e_\xi = \xi_e^d - \xi_e$. The system under control will be at the dynamic equilibrium when errors e_q and e_ξ vanish.

For some applications, we may want the error dynamics of a control law applied to an open kineto-dynamic chain, such as a robot's limb, to behave like a mass-spring-damper system. In other words, the error dynamics of a control law is expected to be governed by

$$M^d \bullet \ddot{e}_p + D^d \bullet \dot{e}_p + S^d \bullet e_p = \xi_e \quad (5.115)$$

where:

- (M^d, D^d, S^d) are the desired mass, damping and stiffness matrices
- $e_p = P_e^d - P_e$ (i.e. the error or deviation of the actual posture of the end-effector's frame from its nominal or desired posture)
- ξ_e is the contact force/torque between the end-effector's frame and the environment. ξ_e is measured by the force/torque sensor mounted onto the end-effector

A control scheme which can achieve this objective is called an *impedance control*. By choosing $M^d = \{0\}$ and $D^d = \{0\}$, the impedance control becomes a *stiffness control*.

Refer to Eq. 5.108. We can express torque vector τ_m as

$$\tau_m = B_m \bullet y + N_m + J_m^t \bullet \xi_e \qquad (5.116)$$

with

$$y = \ddot{q}_m. \qquad (5.117)$$

Refer to Eq. 5.100. We have

$$\dot{P}_e = J_m \bullet \dot{q}_m \qquad (5.118)$$

where $J_m = J \bullet K_r^{-1}$.

Differentiating Eq. 5.118 with respect to time gives

$$\ddot{q}_m = J_m^{-1} \bullet (\ddot{P}_e - \dot{J}_m \bullet \dot{q}_m). \qquad (5.119)$$

As a result, Eq. 5.117 becomes

$$y = J_m^{-1} \bullet (\ddot{P}_e - \dot{J}_m \bullet \dot{q}_m). \qquad (5.120)$$

And the desired output of y when the end-effector's frame is following a desired trajectory is

$$y^d = J_m^{-1} \bullet (\ddot{P}_e^d - \dot{J}_m \bullet \dot{q}_m^d). \qquad (5.121)$$

Now, we linearize Eq. 5.120 around desired output y^d in the following way:

$$y = y^d + J_m^{-1} \bullet \left\{ (M^d)^{-1} \bullet \left[D^d \bullet \dot{e}_p + S^d \bullet e_p + \xi_e \right] + \dot{J}_m \bullet (\dot{q}_m^d - \dot{q}_m) \right\}. \qquad (5.122)$$

The difference between Eq. 5.122 and Eq. 5.120 will result in the expression of Eq. 5.115. In other words, the control law

$$\begin{cases} \tau_m = B_m \bullet y + N_m + J_m^t \bullet \xi_e \\[2mm] y = y^d + J_m^{-1} \bullet \left\{ (M^d)^{-1} \bullet \left[D^d \bullet \dot{e}_p + S^d \bullet e_p + \xi_e \right] + \dot{J}_m \bullet (\dot{q}_m^d - \dot{q}_m) \right\} \end{cases} \qquad (5.123)$$

will make the open kineto-dynamic chain, such as a robot's limb, behave like a mass-spring-damper system while executing its own task (i.e. following a trajectory).

5.5.8 *Image-Space Control*

It is easy to see that the majority of our daily activities are visually guided. Without a visual-sensory system, we would have tremendous difficulty in adapting ourselves to the real-world for any physical activity. Therefore, it is interesting to study how a robot's vision system interacts with its motion control system. However, it is necessary to have a solid understanding of how a robot's vision system works in order to understand the image-space control scheme. As a result, it makes sense to save the study of image-space control until the end of this book.

In Chapter 1, we stated that the intelligence is a measurable characteristic of both human and artificial system. Intelligence is quantitatively measurable and is inversely proportional to the effort spent on the achievement of a predefined goal/target. Naturally, one way to measure the intelligence level of an artificial system is to test its ability to perform visually-guided activities. In Chapter 9, we will study various schemes of limb-eye coordination for a humanoid robot.

For the purpose of cultivating interest in the study of image-space control scheme, let us briefly highlight some important issues related to it.

5.5.8.1 *Scenarios*

A humanoid robot may look like a human being. Accordingly, the combination of the binocular vision and the head-neck-body mechanism will form a typical head-eye system. The behaviors from a head-eye system are called *head-eye coordination*

If we model the mechanism underlying the head-neck-body as an open kineto-dynamic chain, a robotic head-eye coordination system can be formed by putting one or two cameras at the end-effector of an arm manipulator. This is known as the *eye-in-hand* configuration.

On the other hand, a humanoid robot has two arms with hands, and two legs with feet. The motion of these limbs can be guided by visual information in image space of a humanoid robot's binocular vision system. Consequently, another important configuration is the *eye-to-hand*, or *eye-to-feet*) configuration which describes the typical set-up of a robotic *hand-eye coordination*.

5.5.8.2   *Objectives*

For a task-space control scheme, we know that the desired output is the motion at the end-effector's frame in an open kineto-dynamic chain. In order words, the desired output is specified in terms of the posture P_e, its velocity $\dot{P}_e$ and its acceleration $\ddot{P}_e$. (NOTE: By default, all vectors are expressed in a common reference frame).

After we study the robot's visual sensory system in Chapter 8, we will understand that an image is the projection of a three-dimensional space onto a two-dimensional image plane. Conceptually, we can consider that these projections of an end-effector frame's posture, velocity and acceleration onto an image plane are measurable. If we denote p_e, $\dot{p}_e$ and $\ddot{p}_e$ the projections of an end-effector frame's posture, velocity, and acceleration onto an image plane, the objective of the image-space control scheme is to specify the desired output

$$(p_e^d, \dot{p}_e^d, \ddot{p}_e^d)$$

in image space, and design a control law which will command the motion of an open kineto-dynamic chain so that the actual projections

$$(p_e, \dot{p}_e, \ddot{p}_e)$$

converge to the desired ones.

Obviously, the first benefit of the image-space control scheme is the ability to automate the planning of desired output based on visual sensory information. It is a crucial step towards autonomy of a humanoid robot. The second notable benefit is that the image-space control scheme is also applicable to the case in which desired output $(p_e^d, \dot{p}_e^d, \ddot{p}_e^d)$ can be the projections of any object's posture, velocity and acceleration in a scene. This extension will result in interesting solutions to the problem of visually-guided trajectory or target followed by a vehicle or humanoid robot.

5.5.8.3   *Methodologies*

It is not surprising that human vision is qualitative (i.e. not metric) with regard to geometric quantities such as shapes and dimensions. Nonetheless, human beings are skilled at performing visually-guided activities. Naturally, one way of developing an image-space control scheme is to imitate human vision. In this way, we can develop solutions which do not require any metric 3-D reconstruction of the three-dimensional objects or scene.

Alternatively, we can also explore the quantitative aspect of a robot's visual-perception system. Accordingly, we will be able to develop solutions which can outperform human beings in visually-guided activities. We will study these methodologies further, in Chapter 9.

5.6 Summary

A robot is intrinsically a non-linear, multiple variable system with a variable-configuration. It is almost impossible to precisely and completely model the dynamics of a robot system. As a result, a robot's motion control must be operated in a closed-loop feedback manner. And a robot should be treated as a combination of mechanical, electromechanical, and control systems at this point.

An automatic-feedback control system has many advantages over an open-loop control system. We learned that the notable advantages include: a) the ability to achieve the desired output, regardless of unmodelled dynamics of a system under control, b) the resistance to noise and disturbance, c) the ability to achieve fast responses, and d) the ability to withstand all kinds of uncertainty and variations.

There are many mathematical tools for the design and analysis of a closed-loop control system. We leaned that the most important concerns regarding a closed-loop control system are: a) absolute stability (i.e. the steady-state response) and b) relative stability (i.e. the transient response). Although a vast body of knowledge about control has been developed, the most popular control scheme in industry is still the PID control law. And there are many techniques for the manual or automatic tuning of a PID control law. However, there is no proof about a PID control law's absolute stability. Nevertheless, a powerful mathematical tool is the Lyapunov method which is useful for the design of asymptotically stable control systems.

By definition, a system is a set of coherent elements which act together to achieve common objectives and outcomes. Therefore, it is easy to believe that a closed-loop control system is not simply a control law or algorithm. It must have a set of elements which act together. From a systems point of view, we learned that a closed-loop control system must have two sets of elements, namely: a) the control elements (in the forward channel of a closed loop) and b) the sensing elements (in the feedback channel of a closed loop).

A robot is able to execute motion because of the energy applied to the joints of its underlying mechanism. If we use electric motors, energy applied to a robot will come from the electrical domain. There are naturally concerns about: a) how we regulate the amount of electrical energy released by an electric motor to its corresponding joint, and b) how we alter the direction of motion of an electric motor (or its corresponding joint). Thus, we studied the working principles behind the power switch and logic sequence generator for the purpose of altering the electric motor's direction of motions. Subsequently, we also studied the working principles behind the power drive which regulates an electric motor's velocity.

As for sensing elements, we studied the generic sensing principle. While we briefly discussed safety sensors, force/torque sensors, and tactile sensors, we emphasized the study of the popular optical encoders. Almost all the motion sensors found in industrial robots are incremental optical encoders. These require the robot to perform a homing procedure before it is ready to operate. This is not desirable for a humanoid robot because it is dangerous to perform a homing procedure in a nonindustrial environment. As a result, we advocate the use of absolute optical encoders in the design of a humanoid robot's control system.

The most important control element is clearly the digital motion-controller and the control algorithm(s) running on it. A digital motion-controller is part of a robot's information system. Its primary role is to run the control algorithm(s) and interface with the sensors and actuators which are inside the closed-loop control system.

Regarding the design of a robot's control system, we studied the issues of: a) how to plan the desired output, b) how to possibly establish a dynamic model of a robot's system under control, c) how to obtain a timely sensor feedback, and d) how to design suitable control algorithms. The design of suitable control algorithms is not a simple matter. As a result of the complexity of a robot system, there are various control methodologies. Nevertheless, we learned that the PID control and variable-structure control are two popular control methodologies. And, typical control schemes in robotics include joint-space control, task-space control and image-space control.

5.7 Exercises

(1) Explain why a closed-loop control system is better than an open-loop control system.
(2) What is the transient response of a system? And, what is the steady-state response of a system?
(3) Explain the physical meaning of the absolute stability of a closed-loop control system.
(4) Explain the advantages of using Laplace Transform in the study of a closed-loop control system.
(5) The transfer function of a closed-loop control system is

$$\frac{Y(s)}{R(s)} = \frac{K}{(s-1)(s+1)}.$$

Is this a stable system?
(6) Let us use a PID control law to control a system with a transfer function of

$$G_p(s) = \frac{1}{(s+1)(s+2)(s+3)}.$$

Use Simulink of MATLAB to tune the parameters of the PID control law.
(7) What are the control elements inside a robot's control system?
(8) What are the sensing elements inside a robot's control system?
(9) Explain the difference between linear power drive and PWM power drive.
(10) How do you alter an electric motor's direction of motion?
(11) How do you regulate an electric motor's velocity?
(12) Explain why safety is an important issue in the development of a humanoid robot?
(13) What are the considerations for the appropriate placement of a motion sensor?
(14) If an absolute optical encoder has to measure 360^0 with an accuracy of 0.01^0, how many logic states should it have?
(15) Explain why an incremental optical encoder allows us to measure the absolute angular position.
(16) Why should a force/torque sensor's mechanism be stiff?

(17) Refer to Fig. 5.45b. Prove that the input and output relationship is

$$V_i = \frac{R_b}{R_a} \bullet (V_2 - V_1).$$

(18) Explain the difference between a force/torque sensor and a tactile sensor. Why are they important to a robot's motion execution?

(19) Explain the reasons behind the popularity of the PID control law used in the majority of automated equipment in industry.

(20) What is the variable-structure control?

(21) Explain the differences between the joint space control scheme and the task space control scheme?

(22) Draw a block diagram of the closed-loop control system which makes use of the control law in Eq. 5.81.

(23) Draw a block diagram of the closed-loop control system which makes use of the control law in Eq.5.91.

(24) Draw a block diagram of the closed-loop control system which makes use of the control law in Eq. 5.102.

(25) Draw a block diagram of the closed-loop control system which makes use of the control law in Eq. 5.113.

(26) Draw a block diagram of the closed-loop control system which makes use of the control law in Eq. 5.123.

5.8   Bibliography

(1) Astrom, K. J. and T. Hagglund (1995). *PID Controllers: Theory, Design and Tuning*, Instrument Society of America, Research Triangle Park, North Carolina.

(2) Edwards, C. and S. K. Spurgeon (1998). *Sliding Mode Control: Theory and Applications*, Taylor and Francis Ltd.

(3) Horowitz, P. and W. Hill (1989). *The Art of Electronics*, Cambridge University Press.

(4) Koivo, A. J. (1989) *Fundamentals for Control of Robotic Manipulators*, John Wiley and Sons.

(5) Ogata, K. (1997). *Modern Control Engineering*, Prentice-Hall.

(6) Sciavicco, L. and B. Siciliano (1996). *Modelling and Control of Robot Manipulator*, McGraw-Hill.

Chapter 6

Information System of Robots

6.1 Introduction

Without a body, intelligence will not develop. Without intelligence, a body will have limited usage. Although it is an artificial system, a humanoid robot's intelligence should develop. Artificial intelligence is synonymous with computational intelligence because all the mental and physical activities of a humanoid robot rely on data-processing carried out on digital computer(s). As a result, we can compare a humanoid robot's information system with the human brain.

The human brain is undeniably powerful enough to control and coordinate our physical and mental activities in real-time. This ability is an important function of a humanoid robot as well. By definition, *real-time* means that an action is performed by a system within a specific time interval. Thus, an action performed by a system must meet a stipulated *deadline*. When a system performs multiple actions, *real-time* also means the *simultaneity* of actions undertaken by the system.

When a system is capable of performing real-time actions, this means its actions are predictable. The predictability of a system's actions is very important as it is the only way to ensure the system's deterministic behaviors. Therefore, a humanoid robot's information system must be a real-time system.

Today's computer has not yet reached the ability level of the human brain. Consequently, it is not realistic to rely on a single computer to handle the multiple functions of a humanoid robot. It is best to use a cluster of computers to form a distributed system. Thus, a humanoid robot's information system should be a distributed system supported by a real-time network in order to ensure predictable behaviors.

In this Chapter, we will study the basics of the information system underlying a humanoid robot's control, sensing, perception, and decision-making functions. In general, an information system consists of four parts: a) data processing, b) data storage, c) data interfacing and d) data communication. Therefore, in this chapter, we will emphasize the fundamentals of these topics. We will reserve discussion of the computational principles (or algorithms) underlying the data-processing related to control, sensing, perception, and decision-making for other chapters.

6.2   Imitating the Brain

Humans represent the perfect embodiment of mind and body. The mind, which consists of consciousness and intelligence, is undoubtedly a complicated process which physically resides in the brain. From an engineering point of view, we can consider the embodiment of mind and body as elaborate mapping among: a) the body's kineto-dynamic systems, b) the brain, including the decision-making system (see Chapter 9), and c) the body's sensory systems. Fig. 6.1 is a simple illustration of the relationship between the human brain and human body.

The human brain is an extremely complex system, and not a great deal about it is understood. What is certain, however, is that the cerebral structure is highly parallel with the pre-defined divisions of the brain's neural system. Throughout the mental and physical development of the body and mind, the divisions of the brain's neural system are mapped in a highly sophisticated manner to the sensory systems and motor systems along pre-defined pathways (i.e. nerve fibers). From an engineering point of view, this is very similar to the formation of the cluster of basic feedback control loops: sensor-controller-actuator. Because of the complex structure of the brain and body, we are able to perform all kinds of mental and physical activities, such as thought, analysis, knowledge, language, perception, emotion, memorization, sensory-motor coordination, and sensory-integration.

An infant is born with: a) an innate bio-mechanical system, b) an innate bio-kineto-dynamical system (the muscles and body's bio-mechanism), c) innate sensory systems, and d) an innate neural system. A person's mind and intelligence are developed through the whole period of his/her growth. And, the growth of a person's mental and physical capabilities is primarily influenced by interaction with the environment and society.

Now, the question is: What type of innate information system should a

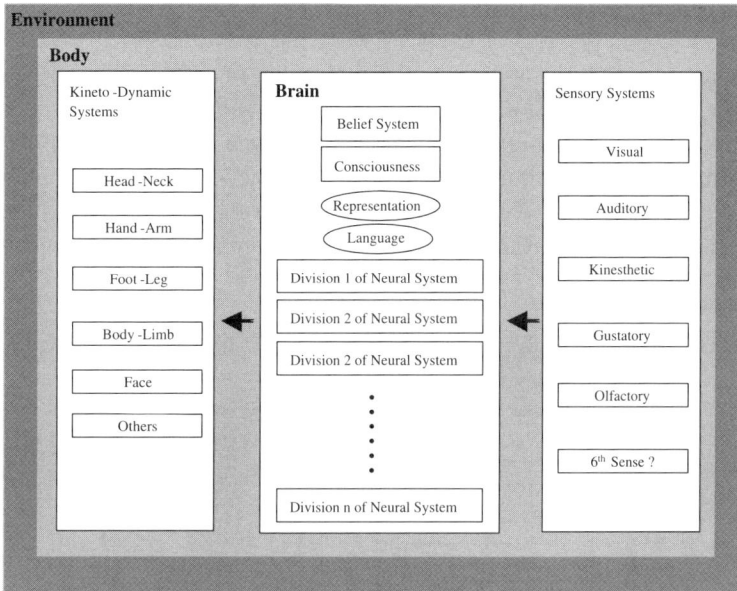

Fig. 6.1 A simple illustration of the relationship between the human brain and human body.

humanoid robot have if we want its intelligence to grow through real-time interaction with the environment and society? It is not easy to give a clear answer to this question because today's information and sensory systems are not yet comparable to those of humans. If we want future humanoid robots to have human-like mental and physical capabilities, we need to improve the available information and sensory systems.

Today, it is easy to form a cluster of networked computers to imitate the parallel structure of the human brain's neural system. However, in order to support the multiple behaviors performed by an artificial body, it is necessary to have the "executive process" (or "operating system") which would harmoniously coordinate the multiple computational tasks running on a cluster of networked microprocessors. However, the challenge in robotics is whether it is possible to develop an innate "operating system" and "software architecture" that can automatically map sensory outputs to actuators' inputs in order to autonomously form *sensor-controller-actuator* loops for specific behaviors. These loops are called *actors or agents*.

6.3   Imitating the Mind

A humanoid robot is not simply a combination of mechanics and control. It is the embodiment of an artificial mind (a complex process for behavior development and learning), and an artificial body, as shown in Fig. 6.2. This embodiment enables a humanoid robot to interact with its environment in a timely manner. The real-time interaction between a humanoid robot and its environment is crucial to the proper development of its physical and mental abilities.

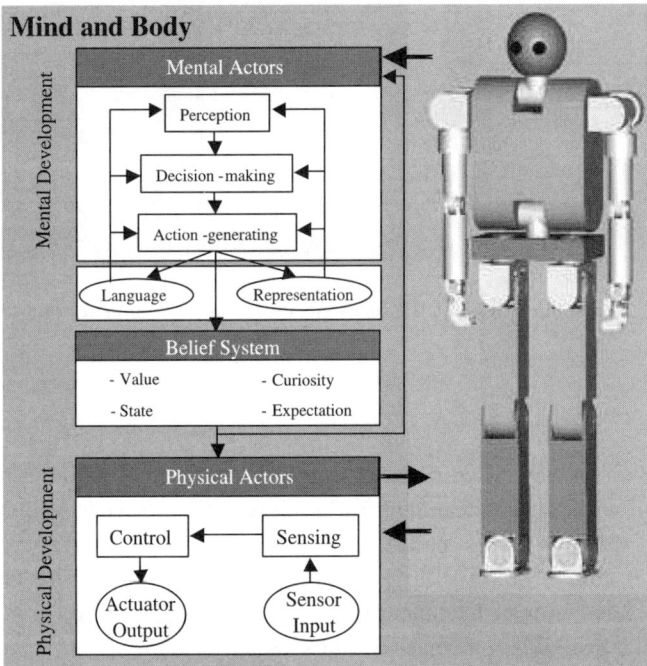

Fig. 6.2   A proposed actor-based framework for the embodiment of artificial mind and artificial body.

6.3.1   *Autonomous Actors or Agents*

As we mentioned already, the human brain is a complex neural system consisting of about 100 billion neurons. The architecture of the human brain's neural system is highly parallel with well-organized partitions dedicated to various physical and mental functions. By physical functions (or abilities), we mean all actions carried out by the sensory-motor systems embedded in the body's mechanism (e.g. limbs, body, neck/head etc). For a sensor-guided action-taking system, there is no doubt that the best control structure is the closed-feedback loop. Naturally, we can consider a sensing-control loop as a *physical actor or agent*, as shown in Fig. 6.2. In this way, we can treat a stand-alone sensor as a degenerated physical actor. Inside an information system, an actor or agent is an autonomous computational process or task. It does not require reprogramming.

Similarly, a mental function refers to a specific type of mental activity which affects the internal representations of the real, external world. A mental activity is intrinsically conditioned by one or multiple sensory input(s). Hence, the basic mechanism that drives a mental activity is the well-known *perception-decision-action* loop. Accordingly, we can consider a perception-decision-action loop as a *mental actor or agent*, as shown in Fig. 6.2. And, we can treat a stand-alone perception as a degenerated mental actor.

The outcome from a mental actor may serve three purposes:

- To enrich language, knowledge, and skills.
- To enrich internal representation.
- To update belief system, a distinct and indispensable component of an autonomous system.

6.3.2 *Autonomous Behaviors*

An actor or agent inside an information system is an autonomous, computational process/task because it requires no reprogramming. Basically, a mental actor responds to input from the sensory systems, acts upon the belief system, and enriches the language and internal representations. On the other hand, a physical actor will mainly be guided by output from the belief system. Note the following belief-guided behaviors:

- Value-driven behavior. (Is it rewarding to undertake certain actions?)
- Curiosity-driven behavior. (Is it interesting and exciting to do, explore, or understand certain new things?)
- State-driven behavior. (Is it different to perform the same tasks under different mental and physiological states?)
- Expectation-driven behavior. (How do we achieve a goal or new activities?)

A physical actor's output will act on the artificial body. For example, if a mental actor's output is to escape a dangerous situation, it will update the belief system to make "safety" the highest priority. This in turn, will trigger a set of physical actors which work together on the artificial body, so that the robot will quickly run away.

Based on the concept of mental and physical actors, we can treat the complex interaction inside a humanoid robot's information system as an *artificial mind system*. We can attempt to define *artificial mind* as follows:

Definition 6.1   An artificial mind is a complex process of behavior development as the result of interaction among the mental and physical actors in a body's information system.

Just as the human mind describes the dynamic characteristics of the human brain, an artificial mind characterizes the dynamics of an information system. The mechanism underlying this complex process is *behavior development and learning*.

By definition, a behavior is a sequence of ordered actions performed by a group of interacting mental and/or physical actors. Therefore, the activation of a behavior depends on the configuration of the interacting mental and/or physical actors. Obviously, the creation of a new mental or physical actor will increase the number of possible configurations of interacting mental and physical actors. Thus, one way to change a humanoid robot's behavior is to form new mental and physical actors to enhance existing mental and physical actors, or to suppress undesirable mental and physical actors. (NOTE: Whether or not a mental or physical actor is undesirable is determined by the belief system).

Accordingly, we can attempt to define *behavior development and learning* in the following way:

Definition 6.2   Behavior development and learning refer to activities dedicated to the formation, enhancement, and suppression of mental and physical actors.

The formation, enhancement and suppression of mental and physical actors cannot occur without input from the society and environment. Interaction in the society, where knowledge is gained from the environment, experience, and others, serves as a catalyst to stimulate the birth, enhancement and suppression of mental and physical actors.

6.3.3   *Computational Tasks*

Refer to Fig. 6.2. The mental and physical actors of a robot's information system are intrinsically computation-dependent because they process sensory feedback and act accordingly. From a data-processing point of view, a humanoid robot's information system must support the following computational tasks:

- Distributed Motion Control:
  The majority of a robot's activities are centered on motion as an action is a sequence of ordered motions. Therefore, motion control is critical to the usefulness of a humanoid robot. As we studied in Chapter 5, a robot's motion control involves sensing elements for feedback, control elements for action-taking, and control algorithm(s) for decision-making. All the sensory feedback and control algorithms run on the *digital motion controllers* which are part of a robot's information system.

- Motion Planning:
  As we studied in Chapter 5, one critical issue in the design of a robot's control system is how to plan the desired motions. There are many computational principles and algorithms for motion planning, all of which must run on a digital computing platform.

- Motion Perception:
  One important control scheme for a robot's motion execution is *image-space control*. There are two paradigms of image-space control: a) visual-servo control and b) visually-guided planning & control. As a result, visual perception is indispensable. Obviously, all principles and algorithms underlying visual perception are computation-dependent.

- Communication:
  A humanoid robot's body includes a complex mechanical system with many degrees of freedom. Intrinsically, a robot is a multiple-input system with a variable configuration. The large number of degrees of freedom implies the necessity to adopt a distributed sensing and control scheme for motion execution and coordination. As we discussed earlier, a robot must be a real-time system with predictable behaviors. Therefore, it is critical for a robot's information system to support real-time communication among its internal motion-control subsystems. And, this communication must rely on a network (e.g. field-bus), which is in proximity to the sensors, actuators, and controllers.

- Inter-robot Communication:
  Obviously, it is desirable to develop humanoid robots which can not only perform tasks as individuals, but also work as a group. Hence, a robot's information system must be able to communicate with other equipment as well as with other robots.
- Man-robot Interaction:
  A humanoid robot's activities should not be confined to a factory. In fact, the humanoid robot has a wider range of applications outside the factory. It is not inconceivable for humanoid robots to co-exist with humans (like an automobile does). Therefore, a humanoid robot must not only be safe, but also friendly. Ideally, a humanoid robot should adapt itself to the ways in which we communicate and interact. And, a humanoid robot's information system should be able to support the computations required for real-time man-robot interaction based on auditory and visual sensory data.
- Behavior Development and Learning:
  A distinct feature of the humanoid robot, as opposed to the industrial robot, is its physical and mental self-development. Just as a human experiences incremental development from infancy to adulthood, a humanoid robot should also be equipped with a self-development mechanism (yet to be discovered) so that it is able to physically and mentally grow through real-time interaction with its environment. Obviously, the principles and algorithms of behavior development and learning for a humanoid robot's physical and mental development are computation-dependent.

Now, we know that the information system is crucial to the development of a humanoid robot. Without an information system, there can be no embodiment of artificial mind and human-like mechanism (body).

6.4   Data-Processing Hardware

Nowadays, we widely use digital computers for data-processing and communication. It is interesting to note that the computer was first built during the Second World War in the 1940s. A computer is a machine which can perform computations (data-processing) and can also handle data storage and communication.

6.4.1 *Digital Computers*

The first generation of computers were based on analog circuits. Digital computer emerged in the early 1950s with the advent of digital circuits (circuits that manipulate pulses). In the early days, digital computers used vacuum tubes to construct the inner logic circuits. This explains why the old digital computers were bulky, expensive, and fragile.

With the rapid development of solid-state circuit technology (especially with the invention of transistors at Bell Laboratory in 1948), microprocessors became a reality in the early 1970s.

Formally, a digital computer can be defined as follows:

Definition 6.3 A digital computer is a machine that undertakes pulse-synchronized calculations through a sequence of ordered instructions called a *program*.

Here, the term *digital* simply means "discrete state". The simplest form of discrete state is the binary state which only has two values: 0 or 1. In electronics, the "0" state can be represented by a low voltage (e.g. 0V), while the "1" state can be represented by a high voltage (e.g. +5V). We also call a binary state a logic signal because we can associate the binary values "0" and "1" with the logic values "true" and "false". Often, we call "1" the logic high and "0" the logic low.

If we plot the logic signal versus the time, we obtain a pulse waveform which is similar to the one shown in Fig. 6.3.

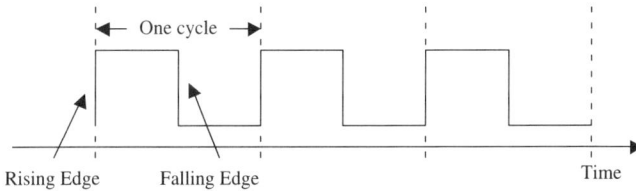

Fig. 6.3   Pulse waveform.

Now, it is easy to understand why a circuit is called a *logic circuit* if it takes a pulse waveform as input and produces a different (or similar) pulse waveform as output. This also explains why the advent of a pulse circuit marked the birth of the digital computer.

As shown in Fig. 6.3, a pulse signal includes two transitions: a) the transition from logic low to logic high, and b) the transition from logic high to logic low. The former is called the *rising edge* and the latter is called the *falling edge*. Because of these transitions, a computer is capable of performing all kinds of calculations. Therefore, a computer can be treated as a machine that controls logic transitions for the performance of intended calculations.

In order to make the results of a computation predictable, all the logic transitions inside a digital computer must be synchronized along the time axis. To do this, it is necessary to use a reference pulse waveform. A device that generates a regular pulse waveform of reference is called the *system clock*. In general, a reference pulse waveform looks like the one shown in Fig. 6.3. For one cycle, there are three events (logic transitions):

- The first rising edge: This indicates the start of the current cycle.
- The falling edge: This indicates the half-cycle.
- The second rising edge: This indicates the end of the current cycle and the start of a new cycle.

In fact, all operations inside a digital computer are synchronized by these three events (logic transitions). As a result, all operations inside a digital computer are called *cycle-by-cycle operations*.

6.4.2 *Microprocessors*

With the advent of solid-state circuits and the invention of the transistor, a digital computer can be made of micro-circuits with transistors which is also called *integrated circuit*. Thus, a new active device, called a *microprocessor*, has become a dominant product in the computer industry. A formal definition of a microprocessor can be stated as follows:

Definition 6.4 A microprocessor is a digital computer which consists of a single integrated circuit (IC, for short).

Like a digital computer, a microprocessor is a data-processing device that manipulates pulse waveforms (i.e. the digital signals) in order to perform calculations.

6.4.2.1 *Memory*

If we look at the following simple expression:

$$a = b + c$$

we easily understand that a is the result of the addition of the two operands b and c. Before this arithmetic operation can take place, we must know where the values of b and c are stored. After this arithmetic operation, we must also know where to store the resulting value a.

Clearly, a data-processing device must have a circuit unit to store data. In a digital computer, this circuit unit is called *memory*, and the data is simply a number.

Memory is a special type of logic circuit. A memory's key features include:

- Its logic output is equal to its logic input.
- The input and output of a memory circuit do not need to be synchronized.
- Input to a memory circuit is called *write*, while output from a memory circuit is called *read*.
- If it is possible to perform both read and write operations while the memory circuit is "on", the memory circuit is called a *random-access memory*, or RAM.
- If it is possible to perform the write operation only once while there is no limit for the read operation, the memory circuit is called a *read-only memory*, or ROM.
- If it is possible to perform the write operation repeatedly only with special equipment or procedures which normally undertake erasing and programming operations, the memory circuit is called an *erasable and programmable read-only memory* or EPROM.

A logic variable that can hold the logic value of either "1" or "0" is called a *bit*. Hence, the basic component in a memory circuit is a logic unit which only stores a bit. A bit is the basic unit in a memory and is also called a *memory cell*. By putting eight such units together, we form a *memory location* which can store a byte (8 bits).

Memory is organized in the form of a list of bytes or memory locations. Each memory location must have an index that indicates its address on the list. If a memory is not divided into a set of sub-lists, the index of a memory location inside the (global) list is called a *memory address*. In summary, a memory is composed of a list of (memory) locations and each location has a unique address.

6.4.2.2 *Number Representation*

A byte is a string of 8 bits that are normally labelled as b_7, b_6, b_5, b_4, b_3, b_2, b_1 and b_0. For any string of n bits, the last bit (b_0) is called the *least-significant bit*, or LSB, while the first bit (b_7) is called the most-significant bit, or MSB. A byte can represent an integer number from 0 to 255. (NOTE: $255 = 2^8 - 1$, the maximum value represented by a string of 8 bits). If we use MSB to represent the sign (0 for positive and 1 for negative), a string of n bits can represent an integer number within the interval:

$$\left[- \left(2^{n-1} - 1 \right), \ \left(2^{n-1} - 1 \right) \right].$$

In real life, many physical quantities are represented with real numbers. A common way to represent a real number with a decimal is to use a *floating-point number* because any real number can be written as the product of a mantissa and the exponent of a base number (e.g. the number 2). In other words, any real number with a decimal can be written as the ratio between two integer numbers. For a string of n bits, a floating-point number is represented as follows:

- The MSB for representing the sign.
- The subsequent n_1 bits for representing the exponent of base number 2.
- The rest of n_2 ($n_2 = n - n_1 - 1$) bits for representing the mantissa.

If we denote F a floating-point number, represented by a string of n bits, we have:

$$F = (-1)^{MSB} \bullet M \bullet 2^{E - E_0} \tag{6.1}$$

where M is an integer representing the mantissa by n_2 bits, E is an integer representing the exponent by n_1 bits, and $E_0 = 2^{n_1 - 1} - 1$.

Example 6.1   We represent a floating-point number with a string of 32 bits. Assume that an exponent is represented by 8 bits ($n_1 = 8$) after the MSB, and the mantissa is represented by the last 23 bits ($n_2 = 23$). Then, $E_0 = 127$ ($2^7 - 1$). The maximum value for mantissa M will be 2^{23} and its minimum value will be 0.

As for exponent E, the maximum value will be 255 (i.e. $2^8 - 1$) and its minimum value will be 0.

Accordingly, the maximum floating-point number will be:

$$F_{max} = (-1)^0 \bullet 2^{23} \bullet 2^{255 - 127}.$$

And the minimum floating-point number will be:

$$F_{min} = (-1)^1 \bullet 2^{23} \bullet 2^{255-127}.$$

◇◇◇◇◇◇◇◇◇◇◇◇◇◇◇◇◇◇◇◇◇◇

6.4.2.3 *Arithmetic Logic Unit (ALU)*

In mathematics, we know that any real and differentiable function can be expanded into a corresponding Taylor series at certain precision. This fact explains why all calculations can be realized with a sequence of ordered arithmetic operations (addition, subtraction, division and multiplication) at certain precision. In theory, all the arithmetic operations of subtraction, division, and multiplication can also be fulfilled with a sequence of ordered additions at certain precision.

Example 6.2 For a sine function $y(x) = \sin(x)$, its corresponding Taylor series is:

$$y(x) = x - \frac{x^3}{2!} + \frac{x^5}{5!} - \text{...} + \frac{(-1)^n \bullet x^{2n+1}}{(2n+1)!} + \text{...} .$$

◇◇◇◇◇◇◇◇◇◇◇◇◇◇◇◇◇◇◇◇◇◇

As a result, the simplest way to design a data-processing unit is to construct a circuit that can perform addition on top of some Boolean logic operations. This explains why the heart of a computer is made of the Arithmetic & Logic Unit or ALU, for short. The ALU is a logic circuitry which is capable of performing the arithmetic and logic operations (e.g. addition, multiplication, and Boolean logic). In general, an ALU can be treated as a logic device which has at least two input (the operands) and one output (the result).

6.4.2.4 *Bus*

So far, we know that the memory is a place to store data (i.e. the numbers) and the ALU is an engine that manipulates data to produce results. The memory and the ALU are two distinct logic circuits in a digital computer. There is an obvious need to connect these two devices. Since these are electronic logic devices, the only way to connect them is to use wires that can conduct pulsed-electrical signals. (NOTE: For long distances, it is advantageous using optic fiber to carry pulsed-signals). In computer engineering, a set of wires that connects electronic logic devices is called a *bus*. In other words, a bus is a vehicle for the transportation of logic signals (or numbers) across electronic logic devices.

As we studied earlier, memory is a list of memory locations, each of which has an unique address. In order to communicate or interface with a memory device, we need to transport the contents (data) as well as their addresses. A common way to do this is to have a bus dedicated to data (contents) and a bus dedicated to addresses. The former is called a *data bus* and the latter is called an *address bus*.

In order to work properly, a digital computer requires other logic signals such as "read", "write", etc. The wires that carry logic signals for controlling operations inside a digital computer are grouped into a bus called a *control bus*. In practice, we seldom refer to this bus because it is more convenient to directly refer to the control-signal lines themselves.

6.4.2.5 *Generic Architecture of Computers*

Now, we know that a digital computer should have ALU, memory, and bus. The next question is: What does a digital computer look like? Nowadays, all digital computers are based on the famous architecture invented by the American mathematician, Von Neumann. The Von Neumann computer architecture, as shown in Fig. 6.4a, is composed of a processor and a memory which are connected by data and address buses. Of course, it is necessary to have control-signal lines. However, for the sake of clarity, these are omitted in the illustration.

As we mentioned earlier, a digital computer performs calculations by executing a sequence of ordered-instructions called a *program*. Since a computation involves at least a program and some data for processing, computers based on the Von Neumann architecture must store the program and data on the same memory unit. And, it is easy to understand that the procedure for an ALU to execute a program should look like the following:

- Step 1: Fetch an instruction.
- Step 2: Fetch the operand(s).
- Step 3: Perform arithmetic and logic operations.
- Step 4: Store the result.
- Repeat Steps 1 to 4 until all the instructions of a program have been completed.

Clearly, it is not possible for the Von Neumann-style computer to perform the operations of Step 1 and Step 2 in parallel. One way to improve

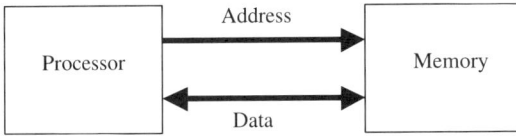

(a) Von Neumann Architecture

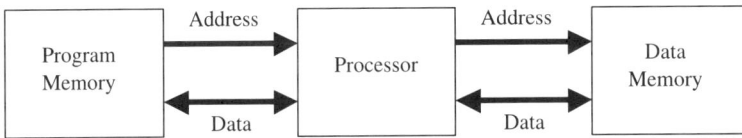

(b) Harvard or DSP Architecture

Fig. 6.4   The Von Neumann and Harvard computer architecture.

the speed of computation is to use the *Harvard architecture*, developed at Harvard University, USA, as shown in Fig. 6.4b. The Harvard architecture is also known as the *DSP architecture* as all the DSPs are based on the Harvard architecture. (NOTE: DSP stands for Digital Signal Processor). As we can see from Fig. 6.4b, the Harvard-style computer separates program storage from data storage. Thus, a Harvard-style computer will have at least two memories: one for programs and one for data. And, the execution of a program will be performed in the following way:

- Step 1: Fetch an instruction and its operand(s).
- Step 2: Perform arithmetic and logic operations.
- Step 3: Store the result.
- Repeat Steps 1 to 3 until all the instructions of a program have been completed.

6.4.2.6   *Cycle-by-Cycle Operations*

As we discussed above, the three basic actions of a processor are: a) read, b) perform arithmetic computation, and c) write. Since all operations in a digital computer must be synchronized in order to produce deterministic results, these three basic actions must also be synchronized.

In fact, a processor's system clock is responsible for producing a regular pulse waveform that serves as a reference for synchronization. Refer to Fig. 6.3 and Fig. 6.4a. A common way to synchronize the read operation is as follows:

- Step 1: When the first rising edge occurs, the processor sends the address to the address bus.
- Step 2: When the falling edge occurs, the memory sends the data stored in the location indicated by the address to the data bus.
- Step 3: When the second rising edge occurs, the processor reads in the data from the data bus.

Similarly, a common way to synchronize the write operation is as follows:

- Step 1: When the first rising edge occurs, the processor sends the address to the address bus.
- Step 2: When the falling edge occurs, the processor writes the data to the data bus.
- Step 3: When the second rising edge occurs, the memory latches in data and stores it in the memory location indicated by the address.

However, a common way to perform arithmetic operations is to allocate a number of cycles of the reference pulse waveform to an instruction. This number depends on the complexity of the instruction. For example, if an instruction requires three cycles of the system clock, then three cycles will be allocated to this instruction. During this period of time, the processor only undertakes the computation of the instruction. In practice, special circuitry called a *control logic unit* automatically handles the allocation of cycles required by the execution of instructions supported by a processor.

Now, let us examine the timing involved in executing one instruction by a processor. Let T be the cycle time of a processor's system clock. We denote:

- n_r the number of cycles required to read in one instruction and its operand(s) from the processor's memory;
- n_i the number of cycles for the processor's ALU to execute the instruction;
- n_w the number of cycles required to write the result in the memory.

It becomes clear that the total execution time for the completion of one instruction will be

$$t_e = (n_r + n_i + n_w) \bullet T. \tag{6.2}$$

For any given instruction, its corresponding execution time is a fixed number (i.e. t_e is predictable). Therefore, a digital computer can be treated as a real-time system and can play the role of a robot's artificial brain. When a processor is executing an instruction, it cannot be interrupted or halted. Hence, a processor's responsiveness depends on the execution time t_e of an instruction. In order to improve time responsiveness, it is desirable to shorten the execution time of an instruction. This can be done in the following ways:

(1) Shorten the cycle time of a processor's system clock (i.e. increase the frequency of a processor's system clock).
(2) Reduce n_r and n_w by increasing the number of bits that a data bus can carry at one time. (For example, a data bus with 32 lines can carry four bytes at a time instant).
(3) Reduce n_i by incorporating more ALUs into a processor.

6.4.2.7 *Generic Architecture of a Microprocessor*

A processor is, basically, a device that can repeatedly perform cycle-by-cycle operations. As a result, a processor should have the following basic elements, as shown in Fig. 6.5:

Register

A register is simply a memory location having 8-bits, 16-bits, or a multiple of 8-bits. For a read or write operation, the processor must have one register, called the address register, to hold the address and one or more registers to hold the data. When the ALU is performing computation, it is very likely that the processor will require registers to store the intermediate results. In fact, for proper operation, a processor requires the following registers:

(1) Address register to hold the address,
(2) Data registers (including *accumulators*),
(3) One, or more, status registers to report the working conditions of the ALU,
(4) A program-counter register to indicate the address of a program's instruction to be subsequently executed,
(5) Index registers to hold the relative location of the sub-lists inside a memory if it is divided into different segments for the separate storage of the operation system, programs, data and stacks etc,

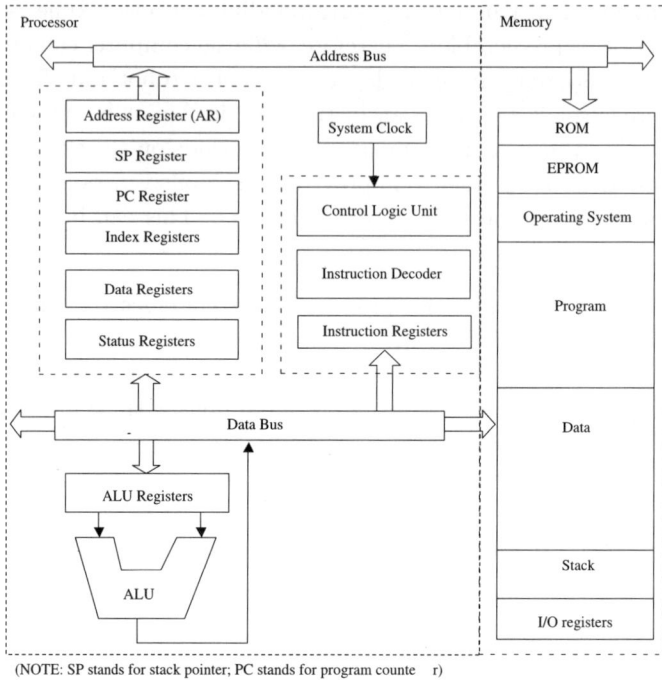

(NOTE: SP stands for stack pointer; PC stands for program counter)

Fig. 6.5   The generic model of a microprocessor.

(6) ALU registers to hold the operands as the input to the ALU,
(7) Instruction register(s) to hold the code of an instruction itself.

Instruction Decoder

Usually, a processor will support a set of pre-defined instructions which is also called an *instruction set*. All the instructions are compactly encoded in order to minimize storage space. When an instruction code is received by a processor after a read operation, it is necessary to have circuitry to decode it, in order to know the arithmetic or logic operation to be performed by the ALU. This circuitry is called *instruction decoder*.

Control Logic Unit

As we discussed earlier, for the proper operation of a processor, it must have a control bus to carry control signals. A typical control signal is a read or write signal which communicates between the processor and its memory.

On the other hand, a processor is a device which performs computations dictated by a program of ordered instructions. It is clear that a processor must handle the sequence logic of a program execution which is, in turn, synchronized by the processor's system clock.

In fact, the circuitry inside a processor which generates the required control-logic signals is called a *control-logic unit*.

Based on the above discussion, we can easily depict a generic model of a processor. Fig. 6.5 shows an example.

6.4.3 *Micro-controllers*

In robotics, we are interested in an information system which can serve as a robot's artificial brain. In order to do this, it is necessary for the information system to be able to interface with various other systems on a robot, or elsewhere. Consequently, a processor with a memory is not sufficient to construct a robot's artificial brain. A set of interfacing systems, commonly called *I/O systems*, must be incorporated into the microprocessor. For the sake of simplicity, let us consider the design solution of memory-mapped I/O systems such as the processor families made by Motorola, Inc., as shown in Fig. 6.6.

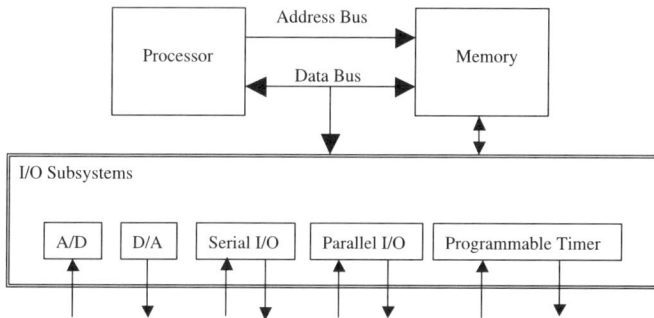

Fig. 6.6   The generic model of a micro-controller.

In general, a microprocessor combined with a set of dedicated I/O systems is called a *micro-controller*. From Fig. 6.6, a micro-controller can be formally defined as follows:

Definition 6.5   A micro-controller is a microprocessor with a set of I/O systems built on a single integrated-circuit or chip.

The I/O systems commonly found on a micro-controller include: a) analog-to-digital conversion system, b) digital-to-analog system, c) serial-communication system, d) parallel-communication system, and e) programmable-timer system. We will study these, in further detail in the later part of this chapter.

There are many types of micro-controllers on the market. Since a micro-controller is built on a single chip, products made from them are normally very cheap (less than US100). As micro-controllers are inexpensive, and DC motors for toys are also inexpensive, toy-type robots are used more and more in places like high schools.

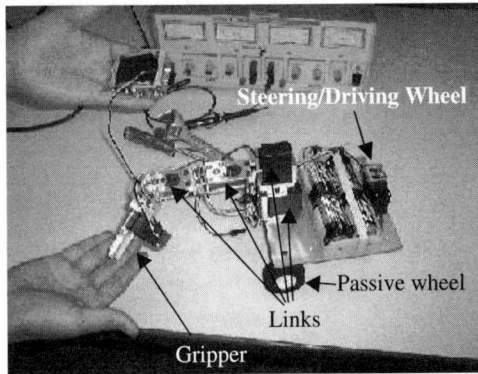

Fig. 6.7   A toy-type mobile manipulator developed with a micro-controller and a low-cost DC motor .

Fig. 6.7 shows an example of a mobile manipulator. This robot has an arm manipulator with four degrees of freedom (DOF). Its hand is a simple gripper with one DOF. Its mobile base has three wheels: two passive wheels and one steering/driving wheel (having two DOFs). All the motions of the joints are controlled by a micro-controller.

6.5   Data-Processing Software

Now, we know that a microprocessor is capable of executing instructions in a cycle-by-cycle manner. In order to make full use of the computational power of a microprocessor, it is necessary to know how to develop software which can effectively process data on a microprocessor.

6.5.1   *Programming*

The development of software requires both programming skill and knowledge of an application's domain. Here, we will study the basic concepts useful in cultivating good programming skills.

6.5.1.1   *Programming Language*

Usually, instructions to be executed by a microprocessor are grouped together to form an entity called a *program*. Different microprocessors from the same or different manufacturers like Intel, Motorola, Texas Instruments etc. usually have their own instruction sets with their own associated syntaxes. A set of instructions with an associated syntax is called the *assembly language*. A program written in an assembly language is known as an *assembly program*.

Since a microprocessor is a digital system, it can only recognize binary numbers. Therefore, each instruction must be encoded with a binary number in order for a microprocessor to interpret and execute it. As a result, an assembly program has to be translated into a series of binary numbers. This results in a series of binary numbers (codes) called a *machine program*. And, the set of binary codes corresponding to the instructions of an assembly language is called a *machine language*.

It is not easy to use an assembly language because it is very different from natural language and is thus difficult to read and understand. Moreover, an assembly program written for one microprocessor cannot be recognized by another microprocessor if their instruction sets are different. This represents a serious drawback for assembly languages.

Since all engineering concepts, principles and mathematical formulas can be expressed in a common natural language (English), one would expect that a microprocessor would be able to undertake and execute a program written in a natural language. However, until today, we have not yet reached that stage. One way to attempt to bridge the gap between the assembly language of a microprocessor and natural language is to use a special programming language. The obvious advantage to introducing such a special programming language is that it enables programs to operate despite different microprocessors.

In engineering, the most popular and high-level programming language is the *C or C++ programming language*. There are many good textbooks which describe the C-programming language in detail. Here, we will focus on some of the important aspects of the C-programming language from a programmer's viewpoint.

6.5.1.2   *Programming Environment*

Today's computers are all equipped with a special program which runs continuously when the power is "on". This special program primarily manages: a) the workload of the processor, b) the usage of memory, c) the data interfacing and communication with peripheral devices (including I/O systems), and d) the execution of *application programs*. This special program is commonly known as an *operating system* (or OS for short). Refer to Fig. 6.5. The OS of a microprocessor is normally loaded into the RAM after the power is switched on. With this addition to the computing hardware, the development of an application program becomes easy and productive.

Fig. 6.8 illustrates an integrated-computing platform for software development. With the support of an OS on a computing platform, many existing programs can be shared among users and programmers. There is no need to "re-invent the wheel" if a program exists already.

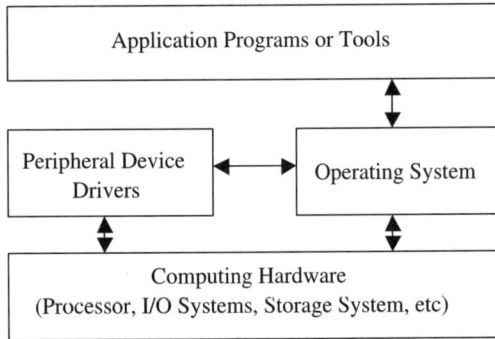

Fig. 6.8   Integrated-computing platform for software development.

For a programmer, the commonly-shared programming tools (i.e. existing proven programs) include:

- An editor for writing and editing a program file,
- A compiler for translating programs in high-level programming languages into assembly programs specific to microprocessors,
- A linker for incorporating shared resources (i.e. libraries) into the assembly program in order to produce an executable file,
- A debugger for examining the execution of a developed program (e.g.

to check for errors and analyze the causes).

6.5.1.3 *Program/Data Files*

In computer engineering, a *file* is a record which stores a string of bits or bytes that may belong to a program or data. A formal definition can be stated as follows:

Definition 6.6 A file is a physical entity which encapsulates the grouping of bits or bytes into a single record.

The content (i.e. bits or bytes) of a file is usually encoded in a specific way to ease file exchange and content display. Since a file's contents is in the form of bits or bytes, it can be directly stored into a memory or any storage device. Inside a memory, the two basic parameters concerning a file are: a) its address (i.e. the address of the location storing the first byte of its content) and b) its length (i.e. the number of bytes it contains). It is much more difficult to implement a file system on a computer. However, the usage of a file is a simple matter. This is because the basic functions for file manipulation are normally provided by a high-level programming language.

6.5.1.4 *Programming Procedure*

A microprocessor is capable of performing computation through the execution of programs. However, a microprocessor or computing system that is capable of directly translating human intention into a corresponding program does not exist. In order to use a microprocessor to perform data processing, a user or programmer must manually do the programming.

A formal definition of *programming* can be stated as follows:

Definition 6.7 Programming is the process of translating a computational task or behavior from the mind of a user (or programmer) into a program.

Programming is a highly creative activity. Fig. 6.9 shows a generic process of programming. Generally speaking, programming with a programming language involves the following steps:

- Step 1:
  Translate a computational task from the mind of a programmer into a corresponding program written in a (high-level) programming lan-

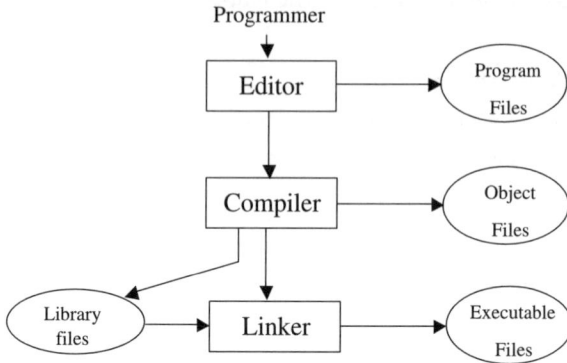

Fig. 6.9   A generic programming procedure.

guage. The outcome of this step is a developed program. For ease in maintenance and editing, it is wise to store a large program into a set of files. This is because it is easier to read and edit a file having a hundred lines than to read and edit a file having several hundred or thousand lines.

- Step 2:
  Compile the program written in a (high-level) programming language into a corresponding machine program encoded in the machine language supported by the microprocessor in use. The outcome of this step is the machine program stored in the *object file*. Usually, each program file will be compiled into a corresponding object file.

- Step 3:
  Put the shareable computational functions into a file called a *library*. The outcome of this step is some library files.

- Step 4:
  Produce the executable program by linking the machine program with the shared computational functions from the libraries. The outcome of this step is the final executable program encoded in machine language and stored in a file called the *executable file*.

The use and sharing of libraries make programming a highly productive activity. A library is a record of the shareable computational functions encoded in a machine language. A formal definition of a library can be stated as follows:

Definition 6.8   A library is a collection of shared program modules or computational functions compiled in a machine language and stored in a file.

While programming is a highly creative and productive activity, a good programmer must observe certain guidelines:

- Readability of Developed Programs:
  As we mentioned earlier, a programming language is not as comprehensive as a natural language. Still, as much as possible, it is important to make a program readable by a human. For example, instead of naming "x" the variable of a motion sensor, one can name it "velocity" directly.
- Reusability of Developed Programs:
  Programming is usually a team effort. Many generic computational functions, like the trigonometric functions, are shared among programmers. Therefore, one should always design a program by making it as modular as possible. In this way, generic modules can be shared with other programmers.
- Inter-operability of Developed Programs:
  Normally, a program written in a high-level programming language can be operated across different computing platforms. However, if a program makes intense usage of shared libraries, one must take special care as the shared libraries may be different across different computing platforms.

6.5.1.5 *The Basics of C-Programming Language*

A good programming language must fulfill two objectives:

- For the sake of simplicity and efficiency, it should be similar to the machine language. In other words, it should allow programmers to explore the characteristics of the microprocessor as directly as possible.
- For ease in readability and programming, it should be similar to a natural language. In this way, a programmer will have less difficulty in translating computational tasks into corresponding programs.

A programming language that satisfies the above two objectives does not exist. So far, the best programming language for engineering simulation and computation is the C-programming language. The C-programming language was invented in the late 1970s. In the early 1980s, this language incorporated *classes*, which implement the concept of object-oriented programming. This has resulted in the *C++ programming language*. The

C-programming language is efficient because it is similar to the machine language. The popular operating system *UNIX* is written in the C-programming language.

Data Types in C

The C-programming language supports basic, primitive data types such as:

- `char`: an 8-bit integer,
- `int`: a 16-bit integer,
- `long`: a 32-bit integer,
- `float`: a 32-bit floating-point number (i.e. a real number),
- `double`: a 64-bit floating-point number (i.e. a real number).

If a variable always takes positive integer numbers, the keyword `unsigned` can be placed in front of a data type for integers. For example, `unsigned char` indicates an unsigned 8-bit integer. An unsigned 8-bit integer varies from 0 to 255.

With the various data types, one can easily declare the variables in C. The following example shows the declaration of variables for the DH parameters of a link inside an open-kinematic chain:

```
float offset_along_x_axis;
float offset_along_z_axis;
float angle_about_x_axis;
float angle_about_z_axis;
```

In engineering, it is inevitable to encounter many constants. In order to enhance the readability of a program in C-programming language, we can associate a constant with a name (i.e. a string of characters) by using the directive `#define`. For example, constant π can be defined as follows:

```
#define PI 3.14;
```

Data Structures in C

To further enhance the readability of a program written in C-programming language, we can group the parameters, or variables, related to a same physical entity to form a data unit called a *data structure*. The directive `typedef struct` allows a programmer to define a data structure.

For example, we can define a data structure, called `LinkData`, to house all the parameters or variables related to a link in an open-kinematic chain:

```
typedef struct {
  float offset_along_x_axis;
  float offset_along_z_axis;
  float angle_about_x_axis;
  float angle_about_z_axis;
  float link_matrix[4][4];
  float link_inertia[3][3];
  float link_mass;
} LinkData;
```

A data structure is treated as a new data type defined by a programmer. We can easily declare a variable of a data structure in the same manner as we declare a variable of an existing data type in C. For example, the following code declares variable link1 of the type LinkData:

```
LinkData link1;
```

Data-Processing Loops in C

Engineering computing may frequently involve repetitive processing of data with a block of instructions. The C-programming language provides two basic mechanisms to repeatedly loop over a block of instructions. This process is called *looping*. The following example shows the use of the directive for(exp1; exp2; exp3) to implement a data-processing loop in C:

```
LinkData left_arm_link[6], right_arm_link[6];

void InitializeAllLinks()
{
  int i;

  for (i=0; i < 6; i++)
  {
    left_arm_link[i].offset_along_x_axis = 0;
    left_arm_link[i].offset_along_z_axis = 0;
    left_arm_link[i].angle_about_x_axis = 0;
    left_arm_link[i].angle_about_z_axis = 0;

    right_arm_link[i].offset_along_x_axis = 0;
    right_arm_link[i].offset_along_z_axis = 0;
    right_arm_link[i].angle_about_x_axis = 0;
```

```
            right_arm_link[i].angle_about_z_axis = 0;
   } ;
}
```

In the above example, exp1 sets the initial condition of the loop and exp2 sets the repetitive condition of the loop. If the condition is true, the loop continues until the condition becomes false. In addition, exp3 sets the default instructions to be executed at the end of each loop (i.e. iteration).

The above example can also be implemented with the directive while in C:

```
LinkData left_arm_link[6], right_arm_link[6];

void InitializeAllLinks()
{
   int i;

   i = 0;
   while (i < 6)
   {
      left_arm_link[i].offset_along_x_axis = 0;
      left_arm_link[i].offset_along_z_axis = 0;
      left_arm_link[i].angle_about_x_axis = 0;
      left_arm_link[i].angle_about_z_axis = 0;

      right_arm_link[i].offset_along_x_axis = 0;
      right_arm_link[i].offset_along_z_axis = 0;
      right_arm_link[i].angle_about_x_axis = 0;
      right_arm_link[i].angle_about_z_axis = 0;
      i++ ;
   } ;
}
```

As we can see, the expression inside the while directive is a condition to test whether or not the loop should be continued. If the condition is true, the loop will be repeated until the condition becomes false.

In C, a programmer can prematurely quit the looping by using the instruction break. This is a very useful and efficient way to quickly exit a loop for whatever reason.

Data-Processing Branches in C

Besides looping, engineering computing may also frequently involve selectively choosing a block of instructions to execute based on the current condition. The process of directing data-processing into different blocks of instructions is called *branching*.

In C, the directive `if-else` provides a simple mechanism to branch between the two blocks of instructions. Obviously, we can nest the `if-else` directive to branch among multiple blocks of instructions. But, this will make a program less comprehensive. An elegant mechanism for branching among multiple blocks of instructions is supported by the directive `switch` in C.

The following example shows how to use the `switch` directive to, selectively, initialize the data structures of the links belonging to a specified open-kinematic chain:

```c
LinkData left_arm_link[6], right_arm_link[6];

void InitializeLinks(int KinematicChainIdentifier)
{
   int i;

   switch(KinematicChainIdentifier)
   {
      case LeftArm:
         for (i=0; i < 6; i++)
         {
            left_arm_link[i].offset_along_x_axis = 0;
            left_arm_link[i].offset_along_z_axis = 0;
            left_arm_link[i].angle_about_x_axis = 0;
            left_arm_link[i].angle_about_z_axis = 0;
         }
         break;

      case RightArm:
         for (i=0; i < 6; i++)
         {
            right_arm_link[i].offset_along_x_axis = 0;
```

```
        right_arm_link[i].offset_along_z_axis = 0;
        right_arm_link[i].angle_about_x_axis = 0;
        right_arm_link[i].angle_about_z_axis = 0;
    }
    break;

  default:
    break;
  }
}
```

As we can see, the expression inside the directive `switch` must be an integer. This expression represents the current condition and is often compared to the predefined cases marked by the keyword `case`. If there is a match, the corresponding block of instructions is executed. Each block of instructions must ended with the `break` instruction. If there is no match for the current condition, the program will enter default, marked by the keyword `default`, and default instructions (if any) will be executed.

Visibility of Data and Functions in C

As we mentioned earlier, for the sake of easy maintenance and editing, it is wise to split a large program into multiple program files. It is now common for a single application program to have multiple program/header files. However, when a single application program is divided into a set of coherent functional modules and each module is stored in its own program file, it raises the issue of how to handle the visibility of data and functions across multiple program files in C.

Fig. 6.10 illustrates an example of a simplified skeleton of a robot-control application. Functionally, we can create one module to handle the forward kinematics and another module to take care of the inverse kinematics. Each module is stored in its own program file (`InverseKinematics.c` and `ForwardKinematics.c`). And, the main body of the application is stored in another file called `RobotControl.c`. As a result, the visibility of data and functions across these files is a concern. In C-programming language, there are some well-defined rules governing the visibility of data and functions.

Rule 1: All the variables declared inside a function are only accessible or visible by the instructions inside this function. These

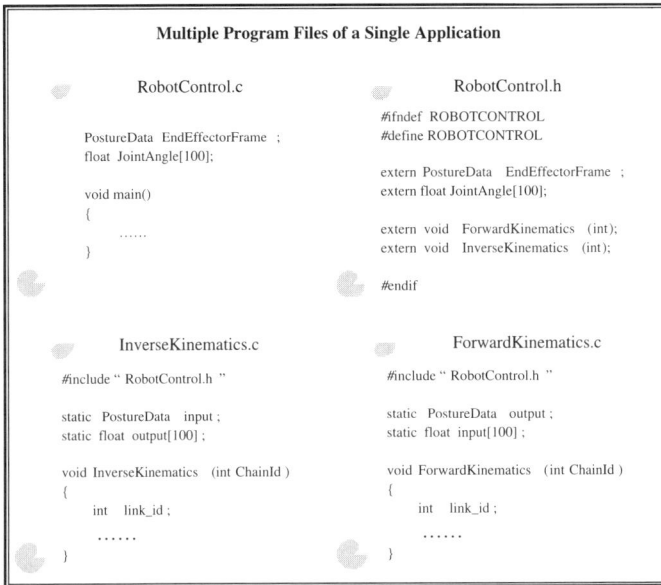

Fig. 6.10   Multiple program/header files of a single application, and the visibility of data (and functions) across the program files.

variables are called *local variables*.

For example, the `InverseKinematics()` function has the variable name `link_id`. It is a local variable and only visible inside this function. If the `ForwardKinematics()` function also requires a local variable with the same name of `link_id`, it has to be declared independently inside the `ForwardKinematics()` function. In fact, these two variables share the same name but will be physically stored in different memory locations.

Rule 2: By default, all the variables declared inside a program file (but not inside any function of the program file) are accessible or visible by all the instructions or functions across all the program files belonging to the same application. These variables are called *global variables*.

For example, the `EndEffectorFrame` and `JointAngle` variables declared inside program file `RobotControl.c` are global variables and visible across all the program files.

Rule 3: The directive `static` placed in front of a global variable will make this variable visible inside its program file only.

For example, the `input` and `output` variables inside program file `InverseKinematics.c` are only accessible by the instructions or functions inside this file.

Rules 2 and 3 are equally applicable to functions. For example, there is no `static` directive in front of the `Inversekinematics()` and `ForwardKinematics()` functions. These two functions are global functions which are visible everywhere across the program files of the application.

To enhance the readability of the developed programs, it is a common practice to notify the global variables and functions in some separate files called *header files*. A global variable or function is notified with the directive `extern`. Refer to Fig. 6.10. The `RobotControl.h` file is a header file. The insertion of the directives:

```
#ifndef ROBOTCONTROL
#define ROBOTCONTROL


#endif
```

is for the purpose of avoiding duplicate notification of a same group of global variables and functions. In this example, the string of characters `ROBOTCONTROL` is user-defined, meaning it can be a string of any characters. And, this string of characters along with the directive `#include`serves as a tag to skip any repetitive notification (e.g. `#include "RobotControl.h"`).

6.5.2   *Multi-Tasking*

In the past, multi-tasking primarily involved: a) time management and b) memory management. Nowadays, however, memory is very cheap. As a result, memory management is no longer a critical issue in multi-tasking. Thus, time management is THE critical issue.

6.5.2.1   *The Basics of Multi-Tasking*

Refer to Fig. 6.4. A digital computer is composed of a processor and memory which are interconnected by buses. Functionally, it can be treated as a system which is capable of undertaking data-processing and storage.

Since the operations inside a processor are all synchronized by the system clock, each instruction of a program is executed by a processor's ALU within a fixed number of clock cycles. In other words, a processor's computing power can be measured in terms of the number of clock cycles per program. But, this measure is not so intuitive because different programs may have different numbers of instructions. Two ways to measure computing power are by MIPS (*million instructions per second*) or MFLOPS (*million of floating-point instructions per second*).

A digital computer has a certain memory capability to store programs and processed data. This memory-storage capacity is commonly measured in terms of bytes. For example, MB stands for "mega-bytes" and GB stands for "giga-bytes".

Now, it is clear that a digital computer can be treated as a system having two important resources: a) computing power and b) storage capacity. If a digital computer is further considered to be an SISO system (single input/single output), there is no issue of resource-sharing. And, there will be no issue of multi-tasking. However, today's computer is becoming more and more powerful in terms of computing power and storage capacity. It is, thus, very common to treat a digital computer as an MIMO system (multiple input/multiple output). In this case, there is concern over resource-sharing. And, it is important to understand how a digital computer handles multiple input and shares the limited resources.

Computational Tasks

The input to a digital computer is known as tasks. These tasks are different from the motion-centric ones performed by a robot. Under the context of an information system or computing system, a task means a computational task related to data-processing. Formally, a computational task can be defined as follows:

Definition 6.9 A computational task, in a real-time system, is a data-processing job with a deadline. It consumes not only the computing power of a digital computer but also memory storage.

As shown in Fig. 6.11, a computational task can be repetitive or nonrepetitive. A repetitive task is a task which can be executed within a set of discrete time intervals. In contrast, a nonrepetitive task must be run within a single time interval.

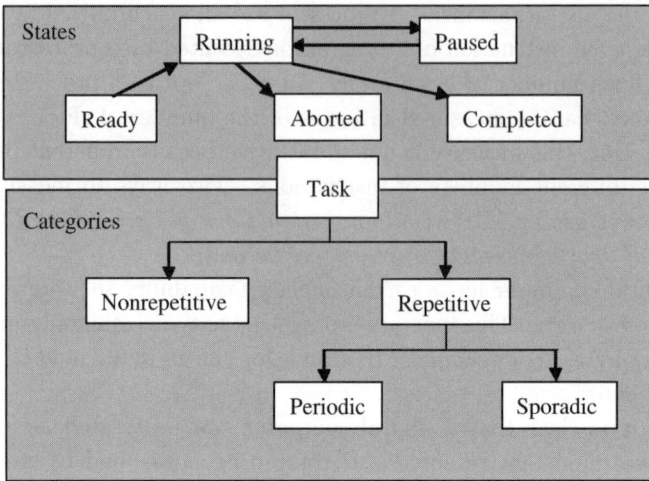

Fig. 6.11　The states and categories related to computational tasks.

Clearly, a nonrepetitive task does not share a digital computer's resources with other tasks because it can neither be interrupted nor preempted. In other words, a nonrepetitive task can only be in one of three states: a) ready (i.e. ready for execution by the processor), b) running (i.e. consuming the processor's resources alone), and c) completed or aborted. Whenever possible, we should avoid nonrepetitive tasks, or transform nonrepetitive tasks into repetitive tasks that satisfy the deadline constraint.

There are two types of repetitive tasks: a) periodic tasks and b) sporadic tasks. Since a repetitive task can be run within a set of discrete time intervals, it can participate in *cooperative multi-tasking* or *preemptive multi-tasking*. This is because a repetitive task can be in one of these states: a) ready, b) running, c) pause (either voluntarily or caused by preemption), and d) completed or aborted. Most importantly, a repetitive task can be periodically switched from "running" to "pause". This provides the basic mechanism for the sharing of a processor's resources.

Time Management in Multi-Tasking

For a complex computing platform with multiple processors (e.g. a parallel computer), there will be no issue of multi-tasking if the number of tasks is equal to or less than the number of processors. The issue of multi-tasking arises only when the number of tasks is greater than the number of processors. For the sake of simplicity, let us consider the scenario of a single processor with multiple computational tasks. In this case, there is an obvious need to manage the processor's time and memory so that the processor's resources can be shared.

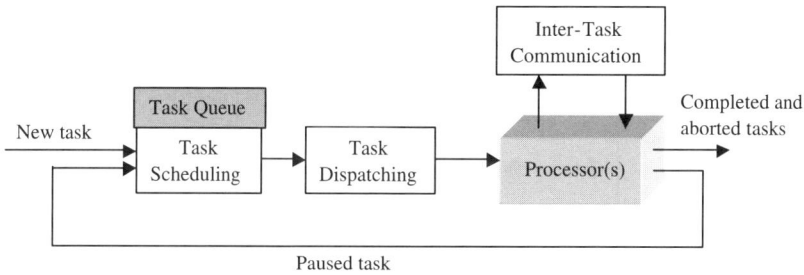

Fig. 6.12 Illustration of multi-tasking process when there is a single processor and multiple computational tasks.

Fig. 6.12 illustrates the process of multi-tasking with a single processor. Since a processor can only execute one task at a time-instant, all other tasks must wait in a place called a *task queue*. A task queue needs to be managed so that the waiting tasks are ordered according to certain rules and priorities. A special program which manages a task queue is called a *task scheduler*.

When a running task has to be stopped for whatever reason, another special program will automatically release the processor's resources from the running task and launch a new task selected by task scheduler. This special program is called a *task dispatcher*.

Memory Management in Multi-Tasking

By definition, a computational task will not only consume a processor's time but also its memory. Thus, we should also manage the sharing of a processor's memory if a large number of tasks are concurrently active, as shown in Fig. 6.13.

A computational task is a computer program for data-processing. Both the program and data require space in the processor's memory. A common way to manage the sharing of this memory is to explicitly divide the memory space into distinct zones which are respectively assigned to the operating system (if any), programs, data and stacks, as shown in Fig. 6.5.

When a task is running, the execution of its instructions requires an

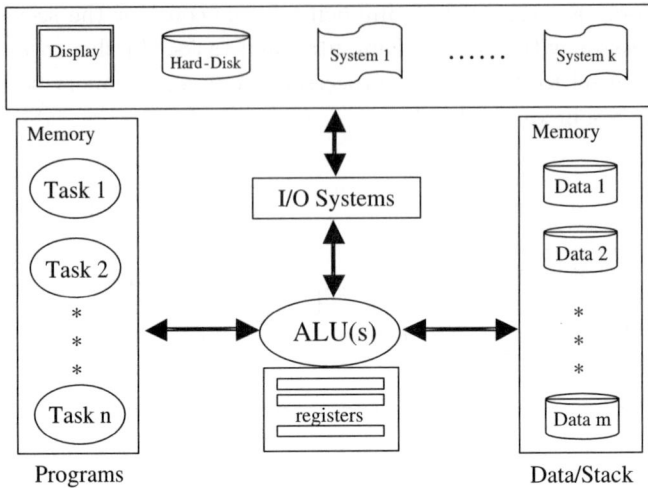

Fig. 6.13   Illustration of memory management required in multi-tasking with a single processor.

intense usage of the processor's registers, such as the program counter, index registers and accumulators. When a running task is voluntarily or preemptively stopped, the contents (also called the *execution context*) of these registers must be saved so that they can be recovered once the task is resumed. A stack is a memory block which holds such contents.

Inter-task Communication

Another requirement of memory management is posed by inter-task communication. If task A wants to exchange information with task B, there should be a mechanism to support this type of inter-task communication. This is commonly achieved by a *shared memory* if the tasks are active inside the same processor.

A shared memory is a reserved memory block in the data zone of a processor's memory. Access is granted to all the tasks that use it for data-exchange.

6.5.2.2   *Cooperative Multi-Tasking*

In practice, there are two strategies of sharing a processor's time among multiple tasks: a) cooperative multi-tasking and b) preemptive multi-tasking.

The basic idea behind cooperative multi-tasking is to treat a set of active tasks as a *state transition network* or *state machine*. Any uncompleted task is an active task as long as it is not aborted. An active task can either be in the running mode or the sleeping mode. Under cooperative multi-tasking, each task voluntarily commutes its running state to a paused (sleeping) state. When one task voluntarily exits the running state, a subsequent task is activated. All the active tasks must be cooperative. The multi-tasking will not function properly if one task selfishly consumes all the resources (time and memory) of a processor.

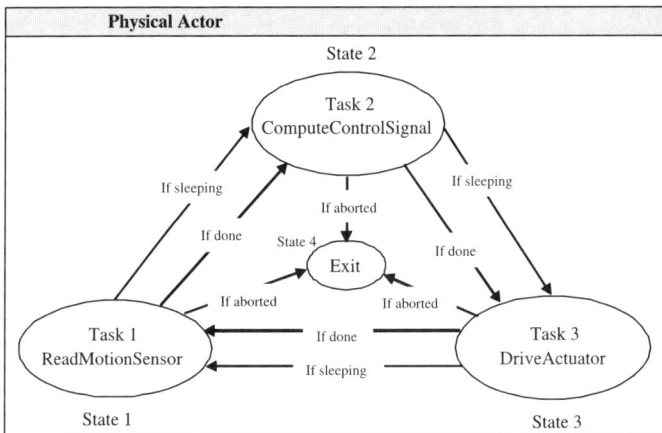

Fig. 6.14 An example of cooperative multi-tasking for a physical actor.

Fig. 6.14 shows an example of cooperative multi-tasking for a physical actor which controls an actuator with an inertial load. The motion controller of the physical actor first reads the sensor feedback. Then, it calculates the control signal according to a control algorithm. Subsequently, it outputs the control signal to the power amplifier which drives the motor. As long as the motor is running, these three operations are repeatedly performed by the physical actor. If we treat the physical actor as a state machine, it will include these three tasks:

- Task 1: Read the motion parameters from the motion sensor,
- Task 2: Compute the control signal,
- Task 3: Output the control signal to the power amplifier.

Since each task is treated as a state, the corresponding state machine will have four states, and the fourth state will describe the situation of exit (e.g. completed motion, sensor failure, actuator failure, etc).

This example can be implemented with a real-time programming language. For example, Dynamic C from Z-World (www.zworld.com) is a C-programming language with multi-tasking. It supports cooperative multi-tasking with a built-in directive `costate`. By using the directive `costate`, we can easily implement multiple tasks, (called *co-statements*, in Dynamic C), which run concurrently on the micro-controller supported by Dynamic C (e.g. Rabbit 2000 micro-controller).

```
CooperativePhysicalActor.c

int   flag_abort ;
void main()
{
    flag_abort = false;
    costate  Task1
    {
        if (ReadMotionSensor () == false)  flag_abort =true;
        waitfor(DelayTicks(20));
    }
    if (flag_abort == true) exit;
    costate   Task2
    {
        if (ComputeControlSignal () == false) flag_abort =true;
        waitfor(DelayTicks(20));
    }
    if (flag_abort == true) exit;
    costate  Task3
    {
        if (DriveActuator () == false) flag_abort =true;
        waitfor(DelayTicks(20));
    }
    if (flag_abort == true) exit;
}
```

Fig. 6.15   Sample program of cooperative multi-tasking in Dynamic C.

Fig. 6.15 shows a sample program in Dynamic C, which implements the state machine described in Fig. 6.14. The function `DelayTicks()` is a built-in function of Dynamic C. For example, a time tick in Dynamic C is equal to 1/2048 second. With the built-in function `waitfor()` in Dynamic C, a task can voluntarily commutate into the sleeping state. As shown in the sample program, these three tasks of the physical actor will voluntarily and periodically sleep for 20 time-ticks.

6.5.2.3 *Preemptive Multi-Tasking*

Cooperative multitasking does not impose a rigid time constraint on concurrently active tasks. Moreover, it is not able to dynamically change the priority or order of task execution. For applications which require stricter time management, it is necessary to use *preemptive multi-tasking*.

Under preemptive multi-tasking, task scheduling is based on priority and order. If the level of priority is the same for all tasks, the order of task execution is determined by a queue which is also known as a *task queue*. A queue is a First-In-First-Out (FIFO) buffer in the memory. A simple way to schedule the execution of tasks having the same level of priority is to assign an equal, predefined time interval to each task. This time interval is called a *quantum* or *time slice*.

Refer to Fig. 6.12. The running task will consume one quantum of the processor's time. If this task is still active after one quantum, it will be preempted and put back in the queue as a paused task. Subsequently, a new task from the queue will be selected as the running task. This process will be repeated until all the tasks are completed or aborted. This process is known as the *round-robin* or *time-slicing* scheduling method.

In order to prioritize task execution, we simply assign a priority level to each task. In this way, active task i will have two important parameters:

- $t_{s,i}$: the assigned quantum or time slice which is dynamically changeable.
- $L_{p,i}$: the assigned priority level which is also dynamically changeable.

Under preemptive multi-tasking, when task i becomes the running task, the amount of the processor's time consumed by this task will be:

$$t_{e,i} = L_{p,i} \bullet t_{s,i}. \tag{6.3}$$

If task i is completed before or within this amount of time, it exits. Otherwise, it is preempted after the elapse of time $t_{e,i}$, and re-enters the task queue to wait until it becomes the running task again.

Assume that a humanoid robot has a built-in distributed information system, as shown in Fig. 6.16, in order to properly coordinate its behaviors in various scenarios of task execution. For example, for dancing or playing a sport, the behaviors of the humanoid robot should include:

- Mental Actor 1: Coordinate visually-guided biped walking,
- Mental Actor 2: Coordinate visually-guided hand gestures or manipulations,

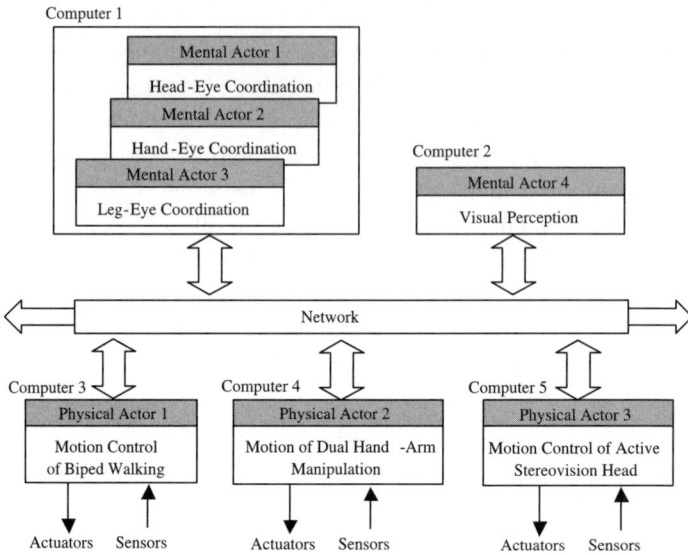

Fig. 6.16 Example of a distributed information system for a humanoid robot.

- Mental Actor 3: Coordinate visually-guided head movements.

If these mental actors run on a single computer, it is necessary to adopt preemptive multi-tasking in order to guarantee the apparent simultaneity of these actors.

Dynamic C from Z-World supports preemptive multi-tasking with the language construct `slice(buffer, time_interval)`. The first argument `buffer` is a parameter which specifies the size of the stack to save the execution context when actor i is preempted. The second argument is time interval $t_{e,i}$, which will be consumed by actor i when it becomes the running actor.

Fig. 6.17 shows a sample program in Dynamic C that implements these behavior coordinations in preemptive multi-tasking.

6.5.3 *Real-Time Operating Systems*

To efficiently handle multi-tasking, it is best to use a real-time operating system. A real-time operating system is usually small, and provides efficient time- and memory-management services such as task scheduling, task dispatching, inter-task communication and real-time interruptions.

```
PreemptiveBehaviorCoordination.c

int   time_slice;
int   priority_actor1, priority_actor2; priority_actor3;

void main()
{
    time_slice = 10;
    slice(500, priority_actor1*time_slice) MentalActor1
    {
        LegEyeCoordination ();
    }

    slice(300, priority_actor2*time_slice) MentalActor2
    {
        HandEyeCoordination ();
    }

    slice(300, priority_actor3*time_slice) MentalActor3
    {
        HeadEyeCoordination ();
    }
}
```

Fig. 6.17   Sample program of preemptive multi-tasking in Dynamic C.

Nowadays, there are many real-time operating systems on the market, such as VxWorks, QNX, RT-Linux. And, many good textbooks provide a comprehensive description of the fundamentals of a real-time operating system.

6.6   Data Storage and Retrieval

Data storage and retrieval is an important part of any information system. This is because the data-processing results must be useful and therefore, reusable. Generally speaking, in robotics, there lacks a standard regarding the storage and retrieval of reusable data (if any). Since a robot incorporates an information system, the issue of how to store useful data-processing results which can be retrieved later for reuse is something that inevitably needs to be dealt with.

If a humanoid robot have to develop its physical and mental abilities through real-time interaction with its environment, it is obvious that a humanoid robot needs a memory to store all of the acquired internal representations of perception and knowledge. Therefore, data storage and retrieval are an indispensable part of a humanoid robot's information system.

6.6.1 *Storage Devices*

As we mentioned earlier, any advanced computing platform will have an operating system. An operating system serves as an interface between the application programs and the computing platform's hardware components/devices. Usually, a general-purpose operating system consists of three major modules: a) the process-management module which manages the allocation of the processor's time for various computing processes or tasks, b) the memory-management module which manages the processor's memory and the storage device, and c) the peripheral-device management module which handles input/output interfacing, and communication between the processor and external devices such as the mouse, keyboard and printer, etc.

The memory-management of an operating system includes management of both the processor's memory (e.g. RAM) and the external storage device (e.g. hard-disk, as shown in Fig. 6.13). The processor's memory, except for ROM and ERPROM, is not a place to store reusable results or data. The common place to store useful and reusable data (including program files) is in an external storage device, such as a hard-disk or a rewritable CD-ROM. But due to the slow access speed, a CD-ROM is more appropriate for data backup than storage.

From the user's point of view, a storage device is a system which takes data as input and produces data as output. Some basic requirements for an external-storage device are:

- It must be able to store a large amount of data.
- It must maintain the integrity of the stored data.
- It must support access (both read and write) to the data by multiple computing processes or tasks.

6.6.2 *A File System*

At the lowest level of a memory, data is a list of ordered bytes. But, at a higher level, data is treated as a record. For example, the DH parameters of an open kinematic chain can form one record. By definition, the entity that stores a record in a storage device is called a *file*. Therefore, from the viewpoint of information, a memory or data storage device is a specifically organized *file system*.

It is true to say that all computing platforms, having an external storage device, have a corresponding file system which is an integral part of an operating system. A file normally has a name and an extension. For example, `InverseKinematic.c` is a file with the name "InverseKinematics" and the extension ".c". The purpose of a file name's extension is to indicate the file's type. For example, the extension ".c" means a file which contains a C program.

6.6.3 *Data Storage Using Unformatted Files*

A computing platform having an external storage device is usually supported by an operating system and/or a real-time kernel. A real-kernel is a mini-operating system which supports, in real-time, the task scheduling, task dispatching, interruption services and inter-task communication. For example, Windows NT (a general-purpose operating system), with the extension of the real-time kernel INTime, becomes a real-time operating system for all personal computers (PCs). Other examples include QNX, VxWorks, RT Linux and Linux with Embedix.

With the support of a real-time operating system, a high-level programming language, like C, normally provides a set of functions for the manipulation of a file. Typical functions include:

- `open`: to open a file for read or write operations,
- `close`: to close a file, so that it is ready to be accessed by other programs or users,
- `read`: to retrieve data from an opened file,
- `write`: to store data to an opened file.

With the help of these functions, a simple way to store and retrieve data is to make use of a file. For example, for a given application, a robot can define a data structure called `RobotData` to organize all the relevant information, parameters and values of the variables related to this application. This data is stored into the `RobotData` structure, referenced by variable `robotdatabase`. During the execution of the application, all data referenced by variable `robotdatabase` is stored in the memory of the processor (RAM). Since the processor's RAM is not a place for the storage and retrieval of any reusable data, at a certain point in time before the application terminates, the data referenced by variable `robotdatabase` must be stored into a file. This can be done with the sample codes, as shown in Fig. 6.18.

In this example, all the bytes inside the record `RobotData`, referenced by variable `robotdatabase`, are stored in an opened file in a single shot

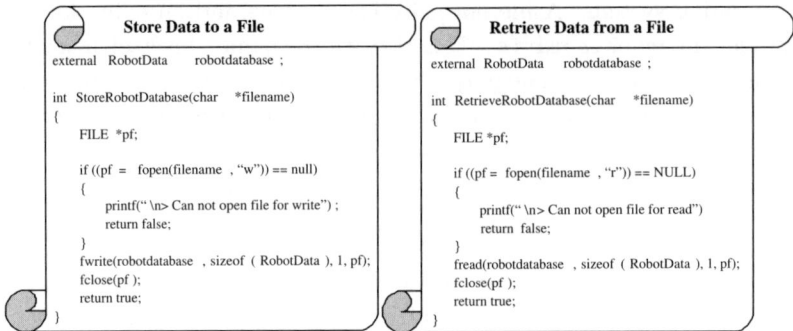

Fig. 6.18 Data storage to an unformatted file.

(with the function `fwrite`). Conversely, all the bytes of the data structure `RobotData` can be retrieved and placed back into the data structure `RobotData`, referenced by variable `robotdatabase`, in a single shot as well (with the function `fread`).

It is clear from this example, that data stored in a file does not require any formatting of the file. Thus, we call this procedure *data storage using an unformatted file*. This method is also applicable for the storage of a list of identical data structures. For example, if there are n units of the data structure `RobotData` (e.g. if we have n robots), we can simply set the third argument of function `fread` or `fwrite` to be n (In Fig. 6.18, it is set to 1).

However, this method has several drawbacks:

(1) If a file is corrupted for whatever reason, all the data stored in the file will be lost.
(2) If a list of data structures is stored, all the data structures in the list must be identical. In other words, the data must be homogeneous.

6.6.4   *Data Storage Using a File System*

Under the file system of an operating system, a set of related files can be grouped together and placed in the same folder, also known as a *directory*. The two basic elements of a file system are: a) the files and b) the directories (or folders).

At the operating system level, a file is a nondivisive entity. But, a directory can be divided into a set of subdirectories. This division of a directory can go into any depth. As a result, a directory can easily be configured into a hierarchical tree-structure of subdirectories and files.

Naturally, one can make use of this type of hierarchy in a file system to create a base for data storage. There are many advantages to this solution:

- It is easy to develop, as a high-level programming language usually provides a set of functions to manipulate the files (e.g. open, close, read, write etc).
- It is easily accessible by any user, as an operating system normally provides a powerful tool (e.g. Windows Explorer) for a user to navigate through the hierarchy of the file system.
- It is safe to store large amounts of data, as a corrupted file will not affect other files.

6.6.5 *Data Storage Using Formatted Files*

If we view a file system to be a hierarchical organization of files in a tree-like structure of directories or folders, a file can also be viewed as a hierarchical organization of records. (NOTE: A record is a sequence of bytes grouped together). Similarly, the records inside a file can be arranged in a tree-like structure of directories or folders. In the C-programming language, the function `fseek` allows a program to store or retrieve a record at any location inside the memory space of a file. With the use of this function, it is a simple matter to create a formatted file. When using a formatted file to store and retrieve data, it is important to consider several issues:

- Should the naming of the records (i.e. the labels assigned to the records inside a formatted file) be standardized so that any user, who follows the standard, can read and write on the formatted file?
- Should library functions be provided for reading and writing records if a new file format is proposed and accepted by a community of users or programmers?

Fig 6.19 shows a generic structure of a formatted file. It is very similar to a file system itself. A formatted file usually has three types of entry:

- Header Entry:
  This is a record that stores some generic and/or historic information about the file. For example, what is the name of the file format (e.g. Postscript, PDF, TIFF etc)? Who invented the file format? Which version is the file format ? Is the data compressed? etc. A header entry will have at least one pointer to a *directory entry*. Here, a pointer means an address within the memory space of a file. In general programming,

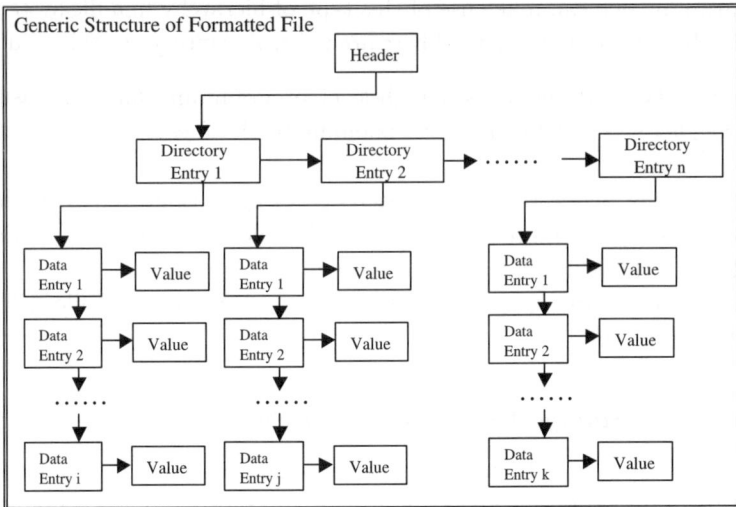

Fig. 6.19   A generic structure of a formatted file.

a pointer means an address within a specific memory space.

- Directory Entries:

  A directory entry is a record which indicates the grouping of related data entries under the same directory or folder. If the directory entries are organized as a chained list, a directory entry can simply contain information about the name of the directory, the number of data entries under this directory, at least one pointer to the first data entry, and a pointer to the next directory entry.

- Data Entries:

  A data entry is a record which stores the associated information to a piece of data. For example, what is the standardized name of the data? What is the size of the data (in terms of bytes)? What is the pointer to the first byte of data? The value record, as shown in Fig. 6.19, is the memory block where data is stored in a file. If the data entries in a directory are organized as a chained list, a data entry will also contain a pointer to the next data entry in that list.

It is easy to share a formatted file among a team of users or programmers. However, information written into a formatted file cannot easily be modified by an ordinary user. This helps to maintain the authenticity of the stored data. In addition, some confidential or proprietary information can be encoded into a formatted file. To a certain extent, it is possible to set up an access control using a password with a formatted file.

6.7   Data Interfacing and Communication

A computing system without a data interfacing or communication module can hardly be useful for the control of any machine, especially a robot. Obviously, a robot's information system must have a comprehensive set of data interfacing and communication subsystems. These subsystems are also called the *I/O communication systems, or devices.*

The primary role of an I/O system, or device, is to receive, send and communicate data. For example, a motion controller is a computing system equipped with some specific I/O systems. In this way, a motion controller can receive data from sensors and send data to actuators through its I/O systems.

A robot is a real-time system, the performance of which depends on the real-time performance of its I/O systems. Hence, an understanding of the working principles of the common I/O systems is important for the design, analysis and control of a robot.

6.7.1   *Basic Concepts*

For the sake of simplicity, let us consider memory-mapped I/O systems.

6.7.1.1   *I/O Registers*

A common way to operate an I/O system is to make use of the I/O registers, as shown in Fig. 6.5. By default, an I/O register is a memory location consisting of 8 bits. A register with 8 bits is comparable to a set of eight electronic switches. You can switch each bit "on" or "off". And, each bit carries one piece of information related to the configuration or status of the I/O system.

One of the advantages of using I/O registers is the ease of programming (this will become clearer later). I/O registers are divided into three groups:

(1) Control registers:
     These registers are for programs to configure an I/O. For example, an I/O system can be configured for input, output, or both.

(2) Status registers:

These registers allow programs to know the status of an I/O system. For example, when a program examines the status register, it will know whether the data communication has been completed successfully or not.

(3) Data registers:

These registers serve as a buffer to hold data which is yet to be transmitted or received.

Sometimes, depending on the complexity of an I/O system, a single register will serve as both the control and status register. However, most of the time, there will be a separate register for the data register, capable of holding one byte at one time-instant.

6.7.1.2 *I/O Ports*

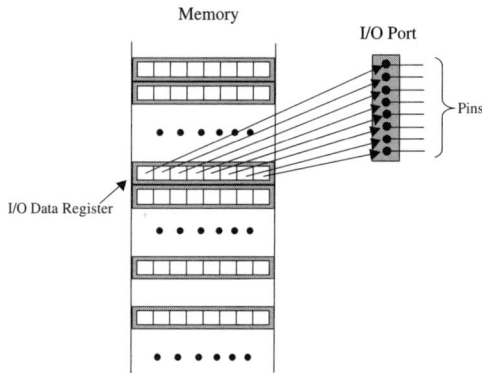

Fig. 6.20   Conceptual illustration of the mapping between an I/O port and its corresponding I/O data register.

We know that an I/O system serves as an interface between two devices or information systems for data exchange. Therefore, there must be some connection pins or lines to physically link these two devices or systems together. Usually, the pins or lines of an I/O system are grouped into units called *I/O ports*. An I/O port has 8 pins which correspond to eight logic bits (or one byte).

For a memory-mapped I/O system, an I/O port is directly mapped to its corresponding I/O data register, as shown in Fig. 6.20.

6.7.1.3 *Communication Networks*

Except in special cases such as sensory data input and data output to actuation devices, an I/O system generally acts as a communication controller for data communication between two devices. The wires which physically connect these two devices together are called the *network*. In a complex system, a network may involve the connection of more than two devices or information systems, which are commonly called *nodes*. When this happens, it is necessary to consider the most appropriate topology.

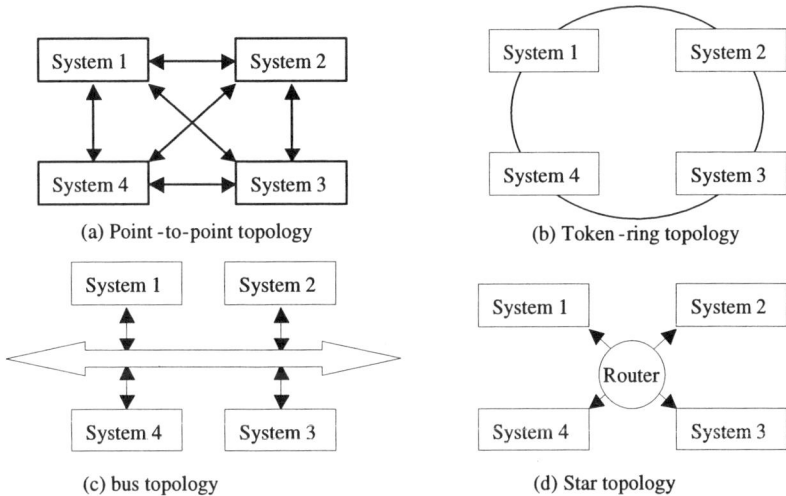

Fig. 6.21   Typical network topologies.

For a distributed real-time controller (i.e. a controller with many computational units connected together by a network), the time responsiveness of the network is as crucial as the time responsiveness of each computational unit inside. A network with a predictable time responsiveness is called a *deterministic network*. Fig. 6.21 shows some typical network topologies, which include:

- Point-to-point Topology:
  In this case, any pair of devices or systems has a dedicated connection line. It is easy to meet the real-time requirement for data communication. However, the network will become complicated if more devices or systems are hooked onto it.

- Token-ring Topology:
  This is a simplified version of point-to-point topology. But, instead of having one connection between any pair of devices, a token-ring topology has a direct connection between two adjacent devices or systems. For this type of network, the data must carry address information so that it is relayed to the proper destination. Depending on the protocol of communication, a network of token-ring topology is usually deterministic.
- Bus Topology:
  A further simplification of point-to-point topology yields the bus topology. From an electrical point of view, bus topology means that all the devices or systems in a network are electrically connected to a common point if we ignore the impedance of the network's wiring. Depending on the protocol of communication, a network of bus topology may be deterministic (e.g. field-bus).
- Star Topology:
  An improvement on bus topology is the star topology with a router to serve as the common connection point. The router allows the communication system to intelligently dispatch data to the proper destination while masking out malfunctioning devices in the network.

6.7.1.4 *Common Issues*

Data communication over a network of a given topology simply means the exchange of a string of bits (serial communication) or bytes (parallel communication) between two devices or systems. Obviously, the first issue of data communication is how to pack data in the form of a string of bits or bytes. The process of packing data for communication is known as *data framing*. The framing process may involve the addition of extra information such as the destination's address and error checking.

Once the data has been framed, the next issue is how to initiate and synchronize data communication. To do this, two common techniques are available:

- Polling Technique (also known as Polled I/O):
  When using this technique, there is one master device on a network and the rest are slave devices. If a slave device wants to communicate data, it must set a flag bit in the status register of its I/O system. In this way, the slave device signals its intention to communicate data. The role of a master device is to periodically check whether a slave device

wants to undertake data communication by detecting the flag bit in the status register of the slave device's I/O system. If the detection is positive, data communication occurs immediately.

• Interrupt-driven Technique (also known as Interrupt-driven I/O):
When using this technique, any device on the network can signal its intention to communicate data with another device by sending out an interrupt signal and taking control of the communication line (if it is free).

After data communication, a third issue is error checking and acknowledgement of data reception. The purpose of error checking is to ensure that the received data is equal to the transmitted data. And, the purpose of acknowledgement of data reception is for the transmitting device to check whether the data was successfully received by the receiving device.

We will examine the details of the above issues when studying serial and parallel I/O systems.

6.7.1.5 *Procedure of Programming I/O Systems*

Although there are many technical issues related to successful data communication, a programmer usually does not need to directly solve these issues because the I/O system's hardware will do it.

The job of a programmer is to know how to operate the I/O system by programming the associated registers. In general, programming an I/O system involves the following steps:

• Step 1:
Configure a chosen I/O system by properly setting up the bits inside the control and/or status registers. For example, a device must indicate whether to transmit or receive data, how to initiate and synchronize data communication, and how to perform data framing (if any) and error checking (if any).

• Step 2:
For a transmitting device, it transmits a data unit (e.g. a byte). For a receiving device, it detects a flag bit on the I/O system's status register in order to know whether a data unit (e.g. a byte) has been received.

• Step 3:
For a transmitting device, it detects a flag bit of the I/O system's status register in order to know whether a data unit has been automatically transmitted by the I/O system's hardware. For a receiving device, it

reads in the data unit and clear the flag bit on the I/O system's status register.

- Repeat Steps 2 and 3 until data communication is completed.

6.7.2   D/A Converters

By definition, if a variable, parameter or signal only takes discrete values encoded in binary numbers, it is called a *digital* variable, parameter or signal. However, if it takes any value encoded in a real decimal number, it is called an *analog* variable, parameter or signal.

Similarly, a system or device which manipulates digital values is called a *digital* system or device. Otherwise, it is called an *analog* system or device.

In nature, most of the values for physical quantities (e.g. temperature, speed, pressure, dimension etc.) are intrinsically analog. However, with the advent of the digital computer, and the use of it for sensing, control, perception and decision-making, it is often necessary to exchange data between an analog system and a digital system. This raises two issues:

- Digital to Analog Conversion (D/A Converter):
  How do we feed the output of a digital system to the input of an analog system (e.g. a digital controller sends control signals to the power amplifier of an actuator)?
- Analog to Digital Conversion (A/D Converter):
  How does a digital system read the output of an analog system (e.g. a digital controller gets motion feedback from a potentiometer)?

In electronics, there are many solutions for D/A conversion. Fig. 6.22 shows one example of a D/A conversion circuit. The input to this circuit is a digital number, for example, a byte: $D_{in} = b_7 b_6 b_5 b_4 b_3 b_2 b_1 b_0$ with $b_i = 0$ or 1 ($i = 0, 1, ..., 7$). Electronically, bit b_i of the input digital number D_{in} can be treated as a logic switch with the resistance value of

$$R_{bi} = \frac{1}{b_i} - 1, \quad \forall i \in [0, 7]. \tag{6.4}$$

It is easy to see that $R_{bi} = \infty$ when $b_i = 0$. This means that the corresponding logic switch is "off". On the other hand, when $b_i = 1$, $R_{bi} = 0$, this means that the corresponding logic switch is "on".

For the example shown in Fig. 6.22, the output of the D/A converter is a voltage signal V_{out}. Depending on the circuit, the following relationship

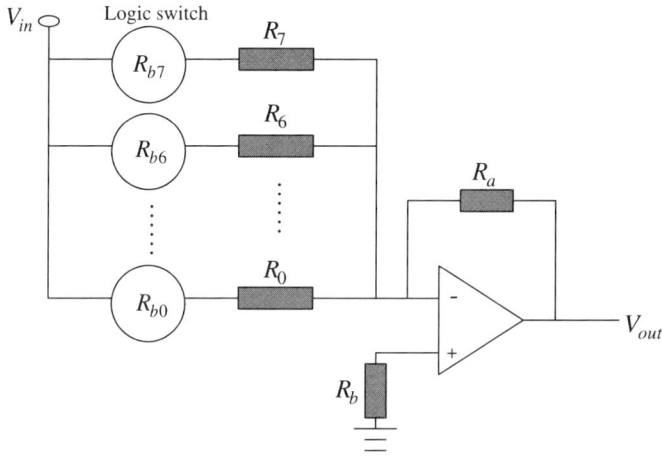

Fig. 6.22   Example of a D/A conversion circuit.

holds:

$$V_{out} = -\left(\frac{V_{in}}{R_{b7} + R_7} + ... + \frac{V_{in}}{R_{b0} + R_0}\right) \bullet R_a. \qquad (6.5)$$

This relationship appears to be nonlinear. However, it will become linear with a specifically selected set of resistor values R_i $(i = 0, 1, 2, ..., 7)$. In fact, two important criteria for evaluating a D/A or A/D converter are:

- Linearity:
  Can the input and output relationship of a D/A or A/D converter be mathematically described by a straight line?
- Sensitivity:
  What is the relationship between variation $\triangle D_{in}$ and variation $\triangle V_{out}$, or vice versa?

From the circuit as shown in Fig. 6.22, we can see that D/A conversion occurs simultaneously. As a result, the time response of a D/A converter is normally deterministic.

6.7.3   *A/D Converters*

In many real applications, it is necessary for a controller to read in the sensory data from an analog device such as a potentiometer, strain gauge, CCD (Coupled-Charge Device) imaging sensor, and microphone etc.

For an A/D converter, the input is an analog signal. In many circumstances, the analog input signal can be represented in the form of a voltage signal, denoted by V_{in}. If the original signal is not in voltage, we can easily convert it to a corresponding voltage signal. And, the output from an A/D converter is a binary number, denoted by D_{out}.

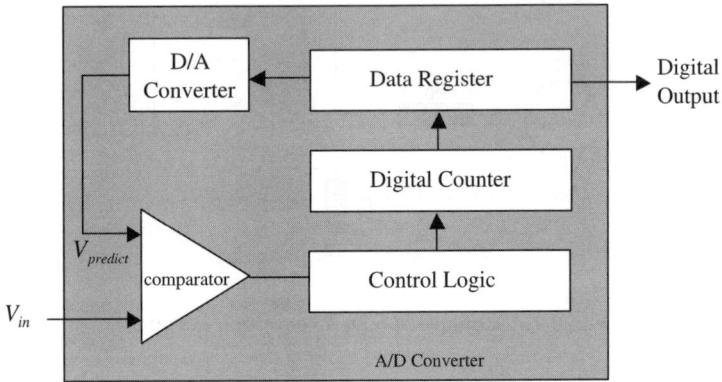

Fig. 6.23   Example of an A/D conversion circuit.

There are many solutions for the design of an A/D conversion circuit. Fig. 6.23 shows a popular one, known as the *progressive-approximation method*. Its working principle is very similar to that of a closed-feedback loop. The basic procedure is as follows:

- Prediction:
  A counter automatically predicts possible output D_{out} from the smallest to the largest number. If the counter has eight bits (one byte), the smallest value is 0 and the largest value is 255.
- Verification:
  For each predicted value of output D_{out}, a D/A converter generates a corresponding predicted voltage $V_{predict}$. As long as the predicted voltage is smaller than the input voltage V_{in}, the cycle of prediction and verification will continue until $V_{predict} \simeq V_{in}$.
- Result Output:
  When $V_{predict} \simeq V_{in}$, the control logic stops the counter, and the actual number of the counter is the output of the A/D converter.

The advantage of this A/D conversion method is its low cost because it is relatively easy to design. The drawback of this method is the variable response time which depends on the length of the counter (e.g. 8 bits, 10 bits 12 bits etc). However, the time response of an A/D converter is still predictable because the conversion time for the largest input value is deterministic.

6.7.4 *Parallel I/O*

An input and output system that has at least eight signal lines is called a *parallel I/O system* because it can input or output at least one byte at a time. Since a parallel I/O system can simultaneously hold at least eight logic signals (bits), it can be used not only for data communication, but also for the control of external devices such as buttons or switches etc.

6.7.4.1 *Device Control with Parallel I/O Systems*

In robotics, there are many buttons, switches and displays which require the attention of a robot's controller. For example, each joint has at least two limit-switches due to the mechanical constraint of a kinematic pair. In other words, if a joint angle varies within the range of $[-45^0, 135^0]$, a limit-switch for the position -45^0 and another limit-switch for the position 135^0 have to be incorporated into a robot's body. On the other hand, if we want to dialogue with a robot equipped with a dedicated keyboard (e.g. the teach pendant), it is necessary for a robot's controller to periodically check the status of the keyboard's push-buttons

In short, all devices which have input and output depending on a set of simultaneous logic signals (i.e. 0 or 1, 0V or +5V, "on" or "off" etc.) can be easily controlled by a parallel I/O system.

Fig. 6.24 shows the use of a parallel I/O system to control a seven-segment LED (Light Emitting Diode) display. This display device has seven segments, labelled as a, b, c, d, e, f, and g. It also has two extra segments for decimal points: LDP (Left Decimal Point) and RDP (Right Decimal Point). A seven-segment display device can show numeric digits from 0 to 9, as well as some simple characters (e.g. A, C, E, F, etc.). An alpha-numerical display is the simplest way for a robot to express its internal states. In this example, assume that the connection between the pins of the I/O port and the segments of the LED display is indicated by Table 6.1. For example, bit b_7 controls the "on" or "off" switch for the LDP while the complement of bit b_7 controls the "on" or "off" switch for the RDP. When $b_7 = 1$, the LDP is "on" while the RDP is "off".

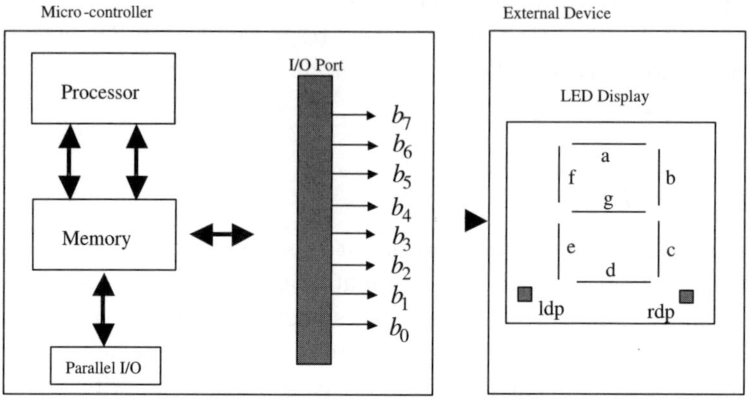

Fig. 6.24   Example of a seven-segment LED-display control by a parallel I/O system.

Table 6.1   Pin Connection.

Pins:	b_7	$\bar{b_7}$	b_6	b_5	b_4	b_3	b_2	b_1	b_0
Segments:	LDP	RDP	a	b	c	d	e	f	g

Example 6.3   For the display control as shown in Fig. 6.24, if we want to display digit "5" and the RDP, the output binary number from the parallel I/O port should be 01001011. In hexadecimal, this number is 0x9B. In order to light up digit "5" and the RDP, the processor simply writes byte 0x9B to the data register of the parallel I/O system.

◇◇◇◇◇◇◇◇◇◇◇◇◇◇◇◇◇

6.7.4.2   Interfacing for Data Communication

By definition, a parallel I/O system is capable of reading in or writing out at least one byte at a time. Obviously, we can use it for data communication because a stream of bytes of any length can be communicated through a parallel I/O system.

Normally, data communication involves two devices: a transmitting device and a receiving device. For the example shown in Fig. 6.16, computer 1 will receive visual sensory information from computer 2. Therefore, computer 1 is a receiving device while computer 2 is a transmitting device. In parallel, for hand-eye coordination behavior, computer 1 will send commands and data to computer 4. In this case, computer 1 is a transmitting device while computer 4 is a receiving device.

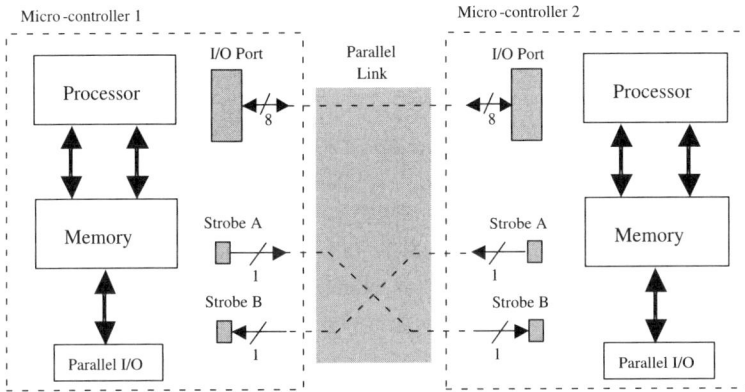

Fig. 6.25   Interfacing for data-communication with a parallel I/O system.

The process of connecting two devices for data communication is called *interfacing*. When two devices communicate with each other using their parallel I/O systems, the general connection looks like the one, as shown in Fig. 6.25.

The slash with a digit placed on a data line, which represents a physical cable, is a graphic representation of the wiring in a computing and communication system. The digit indicates the number of wires inside the cable. For example, the number "8" means that the cable, connecting the pins of the two I/O ports, has eight wires or signal lines inside.

In general, data communication, using a parallel I/O system, requires at least 10 signal lines to form a *parallel link*: a) eight lines for the data bits and b) two lines for the control signals. Obviously, for long-distance communication, this is a costly solution. As a result, data communication using a parallel I/O system is only appropriate for short-distance data communication (e.g. between a computer and a printer; between two devices inside a computing system etc).

6.7.4.3   *Strobed Parallel I/O*

As we studied earlier, the three important issues of data communication are data framing, communication initiation & synchronization, and error checking. An integrated method which handles these issues is generally called a *protocol* of communication. Many standards exist in literature.

However, here we will only cover the working principles at the physical level behind how data communication is initiated and then synchronized for both parallel and serial I/O systems.

By default, it is reasonable to assume that the initiator of data communication is the device or system which outputs a stream of data first. Therefore, a parallel I/O system must have a signal line to output a control signal that initiates a first session of data communication. Here, let us assume that this output signal line is called *strobe A*. On the other hand, a parallel I/O system must have a signal line to input the control signal from an initiating device or system. Let us call this input signal line *strobe B*.

Refer to Fig. 6.25. Assume that micro-controller 1 initiates data communication. In this case, micro-controller 1 undertakes output operation, and micro-controller 2 is in the input operation mode. If data synchronization only relies on the initiating device (micro-controller 1), it is called a *strobed I/O*.

Under the concept of a strobed I/O, the output operation works as follows:

- Step 1: Micro-controller 1 writes a byte to the data register which will put one byte onto the data bus connecting the I/O ports of the two communicating devices or systems.
- Step 2: Micro-controller 1 sends a signal to the strobe A line to tell the receiving device or system that a byte is ready on the data bus. This can be done by driving the strobe A line to logic high or logic low. Another way is to send a positive or negative pulse onto the strobe A line.
- Repeat Steps 1 and 2 until the session of data communication is completed (i.e. a stream of bytes has been exchanged).

Accordingly, the input operation at micro-controller 2 works as follows:

- Step 1: Micro-controller 2 continuously monitors the strobe B line.
- Step 2: If an incoming signal is detected at the strobe B line, it latches in a byte from the data bus to the data register.
- Repeat Steps 1 and 2.

6.7.4.4 *Full Handshake Parallel I/O*

The strobed I/O is the simplest way to undertake data communication. It will work properly under two conditions:

(1) The two devices operate at the same speed for each cycle of input and output.
(2) No noise exists on the data and signal lines.

The first condition is relatively easy to meet while the second condition depends on the working environment. In practice, it is not realistic to assume that transmitted data will be error-free regardless of the level of noise. An alternative solution to using the strobed I/O, which does not have this drawback, is the *full handshake I/O*. The basic idea behind this is to notify the proper reception of each byte. This mechanism provides the opportunity to do all kinds of error checking. It also permits the difference in the speeds at which the transmitting and receiving devices or systems are handling the input and output data.

Refer to Fig. 6.25. The output operation under the full handshake I/O mode works as follows:

- Step 1: Micro-controller 1 writes a byte to the data register which will put one byte onto the data bus connecting the I/O ports of the two communicating devices or systems.
- Step 2: Micro-controller 1 sends a signal to the strobe A line to tell the receiving device or system that a byte is ready on the data bus.
- Step 3: Micro-controller 1 waits for an acknowledgement signal on the strobe B line.
- Repeat Steps 1, 2 and 3 until the session of data communication is completed (i.e. a stream of bytes has been exchanged).

Accordingly, the input operation under the full handshake I/O mode works as follows:

- Step 1: Micro-controller 2 monitors the strobe B line continuously.
- Step 2: If an incoming signal is detected at the strobe B line, it latches in a byte from the data bus and store it in the data register.
- Step 3: Micro-controller 1 sends an acknowledgement signal to the strobe A line if no error is detected.
- Repeat Steps 1, 2 and 3.

6.7.5   *Serial I/O*

We know that the number of wires in the parallel link will be at least 10. If we use a parallel link to form a network, each system or device will have 10 wires. If there are n systems or devices attached to a network, the number of wires (excluding the wires of the network itself) will be $10 \bullet n$. So, obviously, a parallel link is not a good solution to form a network. Nowadays, the serial I/O system and the corresponding serial link have become the standard approach to networked data-communication.

In robotics, the number of wires inside a robot's body is a big concern as a robot is a complex system with many interconnected computational modules. Certainly, in this case, networked data-communication is a good solution.

In this section, we will study the basic working principles behind synchronous and asynchronous serial I/O systems. In fact, all advanced communication protocols, such as the Ethernet, TCP/IP, Field-bus (e.g. CAN bus or DeviceNet) and MODEM, are built on top of these basic working principles.

6.7.5.1   *Interfacing for Data Communication*

Refer to Fig. 6.25. The parallel link has eight signal lines for data exchange and two signal lines for control signals. If we reduce the number of data signal lines to one or two, and the number of control signal lines to zero or one, the parallel link becomes a *serial link* because it can only carry one bit of data at a time. In order to exchange data, the bits of the data have to be sequentially transmitted in series.

The evolution from a parallel link to a serial link sounds simple. But the actual operation of a serial I/O is more complicated than that of a parallel I/O. The control signal line classifies a serial I/O into two categories: a) asynchronous serial I/O, and b) synchronous serial I/O. If a serial link does not have any signal line for synchronization purposes, it is called an *asynchronous serial I/O*. The electronic circuitry which implements the asynchronous I/O mode is known as the *Serial Communication Interface (SCI)*. Obviously, the asynchronous serial I/O is a cost-effective solution for long-distance networked data communication (e.g. Internet and MODEM).

On the other hand, if a serial link requires one signal line for synchronization purposes, it is called a *synchronous serial I/O*. And, the electronic circuitry which implements the synchronous serial I/O mode is known as the *Serial Peripheral Interface (SPI)*. The keyword "peripheral" implies that the synchronous serial I/O is more appropriate for data exchange within a short distance.

Similarly, the data signal line(s) will classify the serial link into one of these three types, as shown in Fig. 6.26:

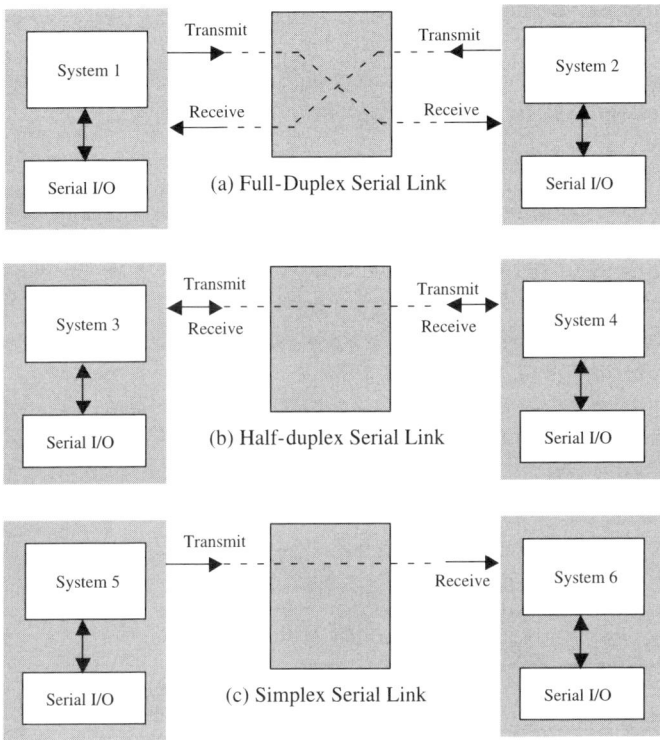

Fig. 6.26 Serial Link Types.

(1) Full-duplex Serial Link:
For this type of serial Link, the two devices or systems engaged in data exchange can simultaneously transmit and receive data. This is because the serial link has two bi-directional data signal lines, as shown in Fig. 6.26a.

(2) Half-duplex Serial Link:
Now, if we reduce the number of data signal lines to one, and if this data signal line is bi-directional, the full-duplex serial link becomes a *half-duplex* serial I/O. For this type of serial link, data can be alternately exchanged between two devices, or systems. During each session of data communication, one device or system is a transmitter and the other is a receiver, or vice versa.

(3) Simplex Serial Link:
For the half-duplex serial link, if we restrict the direction of data com-

munication to one way, then it becomes a *simplex* serial link. For this type of serial link, the transmitter and receiver are pre-specified. They cannot alter their roles. In other words, the data signal line sequentially carries the bits in one direction only.

6.7.5.2 *Bit-oriented Data Framing*

A serial I/O can only communicate one bit at a time. In order to communicate a set of bytes, one must arrange them into a series of bits. The process of packing a set of bytes (i.e. data) into a series of bits is called *data framing*. The result of a data framing process is called a *data frame*.

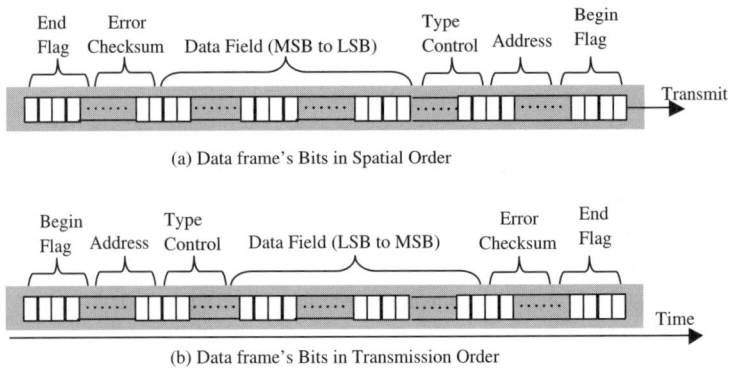

(a) Data frame's Bits in Spatial Order

(b) Data frame's Bits in Transmission Order

Fig. 6.27   Illustration of the concept of a data frame.

Fig. 6.27 illustrates the concept of a data frame. In the spatial order from left to right, a data frame is composed of:

- Begin Flag:
  This usually takes one byte and is used to indicate the start of a data frame.
- Address:
  A serial link is a common solution for networked data-communication. This means that more than two systems or devices can be attached to the same network. Since a network is a bus, there will be only one transmitter at a time. The rest of the systems or devices hooked onto the network are the potential receivers. A simple way to identify the specific receiver is to encode the address (i.e. the identification number) of the receiver into the data frame. This implies that each system or

device involved in a networked data-communication will have its own address. (For example, each computer hooked onto an Internet will have an IP address).

- Type Control:
  This usually takes one byte and is used to indicate the type of data to be transmitted. The types may include: ASCII, Binary, Compressed, Uncompressed, Compression Format, etc.
- Data Field:
  This refers to the set of bytes belonging to the data.
- Error Checksum:
  This may take a few bytes. It contains the results of an error-checking algorithm. Before transmission, this result (error checksum) is calculated and packed into the data frame. At the reception side, the error checksum is recalculated again by the receiver. The calculated result is compared with the received one. If the two error checksums are the same, then there is no error in communication. Otherwise, the receiver may request the transmitter to resend the data.
- End Flag:
  This simply indicates the end of a data frame.

In a serial I/O, the data frame is transmitted by a shift register bit-by-bit. By default, the shift register always moves the bits towards the right-hand side. As a result, the transmission order of a data frame (see Fig. 6.27b) is different from its spatial order (see Fig. 6.27a).

6.7.5.3 *Synchronous Serial I/O*

A serial I/O can only transmit a set of bytes in the bit-by-bit manner. Hence, synchronization is an important issue. In a simple case, this can be done with a dedicated signal line. A serial I/O, with a signal line for synchronization purposes, is called a *synchronous serial I/O*. Fig. 6.28 illustrates data exchange between two micro-controllers using a synchronous serial I/O. Consider the case of a full-duplex serial link. As we can see, the network will have at least three wires: two for data bits and one for a synchronization signal.

Data is a set of bytes. In order to integrally communicate a set of bytes between two systems or devices, a serial I/O's hardware operates two cascaded loops: a) the outer loop on the set of bytes and b) the inner loop on the bits of a byte.

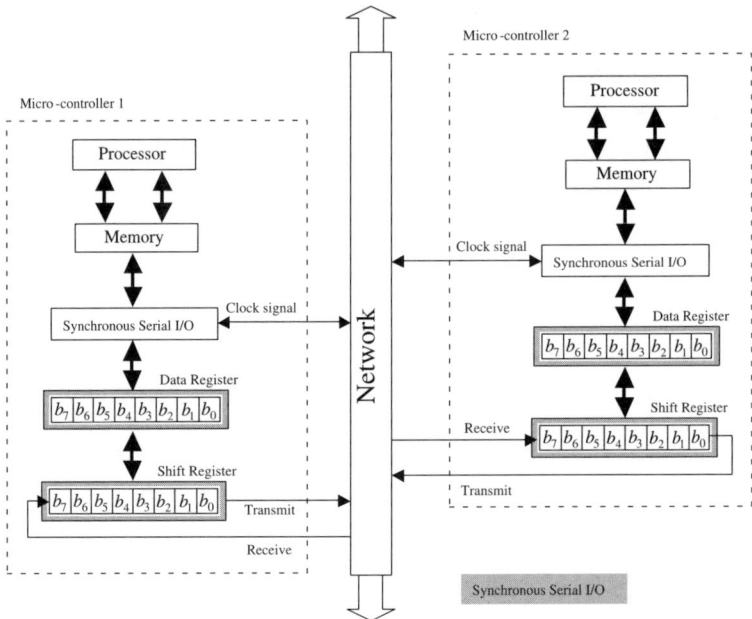

Fig. 6.28 Illustration of data communication with a synchronous serial I/O.

Operational Procedure of Transmitters

Refer to Fig. 6.28. Assume that micro-controller 1 is the transmitter and micro-controller 2 is the receiver. The outer loop of the transmitter's synchronous serial I/O may work as follows:

- Step 1: Configure the synchronous serial I/O for transmission.
- Step 2: When the status register's "transmission complete" flag is set, write a byte to the serial I/O's data register .
- Step 3: Activate the inner loop.
- Repeat Steps 2 and 3 until all bytes have been transmitted.

And, the inner loop of the transmitter's synchronous serial I/O may include the following steps:

- Step 1: Copy the byte from the data register to the shift register.
- Step 2: Send out a synchronization pulse to the synchronization signal line. At the same time, shift out one bit.
- Repeat Step 2 until all bits have been shifted out.

- Step 4: Set the"transmission complete" flag in the serial I/O's status register to inform the outer loop that the transmitter is ready for transmitting another byte.

Operational Procedure of Receivers

The outer loop of the receiver's synchronous serial I/O will work as follows:

- Step 1: Configure the synchronous serial I/O for reception.
- Step 2: Activate the inner loop.
- Step 3: When the status register's "transmission complete" flag is set, read in a byte from the serial I/O's data register.
- Repeat Steps 2 and 3 until all the bytes have been received.

Similarly, the inner loop of the receiver's synchronous serial I/O will include the following steps:

- Step 1: Detect the incoming synchronization pulse.
- Step 2: Shift in one bit for each incoming synchronization pulse.
- Repeat Steps 1 and 2 until all bits have been shifted in.
- Step 4: Copy the byte from the shift register to the data register, and set the "transmission complete" flag in the status register of the serial I/O. This informs the outer loop that one byte has been received.

6.7.5.4 *Asynchronous Serial I/O*

A synchronization signal line in the synchronous serial I/O increases the cost (i.e. one additional signal line) and the complexity (i.e. logic circuitry to generate and detect synchronization pulses). An alternative solution is to remove this line. A serial I/O without a synchronization signal line is known as an *asynchronous serial I/O*. Fig. 6.29 shows data communication between two micro-controllers using an asynchronous serial I/O.

Bit-oriented Byte Framing

When there is a synchronization signal line, it is easy for a device, or system, to take control of a network and transmit data. (Remember, in networked data-communication, there is only one transmitter at a time). So, if we remove the synchronization signal line, we need to find a way to resolve two issues: a) notification of taking control of a network and b) synchronization of the bit-by-bit transmission of a byte by the inner loop of a serial I/O.

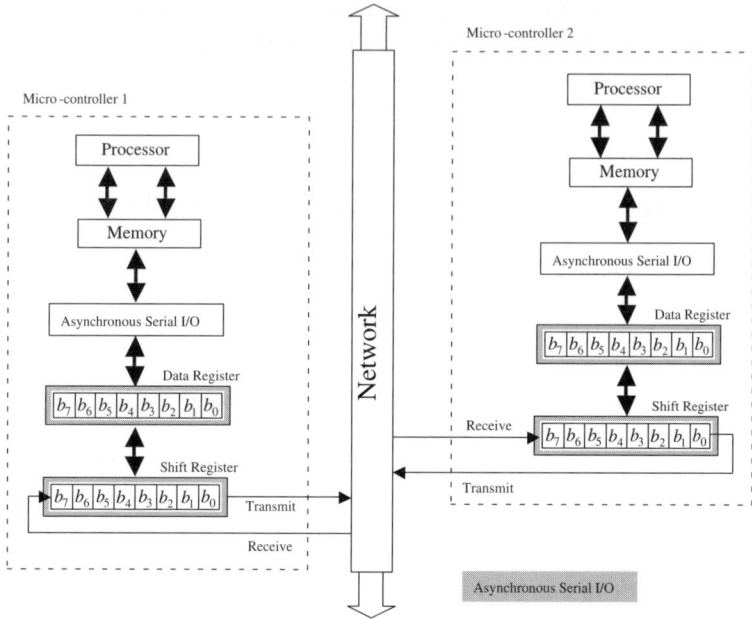

Fig. 6.29 Illustration of data communication using an asynchronous serial I/O.

A common way of solving the first issue for an asynchronous serial I/O is to play with the logic-signal level of the data lines. (There will be two data lines for a full-duplex serial link). For example, when there is no transmission over a network (i.e. no activity), a data line is always at logic high (+5V). This state is called the *idle* state. When a device or system has data to communicate, it first identifies the status of a data line. If the data line is in the idle state for a certain unit of time, the device, or system, drives the data line to logic low (0V) for one unit of time to indicate that it is about to take control of the network. Once a device, or system, has taken control of the data line, the line becomes *busy*. No other device or system can become a transmitter as long as the data line is in the busy state.

Refer to Fig. 6.27. One unit of time along the transmission time axis corresponds to one bit. The solution of driving a data line from the idle state to the busy state for one unit of time is viewed as the insertion of a bit (a bit of logic low, or 0) to the bits of a byte. This bit is called the *start bit*.

A common method for solving the second issue is to predefine a set of transmission speeds. (The RS232 or MODEM have a set of fixed transmission speeds). In other words, the transmitter and the receiver will negotiate one predefined transmission speed during the initialization stage of data communication. In fact, the commonly known *baud rate* of a communication device refers to the transmission speed measured in terms of kilo-bit-per-second, or kbps for short. In a binary number system, one kilo-bit is equal to 1024 bits.

Due to the fact that the transmitter and receiver are two different devices, or systems, of different performance, or quality, it is necessary to consider the discrepancy in the predefined transmission speed of a same value (e.g. 56 kbps). An easy way to accommodate this discrepancy is to introduce some extra units of time after the bit series corresponding to a byte. Since each unit of time, along the transmission time axis, corresponds to a bit, the addition of the extra units of time simply mean the addition of extra bits at the end of the bit series of a byte. These extra bits are known as *stop bits*. Here, we assume that one stop bit is added to the bit series of a byte.

The process of adding a start bit and some stop bits to the bit series of a byte can be called *byte framing*. With the use of byte framing, there is no need to have a physical synchronization signal line. Another advantage is the possibility of simple error checking for each byte. A typical way to perform error checking at the byte level is to introduce a *parity* bit p. With the extra parity bit p, we can make the number of 1s inside a byte be an odd or even number. For example, if we choose the "odd parity", and if the data byte is 0x88 (i.e. 10001000), then the parity bit p is set to "1" because the number of 1s in the byte itself is an even number. If the data byte becomes 0x89 (i.e. 10001001), the parity bit p will be set to "0" because the number of 1s in the byte itself is already an odd number.

The parity of a bit series must be preserved during transmission. If not, this means that there has been an error in communication. Parity checking is a simple way to detect error in data communication. And, Fig. 6.30 summarizes the process of bit-oriented byte framing with the insertion of start, parity and stop bits to the bit series $\{b_i, \ i = 0, 1, 2, ..., 7\}$ of a byte.

Operational Procedure of Transmitters

Similar to the operation of synchronous serial I/O, the hardware underlying an asynchronous I/O will also operate with two nested loops: a) the outer

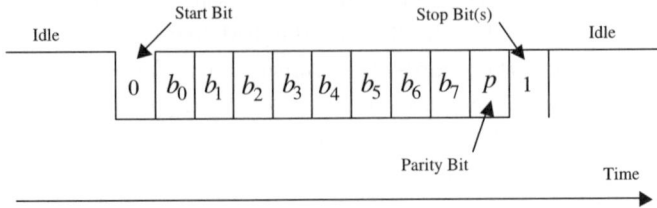

Fig. 6.30   Byte framing for an asynchronous serial I/O.

loop for byte-by-byte transmission of data bytes, and b) the inner loop for bit-by-bit transmission of the bit series of a byte.

Refer to Fig. 6.29. Assume that micro-controller 1 is the transmitter and micro-controller 2 is the receiver. The outer loop of the asynchronous serial I/O at the transmitter side may work as follows:

- Step 1: Configure the asynchronous serial I/O for transmission.
- Step 2: Set the transmission speed (i.e. baud rate) and byte-framing configuration (i.e. type of parity checking, number of stop bits).
- Step 3: When the status register's "transmission complete" flag is set, write a byte to the serial I/O's data register.
- Step 4: Activate the inner loop.
- Repeat Steps 3 and 4 until all bytes have been transmitted.

The inner loop of the transmitter's asynchronous serial I/O may include the following steps:

- Step 1: Do byte framing.
- Step 2: Copy the framed byte from the data register and status register where extra bits are stored to the shift register. (NOTE: In practice, the shift register will have more than 8 bits).
- Step 3: Detect the data line's status.
- Step 4: If the data line is idle, shift out bits from the shift register according to the chosen baud rate.
- Step 5: Set the "transmission complete" flag in the status register of the serial I/O. This informs the outer loop that the transmitter is ready for transmitting another byte.

Operational Procedure of Receivers

On the receiver's side, the outer loop of the asynchronous serial I/O will work in the following way:

- Step 1: Configure the asynchronous serial I/O for reception.
- Step 2: Set the baud rate and byte-framing configuration to be exactly the same as those used for the transmitter.
- Step 3: Activate the inner loop.
- Step 4: When the status register's "transmission complete" flag is set, read in a byte from the data register of the serial I/O.
- Repeat Steps 3 and 4 until all bytes have been received.

The inner loop of the receiver's asynchronous serial I/O will work as follows :

- Step 1: Detect the start bit.
- Step 2: If the receiver is awakened up by the start bit, shift in bits from the data line and perform error checking if any.
- Step 3: Copy the byte from the shift register to the data register.
- Step 4: Set the "transmission complete" flag in the status register of the serial I/O. This informs the outer loop that one byte has been received.

6.7.6 *Programmable Timers*

As we mentioned earlier, a robot's information system must be a real-time system in order to ensure the predictability of its actions or behaviors. With a real-time programming language or operating system, it is possible to perform the data processing and data storage in real-time. This is because all operations inside a computing system are synchronized by a common clock. In addition, we know that data communication between two devices, or systems, must be synchronized in one way or another. Consequently, the time required for the exchange of a set of bytes is also predictable once the two devices or systems have been engaged in data communication. Thus, we can say that all internal operations inside a complex and distributed information system can be achieved in real-time.

Today, robots are used in industry. Tomorrow, they will very likely become part of society as well. As a result, a robot must be able to respond to unpredictable external real-time events in a timely manner.

In society, all social activities are synchronized by a common clock (Greenwich Mean Time). Naturally, a robot should also have a standard sense of time. From our study of the computing system's hardware, we know that a microprocessor has a system clock to synchronize its internal operations. In principle, this system clock is a free-running clock which is not re-programmable by any user or programmer. One way to overcome this restriction is to incorporate an additional clock system which is programmable. (This clock system can simply be derived from the microprocessor's system clock itself). The resulting circuitry is called a *programmable timer*. A programmable timer is an I/O system because it can provide the following generic functions, as shown in Fig. 6.31:

- Active Signal Output,
- Active Signal Capture (input),
- Self-Generating Real-time Interrupt.

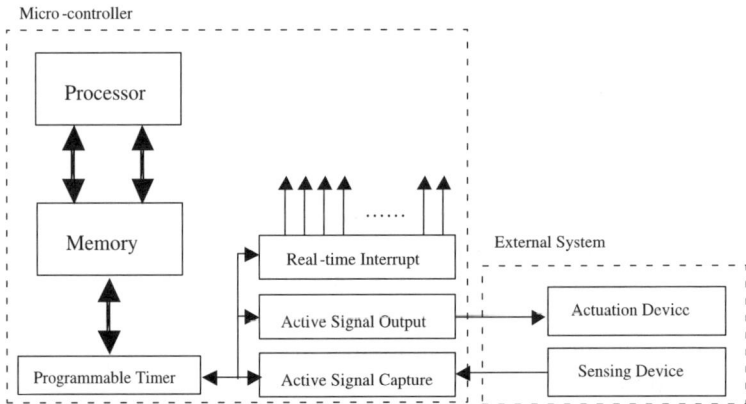

Fig. 6.31   Generic functions of a programmable timer I/O.

A programmable timer has its own free-running timer in the form of a counter. A counter is a register whose value increases at the rate of a clock signal. When the maximum value of the counter is reached, it automatically resets to zero and starts over again at the rate of the clock signal. This free-running timer provides a reference time number t_c to support time-critical actions, sensing or decision-making.

6.7.6.1 *Active Signal Output*

Refer to Fig. 6.31. The function of the active-signal output of a programmable timer is to output an active signal at a programmed time number or instant t_{c1}. If the actual time number of the free-running timer is t_c, an active signal will be sent out when $t_c = t_{c1}$.

An active signal may be in one of these types: a) the transition from logic high to logic low (falling edge), b) the transition from logic low to logic high (rising edge), c) always logic high and d) always logic low. An output active signal's type is also programmable.

The active signal from a programmable timer is useful for many time-critical actions such as turning on a machine, triggering a sensor, stopping an action, etc.

6.7.6.2 *Active Signal Capture*

The second generic function of a programmable timer is the active signal capture. When there is an active signal detected by the programmable time, its job is to record time number t_c into an *input capture data register*. Again, an active signal's type is programmed by the user or programmer.

If we denote t_{c2} the time number held by the input capture data register, t_{c2} indicates the time-instant when an external event occurs.

Example 6.4 Consider the application of a micro-controller controlling an ultra-sonic sensor for range sensing. Fig. 6.32 shows a micro-controller board and three sets of ultra-sonic sensors.

Assume that an ultra-sonic sensor's enabling signal line is driven by the signal from the programmable timer's active signal output. And, the echo signal line from the ultra-sonic sensor is connected to the programmable timer's active signal capture.

At time-instant t_{c1}, the programmable timer generates an active signal output to enable the ultra-sonic sensor. This triggers the sensor to transmit sonar waves towards the environment. If these waves hit an obstacle, they will bounce back as echo. If the ultra-sonic sensor detects an echo at time-instant t_{c2}, it instantly outputs an echo signal to the programmable timer. The active signal capture function of the programmable timer stores the time number corresponding to time-instant t_{c2}.

From the time difference $t_{c2} - t_{c1}$ and the speed of sound in the air, we can derive the distance between the ultra-sonic sensor and the obstacle in the environment.

Fig. 6.32 Example of the use of a micro-controller to control ultra-sonic sensors for range sensing.

◇◇◇◇◇◇◇◇◇◇◇◇◇◇◇◇◇◇

6.7.6.3 *Self-generating Real-time Interrupt*

A third useful generic function of a programmable timer is the generation of real-time interrupts at periodic, programmable time intervals.

In a digital computer, an interrupt signal will automatically halt the execution of the current computational task, and trigger the microprocessor to execute a service routine associated with this signal. An interrupt's service routine is programmable by any user or programmer.

Some diagnostic tasks may be more appropriately programmed as interrupt service routines so that they can be activated at a regular time interval. Some typical diagnostic tasks for a robot include: a) check the level of power supply, b) check the working conditions or alarms of the actuators, c) check the working conditions or alarms of the sensors and d) check the working conditions or alarms of the computing modules, etc.

Due to the low cost of a micro-controller, it may be wise to dedicate one or more micro-controllers to perform diagnostic tasks.

6.8   Summary

It is not particularly important for industrial robots (i.e. arm manipulators) to have advanced information systems for data-processing, storage, and communication. This is because industrial robots depend on manual, off-line programming to schedule their motions. There is no pressing need for industrial robots to be equipped with artificial intelligence.

However, this is not the case with humanoid robots. A humanoid robot has a human-like mechanism, or body. The ultimate purpose of this robot is to have it perform human-like behaviors so that it can be deployed in both industry and society. The research and development of humanoid robots will certainly be considered as marking a new era in the history of robotics.

In this chapter, we first studied the philosophy which stresses the importance of developing a humanoid robot's information system. It is easy to assume that a robot's information system will also be the foundation for an artificial brain. However, it is not that straightforward. And the issue of how to develop the process underlying an artificial mind is still unresolved. Here, we highlighted a possible pathway towards the development of an artificial mind, that is: from autonomous mental and physical actors to autonomous behaviors.

Upon understanding the importance of a humanoid robot's information system, we studied the basic working principles of today's computing hardware and software.

For data-processing hardware, we learned the basic architecture of digital computers. The cycle-by-cycle operations inside the arithmetic and logic unit(ALU) reveal the fact that the program execution of a microprocessor is, for the most part, predictable. This supports the claim that an information system based on microprocessors is intrinsically a real-time system with simultaneous and predictable behaviors.

As for data-processing software, we studied the basic concept of a real-time programming language and real-time operating system. Cooperative multi-tasking and preemptive multi-tasking are two basic solutions for handling the issue of running a pool of multiple, concurrent, computational tasks on one, or a few, microprocessors. However, a humanoid robot's information system will most likely be formed by a cluster of microprocessors or micro-controllers. Thus, it would be desirable to have a programming language, or operating system, which supports multiple behaviors (or multi-behaving) exhibited by a single body. Moreover, one of the challenging questions is: When will it become possible to program a digital computer with a natural language (linguistic programming)?

Finally, we studied the working principles behind data-interfacing and communication. With the help of A/D conversion and D/A conversion, an information system can easily interface with all kinds of analog sensors and actuators. With the support of parallel I/O and serial I/O systems, it is easy to build an information system on top of a cluster of networked microprocessors. For the control of external devices, the presence of a programmable timer will certainly ease the programming of the time-critical sensing and actuation tasks.

6.9 Exercises

(1) Explain why the information system is important to the development of a humanoid robot.
(2) How can you imitate the artificial mind with an information system?
(3) What is a behavior?
(4) What is a mental actor? What is a physical actor?
(5) Rodney Brooks had published an article called, "Intelligence without Representation." To what extent is the statement in the title true?
(6) What is a digital computer?
(7) What is a microprocessor?
(8) What is a micro-controller?
(9) Explain the difference between Von Neumann and Harvard architectures?
(10) Write the Taylor series of the cosine function $y(x) = \cos(x)$, and explain why a microprocessor can calculate such a function.
(11) How is the read or write operation carried out between a processor and its memory?
(12) Explain why the execution time of a program on a microprocessor is predictable.
(13) What is a machine language? What is an assembly language? And, what is a high-level programming language?
(14) Explain the difference between cooperative multi-tasking and preemptive multi-tasking.
(15) Explain the working principle which allows a processor to concurrently handle the execution of multiple tasks.
(16) Explain the importance of data storage and retrieval in robotics.
(17) What are the advantages of using a formatted file for data storage and retrieval?
(18) Search the internet for the TIFF file format specification. Explain why a TIFF-like file format allows for the storage and retrieval of heterogeneous data with a single file.

(19) What are the typical topologies of a data communication network?

(20) Explain the difference between the Polled I/O and the Interrupt-driven I/O.

(21) Prove Eq. 6.5.

(22) Prove that the relationship of Eq. 6.5 is linear if we choose,

$$R_i = \frac{R}{2^i}, \quad i = 0, 1, 2, ..., 7$$

where R is any positive resistance value.

(23) Are the conversion times of the D/A and A/D converters predictable?

(24) Refer to Fig. 6.24. Explain how to display character "F.".

(25) Explain the difference between the synchronous serial I/O and the asynchronous serial I/O.

(26) For networked data communication, how do you identify the receiver?

(27) Why does a network only allow one transmitter at a time?

(28) Search the internet to learn how a pair of MODEM devices work.

(29) Search the internet to learn how a pair of RS232 devices work.

(30) Search the internet to know how the CAN bus works, and explain why it is dedicated to networked industrial equipment control and networked in-vehicle device control.

(31) Search the internet to learn about the human brain's structure and neural system.

6.10 Bibliography

(1) Fuller, J. L. (1991). *Robotics: Introduction, Programming and Projects*, Maxwell Macmillan.

(2) Gilmore, C. M. (1995). *Microprocessor: Principles and Applications*, McGraw-Hill.

(3) Groover, M. P., M. Weiss, R. N. Nagel and N. G. Odrey (1986). *Industrial robotics: Technology, Programming, and Applications*, McGraw-Hill.

(4) Halang, W. A. and K. M. Sacha (1992). *Real-time Systems*, World Scientific.

(5) Krishna, C. M. and K. G. Shin (1997). *Real-time Systems*, McGraw-Hill.

(6) Ray, A. K. and K. M. Bhurchandi (2001). *Intel Microprocessors*, McGraw-Hill.

(7) Skroder, J. C. (1997).   *Using the M68HC11 Micro-controllers*, Prentice-Hall.

(8) Tabak, D. (1995). *Advanced Microprocessors*, McGraw-Hill.

(9) (2002). *Proceedings of IEEE International Conference on Development and Learning*, MIT, USA.

Chapter 7

Visual Sensory System of Robots

7.1 Introduction

They say that, "Seeing is believing." And, among the various sensory systems, the visual sensory system is the one that is able to produce the most accurate description of a scene that is able to "see". They also say that, "A picture is worth a thousand words." This is true in robotics, as while all sensory systems are complementary, the visual sensory system contains the richest information about a scene.

Many physical activities, such as an task execution and social interaction, undoubtedly depend on information-rich and accurate visual perception. A human being's visual sensory system can effortlessly measure the radiation of light rays from a scene, and instantaneously derive the coherent representation of a scene from images (projections of light rays). In robotics, the perception of motion (or action) in a dynamically changing environment is a necessary step towards the development of autonomous behaviors. Therefore, it is important to understand the critical issues underlying a robot's visual sensory and perception systems.

In this chapter, we will study the basic principles and concepts of digital-image acquisition, digital-image modelling and digital-image processing hardware. We will cover the fundamentals of a robot's visual perception system including image processing, feature extraction and perception of geometry in the next chapter.

7.2 The Basics of Light

We live in a colorful world. This sensation of colors is our physiological and psychological interpretation of light. Light is an electromagnetic wave that propagates in space and time. Light carries radiant energy and exhibits both wave-like and (mass-less) particle-like behaviors in space. Accordingly, light has the properties of both physics and optics.

7.2.1 *Physical Properties*

Without light, the world would be invisible to us, and to our robots. Look at the example shown in Fig. 7.1; a robot has a pair of stereo cameras as part of its visual sensory system. Because of the illumination of the scene, the robot is able to visually perceive the stereo images of the scene.

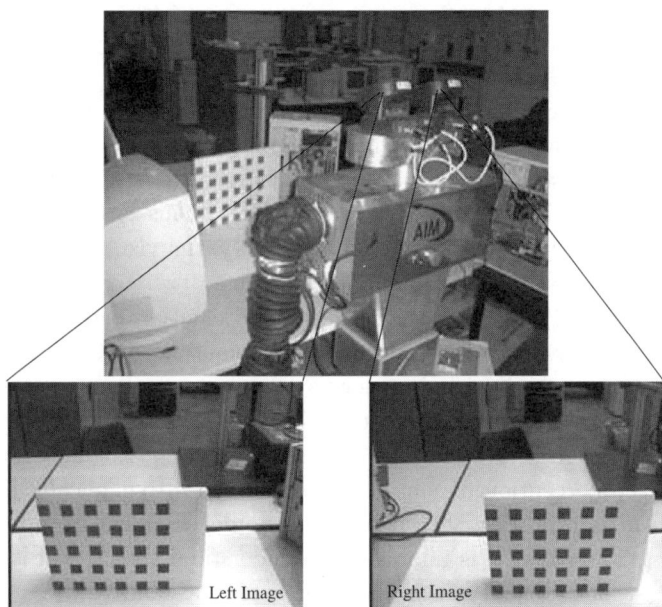

Fig. 7.1   An example of a robot's visual sensory system, and a pair of stereo images.

Light Emission

According to physics, all matters are composed of molecules arranged in specific structures, known as molecular structures. A molecule, in turn, is composed of atoms, also arranged in specific structures (atomic structures). The atom, however, is not the smallest physical entity of matter. In fact, the atom is made up of a nucleus and a cluster of electrons orbiting around the nucleus, as shown in Fig. 7.2a. A nucleus itself is composed of protons and neutrons.

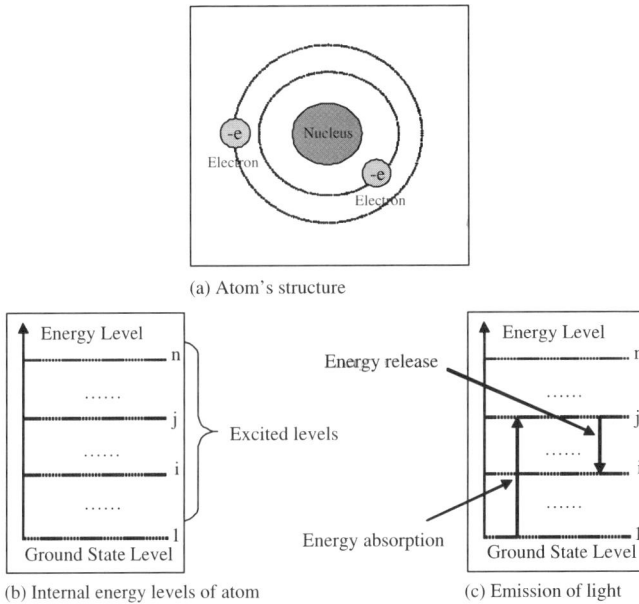

(a) Atom's structure

(b) Internal energy levels of atom

(c) Emission of light

Fig. 7.2   Illustration of light emitting process.

An atom is an actively-vibrating entity because of the energy contained in it. Basically, the total energy contained in an atom comes from: a) the thermal energy (thermal vibration) of the atom and b) the kinetic energy of its orbiting electrons. The latter mainly constitutes an atom's *internal energy*. (NOTE: Electrons and protons have mass and electric charges. They should also contain gravitational and electric potential energies).

According to Danish physicist Bohr's theory (1913), an atom's internal energy is confined to a set of discrete levels. In other words, an atom's internal energy will not make a continuous transition. Instead, it will only transit from one permissible energy level to another permissible energy level, as shown in Fig. 7.2b and Fig. 7.2c. This discrete transition of internal-energy levels can be explained by the stable orbits of the electrons moving around the nucleus. When an atom absorbs or radiates energy, the electrons make the transition across the stable orbits. A stable orbit is a trajectory along which an electron's energy is conserved. When the electrons make the transition, the variation of an atom's internal energy is not continuous but a definite amount.

Additionally, an atom's internal energy tends to stay at the lowest level, also known as the *ground state level*, or level 1. All levels higher than the ground state level are called *excited levels* (see Fig. 7.2b).

When a certain amount of energy is applied to an atom, and absorbed, its internal energy will jump to a higher level, denoted by j. Since an atom's internal energy tends to remain at ground level, this will cause the atom to instantaneously transit its internal energy to a lower level, denoted by i, as shown in Fig. 7.2c. When this transition occurs within a time interval in the order of 10^{-8}s, a definite amount of energy is released in the form of electromagnetic waves that is the origin of *light emission*. This process is known as the *spontaneous emission of light*.

The transition of an atom's internal energy, from a higher excited level to a lower excited level (or ground state level), can also be triggered by an external cause such as an electric field. This process is called the *stimulated emission of light*. The laser (i.e. coherent light of the same wavelength) is produced by a process of stimulated emission.

If we denote

- E_{absorb} the energy applied to an atom,
- $\triangle E_{thermal}$ the variation in the atom's thermal energy,
- $\triangle E_{internal}$ the variation in the atom's internal energy,
- $\triangle E_{emission}$ the energy radiating from the atom,

the application of the principle of energy conservation yields

$$E_{absorb} = \triangle E_{thermal} + \triangle E_{internal} + \triangle E_{emission}. \tag{7.1}$$

Refer to Fig. 7.2c. If E_i and E_j denote the internal energies of the atom at levels i and j, we have

$$\triangle E_{emission} = E_j - E_i. \tag{7.2}$$

Electromagnetic Waves

The energy radiated from an atom, described by Eq. 7.2, is manifested in the form of an electromagnetic wave. An electromagnetic wave is a periodic wave, as illustrated by Fig. 7.3.

Inside a homogeneous medium, the electromagnetic wave travels along a straight line. A wave's spatial period is called a *wavelength*. A wave's wavelength is usually denoted by λ. If there are n periods within a unit

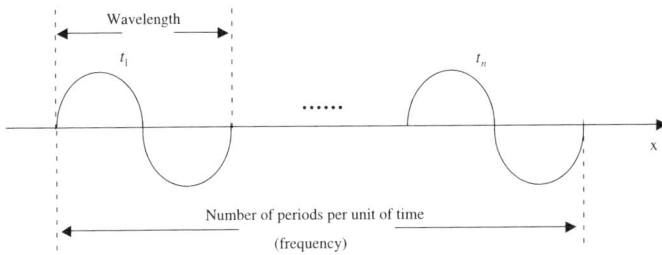

Fig. 7.3   An illustration of a periodic wave in space and time.

of time (one second), this number n is known as the (temporal) frequency, denoted by f.

Refer to Fig. 7.3. A wave's spatial velocity is the travelled distance by the wave within a unit of time. As a result, a wave's spatial velocity (or speed), denoted by v, is

$$v = \lambda \bullet f. \tag{7.3}$$

In a homogenous medium, the speed of an electromagnetic wave is a constant. For example, the speed of an electromagnetic wave in a vacuum is approximately $3 \times 10^8 \mathrm{m/s}$. Usually, we denote $c = 3 \times 10^8 \mathrm{m/s}$ as the speed of light in a vacuum. Therefore, if we know the wavelength, the frequency of an electromagnetic wave in a vacuum is determined by

$$f = \frac{c}{\lambda}. \tag{7.4}$$

Photons

An electromagnetic wave is a carrier of energy. It is not a continuous stream of undulations but is formed by a train of undulating elements called *quanta* or *photons*. A photon is an energy carrier with zero mass. The amount of energy carried by a photon along an electromagnetic wave with frequency f and wavelength λ is

$$E_p = h \bullet f \tag{7.5}$$

where h is Planck's Constant and is equal to 6.626×10^{-34} Newton-meter-second (N·m·s) or Joule-second (J·s). This equation shows that a photon's energy only depends on its frequency. By applying Eq. 7.3, Eq. 7.5 can also

Table 7.1   Colors and their corresponding wavelength ranges

Color	Wavelength Range (nm)
Red	622-780
Orange	597-622
Yellow	577-597
Green	492-577
Blue	455-492
Violet	390-455

be expressed as

$$E_p = h\frac{v}{\lambda}. \tag{7.6}$$

Refer to Eq. 7.2. Eq. 7.6 shows that the wavelength of an electromagnetic wave, emitted by an atom, will be determined by

$$\lambda = h\frac{v}{\triangle E_{emission}}. \tag{7.7}$$

Since $\triangle E_{emission}$ depends on the characteristics of the atom, the wavelength of an electromagnetic wave, radiated from an atom, will also depend on the characteristics of the atom. Because of the difference in wavelengths (or frequencies), different materials will appear in different colors.

Visible Light

According to Eq. 7.1 and Eq. 7.7, all materials excited by external energy will radiate electromagnetic wave(s). The wavelength of an electromagnetic wave depends on the property of the atoms in a material.

Interestingly enough, our eyes are sensitive to electromagnetic waves having wavelengths which fall between 390 nanometers (nm) (or 10^{-9}m) and 780nm. The physiological and psychological responses of our eyes to these wavelengths produce the sensation of colors. And, we commonly call these electromagnetic waves *light* or *visible light*. Table 7.1 lists typical colors and their corresponding wavelength ranges.

Interference

When two electromagnetic waves overlap while travelling in space, this causes the superimposition of the two waves. This process is called *interference*. The interference process can enhance the amplitude of the electromagnetic waves if the phases of the two interfering waves are the same (i.e. in-phase). This phenomenon is called *constructive interference*. But, it can also attenuate the amplitude of the electromagnetic waves if the phases of the two interfering waves are different by 180^0. This phenomenon is called *destructive interference*.

Light Sources

Any object that irradiates a scene is called a *light source*. For example, the Sun is a natural light source that irradiates the Earth. For an object to continuously radiates electromagnetic waves, there must be a balance between energy absorption and energy emission.

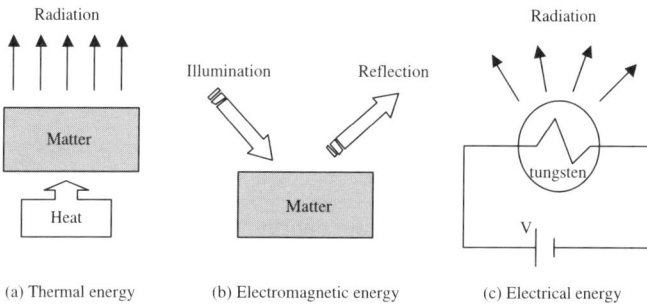

Fig. 7.4   Common effects of radiation.

As shown in Fig. 7.4, there are three common ways to make an object radiate:

- The Use of Thermal Energy (see Fig. 7.4a):
  When heating an object, the internal energy of the atoms inside the matter will actively transit between the ground state level and excited levels. Consequently, more radiations will be emanated. For example, when a material is in flames, our eyes can perceive visible light.
- The Use of Electromagnetic Energy (Fig. 7.4b):
  When the surface of an object is irradiated by a primary source of electromagnetic waves, the atoms on the surface will absorb and release electromagnetic energy. A typical example of this is the light from the Moon at night. Thus, the surface of an object under illumination radiates electromagnetic waves which have wavelengths that primarily depend on the property of the atoms on the object's surface.

This explains why we can perceive the colors of surfaces illuminated by a light source of white color, which is a mixture of all the colors.

- The Use of Electrical Energy (see Fig. 7.4c):
  Certain materials will radiate when a current is applied to them. For example, a tungsten lamp will emit light when current flows through it. Another typical example is the semiconductor device, LED (Light-Emitting-Diode).

7.2.2   *Geometrical Properties*

The physical properties of light demonstrate its particle-like behaviors. These are important in the understanding of a visual sensory system, because they explain the origin of chromatic information. In addition to chromatic information, a robot's visual sensory and perception systems can also infer geometric information about a scene with the help of light. This is because light also has inherent geometric properties.

Light Rays

As we studied earlier, light carries energy (photons) through space and time. At the macroscopic level, a trajectory of light in a homogeneous medium, such as air, glass, or a vacuum, looks like a straight line. That is why we often use the term *light ray*. However, a light's trajectory is actually an undulating wave centered around a straight line in a homogeneous medium. Still, for the sake of studying the geometric properties of light, it is convenient to consider it as a ray (i.e. a straight line in a homogeneous medium).

Reflection

When a light ray strikes a surface between two different homogeneous media such as air and glass, its direction is altered. If this direction-altered light ray bounces back to the same medium, the phenomenon is called *reflection*, as shown in Fig. 7.5. If the light ray passes through the surface and enters another medium, this phenomenon is called *refraction*.

Depending on the smoothness of the surface separating the two media, the reflection may behave in either of the following ways:

- Diffuse Reflection:
  If a surface is not polished properly, and is not smooth enough, the

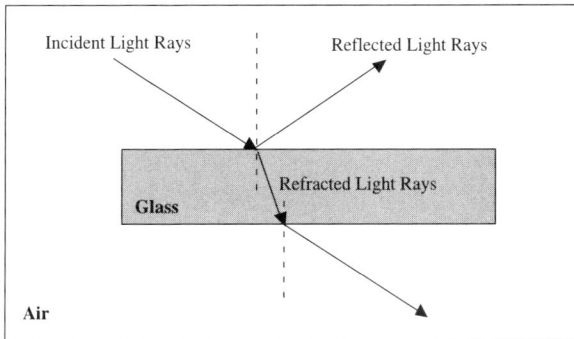

Fig. 7.5 Illustration of reflection and refraction of light rays.

parallel light rays that strike it will be reflected back in all possible directions. This type of reflection is called a *diffuse reflection*.

- Specular Reflection:
  If a surface is smoothly polished and is like a mirror, the parallel light rays that strike it will be perfectly mirrored back as a bundle of parallel light rays. This type of reflection is called a *specular reflection*.

For a specular reflection, the following properties hold:

- The incident light rays, the reflected light rays, and the normal vector of the surface are co-planar.
- The *angle of incidence*, which is the angle between the incident light rays and the surface's normal vector, is equal to the *angle of reflection*, which is the angle between the reflected light rays and the surface's normal vector. If we denote θ_i the angle of incidence and θ_r the angle of reflection, then $\theta_r = \theta_i$.

Refraction

Refraction is the phenomenon where light rays pass from one medium into another.

In a homogeneous medium, light travels at a constant speed. In a vacuum, the speed of light is about $3 \times 10^8 \text{m/s}$ and is denoted by symbol c. In any other medium, the speed v of light is smaller than c. The ratio between c and v is called the *index of refraction*. If we denote n the index

Table 7.2   Index of refraction for yellow light

Medium	Index of refraction
Vacuum	1.0
Air	$\simeq 1.0$
Water	1.3
Glass	1.5
Quartz	1.5
Diamond	2.4

of refraction, we have

$$n = \frac{c}{v}. \tag{7.8}$$

Clearly, the index of refraction is always greater than or equal to 1. This refraction parameter measures the speed change of a light ray when it passes from one medium into another. Table 7.2 lists the indexes of refraction for some medium with respect to yellow light, which has a wavelength in a vacuum that is about 589nm.

When light can continuously travel in the same medium, there is no loss of energy. Alternatively, if the photons are absorbed by the medium, there will be no light. According to Eq. 7.5, a light's frequency f will remain the same if a photon continues travelling in the same medium. Since $v = \lambda \bullet f$ (i.e. Eq. 7.3), this means that a light's wavelength will change when it passes from one medium into another. If we denote λ_0 the wavelength of light in a vacuum and λ the wavelength of light in any other medium, we have

$$\frac{v}{\lambda} = \frac{c}{\lambda_0} \tag{7.9}$$

or

$$\lambda = \frac{\lambda_0}{n}. \tag{7.10}$$

Eq. 7.10 indicates that a wavelength will be shortened when light goes from a vacuum into any other medium. However, for a robot's visual sensory and perception systems, an important property of light is the direction change when a light passes through the surface separating two media, as shown in Fig. 7.5.

Let us denote θ_r the angle between the refracted light rays and the surface's normal vector. This angle is also called the *angle of refraction*.

Then, the following geometric properties governing the refraction will hold:

- The incident light rays, the refracted light rays, and the normal vector of the surface separating two media are co-planar.
- The ratio between the sine functions of the angles of incidence and refraction is equal to the ratio between the index of refraction n_2 of the second medium (e.g. glass) and the index of refraction n_1 of the first medium (e.g. air). In other words, we have

$$\frac{\sin(\theta_i)}{\sin(\theta_r)} = \frac{n_2}{n_1} \qquad (7.11)$$

or

$$n_1 \bullet sin(\theta_i) = n_2 \bullet sin(\theta_r). \qquad (7.12)$$

Eq. 7.11 is useful to guide the design and fabrication of optical lenses widely found in cameras, microscopes, CD-ROM Readers/Writers, CD players, and other optical devices or equipment.

7.2.3 *Refraction of Light Rays by a Thin Lens*

Refer to Fig. 7.5. After the light rays enter the glass, they will continue to travel along a straight line. When they reach the bottom surface of the glass, a new refraction will occur. After the second refraction, the light rays will enter the air again. If the top and bottom surfaces of the glass are parallel, the light rays, exiting from the glass's bottom surface, will be parallel to the incident light rays. This property can easily be verified by Eq. 7.11.

This observation indicates that if the top and bottom surfaces of a transparent material are parallel to each other, it is not possible to converge or diverge the light rays. In other words, it is not possible to create images with scalable dimensions.

However, if the top and bottom surfaces of a transparent material are not parallel to each other, the light rays will converge or diverge. Thus is born the optical lens system.

Geometrically, an optical lens system can be best explained using a *convex thin lens*. A convex thin lens is made of transparent material (e.g. glass), which is shaped into two spherical surfaces, as shown in Fig. 7.6. The thickness of a thin lens is negligible.

For optical imaging, a convex thin lens has the following useful properties:

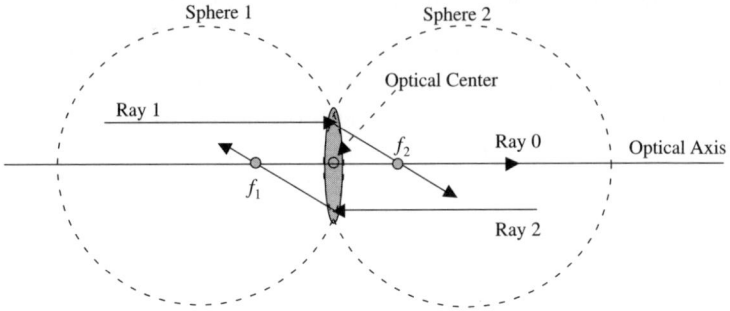

Fig. 7.6 Sectional view of a convex thin lens.

- The center of the lens is called the *optical center*.
- The axis that passes through the optical center and perpendicular to the two spherical surfaces is called the *optical axis*.
- The ray (ray 0 in Fig. 7.6) which coincides with the optical axis, will not be refracted by the lens. In other words, it will go through the lens without altering its direction.
- The ray parallel to the optical axis (ray 1 or ray 2 in Fig. 7.6) which strikes the lens at one side will be refracted and exit from the other side. The refracted ray will intersect the optical axis at a point known as *focal point* (i.e. f_2 or f_1).
- The distance between a focal point and the optical center is called the *focal length*.
- Interestingly, the focal lengths at both sides of a convex thin lens are equal to each other regardless of the radii of the two spherical surfaces.

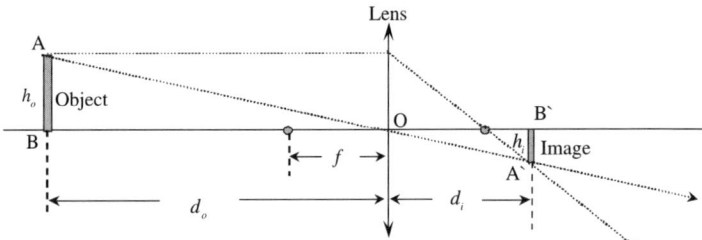

Fig. 7.7 Formation of an optical image with a convex thin lens.

The refractive property of a convex thin lens is useful in the formation of optical images, as shown in Fig. 7.7. When an object is placed in front of a convex thin lens, an optical image will appear at the other side of the lens. As shown in Fig. 7.7, point A will have a corresponding image A'. Similarly, point B will have a corresponding image B'. Let us denote f the focal length (i.e. the distance between the optical center and the focal point at either side of the lens), d_o the distance along the optical axis between an object and the optical center, and d_i the distance along the optical axis between the image and the optical center. From the two similar triangles OAB and $OA'B'$, it is easy to prove the following well-known formula:

$$\frac{1}{d_o} + \frac{1}{d_i} = \frac{1}{f}. \tag{7.13}$$

Example 7.1 Let us place an object at a distance of 1 meter (m) from a convex thin lens with a focal length of 35 millimeters (mm). By applying Eq. 7.13, the distance of the optical image relative to the thin lens will be

$$d_i = \frac{f \bullet d_o}{d_o - f} = \frac{0.035}{0.965} = 0.03627 \ (m).$$

Now, let us move the object back another meter (i.e. $d_o = 2$m). The distance of the optical image relative to the thin lens is now

$$d_i = \frac{f \bullet d_o}{d_o - f} = \frac{0.07}{1.965} = 0.03562 \ (m).$$

The above results indicate that the image shifts toward the focal point (e.g. at 35mm) when the object moves away from the convex thin lens.

◇◇◇◇◇◇◇◇◇◇◇◇◇◇◇◇◇◇◇

Now, let us denote h_o the height of the object in the direction perpendicular to the optical axis, and h_i the height of the image. From the similarity between the two triangles OAB and $OA'B'$, as shown in Fig. 7.7, we have

$$\frac{h_o}{h_i} = \frac{d_o}{d_i} \tag{7.14}$$

or

$$h_i = h_o \bullet \frac{d_i}{d_o}. \tag{7.15}$$

Eq. 7.15 indicates that a convex thin lens is able to form an image of scalable size. The scaling factor depends on the distance of the optical lens from the object and image (i.e. d_o and d_i). In fact, Eq. 7.15 is useful for inferring geometry from images. It forms an important mathematical basis for a robot's visual perception system.

Example 7.2 Assume that an object with a height of 0.5 m is placed at a distance of 1m from a convex thin lens with a focal length of 35mm. From the previous example, we know that the image is formed at

$$d_i = 0.03627 \ (m).$$

By applying Eq.7.15, the height of the image can be calculated as follows:

$$h_i = 0.5 \bullet \frac{0.03627}{1.0} = 0.018135 \ (m).$$

Now, let us move the object back another meter (i.e. $d_o = 2m$). The height of the image is now

$$d_i = 0.5 \bullet \frac{0.03562}{2.0} = 0.008905 \ (m).$$

The above results illustrate that an image becomes smaller when the object is moved further away from the convex thin lens.

$$\diamond\diamond\diamond\diamond\diamond\diamond\diamond\diamond\diamond\diamond\diamond\diamond\diamond\diamond\diamond\diamond$$

7.3 The Basics of the Human Eye

Human eyes (including most animal eyes) magically explore the refractive property of a convex thin lens as the basis for forming optical images. In addition, human vision is intrinsically color vision. Here, we will study some basic facts about image-sensing by human eyes.

7.3.1 *Eyeballs*

In order to form a color image of a reduced scale, the eye or camera must have the following three elements:

- Optical-Lens System:
  The dimension of a scene can be infinite. However, a sensing device has a definite, small size. Therefore, it is necessary to scale down the projection of this infinite scene into a definite, small area. This is only achievable with an optical-lens system.

- Photon-Detector Array:
  A light ray is the locus of a photon moving at a high speed (in the order of 3×10^8 m/s). In order to capture photons and convert their energies into the corresponding electrical signals, an array of photon-detectors must be placed at a focusing plane behind the optical-lens system. This focusing plane is also called *image plane*.
- Iris:
  All sensing elements have a working range. Beyond the limit where a signal starts to saturate, a sensing element will no more respond to any input. Therefore, there must have a mechanism to regulate the amount of photons entering an eye or camera. The device that controls the amount of incident photons is called an *iris* or *diaphragm*.

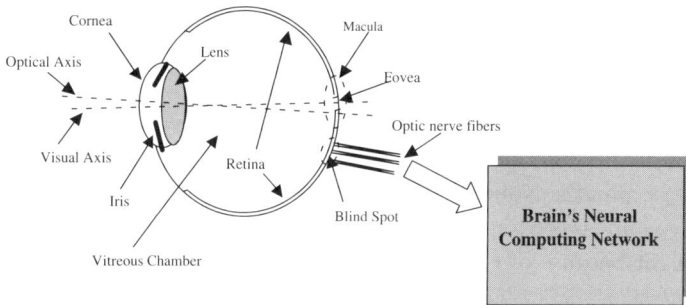

Fig. 7.8   Simple illustration of a human eyeball, sectional view.

Of course, the human eye has the three above-mentioned elements, as shown in Fig. 7.8. More specifically, it is mainly composed of: a cornea, iris, crystalline lens, inner vitreous chamber, retina, macula (including fovea) and optic-nerve fibers. However, the human lens system consists of the cornea, the vitreous substance inside the eyeball's chamber and the crystalline lens. This system's primary function is to refract incoming light rays to sharply focus them onto the *macula*, and to loosely focus them onto the large inner spherical area, called the *retina*.

The retina and macula are made of photon-sensitive cells. They convert the energy, carried by the captured photons, into electrical signals which are then transmitted to the brain's neural computing network through *optic-nerve fibers*. The area where the optic-nerve fibers are connected to the inner spherical surface of the eyeball is known as the *blind spot*. This area is insensitive to incoming light rays.

Refer to Fig. 7.8. The line connecting the center of the crystalline lens and the center of the macula is called the *visual axis*. As we know, any spherical lens has its own optical axis. Interestingly enough, the eye's visual axis does not coincide with the optical axis of the lens. This means that the center of the image at the macula is not the projection of the optical axis of the lens. This phenomenon is also true for a man-made camera as any error in assembling a camera will make it impossible to project the optical axis exactly onto the image center.

7.3.2   *Photosensitive Cells*

The human eye has two sets of photosensitive cells: a) cone cells and b) rod cells. The distribution of the cone cells inside a human eye looks like a very sharp Gaussian distribution which is centered at the macula. But, the rod cells are scattered across the retina and macula. However, at the narrow center of the macula, there is no rod cells.

George Wald, and his team at Harvard University, pioneered the understanding of the functional and molecular properties of rod and cone cells. In 1967, he won the Nobel Prize in Medicine and Physiology for his work on human vision. According to Wald's work, cone and rod cells are sensitive to visible light because of the presence of a protein called *rhodopsin*, or *visual pigment*.

Structurally, a cone cell is much smaller than a rod cell. So, the amount of rhodopsin inside a cone cell is much less than the amount of rhodopsin inside a rod cell. As a result, a rod cell is almost a thousand times more sensitive to light than a cone cell. This high sensitivity of rod cells explains why rod cells will only respond to the variation of energy carried by the captured photons. To a certain extent, a single photon is sufficient to trigger a response from a rod cell. Because of the high sensitivity, rod cells are not able to differentiate among various wavelengths of incoming light, and thus, insensitive to colors. Since the amount of energy passing through a unit area is called *light intensity*, we can say that rod cells are only sensitive to light intensity.

On the other hand, a cone cell can only moderately respond to light intensity. Because of the low sensitivity, a cone cell is able to respond differently to the wavelengths of incoming light. This explains why a cone cell is able to selectively respond to colors. This selectivity, in response to colors, forms the basis for human color vision. Artificial color vision works on a similar principle. According to the selective response to colors, the cone cells of the human eye are magically divided into three types:

- Red-sensitive cone cells, which are only sensitive to red,
- Green-sensitive cone cells, which are only sensitive to green,
- Blue-sensitive cone cells, which are only sensitive to blue.

7.3.3 *Central Vision*

Together, the cornea, the crystalline lens and the vitreous substance are responsible to refract and focus the incoming light rays to an eye. However, a focused optical image will fall into a large inner spherical area inside an eye. As shown in Fig. 7.8, the inner spherical surface of the eye is composed of: a) the retina, b) the macula and c) the blind spot. By definition, the sensing and processing of images from the macula area is called the *central vision*, while the sensing and processing of images from the retinal area is called the *peripheral vision*.

Human central vision makes use of both cone cells and rod cells to convert light into the corresponding electrical signals. The following are key characteristics of human central vision:

- Color Vision:
  As cone cells are sensitive to colors, the central vision is intrinsically color vision.
- Daylight Vision:
  Cone cells can only moderately respond to light intensity. Thus, the light intensity has to be above a certain level in order to trigger any response from a cone cell. As a result, it is necessary to illuminate a scene so that the central vision can perform properly.
- High-resolution Vision:
  The human eye has around 6-7 million cone cells. However, the spherical angle of the macula is about $\pm20^{o}$ while the spherical angle corresponding to the fovea is about $\pm2^{o}$. Since the distribution of cone cells follows a very sharp Gaussian distribution centered at the fovea, the center of the macula has the highest density of cone cells. On the other hand, there are about 120 million rod cells scattered across the retina and macula (except for the fovea). Relatively speaking, the combined density of both cone and rod cells at the macula is higher than at other areas of the inner spherical surface of the eye.

Functionally, our eyes always focus the sharp optical image onto the macula. In other words, central vision is very sharp as well. Thus, all important visual information will come from the central vision of our eyes.

7.3.4 *Peripheral Vision*

The visual attention of our eyes is mainly on the macula. This is why visual-sensing and processing for signals from the zone outside the macula is called *peripheral vision*. The photosensitive cells of peripheral vision are mainly composed of rod cells. Only a very small percentage of cone cells is present in peripheral vision. Therefore, the key characteristics of human peripheral vision are:

- Dim-light VisionDim-light vision:
  Since a rod cell is able to respond to a single photon, peripheral vision works even in a poorly-illuminated environment.
- Motion Detection:
  Signals from peripheral vision mainly contribute to motion detection. From an engineering point of view, this can be explained by the fact that the brain's neural computing network performs very simple processing on signals from these cells because the main attention is devoted to signals from the macula's cone cells. One of the simplest image-processing operations, called image substraction, is a good, simple, and fast motion-detector in image space. We will study some basic image-processing techniques in the next chapter.

7.4 Digital Image Acquisition

Refer to Fig. 7.8. The human eye is composed of a lens system, an iris, and an image-sensing area. The output from the eye is a series of electrical signals which are transmitted to our brain's neural-computing network through optic-nerve fibers. In the image-sensing area, there are two arrays of photosensitive cells: a) an array of cone cells and b) an array of rod cells. In addition, the eye can perceive the three primary colors (red, green and blue) because the array of cone cells can further be divided into three sub-arrays which are sensitive to these colors. It goes without saying that human vision is a very sophisticated system.

In order for a robot to see a colorful world, it is necessary to develop an artificial eye, commonly called an *electronic camera* (or *camera* for short).

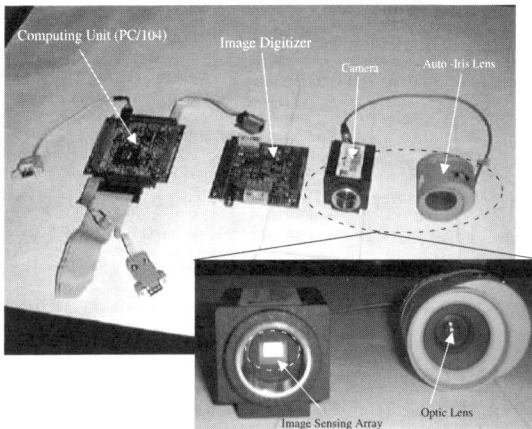

Fig. 7.9 The minimum set of hardware components necessary for a robot's visual sensory system.

Functionally, a camera is similar to a human eye. Due to advances in semiconductor technology, it is easy to construct a visual sensory system composed of an optical lens, artificial iris, light splitter & color filter, image-sensing array(s), signal-conversion circuit, image digitizer, and computing unit. Fig. 7.9 shows the minimum set of hardware components necessary for a robot's visual sensory system.

Certainly, the most important item in a robot's visual sensory system is the electronic camera. Without a camera, there is no vision. Just as in the human eye, a camera consists of a lens, an iris, light splitter & color filter, an image-sensing array, and a signal-conversion circuit. In this section, we will study these items in detail.

7.4.1 *Formation of Optical Images*

As we already mentioned, while a scene can be infinitely large, the human eyes have a definite, small size. In order to perceive a large scene with a small sensing device, it is necessary to first focus the incoming light rays into a small area, similar to the macula in an eye. This process is known as the *formation of optical images*. For an electronic camera designed for color-image sensing, the process of forming optical images will involve these steps:

- Optical-image focusing,
- Control of incoming light,
- Light splitting & color filtering.

7.4.1.1   *Optical-Image Focusing*

The first step in forming an optical image is to focus incoming light rays into a small area. This can be done with an optical lens system. For the sake of simplicity, let us consider an optical lens system to be a convex thin lens. According to Example 7.1, an object placed at different distances, relative to one side of a convex thin lens, will have corresponding images at different distances relative to another side of the convex thin lens. If an object has an arbitrary shape which is not planar, the focused image will not be planar. However, in practice, we can consider that a focused image falls approximately into a planar surface. In this way, an image-sensing device can be a planar device. As shown in Fig. 7.10, the planar surface of an image-sensing device is called an *image plane.*

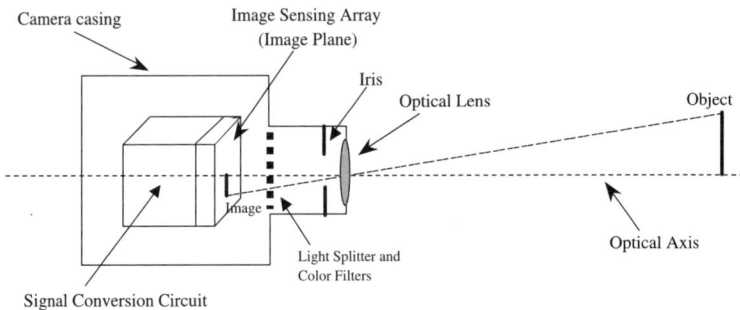

Fig. 7.10   Generic structure of an electronic camera.

Since a scene has a certain depth which is the permissible variation range in the direction parallel to the optical axis, we should not fix the distance between an image plane and an optical lens. In fact, all optical lenses have a mechanism which allows us to manually (or automatically) adjust the position of the optical center along the optical axis. The process of adjusting the position of the optical center is called *image focusing.* Adjusting the position of the optical center is equivalent to adjusting distance d_i between an image plane and an optical lens. For the sake of convenience, d_i is commonly called a *focal length.* d_i does not refer to the focal length f of an optical lens. To avoid any confusion, we explicitly state the definition of *focal length* as follows:

Definition 7.1 *Focal length* in a robot's visual sensory system refers to the distance between the image plane and the optical lens. For the sake of clarity, we call it the *focal length of a camera* and denote it with f_c.

Given a value for f_c, the sharpness of the images formed on an image plane depends on distance d_o between an object and the optical lens. Refer to Eq. 7.15. Differentiating this equation with respect to time allows us to obtain the following relationship between the variation of h_i in the image plane and the variation of the object's distance d_o relative to the optical lens:

$$\|\triangle h_i\| = h_o \bullet \frac{d_i}{d_o^2} \bullet \|\triangle d_o\|. \tag{7.16}$$

Eq. 7.16 clearly illustrates the *out-of-focus* phenomenon as well as the concept of the *circle of confusion*. When a point on an object is in focus, its image will be a point image. This is illustrated by Fig. 7.11, where the image at point B is in focus. However, any change in the object's distance relative to the optical lens will cause the image to become a spot. Eq. 7.16 determines the dimension of an image spot. If point B is on the optical axis, the image spot will be a circle. In general, when a point on an object is out of focus, the circle enveloping the image spot is called a *circle of confusion*.

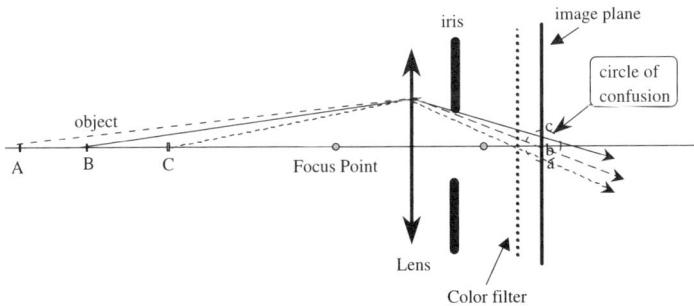

Fig. 7.11   Illustration of an optical image focused onto a planar surface.

Since an image-sensing array is composed of discrete sensing elements (or cells) which are arranged in a regular pattern, the image of a point will appear to be in focus as long as the circle of confusion is not bigger than the size of the sensing element (or cell). Otherwise, the image spot of a point will appear to be blurry (out of focus).

Since the size of the image-sensing cells is fixed, the largest permissible circle of confusion, described by $\|\triangle h_i\|_{max}$, is fixed as well. Therefore, Eq. 7.16 allows us to determine the largest permissible variation of $\triangle d_o$ as follows:

$$\|\triangle d_o\|_{max} = \frac{d_o^2}{h_o \bullet d_i} \bullet \|\triangle h_i\|_{max}. \qquad (7.17)$$

The value of $\triangle d_o$ is called the *depth of field* of a visual sensory system. Interestingly, when d_o increases, the depth of field increases as well. This clearly explains how we can perceive a large scene as well-focused images.

7.4.1.2 *Control of Incoming Light*

A *passive* sensing element always has a limited working range. When input to a passive sensing element goes beyond that working range, the output saturates. Therefore, it is necessary to have a control mechanism, such as a photosensitive cell, which can regulate the amplitude of input. As shown in Fig. 7.11, all cameras have an iris device which can be manually or automatically controlled in order to regulate the amount of incoming light.

If a scene's lighting is dynamically changing, it is necessary to automatically adjust the iris device. However, this can be avoided if we are able to adaptively adjust the working range of the sensing elements. With CMOS (Complementary Metal Oxide Semiconductor) technology, it is easy to develop active image-sensing cells which have working ranges that can be dynamically reprogrammed. We will study CMOS-based imaging sensors in further detail later in this chapter.

When an iris device is present in a camera, the amount of incoming light is controlled by adjusting the diameter of the circular hole of that iris. This hole is called an *aperture*. Interestingly enough, an iris device will allow us not only to control the amount of incoming light but also to adjust the permissible dimension of the circle of confusion. This second advantage can be explained as follows:

If we have a point at a fixed location in front of an optical lens, its image spot on the image plane has a fixed size. However, when the diameter of the iris decreases to a value less than the diameter of the optical lens, the image spot on the image plane decreases. Since the size of the sensing elements is fixed, the circle of confusion has a fixed permissible size as well. When an iris reduces the size of a point's image spot, this means that the permissible variations of an image spot can be increased. With a simple drawing, we can easily verify this fact.

Now, let us define k_i as an *iris effect factor*. When the aperture of the iris is bigger than or equal to the dimension of the optical lens, $k_i = 1$. When it becomes smaller than the diameter of the optical lens, $k_i > 1$. If we denote $\triangle h_i$ the variation of an image spot without an iris, and $\triangle h_i^*$ the variation of an image spot with an iris, then we have

$$\triangle h_i = k_i \bullet \triangle h_i^*. \tag{7.18}$$

Eq. 7.18 simply illustrates that $\triangle h_i^*$ is not bigger than $\triangle h_i$.

Substituting Eq. 7.18 into Eq. 7.17 yields

$$\|\triangle d_o\|_{max} = \frac{d_o^2}{h_o \bullet d_i} \bullet k_i \bullet \|\triangle h_i^*\|_{max}. \tag{7.19}$$

Eq. 7.19 indicates that a smaller aperture permits a larger depth of field because $k_i > 1$.

7.4.1.3 *Light Splitting & Color Filtering*

Human vision is intrinsically color vision. Human eyes can perceive colors because of the presence of three types of cone cells:

- Cone cells which are sensitive to red light,
- Cone cells which are sensitive to green light,
- Cone cells which are sensitive to blue light.

A simple way to perceive colors with an electronic camera is to use a light splitter and color filter.

A color filter is a transparent material which permits lights having a wavelength within a very narrow range to pass through. With color filters, we can selectively channel lights of a specific wavelength into an image-sensing area. There are two common ways of performing light splitting and color filtering.

Independent Light Splitting and Color Filtering

We normally place color filters behind an optical lens. In this way, we can split incoming light into three separate light beams, as shown in Fig. 7.12. Each beam is then channelled into a specific color filter (red, green or blue), which filters the separated light beams. This is shown in Fig. 7.12.

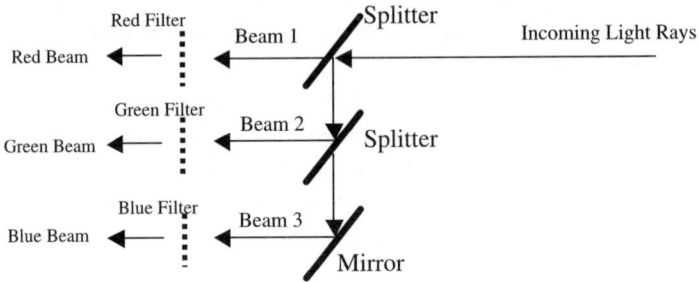

Fig. 7.12   Illustration of independent light splitting and color filtering.

Since each beam after light splitting can be sensed by a dedicated image-sensing array (chip), it is necessary to use a set of three Charge-Coupled Device (CCD) chips to independently convert the three separated light beams into the red, green and blue component color images. A color camera which does this is called a *3CCD camera*. One advantage of this method of light splitting and color filtering is that it preserves the spatial resolution of incoming light.

However, an obvious drawback is the cost and complexity, because 3CCD camera consists of a light splitter, mirror, color filters and three CCD chips (including associated electronic circuits). Under the scheme of independent light splitting and color filtering, the camera output will be a set of three RGB component color images.

Correlated Light Splitting and Color Filtering

An alternative solution instead of splitting the incoming light into three separate beams is to design a special color filter which is composed of an array of filtering patterns, as shown in Fig. 7.13. This idea was first proposed by Bryce Bayer from Kodak Inc. and is called the Bayer Pattern. He proposed a filtering pattern formed by a 2×2 block of GR/BG filtering elements. He included two filtering elements for the green color because human eyes are more responsive to this color. However, for a robot's visual sensory system, we have the freedom to choose any combination of R, G, and B filtering elements to form a filtering pattern.

As it is sufficient to use a single image-sensing array (chip) to convert incoming light into corresponding electrical signals, the notable advantage of a filtering pattern is its simplicity and low-cost. Unfortunately, this

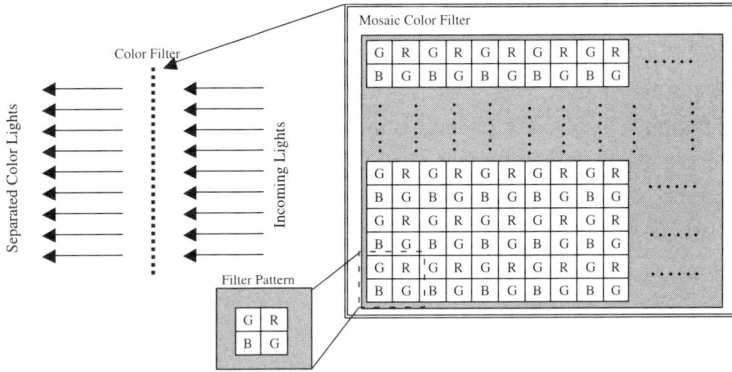

Fig. 7.13   Illustration of correlated light splitting and color filtering.

advantage is achieved to the detriment of the spatial acuity of colors. In addition, it is necessary to perform a color-correction procedure, or decorrelation, in order to retrieve the three RGB component color images from the output of a single image-sensing chip. The process of color correction can be explained as follows:

Assume that V_r', V_g' and V_b' are the direct output of R, G, and B values from the sensing elements placed behind a filtering pattern. The actual output of R, G, and B values will be the weighted sum of the direct output. A simple way to compute weighted sums if there are four elements in a filtering pattern is as follows:

$$\begin{cases} V_r = C_{r1} \bullet V_r' + C_{r2} \bullet V_g' + C_{r3} \bullet V_b' \\ V_g = C_{g1} \bullet V_r' + C_{g2} \bullet V_g' + C_{g3} \bullet V_b' \\ V_b = C_{b1} \bullet V_r' + C_{b2} \bullet V_g' + C_{b3} \bullet V_b' \end{cases} \qquad (7.20)$$

where correction matrix C is

$$C = \begin{pmatrix} C_{r1} & C_{r2} & C_{r3} \\ C_{g1} & C_{g2} & C_{g3} \\ C_{b1} & C_{b2} & C_{b3} \end{pmatrix}.$$

Correction matrix C can be determined by a color-calibration process. After color correction, a color camera's output is still a set of three RGB component color images.

7.4.2　Formation of Electronic Images

Refer to Fig. 7.10. After the formation of an optical image on a planar surface (image plane), the critical issue is how to develop an image-sensing device which will convert optical images into corresponding electronic images. A common way to do this is to use photosensitive cells made of semiconductors. A camera which uses photosensitive cells is called an *electronic camera* or *film-less camera*.

7.4.2.1　Photoelectric Effect

In 1887, Heinrich Hertz accidentally discovered that the surface of certain materials will emit electrons when exposed to a bombardment of light rays. This phenomenon is called the *photoelectric effect* and can be explained with Bohr's theory, as illustrated in Fig. 7.2. The photoelectric effect works as follows:

When light rays strike the surface of a material, the internal energies of the atoms at the surface will jump to higher levels. The electrons orbiting the atoms will transit to outer stable orbits. For some specific materials, the orbiting electrons of the atoms will even jump out of the largest stable orbit. These electrons become free moving electrons in a material. As a result, the incoming light striking the surface will trigger the release of electrons.

With the advent of semiconductors, the photoelectric effect has become a practical way of developing image-sensing chips for electronic cameras. Let us take a look at two examples: the CMOS and CCD imaging sensors.

7.4.2.2　CMOS Imaging Sensors

As a result of the invention of active image-sensing cells at NASA's Jet Propulsion Laboratory in the mid-1990s, the CMOS imaging sensor has become increasingly popular. CMOS is the basic element behind all microprocessors, memory chips and other digital devices. When the CMOS is produced in large quantities, its manufacturing cost goes down. As a result, the CMOS imaging sensor is a commercially viable and attractive solution for electronic imaging applications, such as video entertainment, visual inspection and visual guidance.

P-type and N-type Semiconductors

The term *semiconductor* is composed of two words: a) semi, meaning "partial" and b) conductor, meaning "something that conducts electricity". A semiconductor's (electric) current conductivity is between a perfect conductor and a perfect isolator. By adding an *impurity*, or *doping material*, to pure semiconductor material, the conductivity can be controlled externally. Two typical external causes are: a) an electric field, and b) exposure to lights.

The process of adding a different material to pure semiconductor material is called *doping*. The material used for doping purposes is known as an *impurity, or doping material*.

A special method of doping is to add a material which has an atom with one electron more or less than the semiconductor material's atom. For example, we can choose Germanium as the pure semiconductor material. Germanium's atom has 32 orbiting electrons. If pure Germanium material is mixed with a small percentage of doping material, one of these two things may happen:

- Case 1:
  An atom of doping material has one electron less than that orbiting the atom of pure semiconductor material. When the doping material is evenly diffused inside the pure semiconductor material, the atoms of doping material will be surrounded by the atoms of pure semiconductor material. The addition of one external electron to an atom of doping material will make that atom behave like an atom of pure semiconductor material. In other words, the pure semiconductor doped with an impurity material will readily act as a receptor of external electrons. This type of man-made semiconductor is called a *P-type semiconductor*.

- Case 2:
  An atom of doping material has one electron more than that orbiting the atom of pure semiconductor material. In this case, the release of one electron from an atom of doping material will make that atom behave like an atom of pure semiconductor material. In other words, the pure semiconductor doped with an impurity material will readily act as a donor of electrons. This type of man-made semiconductor is called an *N-type semiconductor*.

Example 7.3 In a material made of Germanium, an atom has 32 orbiting electrons. If the Germanium material is doped with a small percentage of material made of Arsenic, the result will be an N-type semiconductor because an atom of Arsenic material has 33 orbiting electrons.

◇◇◇◇◇◇◇◇◇◇◇◇◇◇◇◇◇

Example 7.4   Now, if we mix a pure semiconductor material, made of Germanium, with a small percentage of material made of Gallium, we will obtain a P-type semiconductor. This is because an atom of Gallium material has 31 orbiting electrons.

◇◇◇◇◇◇◇◇◇◇◇◇◇◇◇◇◇

Photodiodes and Phototransistors

The invention of P-type and N-type semiconductors in the late 1940s accelerated development of important electronic devices such as *diodes* and *transistors*. In fact, today's microprocessors and solid-state digital devices are all made of diodes and transistors.

If we specifically arrange P-type and N-type semiconductors together, we can develop an element called a *photosensitive cell*. Fig. 7.14a shows one design solution for photosensitive cells.

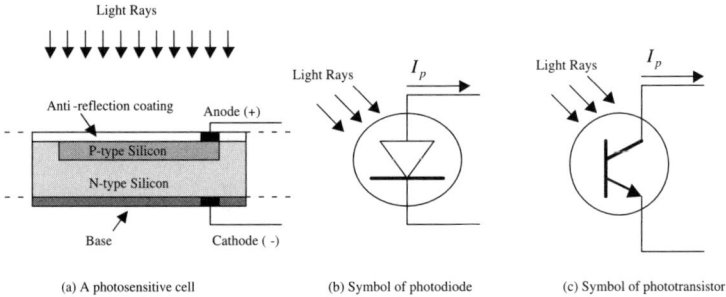

(a) A photosensitive cell   (b) Symbol of photodiode   (c) Symbol of phototransistor

Fig. 7.14   Illustrations including a sectional view and symbolic representation of a photodiode and phototransistor.

For the example shown in Fig. 7.14a, we can see that the main body of a photosensitive cell is made of an N-type semiconductor. The main body sits on a base which can be either metallic or a layer of isolator. On the top surface of the main body, a thin layer of P-type semiconductor is created through a diffusion process. After the P-type semiconductor is in place, the surface is coated with a layer of anti-reflection (i.e. light absorption) material. The device, as shown in Fig. 7.14a, is a photosensitive cell (or element). It works as follows:

When light strikes the top surface of this device, the P-type semiconductor releases electrons as a result of the photoelectric effect. After this, the electric charges (the accumulation of electrons) build up. As the layer of P-type semiconductor is very thin, electric charges will flow into the N-type semiconductor. If a pair of anode/cathode is placed on the device, as shown in Fig. 7.14a, an external closed-circuit will permit current to flow from the anode to the cathode. The direction of the current is opposite to the direction of the moving electrons. Clearly, a photosensitive cell acts more like a current generator than a voltage generator.

In electronics, if a main body made of an N-type semiconductor is deposited with a layer of P-type semiconductor, it is called a *P-N junction*. The P-N junction was invented in 1949 and is the basic element in diodes and transistors. Because of the P-N junction, a photosensitive cell is also referred to as a *photodiode* (see Fig. 7.14b) or *phototransistor* (see Fig. 7.14c). Here, we call a photosensitive cell a photodiode.

The Equivalent Circuit to a Photodiode

In order to better understand the electrical properties of a photodiode, we can draw an equivalent electronic circuit, as shown in Fig. 7.15. The basic elements inside this circuit include:

- I_p: A current generator with an output proportional to the amount of photons striking the surface of the photodiode. Depending on light intensity, I_p falls in the range of milli-Amperes or micro-Amperes.
- C_J: A capacitor of the P-N junction. Depending on design and intended application, C_J can be big or small. For a Charge-Coupled Device (CCD) sensing element, C_J should be big.
- R_J: Shunt resistor of the P-N junction. In general, the value of R_J falls in the range of mega-Ohms.
- R_S: Series resistor of the P-N junction. In general, the value of R_S is usually small.

Active-Sensing Cells

Refer to the circuit equivalent to a photodiode, as shown in Fig. 7.15. A photodiode may behave like a current generator, a voltage generator or a

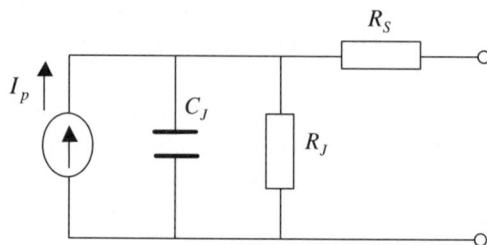

Fig. 7.15   Circuit equivalent to a photodiode.

capacitor. By adjusting the design parameters of the photodiode, we can make it behave more like a current generator or more like a capacitor. The former is the design objective for CMOS image-sensing cells while the latter is the design objective for CCD image-sensing cells.

In order to convert an optical image into its corresponding electronic image, it is necessary to have an array of image-sensing cells. These cells are also called *image elements* or *pixels*, for short. Obviously, a photodiode can serve as a pixel.

However, the current (or voltage) produced by a photodiode is usually small, falling in the range of milli-Amperes or micro-Amperes. Thus, it is necessary to enhance the electrical signal of photodiode in one way or another. Since the photodiode is made of the semiconductor, one possible method is to increase the number of transistors in the photosensitive cell. This possibility led to the invention of *active pixels* in the mid-1990s at the Jet-Propulsion Laboratory in California.

As shown in Fig. 7.16a, an active pixel is composed of the following elements:

(1) A Photodiode:
   This transistor is responsible for converting light intensity into a corresponding current.
(2) A Current-to-Voltage Conversion Transistor:
   This transistor takes current from the photodiode as input. Output voltage V_p from the transistor is proportional to I_p. If R is the electrical resistance across the collector and emitter of the transistor, and K is the amplifying gain, then we have

$$V_p = (1 + K) \bullet R \bullet I_p. \tag{7.21}$$

(NOTE: A transistor has three terminals: Base, Collector and Emitter).

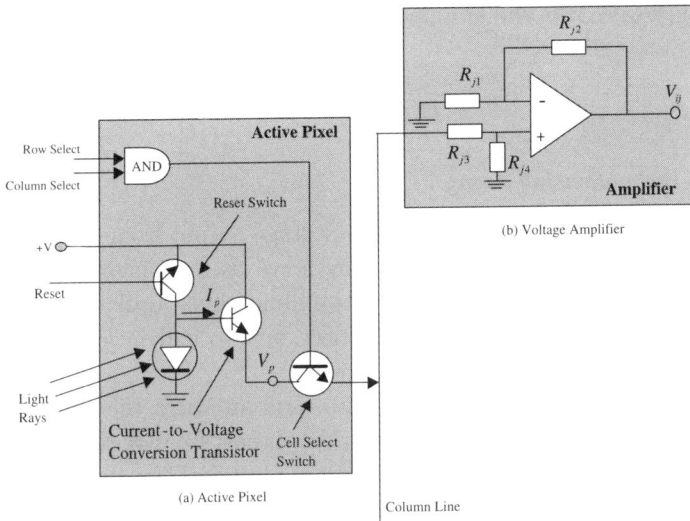

Fig. 7.16   Illustration of a conceptual design of an active-sensing cell, or active pixel.

(3) A Reset Switch:

This is a transistor which serves as an electronic switch. Its purpose is to discharge the photodiode after the output has been read. In other words, this switch resets the photodiode so that it can start a new sensing cycle. From Fig. 7.16a, we can see that the power supply $(+V)$ is connected to the ground when the reset switch is "on".

(4) A Logic Circuit for Cell Selection (optional):

In order to individually select a cell from an array of sensing cells, a logic circuit of cell selection can be added to an active pixel. This logic circuit consists of a logic "AND" device as well as a transistor which serves as an electronic switch. The logic "AND" device has two input lines: a) row select and b) column select.

The above elements make an active pixel look like a controllable voltage generator. Although the output voltage signal from an active pixel is still weak, this is not a drawback because the voltage signal can be easily amplified by a circuit, as shown in Fig. 7.16b. Assume that the active pixel in Fig. 7.16a is located at the (i, j) position (ith row and jth column) of an image-sensing array. If the voltage signal is amplified by the circuit, as shown in Fig. 7.16b, the amplified output will be

$$V_{ij} = \frac{R_{j2} \bullet R_{j4}}{R_{j1} \bullet (R_{j3} + R_{j4})} \bullet V_p. \tag{7.22}$$

CMOS Image-Sensing Array

At any time instant, there is only one voltage output from an active pixel. Hence, a single, active pixel can only serve as an imaging sensor for a single point. In order to convert a two-dimensional optical image into a corresponding electronic image, it is necessary to make use of an array of active pixels.

If we place active pixels together to form an array, the critical issue is how to simultaneously read the electrical signals from these active pixels. In other words, is it necessary to read the electrical signals from an array at one time instant, or can it be done within a short time interval?

If it is necessary to simultaneously read the electrical signals from an array of active pixels at one time instant, this will be costly and impractical. This is because the size of an array may be prohibitive. For example, the number of active pixels in an image-sensing array may be on the order of 10^6 pixels. A more practical solution, therefore, is to read the electrical signals from an array of active pixels within a short period of time. For example, if one image is read within a time interval of 40 milliseconds (ms), an imaging sensor will be able to output 25 images per second.

Refer to Fig. 7.16. If an active pixel has a cell-selection circuit, it is similar to a memory cell which is an integrated-circuit for holding the logic value "0" or "1". In this case, it is easy to form an image-sensing array with a set of active pixels. Fig. 7.17 shows a possible design solution for constructing a CMOS imaging sensor based on an array of $n \times m$ active pixels. Each active pixel can be treated as a memory location with a dedicated address (i, j) (ith row and jth column). In this way, we can individually read the voltage signal from an active pixel at location (i, j). If the electrical signals are read out pixel-by-pixel, it will be too slow. A better solution is to read out the voltage signals row-by-row, as shown in Fig. 7.17. The procedure of reading voltage signals and converting them into a continuous video format is as follows:

- Step 1: Reset all active pixels.
- Step 2: Expose active pixels to incoming light rays for a certain period of time. The time of exposure can automatically be determined, based

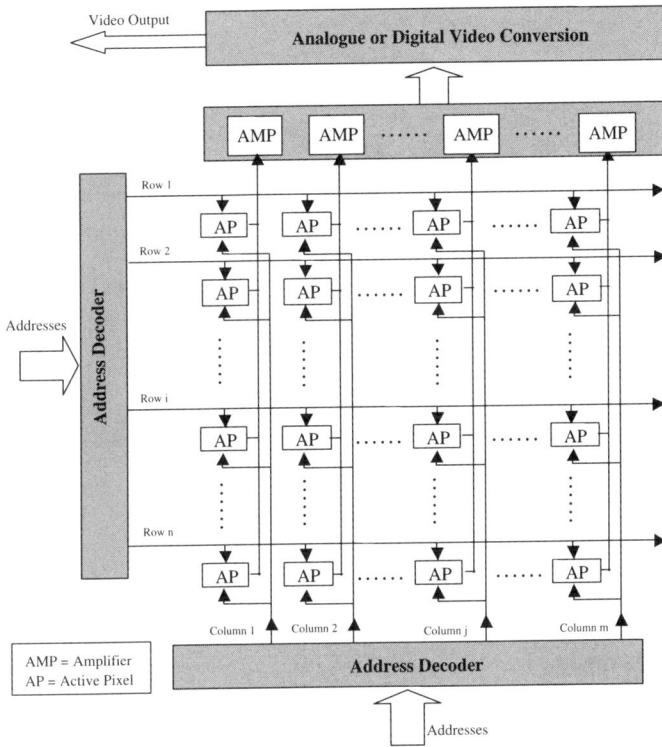

Fig. 7.17   Illustration of the generic architecture of a CMOS imaging-array having $n \times m$ active pixels.

on light intensity and the characteristics of the active pixels.

- Step 3: Select columns, and set row index i to be 0.
- Step 4: Select row i, and amplify the voltage signals from it. Each column will have its own dedicated voltage amplifier, as shown in Fig. 7.17.
- Step 5a: Pack the amplified voltage signals from row i into the current image, which is an image inside a video. An image can be either analogue or digital.
- Step 5b: At the same time, increase row index i ($i = i + 1$), and repeat Steps 4 and 5 until an image has been read.
- Repeat Steps 1, 2, 3, 4, and 5 to continuously generate a video.

Fig. 7.18 shows an example of a CMOS color camera. A CMOS imaging sensor's notable advantages include:

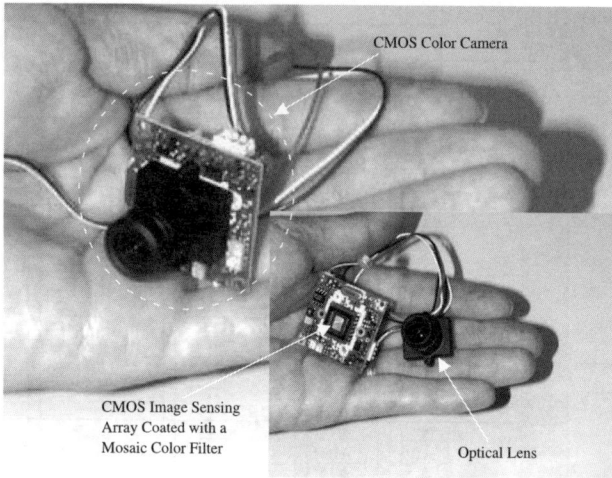

Fig. 7.18   Example of a CMOS Color Camera.

- Low Cost:
  The elements inside a CMOS imaging sensor are transistors or transistor-like devices. CMOS imaging sensors are manufactured in the same facilities as microprocessors and memory chips. Due to the mass production, the unit cost of a CMOS imaging sensor is very low.
- Low Power Consumption:
  The voltage signal in a CMOS device is not higher than +5V. As a result, a CMOS imaging sensor is energy efficient. This is a very important feature for embedded and/or mobile applications where power consumption is a major concern.
- Random Pixel Access:
  Refer to Fig. 7.17. A CMOS imaging sensor is similar to a memory chip. As a result, one can directly access an individual pixel if a full-address decoding circuit is present.
- Wide Dynamic Range of Active Pixels:
  A passive sensing cell has a limited dynamic range (working range). Because of the reset switch inside an active pixel, it is easy to increase this working range. In fact, we can treat a reset action as an overflow, similar to the overflow of a numerical counter. Let n_{ij} denote the number of overflows which occur at the active pixel located at (i, j). If V'_{ij} is the actual output of the voltage signal without considering

overflows, then the true output of the voltage signal will be

$$V_{ij} = V'_{ij} + n_{ij} \bullet V_{ij}^{max}$$

where V_{ij}^{max} is the maximum voltage signal which triggers a reset action.

- Automatic Compensation of Light Intensity:
  We can also make use of the reset switch to automatically adjust the response of an active pixel to lighting conditions in a scene. For example, the frequency of reset can be high if light intensity is strong, and low if the lighting is dim. Because of this, it is not necessary to have an automatic iris (or auto-iris). The camera shown in Fig. 7.18 has an optical lens with an iris that is not automatic.

- On-chip Processing:
  Since CMOS imaging sensors use the same manufacturing process as microprocessors and memory chips, it is easy to incorporate signal-processing hardware into an imaging sensor. Because of this, it is simple to develop smart cameras with built-in programmable functions.

7.4.2.3 *CCD Imaging Sensors*

An important milestone for electronic imaging was the invention of the Charge-Coupled Device (CCD)) by Boyle and Smith from Bell Labs in 1970. After this, the first black-and-white CCD camera was announced in 1971. And, the first color CCD camera was announced in 1972. Both were from Bell Labs.

Most electronic cameras are based on CCD imaging sensors (or chips). However, since the late 1990s, the market share of electronic-imaging products was made up of CMOS cameras. Today, however, CCD cameras still dominate the high-end (high resolution and quality) electronic-imaging products.

Charge-Coupled Devices (CCD)

A CCD can be treated as a pair of coupled capacitors: one for light sensing and the other for charge storage. The first capacitor is a photodiode that can serve as a photosensitive cell which will convert incoming light into corresponding electrical signals. Thus, a CCD can also be called a *CCD imaging element* or *CCD pixel*.

Fig. 7.19 shows a conceptual design solution for a CCD pixel. The photodiode in Fig. 7.19 behaves more like a capacitor than a current generator,

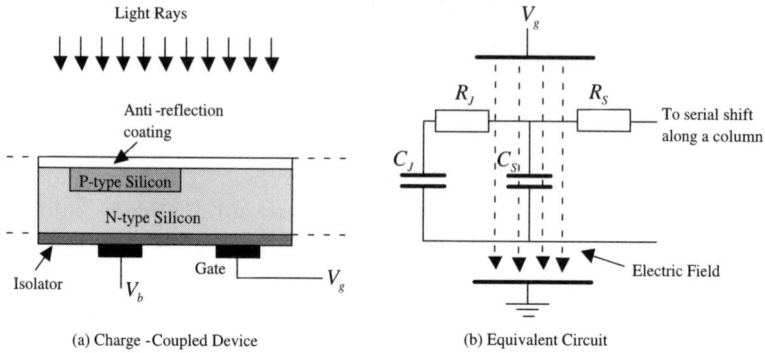

Fig. 7.19   Illustration of a CCD pixel.

as shown in Fig. 7.14. When incoming light strikes the surface of a CCD
pixel, the P-type semiconductor will release electrons. Since the layer of
P-type semiconductor is very thin, the free-moving electrons will flow into
the main body (N-type semiconductor). This will result in an accumulation
of electric charges. In order to enhance this accumulation, we can apply
bias voltage V_b which will in turn increase the signal-to-noise ratio.

Next to a CCD-pixel's photodiode, there is a capacitor. This capacitor
serves as a reservoir to store electric charges from the photodiode. When
the photodiode is exposed to incoming light for a certain period of time, a
charge transfer occurs. Then, the gate voltage V_g is turned "on" to generate
an electric field. This electric field induces electric charges to flow from the
photodiode into the capacitor.

A CCD-pixel's function can better be described by its equivalent circuit,
as shown in Fig. 7.19b. The basic elements inside a CCD pixel include:

- C_J: Junction capacitor of the photodiode,
- C_S: Capacitor for storage of electric charges from C_J,
- R_J : Electrical resistance along the path connecting C_J to C_S,
- R_S: Electrical resistance when transferring electric charges out of C_S.

Refer to Fig. 7.19b. A charge transfer from C_J to C_S will have no
signal loss as the thermal effect caused by R_J is negligible. This is not the
case, however, if the photodiode behaves like a current generator (e.g. an
active pixel made of CMOS). This explains why a CCD imaging sensor can
produce high-quality electronic images.

CCD Imaging Arrays

Just as with a CMOS imaging sensor, we can place a set of CCD pixels into the shape of an array. With an array of CCD pixels, we face the same issue of how to read the electric charges.

As we discussed earlier, a CCD pixel consists of a pair of coupled capacitors: one for sensing light rays, and the other for storing electric charges. Since a CCD pixel is more complicated than a transistor, manufacturing of CCD imaging sensors is not the same as it is for microprocessors and memory chips. Thus, CCD imaging sensors are more expensive than CMOS imaging sensors. In addition, the manufacturing is even more expensive if we attempt to build an "active pixel" based on a CCD. As a result, an existing CCD imaging sensor is not similar to a memory chip, and we cannot access individual CCD pixels. A common solution to read the electric charges is to treat the charge-storage capacitors within a column of a CCD imaging array as a shift register. In this way, the electric charges of CCD pixels in a column can be sequentially shifted out to a signal-amplification circuit for post-processing.

Fig. 7.20 shows a conceptual design solution for a CCD imaging sensor made of an array of $n \times m$ CCD pixels. The CCD image sensing array works as follows:

- Step 1: Discharge all CCD pixels. This can be done either by transferring the CCD pixels' electric charges out or by discharging the CCD pixels using a reset circuit.
- Step 2: Expose the CCD pixels to incoming light for a certain time interval.
- Step 3: Transfer electric charges from the photodiodes to the charge-storage capacitors. This transfer is done horizontally (for all columns) at the same time, and is called *parallel shifts*.
- Step 4: Transfer the electric charges vertically (row-by-row). This operation treats the charge-storage capacitors in each column as a single vertical shift register, and is called *serial shifts*. The serial shift in each column must be synchronized by a serial shift clock. Refer to Fig. 7.20. Each cycle of serial shifts will transfer a row of electric charges from row 1.
- Step 5. Amplify the electric charges which are vertically shifted from row 1. This is done by a series of charge-to-voltage amplifiers.
- Step 6: Pack the amplified electrical signals into the current image which is either analogue or digital.

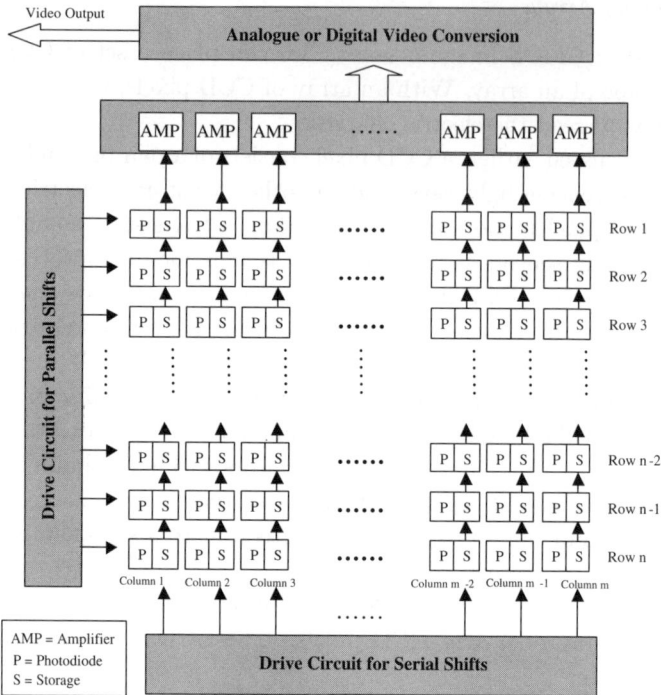

Fig. 7.20 Illustration of the generic architecture of a CCD-image sensing array having $n \times m$ pixels.

- Step 7: Repeat Steps 4, 5, and 6 until one image has been read.
- Repeat Steps 1, 2, 3, 4, 5, 6, and 7 for a continuous video.

To perform parallel and serial shifts of electric charges, it is necessary to make use of clock signals and high voltages. Thus, the drive circuits for CCD imaging sensors are complicated, and the cameras using CCD imaging sensors are bulky. The two cameras on top of the robot, as shown in Fig. 7.1, are CCD-color cameras. Additionally, because it is necessary to have high voltage to transfer charges, the CCD imaging sensor is not as energy-efficient as the CMOS imaging sensor.

7.4.3 *Formation of Digital Images*

Refer to Fig. 7.17 and Fig. 7.20. The output from a camera (either CMOS or CCD) is a continuous stream of images (video). In the case of a color camera, however, the direct output from an imaging sensor consists of three streams of images: red component color images, blue component color images and green component color images.

If a camera's output has an analogue signal, the camera is called an *analogue video camera*. Output from an analogue video camera will be a continuous one-dimensional signal, representing a stream of two-dimensional analogue images. This analogue signal is also called an *analogue video*.

Digital cameras have been on the market since the late 1990s. Output from a digital camera can be a stream of two-dimensional digital images, called a *digital video*. If a digital camera outputs digital video, then it is called a *digital video camera*.

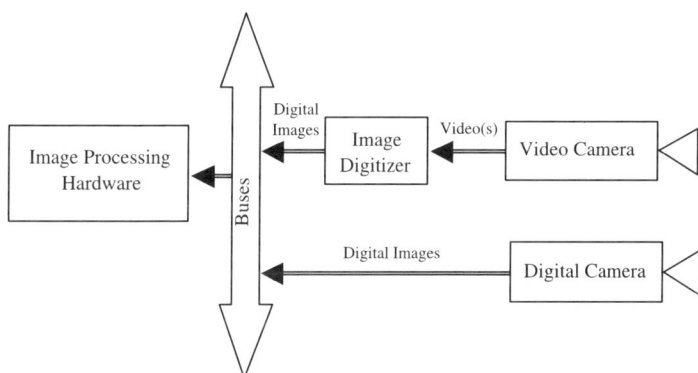

Fig. 7.21 Visual sensory system, composed of a camera, image digitizer and computing hardware.

As shown in Fig. 7.21, the computing hardware in a robot's visual sensory system is digital. Thus, if a camera outputs analogue video, it is necessary to convert this into a stream of digital images. Since analogue video cameras are still popular, a robot's visual sensory system needs an image digitizer which converts analogue video into the corresponding digital video. (See Fig. 7.21).

7.4.3.1 *Analogue Videos*

Everyday, we receive a large amount of information, images and news from a variety of TV programs, broadcasting through air or over cable. A TV program consists of a continuous stream of images encoded with the associated audio signals. Another popular and important source of images are videos. Because of the huge consumer market for videos, it is important that video production equipment ensures a certain standard, guaranteeing its compatibility with display equipment such as TV receivers, video monitors, VCR etc.

A video is a continuous stream of images. Hence, it is dynamic data which can be specified (or described) by a set of parameters related to the color, intensity and timing of images. So, while it is possible to specify the parameters, there is concern over how to standardize them.

Video Standards

A standard is a set of specifications for a product or process. When manufacturers follow a common standard, products or processes for a similar application (or function) can be interchanged. In fact, a standard creates a platform for product and service providers (manufacturers of a similar product) to compete against one another in order to better serve the end-users (consumers).

In the TV industry, there are two dominant standards: a) the National Television Standards Committee (NTSC) from the U.S., and b) Phase-Alternating Line (PAL) from Western Europe.

The technical name for NTSC video is RS-170A or Electronic Industry Association (EIA). The following are specifications of NTSC:

- Image Rate:
  The image rate is 30 images per second, allowing an image to last for 33.33 milliseconds (ms). This time interval is also known as an *image period*. In electronic imaging, an image is commonly called a *frame*. In robotics, a frame usually refers to a coordinate system. So, in order to avoid confusion, let us stick to the term *image*.
- Number of Lines per Image:
  The number of lines per image is 525. Only 484 of these lines are used to carry the signals of a video image. These lines are called *signal lines*. Since the image period lasts 33.33ms, one signal line will last 63.5 microseconds (μs). This time interval is called a *line period*.
- Interlaced Image:
  As we mentioned earlier, an analogue video is a one-dimensional signal. When an image is read row-by-row, as shown in Fig. 7.17 and Fig. 7.20, it is packed into a video in the form of continuous lines (rows). If we pack the lines by following line index i which varies from 1 to n, these lines will appear in the same order when displayed on the TV screen.

Our eyes' latency to the image change-rate is 25 images per second. Therefore, the image-by-image display mode, at a rate of 30 images per second, is acceptable. However, for a long video, a higher display-rate is desirable in order to ease the stress on our eyes. A common way to double the display rate is to split an image into two fields: a) the odd-line field and b) the even-line field. The odd-line field is composed of the odd lines of an image while the even-line field is composed of the even lines of an image. This split image is called an *interlaced image*. The field rate of NTSC video is 60 fields per second.

- Black and White Signal Levels:
  The black signal level is 0.0 volts while the white signal level is 1.0 volt.
- Horizontal-Synchronization Pulse:
  When packing a series of lines together, it is necessary to have a "marker" to indicate the beginning or end of a line. One way to do this is to use a negative pulse. This negative pulse is known as a *horizontal-synchronization signal* or, *H-Sync* for short. For NTSC videos, H-sync is a negative pulse of 0.4 volts.
- Vertical-Synchronization Pulse:
  A video is a series of interlaced odd and even fields. Thus, it is necessary to have a "marker" to indicate the beginning or end of each field. In order to do this, we can use a negative pulse which is different from H-Sync, or we can use a series of negative pulses which are the same as H-Sync. The NTSC adopts this latter solution to mark the end of a field. As a result, the vertical-synchronization pulse, or V-Sync, consists of a series of negative 0.4-volt pulses which may last for a certain number of line periods. If n_1 is the number of signal lines in a field, and n_2 is the corresponding number of line periods taken up by V-Sync, the following condition must hold:

$$2 \bullet (n_1 + n_2) = 525.$$

NTSC is not the most popular video standard in the world but Phase-Alternating Line (PAL) is. The technical name for PAL is the Consultative Committee for International Radio (CCIR). The following are the specifications of PAL:

- Image Rate:
  The image rate is 25 images per second. The corresponding *image period* is 40ms.

- Number of Lines per Image:
  The number of lines per image is 625. The corresponding line period is $64\mu s$.
- Interlaced Image:
  A PAL video is also composed of a series of interlaced images. However, the field rate of a PAL video is 50 fields per second.
- Black and White Signal Levels:
  The black signal level is 0.0 volts while the white signal level is 0.7 volts.
- Horizontal-Synchronization Pulse:
  The H-Sync is a negative pulse of 0.3 volts.
- Vertical-Synchronization Pulse:
  The vertical-synchronization pulse also consists of a series of negative 0.3-volt pulses. If n_1 is the number of signal lines in a field, and n_2 is the corresponding number of line periods taken up by V-Sync, the following condition must hold:

$$2 \bullet (n_1 + n_2) = 625.$$

In NTSC or PAL , the H-Sync and V-Sync are also known as *timing signals* (or sync signals). These signals are encoded into an analogue video so that there is only a single, one-dimensional pulse waveform which represents the video. For the sake of simplicity, the following discussions use PAL as a reference.

Component Videos

A color-imaging sensor's direct output consists of three streams of video: a) the video of red component images, b) the video of green component images and c) the video of blue component images. Each of these video streams is called a *component video*.

For a color camera's three component videos, the timing signals are the same. Therefore, it is sufficient to pack the timing signals into one component video. It is common to use the green component video, as shown in Fig. 7.22. Alternatively, we can carry the time signals in a separate signal line, known as a *sync signal line*.

A one-dimensional signal wave can be transmitted through a cable with two wires: a) one wire for signal, and b) one wire for ground. If a color camera directly outputs a set of RGB component videos with the timing signals packed into the green channel, it is necessary to use three cables,

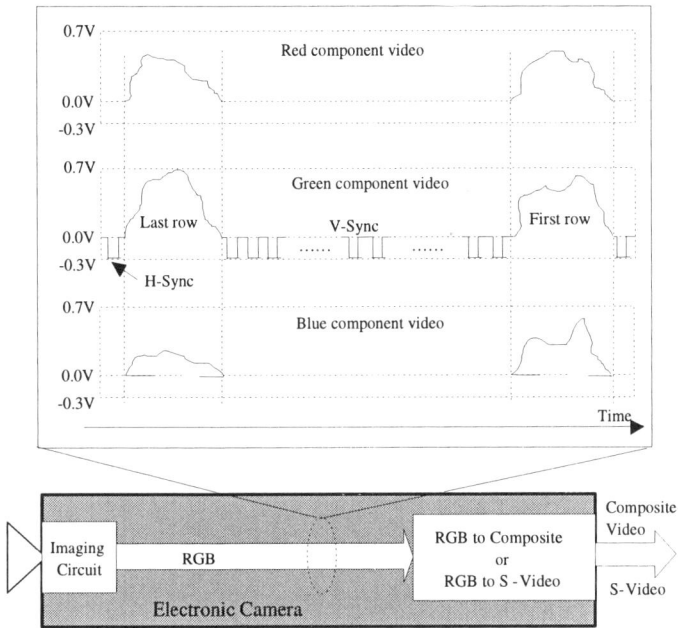

Fig. 7.22   Illustration of an electronic camera, including a set of RGB component videos.

such as BNC cables, to interface the camera with an image digitizer. If the timing signals are in a separate signal line, then it is necessary to have four cables.

Composite Videos

As it is necessary to have at least three cables in order to transmit a set of RGB component videos, it is very costly to broadcast them over the air by modulating them into a band of radio frequencies (RF). It would also be costly for a TV receiver to directly receive RGB component videos. Thus, in order to reduce the cost, the TV industry commonly uses a *composite video* for broadcasting and reception.

A composite video is obtained by encoding the RGB component videos in a specific procedure which involves two steps:

- Step 1: RGB to Luminance/Chrominance Conversion:
  The values of R, G and B define a three-dimensional color space. This color space can be projected onto any other three-dimensional space.

An interesting way of projecting the color space is to separate the luminance (intensity-related information) from the chrominance (color-related information). For PAL videos, this adopted conversion is called an *RGB-to-YUV* conversion. The following equation describes an RGB-to-YUV conversion:

$$\begin{cases} Y = 0.3 \bullet R + 0.59 \bullet G + 0.11 \bullet B \\ U = 0.493 \bullet (B - Y) \\ V = 0.877 \bullet (R - Y) \end{cases} \tag{7.23}$$

where R, G, and B are the values of red, green and blue component colors at a pixel. In Eq. 7.23, Y represents the luminance value, U represents the difference between blue component color B and luminance Y, and V represents the difference between red component color R and luminance Y. In fact, variables U and V represent the chromaticity of a color. The plot $U - V$ is also called a *chromaticity diagram*. We can see that luminance Y is the weighted average of R, G, and B because the sum of the weighting coefficients is equal to 1 $(0.3 + 0.59 + 0.11 = 1.0)$. As for NTSC videos, the adopted conversion is called an *RGB-to-YIQ* conversion. I and Q describe both the *hue* and *saturation* of a color. If U and V are known, then I and Q are determined as follows:

$$\begin{cases} I = -U \bullet \sin(33^0) + V \bullet \cos(33^0) \\ Q = U \bullet \cos(33^0) + V \bullet \sin(33^0). \end{cases} \tag{7.24}$$

- Step 2: Generation of a Composite Video:
  For PAL, a composite video is obtained by combining Y, U and V as follows:

$$C = Y + U \bullet \sin(2\pi f_1 t) + V \bullet \cos(2\pi f_1 t) \tag{7.25}$$

where $f_1 = 4.43$MHz. For NTSC, a composite video is obtained by combining Y, I and Q as follows:

$$C = Y + I \bullet \sin(2\pi f_2 t) + Q \bullet \cos(2\pi f_2 t) \tag{7.26}$$

where $f_2 = 3.58$MHz.

In theory, it is not possible to recover the original values of R, G and B if the values of a one-dimensional composite video C are known. In practice, however, we can estimate the R, G and B values from a set of at least three consecutive C values. While this estimation results in some loss of color information, our eyes can easily tolerate it, so it is not a concern for the

TV industry. However, it is a serious drawback for a robot's visual sensory system because exact color information is very important in image-feature detection and object identification.

S-Videos

One way to minimize the loss of color information is to remove luminance Y from Eq. 7.26. As a result, R, G, and B will be projected onto Y and C as follows:

$$\begin{cases} Y = 0.3 \bullet R + 0.59 \bullet G + 0.11 \bullet B \\ C = U \bullet \sin(2\pi f_1 t) + V \bullet \cos(2\pi f_1 t) \end{cases} \tag{7.27}$$

or

$$\begin{cases} Y = 0.3 \bullet R + 0.59 \bullet G + 0.11 \bullet B \\ C = I \bullet \sin(2\pi f_2 t) + Q \bullet \cos(2\pi f_2 t). \end{cases} \tag{7.28}$$

Now, Y and C become two separate one-dimensional signal waves. A video's timing signals can be packed into either signal Y or signal C. Together, Y and C signals are called an *S-Video*. For the wired transmission of an S-video, a cable must have two wires for signals Y and C. An S-video can reconstruct R, G, and B better than a composite video. So far, however, the S-video format is not intended for broadcasting TV programs. Rather, it is widely-used in video products, such as camcorders.

7.4.3.2 *Digitization of Analogue Videos*

Most electronic cameras are used for news or video-entertainment industries. As a result, an electronic camera's output is generally not in a digital video format. As shown in Fig. 7.21, an *image digitizer* is still a necessary part of a robot's visual sensory system. The purpose of an image digitizer is to convert analogue videos into corresponding digital videos which can be directly stored in the memory, or communicated to a microprocessor.

Fig. 7.23 shows the generic architecture of an image digitizer. If the input is not a set of RGB component videos, there will be a color decoder which will decode an S-video or composite video into such a set. Subsequently, the sync signals will be separated from the RGB component videos (or the green channel). And then, three component video streams will be sent to three A/D converters respectively. Each A/D converter will turn a component video into a series of digital images.

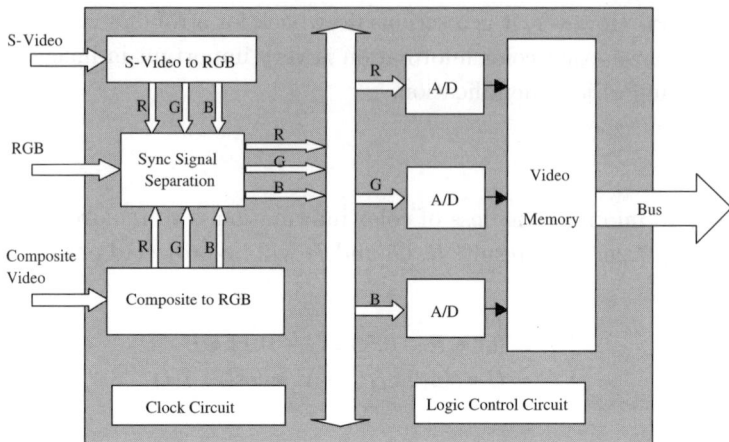

Fig. 7.23   Illustration of the generic architecture of an image digitizer.

The process of image digitization involves two cascaded loops. The outer loop works as follows:

- Step 1: Configure the image digitizer. We must set the resolution of the digital image. The image resolution indicates the number of rows and columns that a digital image will have. Here, we denote (rx, ry) as the number of columns and the number of rows.
- Step 2: Detect V-Sync pulse and log onto the odd-field.
- Step 3: For line index i_o which varies from 1 to $ry/2$ in the odd field, perform an A/D conversion with the inner loop. (See below).
- Step 4: Log onto the even field.
- Step 5: For line index i_e which varies from 1 to $ry/2$ in the even field, perform the A/D conversion with the inner loop. (See below).

The inner loop works as follows:

- Step 1: Set the pixel clock.
  For PAL, the period of H-Sync is $10\mu s$. Therefore, the period of an effective signal line will be $54\mu s$ ($= 64 - 10$). And, the period of a pixel clock can be calculated as follows:

$$t_p = \frac{54}{rx} \ \mu s. \tag{7.29}$$

In fact, the frequency $(1/t_p)$ of a pixel clock is the sampling rate when an effective signal line is digitized.

- Step 2: For pixel clock number j which varies from 1 to rx, perform an A/D conversion and save the results (pixel value) in the memory which stores digitized images. The pixel value will be stored at location (i, j) where index i is determined by

$$i = \begin{cases} 2 \bullet i_o - 1 & \text{if digitizing the odd field,} \\ 2 \bullet i_e & \text{if digitizing the even field.} \end{cases}$$

From studying the process of image digitization, it is clear that resolution (rx, ry) of a digital image is not equal to resolution (n, m) of an imaging sensor, shown in Fig. 7.17 or Fig. 7.20. In a robot's visual sensory system, the typical values for (rx, ry) are (256, 256), (512, 512), or (1024, 1024), etc.

Example 7.5   An image digitizer's output is a set of digitized red, green and blue component color images. Fig. 7.24 shows one such set. From the highlighted areas, we can see that the input color image contains red, blue and orange.

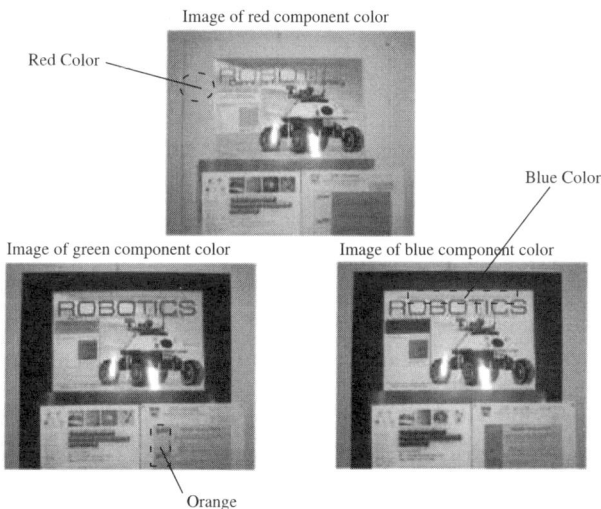

Fig. 7.24   Example of a set of digitized red, green and blue component color images.

7.5   Modelling of Digital Images

So far, we have studied the physical properties and working principles behind the optical lens, the imaging sensor, and the image digitizer. An image is a two-dimensional array and each of its elements is called a *pixel*. Within this array, a pixel has a location and three integer values corresponding to the component colors: red, green and blue. Therefore, an image contains both chromatic and geometric information about a scene. In order to infer the geometry of a scene from images, it is necessary to know how the color and geometry in the images are related to the color and geometry in the scene.

7.5.1   *Chromatic Modelling*

Perceptually and numerically, color is a combination of chrominance and luminance. Any color can be reproduced by a combination of three primary colors: red, green, and blue. Therefore, any color can be represented by a point in a three-dimensional color space. Interestingly enough, one color space can be projected onto another color space. Because of this, colors will have different representations in different color spaces.

7.5.1.1   *Representation in RGB Color Space*

We know that a color-imaging sensor's direct output is a set of three RGB component color images. Therefore, a color image can be represented in an RGB color space.

After digitizing an analogue color image, the output will be a set of three matrices of pixels: one for the red component color, one for the green component color and one for the blue color. Let I_R, I_G and I_B denote these three matrices respectively. Then, we can represent a color image by

$$\begin{cases} I_R = \{r(i,j), & 1 \le i \le ry \text{ and } 1 \le j \le rx\} \\ I_G = \{g(i,j), & 1 \le i \le ry \text{ and } 1 \le j \le rx\} \\ I_B = \{b(i,j), & 1 \le i \le ry \text{ and } 1 \le j \le rx\} \end{cases} \tag{7.30}$$

where (rx, ry) is the image resolution.

The three component color images, as shown in Fig. 7.24, are an example of the display of (I_R, I_G, I_B). Given a pixel at location (i, j), the component colors at this location will be $[r(i,j), g(i,j), b(i,j)]$, which define the color coordinates of a point in RGB color space. If a color coordinate is encoded with 8 bits (one byte), a pixel's color coordinates $[r(i,j), g(i,j), b(i,j)]$ at location (i,j) will require 24 bits (three bytes) to store the RGB values.

Eq. 7.30 indicates that we are only able to describe component color images in the form of two-dimensional matrices. In other words, by default, there is no generic parametric representation of color distribution in an image.

RGB color space does not explicitly represent a color's chrominance and luminance. If it is necessary to explicitly represent a color's chrominance and luminance, the only solution is to project the RGB color space onto another appropriate color space. We will discuss this further in the next chapter.

It is worth noting that RGB color space is not perceptually uniform. This means that the same numerical differences in RGB values do not produce the same physiological sensation of color difference. For example, the difference ($\triangle R = 100 - 80, \triangle G = 50 - 20, \triangle B = 150 - 100$) is numerically equal to the difference ($\triangle R = 120 - 100, \triangle G = 70 - 40, \triangle B = 170 - 120$). However, the perceived difference in colors by the brain is not the same. This phenomenon is called *perceptual non-uniformity*.

7.5.1.2 *Representation of Intensity Images*

The processing and computation of three two-dimensional matrices will be expensive. However, a lot of useful visual information, such as uniformity and discontinuity, is encoded into luminance value Y. Thus, it is sufficient to process *intensity images* in order to derive geometric information about a scene. If we know the value of a set of RGB component color images, and compute the weighted average, we can obtain the corresponding intensity image as follows:

$$I_I = \{I(i,j), \ 1 \leq i \leq ry \ \text{ and } \ 1 \leq j \leq rx\} \tag{7.31}$$

where

$$I(i,j) = 0.3 \bullet r(i,j) + 0.59 \bullet g(i,j) + 0.11 \bullet b(i,j). \tag{7.32}$$

As human brain interprets the three primary colors (red, green and blue) differently, the weighting coefficients are chosen to be different. When the numerical values of the three primary colors are the same, the perception of green is the strongest. Moreover, the sensation of red is stronger than that of blue. Thus, the coefficient for green is larger than the coefficient for red, which in turn, is larger than the coefficient for blue.

Example 7.6   For the color image in Example 7.5, the application of Eq. 7.32 yields the intensity image, as shown in Fig. 7.25. We can see that there is almost no loss of visual information such as continuity and discontinuity.

Fig. 7.25   Example of an intensity image.

◇◇◇◇◇◇◇◇◇◇◇◇◇◇◇◇◇

From this example, we can see that an intensity image largely, if not totally, preserves the uniformity and discontinuity of the chrominance distribution across an image plane.

7.5.2   *Geometric Modelling*

The geometry of a scene, in the form of posture, motion, shape and curved surface, is not explicitly measured by a visual-sensory system. The ultimate goal of a robot's visual perception system is to infer the geometry of a scene from digital images. We will study this topic further in the next chapter.

However, let us consider the simplest case: a single point. There is a deterministic relationship between the image coordinates of a point and its coordinates in a three-dimensional space. In fact, Eq. 7.15 indicates that the position of a point in a camera's image plane depends on the position of this point in a three-dimensional space. Therefore, it is important to know how the coordinates (X, Y, Z) in a three-dimensional space are related to the coordinates (u, v) in a camera's image plane.

7.5.2.1 *Pin-hole Model and Perspective Projection*

Refer to Fig. 7.10. The area in focus on the image plane can be treated as a planar surface. In addition, the aperture of the iris is usually very small compared to the physical dimension of the objects in a scene. Thus, we can treat an optical lens as a small hole. Accordingly, the optical lens of a camera is described by a *pin-hole*. In this way, the formation of an optical image follows a perspective projection, as shown in Fig. 7.26a.

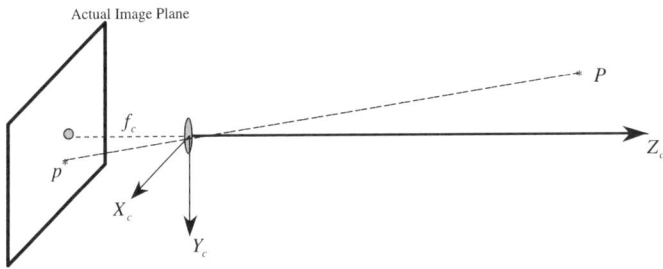

(a) Pin-hole camera model

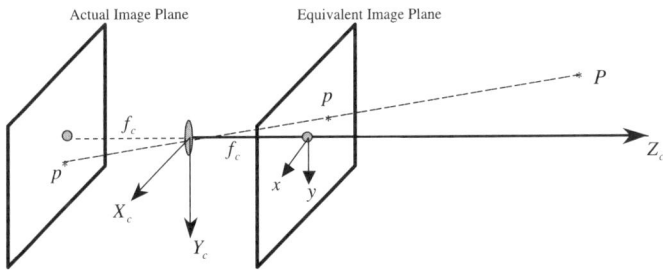

(b) Equivalence of perspective projection

Fig. 7.26   Illustrations of pin-hole camera model and perspective projection, including the assignment of coordinate systems.

Mathematically, perspective projection is a projection under which all points are projected onto a common plane by following straight lines which intersect at a common point known as the *center of projection*. When a camera is described by a pin-hole model, the center of projection is the center of the optical lens.

7.5.2.2   *Assignment of Camera Coordinate Systems*

For the study of robot kinematics, we adopted the strategy of assigning a coordinate system (frame) to a rigid body. A camera is a physical entity (rigid body) as well. Thus, a coordinate system (or frame) can be assigned to it. In this way, the camera's posture is described by the posture of its frame.

When we assign a frame to a camera, it is common to define the Z axis along the optical axis and the origin of the camera frame at the center of projection, as shown in Fig. 7.26a. By default, a frame is a right-handed coordinate system. However, we can choose the X axis to be in the direction parallel to the imaging sensor's rows, and the Y axis to be in the direction parallel to the imaging sensor's columns.

7.5.2.3   *Assignment of Image Coordinate Systems*

Physically, a camera's image plane is behind the center of projection, as shown in Fig. 7.26a. Distance f_c between the image plane and the center of projection is known as the focal length of a camera. In a camera frame, the value of this focal length is negative because the image plane is on the negative Z axis.

For the sake of convenience, we can make use of the equivalent image plane on the positive Z axis, as shown in Fig. 7.26b. The equivalent image plane and the actual image plane are symmetric about the XY plane of a camera frame. Under a perspective projection, the geometric properties of the objects on these two planes are strictly equivalent except for the signs of the values.

By default, the equivalent image plane is also called an *image plane*. Since an image plane can be treated as a physical object, it is necessary to assign a coordinate system to it. A common way to do this is to define the origin of the image frame at the intersection point between the image plane and the optical axis (the Z axis). The x axis is parallel to the X axis of the camera frame, and the y axis is parallel to the Y axis of the camera frame. Fig. 7.26b illustrates the assignment of an image frame.

7.5.2.4   *Determination of Image Coordinates*

As we discussed above, a camera will contain two frames: one frame for the camera, and the other for its image plane. If P is a point in the camera frame, and p is the projection of P onto the image plane, the question is:

What is the relationship between the coordinates of P and the coordinates of p?

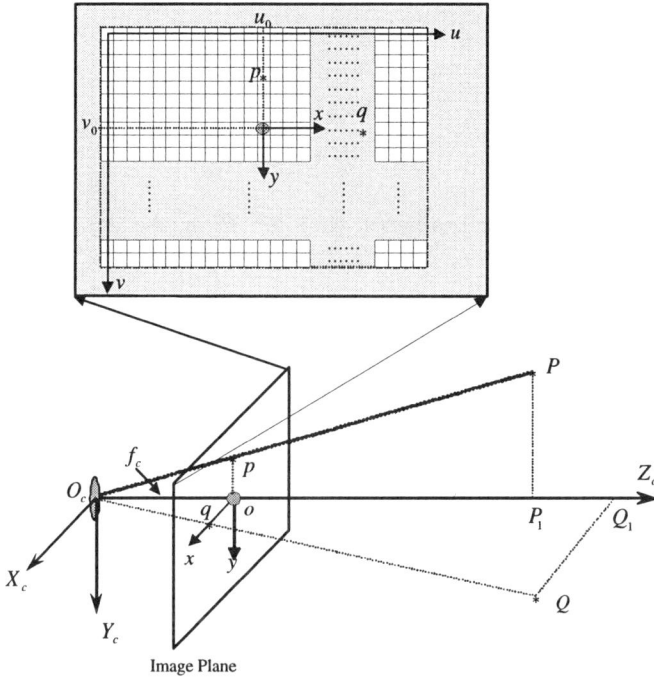

Fig. 7.27   Illustrations, including the projection of coordinates from camera space onto an image plane, and an image array.

Refer to Fig. 7.27. Assume that P is located on the YZ plane of the camera frame. In this case, $P = (0, {}^cY, {}^cZ)^t$. According to the perspective projection, p will fall onto the y axis of the image frame. Let us denote $p = (0, y)$. From the fact that triangle PP_1O_c is similar to triangle poO_c, we have

$$\frac{{}^cY}{y} = \frac{{}^cZ}{f_c} \tag{7.33}$$

or

$$y = f_c \bullet \frac{{}^cY}{{}^cZ}. \tag{7.34}$$

Similarly, let us assume that point Q is located on the XY plane of the camera frame. Under a perspective projection, image q of point Q will fall onto the x axis of the image frame. Since triangle QQ_1O_c is similar to triangle qoO_c, we have

$$\frac{^cX}{x} = \frac{^cZ}{f_c} \tag{7.35}$$

or

$$x = f_c \bullet \frac{^cX}{^cZ}. \tag{7.36}$$

In general, if the coordinates of point P in the camera frame are $(^cX, \ ^cY, \ ^cZ)^t$, and the coordinates of its image p are (x, y), the following relation will hold:

$$\begin{cases} x = f_c \bullet \frac{^cX}{^cZ} \\ \\ y = f_c \bullet \frac{^cY}{^cZ}. \end{cases} \tag{7.37}$$

Eq. 7.37 is the mathematical description of a perspective projection from a three-dimensional space into a two-dimensional plane. Eq. 7.37 can also be rewritten in matrix form as follows:

$$s \bullet \begin{pmatrix} x \\ y \\ 1 \end{pmatrix} = \begin{pmatrix} f_c & 0 & 0 & 0 \\ 0 & f_c & 0 & 0 \\ 0 & 0 & 1 & 0 \end{pmatrix} \bullet \begin{pmatrix} ^cX \\ ^cY \\ ^cZ \\ 1 \end{pmatrix} \tag{7.38}$$

where s is a scaling factor, $(x, y, 1)^t$ the equivalent projective coordinates of a point in the image frame (see Chapter 2), and $(^cX, \ ^cY, \ ^cZ, \ 1)^t$ the equivalent projective coordinates of a point in the camera frame.

From Eq. 7.38, we can see that the coordinates (x, y, s) will be uniquely determined if a set of $(^cX, \ ^cY, \ ^cZ, \ 1)^t$ is given as input. However, the coordinates $(^cX, \ ^cY, \ ^cZ)^t$ in the camera frame cannot be determined if the coordinates (x, y) in the image frame are known. This is because the scaling factor s is unknown.

If we define Eq. 7.38 as the *forward projective-mapping* of a robot's visual sensory system, the recovery of $(^cX, \ ^cY, \ ^cZ)^t$ from (x, y) can be defined as the *inverse projective-mapping*. With a single camera and no *a priori* knowledge about an object or scene under consideration, there will be no unique solution to inverse projective-mapping. We will study this problem as well as some solutions in the next chapter.

7.5.2.5 *Determination of Index Coordinates*

For the sake of programming, it is convenient to use the indexes to indicate a pixel's location in a two-dimensional array. These indexes are called the *index coordinates* of a pixel.

Refer to Fig. 7.27. The row and column indexes indicate that there is an *index coordinate system* associated with an image array. In the memory which stores digital images, the row index of a two-dimensional matrix is counted from top to bottom while the column index is counted from left to right. If we denote u the axis for the column index, and v the axis for the row index, the index coordinate system uv is assigned to an image array, as shown in Fig. 7.27. For example, the index coordinates for the upper-left pixel will be either $(0,0)$ ($u = 0$ and $v = 0$ in C-programming language) or $(1,1)$ ($u = 1$ and $v = 1$ in the MATLAB programming environment).

Now, the question is: What is the relationship between image coordinates (x, y) and index coordinates (u, v)?

From our study of digital-image formation, we know that a digital image can be obtained in two ways:

- Direct Output from an Imaging Sensor:
  It is best to directly perform A/D conversion with the output from an imaging sensor as there will be no loss of chromatic information caused by the back-and-forth conversion from analogue images to video, and from video back to digital images. If a digital image is directly obtained from the A/D conversion of an imaging-sensor's output, a pixel will correspond to an image-sensing cell. And, an image-sensing cell can be treated as a rectangular surface. If (C_x, C_y) defines the dimension of an image-sensing cell in the x and y directions of the image plane, the size of a pixel, represented by (D_x, D_y), will be

$$\begin{cases} D_x = C_x \\ D_y = C_y. \end{cases} \tag{7.39}$$

- Direct Output from Image Digitizer:
  Today, it is still common to use an image digitizer to convert an analogue video into a stream of digital images. When a digital image is obtained from digitizing an analogue video, a pixel's size depends on the sampling steps along the columns and rows of an imaging sensor. If (S_x, S_y) are the sampling steps for digitizing an analogue video, then

we have

$$
\begin{cases} D_x = S_x \\ D_y = S_y. \end{cases} \tag{7.40}
$$

As we studied earlier, the resolution of a digital image is different from the resolution of an imaging sensor. In general, (S_x, S_y) is not equal to (C_x, C_y).

Now, let us assume that the origin of an image frame is at (u_0, v_0), with respect to the index coordinate system assigned to the imaging sensor. Let (D_x, D_y) be the physical dimension of the pixels in the image array. If the index coordinates corresponding to image coordinates (x, y) is (u, v), then we have

$$
\begin{cases} \frac{x}{D_x} = u - u_0 \\[2ex] \frac{y}{D_y} = v - v_0 \end{cases} \tag{7.41}
$$

or

$$
\begin{cases} u = u_0 + \frac{x}{D_x} \\[2ex] v = v_0 + \frac{y}{D_y} \end{cases} \tag{7.42}
$$

where $\frac{x}{D_x}$ and $\frac{y}{D_y}$ are the number of digitizations which occurred along a row and column respectively.

Eq. 7.42 can also be rewritten in matrix form as follows:

$$
\begin{pmatrix} u \\ v \\ 1 \end{pmatrix} = \begin{pmatrix} \frac{1}{D_x} & 0 & u_0 \\ 0 & \frac{1}{D_y} & v_0 \\ 0 & 0 & 1 \end{pmatrix} \bullet \begin{pmatrix} x \\ y \\ 1 \end{pmatrix}. \tag{7.43}
$$

7.5.2.6 *Intrinsic Parameters of Cameras*

Consider a camera represented by the pin-hole model. Eq. 7.38 describes the relationship between the coordinates in the image frame and the coordinates in the camera frame. And, Eq. 7.43 describes the relationship between the image coordinates and the corresponding index coordinates. If we eliminate image coordinates $(x, y, 1)^t$, the combination of Eq. 7.38 and Eq. 7.43 yields

$$s \bullet \begin{pmatrix} u \\ v \\ 1 \end{pmatrix} = {}^{I}P_c \bullet \begin{pmatrix} {}^{c}X \\ {}^{c}Y \\ {}^{c}Z \\ 1 \end{pmatrix} \tag{7.44}$$

with

$$^{I}P_c = \begin{pmatrix} f_x & 0 & u_0 & 0 \\ 0 & f_y & v_0 & 0 \\ 0 & 0 & 1 & 0 \end{pmatrix} \tag{7.45}$$

and

$$\begin{cases} f_x = \frac{f_c}{D_x} \\ f_y = \frac{f_c}{D_y}. \end{cases} \tag{7.46}$$

Matrix $^{I}P_c$ describes the perspective projection of coordinates from the camera frame to the corresponding index coordinate system. We call such a matrix a *projection matrix*.

In Eq. 7.45, f_x and f_y are two new variables. They can be interpreted as the focal lengths of the camera in the directions parallel to the x and y axes respectively. The four parameters f_x, f_y, u_0 and v_0 are commonly called the *intrinsic parameters* of a camera.

In general, it is difficult, if not impossible, to know a camera's intrinsic parameters. This is because the parameter specifications of the camera and lens are not accurate. Interestingly enough, a camera's intrinsic parameters can be estimated by a process known as *camera calibration*. We will study camera calibration further in the next chapter.

If we know a camera's intrinsic parameters, the index coordinates $(u, v)^t$ can be uniquely determined if $({}^{c}X, {}^{c}Y, {}^{c}Z)^t$ is given as input. However, $({}^{c}X, {}^{c}Y, {}^{c}Z)^t$ cannot be determined for any given $(u, v)^t$ because the scaling factor s is unknown.

7.6 Digital Image-Processing Hardware

"An image is worth a thousand words" implies that a large amount of information can be extracted from images. However, human brain's neural computing network is highly parallel and is able to process, in real-time, a large amount of information captured by all the five sensory systems of the body.

Now, if a robot's visual perception system must respond in real-time to streams of video images captured by a visual sensory system, it is necessary to have an appropriate image-processing hardware to support this need.

7.6.1   *Host Computers*

Today's popular computing platform is the personal computer (PC), which is made from Intel's microprocessors (e.g. the Pentium). Due to the proliferation of web-based applications, such as the video-phone and net-meeting, today's PC's are capable of interfacing with all kinds of digital cameras and web-cams through the Universal Serial Bus (USB), or FireWire. Because of the speed limitations, USB and FireWire are suitable for producing low-resolution digital videos. If an application requires high-resolution digital video, it is necessary to form a visual-sensory system which consists of an electronic camera, image digitizer and host computer, as shown in Fig. 7.28.

Refer to Fig. 7.28. A host computer normally includes:

- A microprocessor for all computational tasks,
- An external bus, such as Peripheral Component Interconnect (PCI), for interfacing with external devices or systems, such as an image digitizer,
- A hard-disk for nonvolatile storage of images and results,
- A graphic card, which packs digital data (including images) into a video of a specific graphic format, such as VGA, SVGA or XGA,
- A monitor which displays analogue videos in a graphic format which is different from PAL or NTSC,
- Mouse and keyboard for interfacing with the user or programmer.

In general, a robot's host computer can also serve as an interface between the robot and the human master. Through the host computer, a human master can develop application programs for the robot to execute.

7.6.2   *DSP Processors*

Normally, a pixel's component color is encoded with one byte. For a color image of the size of 512×512, the data size, in terms of the number of bytes, will be $3 \times 512 \times 512$ or 768 kilobytes (KB). (NOTE: 1KB is equal to 1024 or 2^{10} bytes). Therefore, it is quite a heavy computational load for a robot to process a sequence of color images in real-time.

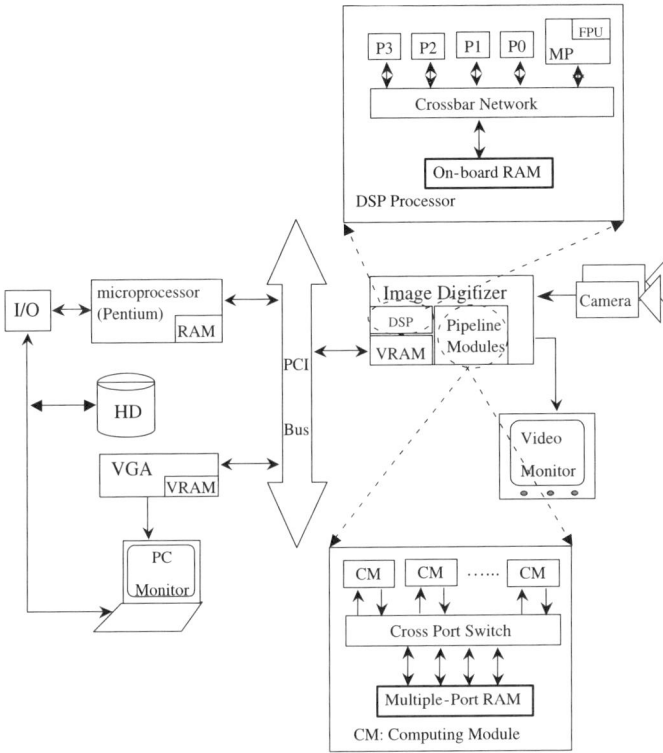

Fig. 7.28   Illustration of the generic image-processing hardware, including the host computer, image digitizer, DSP processor and pipelined image-computing modules.

One common way to reduce the computational load of the host computer is to add a dedicated microprocessor on the board of the image digitizer. A popular type of on-board microprocessors are the Digital Signal Processors (DSP), such as C80 from Texas Instruments, Inc.

As shown in Fig. 7.28, a DSP microprocessor is made up of five processors, each of which has its own Arithmetic and Logic Unit (ALU). One of them is the master processor (MP), with a floating-point processing unit (FPU). And, the other four are co-processors, which can only perform computations with integers.

A DSP microprocessor usually has its own on-board memory (RAM) which can store a large amount of image data. An internal bus, known as the *cross-bar network*, supports the fast exchange of large streams of bytes among the processors and on-board memory (RAM).

Example 7.7 Fig. 7.29 shows an example of a color image digitizer with an on-board DSP microprocessor. This DSP microprocessor is programmable through the host computer.

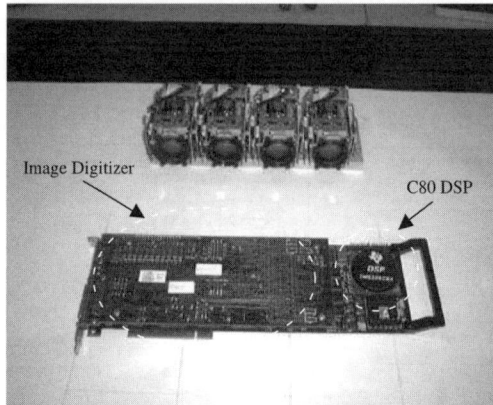

Fig. 7.29   Example of an image digitizer equipped with a DSP microprocessor.

◇◇◇◇◇◇◇◇◇◇◇◇◇◇◇◇◇

7.6.3   *Pipelined Computing Modules*

Sometimes, it is more efficient to implement some computational tasks with well-established algorithms using dedicated hardware instead of dedicated processors (e.g. certain convolution filters for image feature-enhancement or pre-processing).

A filter, or algorithm, implemented on a chip is called a *computing module* (CM). Since a filter's output can be another filter's input, a series of computing modules can be linked together to form a *pipe*. This pipe is known as an *image-processing pipeline*. (See Fig. 7.28).

The configuration of an image-processing pipeline is user-programmable. This means it is easy to form a sub-pipeline for a given computational task. For example, we can form the sub-pipeline: CM1 → CM3 → CM2 or CM2 → CM3 → CM1 even if CM1, CM2, CM3 and CM4 are all present in a pipelined image-processing hardware. This flexibility for reconfiguring an image-processing pipeline is achieved with a *cross-port switch* which ties the pipeline's computing units together. In fact, a cross-port switch makes the pipelined image-processing hardware behave like the network in a star-topology.

Example 7.8 Fig. 7.30 shows an example of a color image digitizer with pipelined image-processing hardware. This image-processing pipeline is programmable through the host computer.

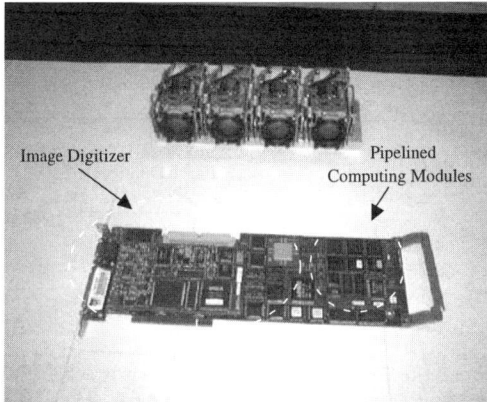

Fig. 7.30 Example of an image digitizer equipped with an image-processing pipeline.

◇◇◇◇◇◇◇◇◇◇◇◇◇◇◇◇◇◇◇

7.6.4 *Parallel Computing Platforms*

For advanced computations in real-time such as 3-D reconstruction, face recognition, hand-gesture recognition, and conceptual symbol-learning, it is necessary to use a cluster of microprocessors in order to supply the computational power required. In general, a robot's visual sensory and perception systems should be on a parallel computing platform because a robot normally has two or more electronic eyes.

Depending on the configuration of the processors and memories, parallel computing platforms can be divided into two categories: a) Multiple Instructions Single Data (MISD), and b) Multiple Instructions Multiple Data (MIMD). For an MISD platform, all processors have equal access to the shared memories. A DSP microprocessor is a mini-MISD computing platform, because its memory is shared among its five processors. In general, however, an MISD computing platform has 16 or more processors. It

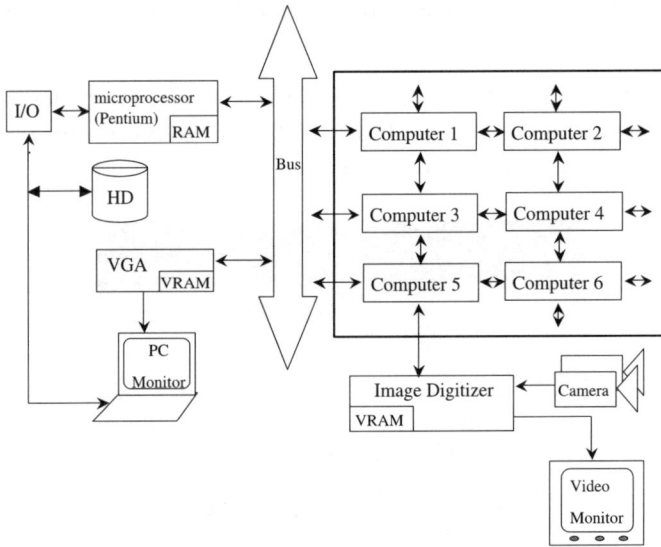

Fig. 7.31   Illustration of the generic image-processing hardware, including the host computer, image digitizer, and an array of parallel processors.

is a stand-alone computing system which is very different from a personal computer. As a result, it is expensive and unpopular.

An MIMD computing platform is composed of a cluster of networked processors. With the availability of USB and FireWire ports in today's personal computers, it is easy to form an array of personal computers, as shown in Fig. 7.31. This architecture is similar to a transputer-based parallel computing platform (no longer popular). The USB is a serial communication standard, and a high-speed USB can transmit 12 Megabits per second (Mbits).

IEEE-1394 (or Firewire) is a new serial communication standard for computers and peripheral devices. It is a universal standard for interconnecting all digital devices such as computers and digital cameras. And, the basic working principle behind it is asynchronous serial communication (see Chapter 6). The transmission speed of the FireWire falls within the range of 100Mbits to 1000Mbits, making it a good way to communicate digital images among an array of microprocessors.

Instead of using serial communication links such as the USB or FireWire, it is possible to make use of the Internet because today's personal computers are all Internet-enabled.

As we mentioned in Chapter 6, the biggest challenge now is how to develop a real-time programming language (or real-time operating system) which can effectively manage the distributed processing powers and memories, in the form of a cluster of networked microprocessors.

7.7 Summary

Although the electronic camera is one of the most important components in the robot, it is a common fact that the robotics- and vision-research communities have made little effort in its development. Rather, this has been fuelled by competition in consumer products (e.g. digital cameras, camcorders) and the media industry.

Based on our studies in this chapter, we should realize that the visual sensory system is of fundamental importance to the humanoid robot or other autonomous robots. Without visual sensory input, the important perception-decision-action loops cannot function properly at all. Therefore, we should not neglect the importance of the continuous improvement of electronic cameras.

In this chapter, we learned that an electronic camera is a device which is part of a robot's visual sensory system and imitates the human eye. Functionally, the optical lens of a camera focuses incoming light into a small planar area, known as an *image plane*. The focused light's intensity is then converted into corresponding electrical signals by an imaging sensor. Subsequently, the electrical signals are packed with the timing signals to form videos which can be either analogue or digital.

We also learned that when an optical lens focuses incoming light onto an image plane, it is necessary to control the light intensity with the help of a device called an *iris*. In addition, we learned that one way of sensing colors selectively is to make use of a light-splitting and color-filtering mechanism. In this case, the output from an imaging sensor will be three streams of RGB component color images.

Since 1970s, electronic imaging has been dominated by CCD imaging sensors. A CCD image-sensing cell looks like a pair of coupled capacitors: one for sensing incoming light, and the other for storing electric charges. It is necessary to form an array of CCD sensing-cells in order to convert an optical image into a corresponding electronic image.

Since 1990s, the CMOS imaging sensor has become the new device for electronic imaging. A CMOS image-sensing cell looks like a current generator. Output from a CMOS image-sensing cell is proportional to the lights striking its surface. As CMOS imaging sensors are produced with the same equipment as that used in the manufacturing of microprocessors and memory chips, they are more cost-efficient. And this broadens the horizon for the development of the smart camera on a chip. This will be a tremendous advantage for the future of embedded & vision-guided applications such as the humanoid robot.

As electronic cameras cater to the video entertainment and media industry, electronic camera output must follow a certain video standard (e.g. PAL and NTSC). In order to conform to video standards, an imaging sensor's direct output must undergo signal conversion. This conversion will result in the loss of chromatic information, and thus, it is undesirable for a robot's visual sensory and perception systems.

Before a computer can process an image, it is necessary to convert analogue video into a stream of digital images. This can be done by using an image digitizer. If a color camera's output is an image digitizer's input, the image digitizer's output will be three streams of RGB component color images. These digital images can be represented by three two-dimensional matrices. In order to reduce the computational load caused by three two-dimensional matrices, it is sufficient to process an intensity image which preserves most visual information and can easily be obtained by a weighted sum of the RGB component color images.

Finally, the geometric aspect of an image formation can be treated approximately as a perspective projection. If we know a camera's intrinsic parameters, the coordinates in the camera frame can be uniquely projected onto the index coordinates in the image plane. However, when the coordinates are projected from a three-dimensional camera space onto a two-dimensional image space, a dimension is lost. For a robot's visual perception system, the ultimate challenge is to recover this missing dimension. We will discuss this challenge further, in the next chapter.

7.8   Exercises

(1) What is light?
(2) Does a photon's energy depend entirely on a light's wavelength?
(3) What is the index of refraction?
(4) Prove Eq. 7.13.
(5) Explain the difference between the human eye's central vision and its

peripheral vision.

(6) What is a circle of confusion?

(7) Draw a graph to illustrate the fact that a smaller aperture will reduce the dimensions of the confusion circles.

(8) Human vision is intrinsically color vision. But, is it possible for a robot to see a colorful world?

(9) Explain how to design an active pixel.

(10) Prove Eq. 7.22.

(11) Explain the working principle of a CMOS imaging sensor.

(12) Explain the working principle of a CCD imaging sensor.

(13) Explain the advantages and disadvantages of a CMOS imaging sensor vs. a CCD imaging sensor.

(14) What is a video camera's output?

(15) What are the typical types of video?

(16) Explain why video camera's output is usually not digital.

(17) Explain why an image digitizer is necessary for a robot's visual sensory system.

(18) Is a digital image's resolution equal to that of an imaging sensor's?

(19) When digitizing a PAL video, the resolution of the digital images is set to be (512, 512). What is the frequency of the pixel clock inside the image digitizer?

(20) How is a color image represented inside a robot's "brain"?

(21) How do you obtain the intensity image from the component color images?

(22) Prove that the projection matrix from a camera frame's coordinates to the index coordinates with respect to the image array is:

$$^I P_c = \begin{pmatrix} f_x & 0 & u_0 & 0 \\ 0 & f_y & v_0 & 0 \\ 0 & 0 & 1 & 0 \end{pmatrix}.$$

(23) Explain why a robot should have dedicated computing hardware as part of the visual sensory system.

(24) What are the possible ways to form a parallel computing platform for image-and-vision computing.

(25) What issues are related to the use of a parallel computing platform?

7.9 Bibliography

(1) Hecht, E. (1987). *Optics*, Addison-Wesley.

(2) Galbiati, Jr. L. J. (1990). *Machine Vision and Digital Image Processing Fundamentals*, Prentice-Hall.

(3) Gonzalez, R. C. and R. E. Woods (1992). *Digital Image Processing*, Addison-Wesley.

(4) Ng, K. K. (1995). *Complete Guide to Semiconductor Devices*, McGraw-Hill.

(5) Pratt, W. K. (1991). *Digital Image Processing*, John Willey and Sons.

(6) Umbaugh, S. E. (1998). *Computer Vision and Image Processing*, Prentice-Hall.

Chapter 8

Visual Perception System of Robots

8.1 Introduction

We constantly perform all kinds of activities and this dynamism depends on our ability to act and interact with our environment. This autonomy and freedom of mobility in turn depends on our powerful vision.

While blind people can live very rewarding lives, it is true that without vision, people's activities are highly limited, their knowledge acquisition is tremendously compromised, human interaction is difficult and, in general, mental and physical development is slowed.

This is also true for a humanoid robot or any intelligent artificial system. Vision is an ability which enables autonomy in action-taking, knowledge acquisition, and social interaction with humans. In fact, there can be no humanoid robot if there is no competent (artificial) vision.

Image processing, computer vision, and pattern recognition are three related but distinct subjects. These three subjects form the important pillars of machine perception and we have accumulated a great deal of knowledge about these fields over the past 30 or 40 years. I do not intent to cover all this knowledge here, but rather to study those topics of primary importance to motion execution.

In this chapter, we will discuss the basic computational principles and methodologies of image processing, feature extraction and computational vision, which are considered to be relevant to the development of a humanoid robot. We will place special emphasis on the computations of geometrical features which are useful for determining an object's posture and motion in a scene.

8.2   The Basics of Visual Perception

From Chapter 7, we know that a visual-sensory system's output is digital images. This output will be the visual-perception system's input. One may question why we are concerned about a visual-perception system which takes digital images as input?

8.2.1   *A Process of Visual Perception*

From a systems point of view, artificial vision is an information process that takes digital images as input, and produces descriptions of the objects in a scene as output. This input-output relationship is illustrated in Fig. 8.1. Those who are familiar with computer graphics can easily see that the visual-perception process is the inverse of the computer-graphics process. In computer graphics, the goal is to produce realistic images or image sequences from descriptions of objects and characters in a (virtual) scene.

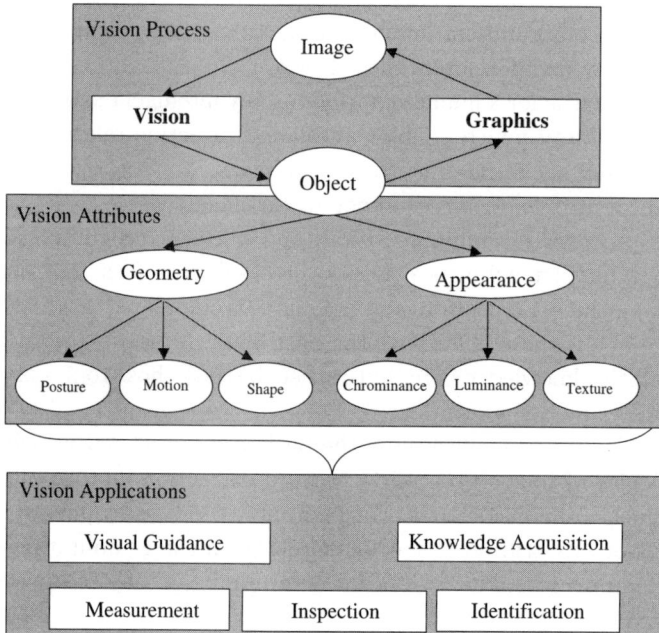

Fig. 8.1   Illustration of the process, attributes and applications of a visual-perception system.

Formally, we can define the process of visual perception as follows:

Definition 8.1   Visual perception is an information process which takes digital images as input, and produces descriptions of the objects in a scene as output.

In robot vision (and computer vision), a scene means the environment in proximity to the robot. We can formally define the term *scene*, as follows:

Definition 8.2   A scene is a collection of objects and physical entities which are in proximity to an observer such as a robot.

From our study of the robot's visual-sensory system, we know that the process underlying it is very similar to the process underlying computer graphics. In fact, outputs from these two processes are of the same nature (digital images or image sequences). However, there is a fundamental difference between them:

- In computer graphics, input is a precise description of objects in a (virtual) scene.
- In a visual-sensory system, input is not known in advance and the scene is real (not virtual).

Moreover, a robot's visual-sensory system does not explicitly output descriptions of objects in a real scene, and therefore the visual-perception system must infer them from digital images. In this way, the visual-perception and visual-sensory systems complement each other.

The inference of geometrical features from images is a decision-making process. We will study this topic further in Chapter 9. However, it turns out that today's computers perform poorly when it comes to complex decision-making. Whereas, humans can perform visual perception effortlessly, it is a challenging task for a machine (including a robot) to do it with comparable quality.

8.2.2   *Attributes of Visual Perception*

A physical object generally exhibits two types of properties:

- Physical properties such as smell, color, stiffness, electrical conductivity, and thermal conductivity, etc.
- Geometric properties such as shape (implicitly dimension), posture (position and orientation), and motion (velocity and acceleration).

As shown in Fig. 8.1, a digital image only preserves the appearance of an object in terms of chrominance (color), luminance (intensity), and texture (specific and regular distribution of chrominance and/or luminance). This information combined with geometric features is useful for visual perception because an object's most important attributes are its geometric properties. In summary, a robot's visual-perception system will consider the following attributes:

(1) Geometric properties, in terms of posture, motion, and shape,
(2) Appearance, in terms of chrominance, luminance, and texture.

An image is the projection of a scene onto a camera's image plane. In a spatial domain, a scene is a three-dimensional (3-D) space, and an image is a two-dimensional (2-D) plane. (NOTE: Time adds another dimension). As one spatial dimension is lost, an image does not explicitly measure a scene's geometric properties.

On the other hand, a digital image is a 2-D matrix of pixels, which, as it is, doesn't contain any geometric information. Since a digital image only preserves the appearance of objects in a scene, their geometric properties must be inferred through the processing of chrominance, luminance, or texture information. Thus, a visual-perception system always starts with image processing and feature extraction which computationally manipulate chrominance, luminance, or texture information in order to obtain useful geometric information.

8.2.3 *Applications of Visual Perception*

A visual-perception system is capable of recovering the geometry and appearance of a scene from images. This capability is indispensable for a system such as the humanoid robot which should autonomously act and interact in a dynamically changing environment.

Refer to Fig. 8.1. Vision-enabled, autonomous behaviors can be classified into these categories:

(1) Visual Guidance:
    Walking, manipulating, and grasping are three types of daily human activities. The autonomy we gain from visual guidance gives us a sense of freedom in performing these activities. For a humanoid robot, it is necessary for its visual perception system to link tightly with the execution of activities, such as walking, manipulating, and grasping. From

the viewpoint of control, a humanoid robot should adopt as much as possible the image-space control scheme for most of its motion executions.

(2) Knowledge Acquisition:

Our mental and physical abilities develop through real-time interaction with a dynamically changing environment. One important aspect of mental development is knowledge acquisition which results from the comprehension of natural languages (e.g. English). With the help of a visual-perception system, it is easy for humans to read, observe and associate meanings with conceptual symbols. A humanoid robot's visual-perception system will also enable it to develop its ability to acquire knowledge through the use of a language.

(3) Visual Inspection:

A successful vision application used in industries is visual inspection for quality control. The quality of a product is measured with respect to the product's design specifications. The purpose of quality control is to minimize variation in a product's parameters with respect to the reference values stipulated by the design (the mean and standard deviations). A subgroup of the product's parameters are usually related to the uniformity of geometry and appearance (i.e. shape, alignment, dimension, color, smoothness of surface, etc). Thus, automatic visual inspection is a must, especially given the increase in the rate of production and the decrease in product size (e.g. semiconductors).

(4) Visual Identification:

Geometry and appearance may also provide other useful information, such as identity, emotion, and language symbols. Two typical examples of visual identification in the domain of bio-metrics are finger-print and face identifications. In the area of man-machine interaction, the ability to decode meaning and emotion from body movements and gestures is very important in the development of a sociable humanoid robot.

(5) Visual Measurement:

It is worth noting that human vision is not metric. In terms of appearance, color perception in human vision is the physiological and psychological interpretation of lights by our brain. Moreover, color perception depends on the lighting conditions of a scene. On the other hand, human vision does not metrically infer the geometry of a scene or object. For example, human vision is unable to guess the length of an object with a reasonable level of precision. This is not the case with a robot's visual-perception system. In fact, an electronic camera can serve as a scientific instrument for the measurement of color (colorimeter) and geometry (photogrammetry or range finder).

In this chapter, we will study the principles and methodologies relevant to the process of inferring geometry from digital images.

8.2.4　*Information Processing in Visual Perception*

We know that a color image inside a robot's visual-sensory or perception system is represented by three image arrays (2-D matrices) as follows:

$$\begin{cases} I_R = \{r(v,u), \; 1 \le v \le r_y \text{ and } 1 \le u \le r_x\} \\ I_G = \{g(v,u), \; 1 \le v \le r_y \text{ and } 1 \le u \le r_x\} \\ I_B = \{b(v,u), \; 1 \le v \le r_y \text{ and } 1 \le u \le r_x\}. \end{cases} \tag{8.1}$$

These represent the Red, Green and Blue (RGB) component color images. In Eq. 8.1, (r_y, r_x) is the image resolution (i.e. the number of columns and rows in an image array). And, (u,v) are the index coordinates. A set of RGB component color images contain both the chrominance and luminance information about a scene being projected onto an image plane.

By computing the weighted average of RGB component color images, an intensity image can be obtained as follows:

$$I_I = \{I(v,u), \; 1 \le v \le r_y \text{ and } 1 \le u \le r_x\} \tag{8.2}$$

where

$$I(v,u) = 0.3 \bullet r(v,u) + 0.59 \bullet g(v,u) + 0.11 \bullet b(v,u). \tag{8.3}$$

As we already mentioned, an intensity image encodes the luminance of a scene being projected onto an image plane. Thus, a digital image explicitly measures the chrominance, luminance, and texture of a scene, but not its geometry. Since the most useful information relevant to motion execution is the geometry of a scene, it is necessary to find a way to infer this from the available information.

Refer to Fig. 8.2. The process of inferring geometry from digital images starts with image processing, followed by feature extraction and finally, geometry measurement. Let us consider intensity images as input to image processing. As shown in Fig. 8.2, the flow of information in a visual-perception system generally involves the following modules:

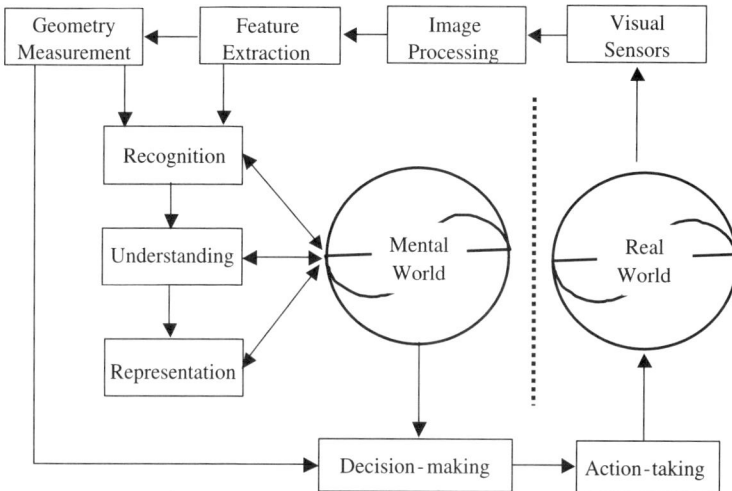

Fig. 8.2   Illustration of the flow of information in a visual-perception system.

Image Processing

A digital (color) image explicitly contains the chrominance, luminance, and texture of a scene. The corresponding geometric properties are implicitly encoded into the uniformity, continuity, and discontinuity of the chrominance, luminance, and texture. Therefore, the primary goal of image-processing, with regard to the perception of geometry, is to detect the uniformity, continuity, and discontinuity in a digital image's chrominance, luminance, and texture. This can be achieved with some specific linear and/or nonlinear filters.

Feature Extraction

Image features are closely related to the uniformity, continuity, and discontinuity in terms of chrominance, luminance, or texture. The presence of image features can be caused by various physical reasons. But, the main causes are the object's geometry and color. For example, the discontinuity of chrominance or luminance may be due to the discontinuity of an object surface's normal vectors, depth, and color. Therefore, image-feature extraction is a first and necessary step towards geometry inference.

Geometry Measurement

The spatial locations of image features are measured with respect to the index coordinate system *uv*, which has no unit of measurement. Therefore, it is necessary to know the relationship among the index coordinates (u, v), the image coordinates (x, y), and the three-dimensional coordinates $(^cX, \ ^cY, \ ^cZ)$ in a scene. In this way, it is possible to compute the actual dimensions of image features. However, there is a loss of one spatial dimension when three-dimensional coordinates in a scene are projected onto an image plane. And, it is still a question on how to reliably determine the three-dimensional coordinates from images.

Object Recognition

Object recognition refers to the identification of objects in terms of geometry, appearance, and meaning, from images. This is relatively easy in a well-controlled and simple environment such as a production line, where there are only a few types of products which are known in advance. However, it is difficult for a robot to perform object recognition. A crucial step towards object recognition is to group image features into distinct sets, each of which corresponds to only one physical object. This will be difficult unless the robot is capable of extracting image features and making a reliable estimation of the geometry of a three-dimensional scene. Thus, these remain the critical issues for a visual-perception system.

Image Understanding

Image understanding refers to the process of generating an ordered sequence of descriptions (or interpretations) of a scene, or the content in an image. The success of image understanding depends on the robot's ability to perform object recognition and description at a semantic level.

Knowledge Representation

The outcome of the linguistic description (or programming) of a scene is naturally a representation of knowledge. Therefore, the visual-perception system is important for knowledge acquisition which is based on real-time interaction with the environment and others. The results of the linguistic description or interpretation of the external world will largely contribute to the formation of the robot's *mental* or *virtual* world.

Scientists around the world are actively researching object recognition, image understanding, and knowledge representation. A systematic discussion of these topics is beyond the scope of this book. Instead, we will focus our study on the basics of image processing, feature extraction, and geometry measurement, which the results can directly contribute to the perception-decision-action loop, as shown in Fig. 8.2.

8.3 Image Processing

Refer to Eq. 8.2. By default, a digital image means an intensity image and is a function of two spatial variables (u, v). However, a robot's visual perception is a continuous dynamic process. The input to a robot's visual-perception system is a sequence of digital images (or digital video, for short). Thus, the exact representation of a digital image should be a function of three variables: a) two spatial variables (u, v) and b) one temporal variable t (time). In general, we can denote a digital image by $I_I(v, u, t)$.

Accordingly, image processing will include all possible computational techniques for the manipulation of image function $I_I(v, u, t)$, with respect to variables (u, v, t) and their neighborhood. Of all the possible computational techniques, the two special types which are important to a robot's visual-perception system are: a) image transformation and b) image filtering.

8.3.1 *Image Transformation*

Image transformation refers to all operations which aim at obtaining alternative representations of an input image. The result of an image transformation is called an *image transform*. In other words, an image transform is an alternative representation of the input image. If there is no loss of information after image transformation, the original input image can be recovered exactly by an inverse transformation applied to the image transform.

8.3.1.1 *Frequency-domain Transforms*

One well-known image transform is the Fourier Transform which is the representation of an image in a frequency domain. If an input image at time-instant t is represented by

$$I_I(v, u), \quad 1 \leq v \leq r_y \quad \text{and} \quad 1 \leq u \leq r_x,$$

the corresponding discrete Fourier Transform will be

$$F(m,n) = \sum_{v=1}^{r_y} \sum_{u=1}^{r_x} \left\{ I_I(v,u) \bullet e^{-j\omega_m v} \bullet e^{-j\omega_n u} \right\} \tag{8.4}$$

where

$$\begin{cases} \omega_m = \frac{2\pi}{r_y} \bullet (m-1), & \forall m \in [1, r_y] \\[2mm] \omega_n = \frac{2\pi}{r_x} \bullet (n-1), & \forall n \in [1, r_x]. \end{cases} \tag{8.5}$$

The discrete Fourier Transform has an inverse, which is

$$I_I(v,u) = \frac{1}{(r_y - 1)(r_x - 1)} \bullet \sum_{v=1}^{r_y} \sum_{u=1}^{r_x} \left\{ F(m,n) \bullet e^{j\omega_m v} \bullet e^{j\omega_n u} \right\}. \tag{8.6}$$

Since the Fourier Transform is an alternative representation of an input image in a frequency domain, we can perform image feature extraction either with the original image or with its Fourier Transform. However, for geometric feature extraction, it is not wise to do the latter.

8.3.1.2 *Time-domain Transforms*

If the result of an image transformation is of the same nature as the input image, it is called *time-domain image transform*.

Example of Color-Image Transformation

The result of the transformation from an RGB color image, represented by Eq. 8.1, to the corresponding intensity image, represented by Eq. 8.2, is a time-domain transform.

Mathematically, the red, green, and blue component colors define a three-dimensional color space, commonly known as the *RGB color space*. Any 3×3 matrix will transform the RGB color space into another space. Of all the possible 3×3 matrices, the one below, used by NTSC TV, defines the transformation from RGB color space to *YIQ color space*:

$$\begin{pmatrix} R \\ G \\ B \end{pmatrix} = \begin{pmatrix} 0.30 & 0.59 & 0.11 \\ 0.60 & -0.27 & -0.32 \\ 0.21 & -0.52 & -0.31 \end{pmatrix} \bullet \begin{pmatrix} Y \\ I \\ Q \end{pmatrix}. \tag{8.7}$$

Since the matrix in Eq. 8.7 has an inverse, the transformation from YIQ to RGB will be

$$\begin{pmatrix} Y \\ I \\ Q \end{pmatrix} = \begin{pmatrix} -4.8084 & 7.3086 & -9.2505 \\ 6.9074 & -6.7504 & 9.4192 \\ -14.8439 & 16.2742 & -25.2922 \end{pmatrix} \bullet \begin{pmatrix} R \\ G \\ B \end{pmatrix}. \qquad (8.8)$$

Example of Intensity-Image Transformation

Image thresholding which acts on an intensity-image's individual pixels, is the simplest and most useful transformation. If we denote $I_I^{in}(v,u)$ the input image, and $I_I^{out}(v,u)$ the corresponding image transform, the mathematical description of the image thresholding transformation is

$$I_I^{out}(v,u) = \begin{cases} 1 & \text{if } I_I^{in}(v,u) > v_0 \\ 0 & \text{otherwise} \end{cases} \qquad \forall v \in [1, r_y] \text{ and } \forall u \in [1, r_x] \quad (8.9)$$

where v_0 is a threshold value. If a pixel's luminance is encoded with one byte, v_0 will fall within the range of 0 to 255. In fact, image thresholding transformation divides an intensity-image's pixels into two groups: a) the foreground pixels and b) the background pixels.

Threshold value v_0 can be manually set if the nature of the input images is known in advance. One useful bit of prior knowledge is the *intensity histogram*. This is a vector that each element encodes a luminance-value's frequency of occurrence. If v denotes the luminance value which varies from 0 to 255, a histogram is simply described by

$$H_f = \{h(v), \ \forall v \in [0, 255]\} \qquad (8.10)$$

where v serves as the index from which to retrieve the frequency of occurrence $h(v)$ of luminance value v.

Example 8.1 A common process for estimating the camera's parameters in a robot's visual sensory system is calibration. This process normally makes use of a special object called a *calibration rig*. This rig can be as simple as a box with a grid or an array of regular patterns painted on one or more of the box's facets. Fig. 8.3a shows an image of a calibration rig, as seen by a robot's camera. Before the calibration process can begin, it is necessary to precisely locate, in the image plane, the array of patterns on the calibration rig.

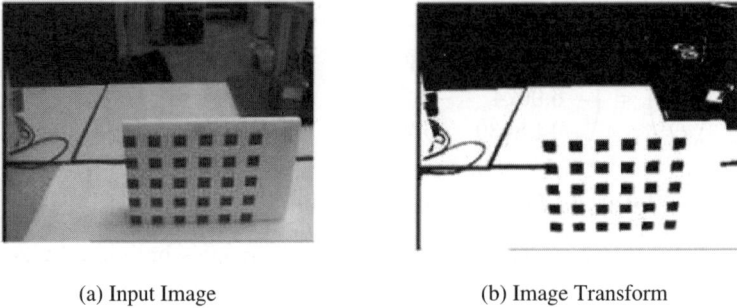

| (a) Input Image | (b) Image Transform |

Fig. 8.3   Example of image-thresholding transformation: a) input image and b) image transform.

In this example, we first compute the intensity histogram. Refer to Fig. 8.4. We can see that there are two dominant peaks in the histogram, which are separable around the index value 80. So, we choose $v_0 = 80$. Fig. 8.3b illustrates the result of image-thresholding transformation when $v_0 = 80$. Clearly, the array of patterns on the calibration rig has been isolated from the other objects. (There is only one exception to the pattern at row 2, column 1).

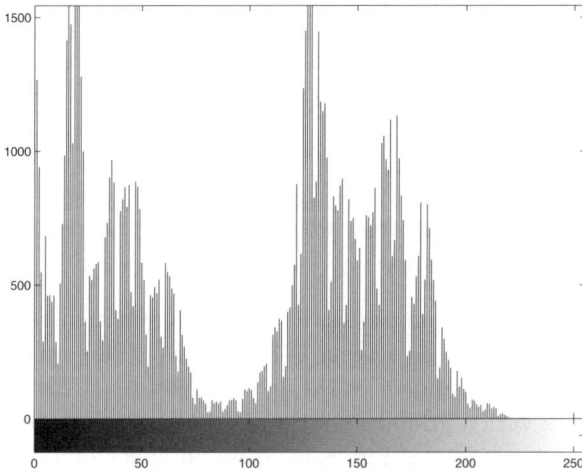

Fig. 8.4   Example of a histogram.

◇◇◇◇◇◇◇◇◇◇◇◇◇◇◇◇◇◇◇

As shown in the above example, it is possible to automatically determine the threshold value by analyzing the intensity distribution in an image. A simple and practical algorithm automatically determining the threshold value is Otsu's method. This method evaluates all possible threshold values ranging from 0 to 255, and chooses the one which minimizes the variances of luminance of both the foreground and background pixels.

8.3.1.3 *Spatial Transforms*

Mathematically, we can also manipulate spatial variables (u, v) without altering the luminance value. The outcome of a transformation acting on spatial variables (u, v) is called a *spatial transform*. If we denote $I_I^{in}(v, u)$ the input image and $I_I^{out}(v_1, u_1)$ the output image, a general spatial transformation can be described by

$$I_I^{out}(v_1, u_1) = I_I^{in}(v, u), \quad \forall v \in [1, r_y] \text{ and } \forall u \in [1, r_x] \tag{8.11}$$

where

$$s \bullet \begin{pmatrix} u_1 \\ v_1 \\ 1 \end{pmatrix} = \begin{pmatrix} a & b & c \\ d & e & f \\ g & h & i \end{pmatrix} \bullet \begin{pmatrix} u \\ v \\ 1 \end{pmatrix}. \tag{8.12}$$

In fact, Eq. 8.12 represents the following types of spatial transformation:

- Projective Transformation, if (g, h, i) is not equal to $(0, 0, 1)$,
- Affine Transformation, if (g, h, i) is equal to $(0, 0, 1)$,
- Motion Transformation, if (g, h, i) is equal to $(0, 0, 1)$, and the four elements (a, b, d, e) form a rotation matrix.

8.3.2 *Image Filtering*

"Filtering" literally means to remove impurities or undesirable substances from something (e.g. water, air, or even signals). For image processing, "filtering" means to remove a subset of (undesirable) signals from an input image. In a frequency domain, a signal is the sum of harmonic functions, such as sinusoids and exponentials which have different frequencies, magnitudes and phases. Depending on the application, one may wish to remove the harmonic functions above, below, or outside a certain frequency. In general, image filtering causes a loss of information. Consequently, it does not make sense to recover the original input image from a filtered image in the frequency domain.

Just as the purpose of a control law is to alter the dynamics of the system under control, the purpose of filtering is to alter the dynamics of a signal. From this point of view, it is possible to unify the philosophy underlying the design of a mechanical system, a control system, and a computational algorithm for image feature extraction.

8.3.2.1   Convolutions

In Chapter 5, we learned that a differential equation describes a linear system's dynamics. And, the Laplace transform of a differential equation is in algebraic form. For a linear system, the ratio between the output's Laplace transform and the input's Laplace transform is called a *transfer function*. If the input to a system is an impulse, its Laplace transform is 1 and the Laplace transform of the output is exactly equal to the transfer function. Because of this, the inverse Laplace transform of the transfer function when the input is an impulse is called an *impulse response*.

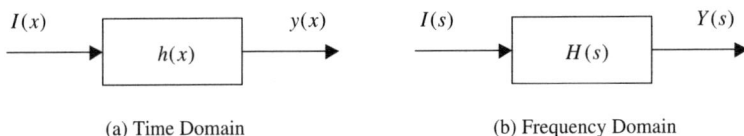

(a) Time Domain   (b) Frequency Domain

Fig. 8.5   Dynamics of a linear system: a) the impulse-response in the time domain, and b) the transfer function in the frequency domain.

As shown in Fig. 8.5, a system's output in the frequency domain is expressed as the product of the input's Laplace transform and the system's transfer function, that is,

$$Y(s) = H(s) \bullet I(s). \tag{8.13}$$

If we treat $I(s)$ as the Laplace transform of an input signal or image, $Y(s)$ is the filtered signal or image, the dynamics of which have been altered by the filter's transfer function $H(s)$. Clearly, it is easy to change the dynamics of an input signal or image by manipulating the filter $H(s)$.

In the time domain, the inverse Laplace transform of Eq. 8.13 is

$$y(x) = h(x){*}I(x) = \int_{-\infty}^{x} h(\alpha) \bullet I(x-\alpha)\, d\alpha = \int_{-\infty}^{x} I(\alpha) \bullet h(x-\alpha)\, d\alpha \tag{8.14}$$

where α is a dummy variable for integration.

By definition, the operation in Eq. 8.14 is called *convolution*, and the impulse response $h(x)$ is known as the *convolution kernel* for signal or image filtering.

The above results are obtained in a continuous time domain. In a discrete image plane, an input digital image at time-instant t is represented by

$$I_I^{in} = \{I_I^{in}(v,u), \ \forall v \in [1,r_y] \text{ and } \forall u \in [1,r_x]\}$$

and the output digital image is represented by

$$I_I^{out} = \{I_I^{out}(v,u), \ \forall v \in [1,r_y] \text{ and } \forall u \in [1,r_x]\}.$$

If a discrete convolution kernel is

$$\{h_k\} = \{h_k(m,n), \ \forall m \in [1,ky] \text{ and } \forall n \in [1,kx]\}$$

with

$$\begin{cases} h_k(m,n) \neq 0 & \text{if } \forall m \in [1,ky] \text{ and } \forall n \in [1,kx] \\ \\ h_k(m,n) = 0 & \text{otherwise,} \end{cases} \tag{8.15}$$

the result of discrete convolution will be

$$I_I^{out}(v,u) = h_k(v,u) * I_I^{in}(v,u)$$
$$= \sum_{v_1=1}^{v} \sum_{u_1=1}^{u} \left\{ h_k(v-v_1, u-u_1) \bullet I_I^{in}(v_1, u_1) \right\}. \tag{8.16}$$

By applying Eq.8.15, Eq. 8.16 can be rewritten as

$$I_I^{out}(v,u) = \sum_{v_1=v-ky}^{v-1} \sum_{u_1=u-kx}^{u-1} \left\{ h_k(v-v_1, u-u_1) \bullet I_I^{in}(v_1, u_1) \right\}. \tag{8.17}$$

If we substitute $v - v_1$ with m, and $u - u_1$ with n, Eq. 8.17 will become

$$I_I^{out}(v,u) = \sum_{m=1}^{ky} \sum_{n=1}^{kx} \left\{ h_k(m,n) \bullet I_I^{in}(v-m, u-n) \right\}. \tag{8.18}$$

Consider an image's boundary condition. Indexes (v,u) in Eq. 8.18 must vary within the range of $[ky+1, r_y]$ and $[kx+1, r_x]$. This indicates that the output is not centered on the original input image. In other words, the information in the first ky rows and kx columns is not considered in the convolution. To make full use of the information centered on an image plane, we should shift the input image using displacement vector $(kx/2 + 1, ky/2+1)$ so that the output will be centered on the input image. (NOTE: If $kx = 3$, $kx/2$ is equal to·1; and if $ky = 3$, $ky/2$ is equal to 1). Accordingly, Eq. 8.18 becomes

$$I_I^{out}(v, u) = \sum_{m=1}^{ky} \sum_{n=1}^{kx} \left\{ h_k(m, n) \bullet I_I^{in}(v - m + ky/2 + 1, u - n + kx/2 + 1) \right\}.$$

$$(8.19)$$

If a convolution kernel is symmetric, Eq. 8.20 can also be expressed as

$$I_I^{out}(v, u) = \sum_{m=1}^{ky} \sum_{n=1}^{kx} \left\{ h_k(m, n) \bullet I_I^{in}(v + m - ky/2 - 1, u + n - kx/2 - 1) \right\}.$$

$$(8.20)$$

However, a convolution kernel can be asymmetric. In this case, it is still possible to use Eq. 8.20 to perform the convolution if the results of the convolution with either a convolution kernel or its symmetrically-swapped version are equally acceptable. For example, the symmetrically-swapped version of the convolution kernel

$$\begin{bmatrix} 1 & 0 & -1 \\ 1 & 0 & -1 \\ 1 & 0 & -1 \end{bmatrix}$$

is

$$\begin{bmatrix} -1 & 0 & 1 \\ -1 & 0 & 1 \\ -1 & 0 & 1 \end{bmatrix}.$$

Example 8.2   Fig. 8.6a shows an image array which the size is 10×10. Assume that the convolution kernel is

$$\{h(m, n)\} = \frac{1}{6} \bullet \begin{bmatrix} 1 & 1 & 0 \\ 1 & 1 & 0 \\ 1 & 1 & 0 \end{bmatrix}.$$

The result of the convolution after applying Eq. 8.20, is shown in Fig. 8.6b. The convolution is computed using integers. Therefore, $100/6$, which would be 16.6667 if working with floating-point numbers, is 17.

◇◇◇◇◇◇◇◇◇◇◇◇◇◇◇◇◇◇◇

1	0	0	0	0	0	0	0	0	0	0
2	0	0	0	0	100	0	0	0	0	0
3	0	0	0	100	100	100	0	0	0	0
4	0	0	100	100	100	100	100	0	0	0
5	0	100	100	100	100	100	100	100	0	0
6	0	0	100	100	100	100	100	0	0	0
7	0	0	0	100	100	100	0	0	0	0
8	0	0	0	0	100	0	0	0	0	0
9	0	0	0	0	0	0	0	0	0	0
10	0	0	0	0	0	0	0	0	0	0
11		2		4		6		8		10

(a) Input

1	0	0	0	0	0	0	0	0	0	0
2	0	0	0	17	50	50	17	0	0	0
3	0	0	17	50	83	83	50	17	0	0
4	0	17	50	83	100	100	83	50	17	0
5	0	17	67	100	100	100	100	67	17	0
6	0	17	50	83	100	100	83	50	17	0
7	0	0	17	50	83	83	50	17	0	0
8	0	0	0	17	50	50	17	0	0	0
9	0	0	0	0	17	17	0	0	0	0
10	0	0	0	0	0	0	0	0	0	0
11		2		4		6		8		10

(b) Output

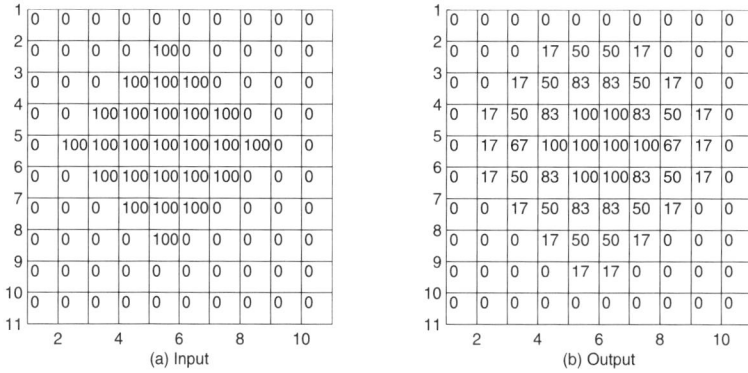

Fig. 8.6   Example of convolution with a convolution kernel: a) the input image and b) the result of the convolution.

8.3.2.2 *Derivative of Convolutions*

If the initial conditions of linear function $f(x)$ are all zeros, the Laplace transform of its derivative $\frac{df(x)}{dx}$ will be $s \bullet F(s)$, in which $F(s)$ is the Laplace transform of function $f(x)$. Now, let us multiply the complex variable s to both sides of Eq. 8.13. This yields

$$s \bullet Y(s) = s \bullet H(s) \bullet I(s). \qquad (8.21)$$

Eq. 8.21 can also be written as

$$\{s \bullet Y(s)\} = H(s) \bullet \{s \bullet I(s)\}. \qquad (8.22)$$

The computation of the inverse Laplace transform of Eq.8.22 results in

$$\frac{dy(x)}{dx} = h(x) * \frac{dI(x)}{dx}. \qquad (8.23)$$

Eq. 8.23 indicates that a linear system's response to an input's derivative is equal to the derivative of the system's response to the input itself. Therefore, it is possible to compute derivative $\frac{dy(x)}{dx}$ directly without invoking the input's derivative if response $y(x)$ is already known. This is a well-known result in control engineering. Here, we call it the *principle of dynamics conservation*. In control engineering, we often use a unit-step function as a linear system's input. If we know a system's response to a unit-step input, then its derivative is exactly equal to the system's response to a unit-impulse input.

Interestingly enough, there is a different interpretation to the above principle when dealing with image or signal filtering. In fact, Eq. 8.21 can alternatively be expressed as

$$\{s \bullet Y(s)\} = \{s \bullet H(s)\} \bullet I(s). \tag{8.24}$$

Computation of the inverse Laplace transform of Eq.8.24 yields

$$\frac{dy(x)}{dx} = \frac{dh(x)}{dx} * I(x). \tag{8.25}$$

Eq. 8.24 indicates that the response of a linear system (or filter) replicates the dynamics of the system (or filter) even though the input remains the same. In other words, input $I(x)$ is able to resonate not only with a system's impulse-response $h(x)$, but also with its derivative $\frac{dh(x)}{dx}$. We call this phenomenon the *principle of dynamics resonance*.

With regard to feature extraction from signals, it is possible to interpret a feature as a specific type of dynamics exhibited by the signal itself, in terms of uniformity, continuity, and discontinuity. If a filter is sensitive to a specific type of dynamics and the resonance stimulated by the input signal does not vanish, this indicates that the signal contains features corresponding to this type of dynamics. From this point of view, feature extraction from a signal can easily be achieved in two steps: a) resonance generation and b) resonance detection. We will discuss this in more detail in the section on image-feature extraction.

Example 8.3　　Fig. 8.7a shows an image array that the size is 10×10. If we compute the derivative of the convolution kernel from Example 8.2 in the horizontal direction, we will obtain

$$\left\{\frac{dh(m,n)}{dn}\right\} = \frac{1}{6} \bullet \begin{bmatrix} 1 & 0 & -1 \\ 1 & 0 & -1 \\ 1 & 0 & -1 \end{bmatrix}.$$

If we apply Eq. 8.20, the result of the convolution between the input image and convolution kernel $\left\{\frac{dh(m,n)}{dn}\right\}$ is shown in Fig. 8.7b. Clearly, there is a resonance caused by the presence of vertical edges in the input image.

◇◇◇◇◇◇◇◇◇◇◇◇◇◇◇◇◇◇◇

0	0	0	0	0	0	0	0	0	0
0	100	100	100	100	100	100	100	100	0
0	100	100	100	100	100	100	100	100	0
0	100	100	100	100	100	100	100	100	0
0	0	0	150	150	150	150	0	0	0
0	0	0	150	150	150	150	0	0	0
0	0	0	150	150	150	150	0	0	0
0	0	0	150	150	150	150	0	0	0
0	0	0	150	150	150	150	0	0	0
0	0	0	0	0	0	0	0	0	0

(a) Input

0	0	0	0	0	0	0	0	0	0
0	33	0	0	0	0	0	0	33	0
0	50	0	0	0	0	0	0	50	0
0	33	25	25	0	0	25	25	33	0
0	17	50	50	0	0	50	50	17	0
0	0	75	75	0	0	75	75	0	0
0	0	75	75	0	0	75	75	0	0
0	0	75	75	0	0	75	75	0	0
0	0	50	50	0	0	50	50	0	0
0	0	0	0	0	0	0	0	0	0

(b) Output

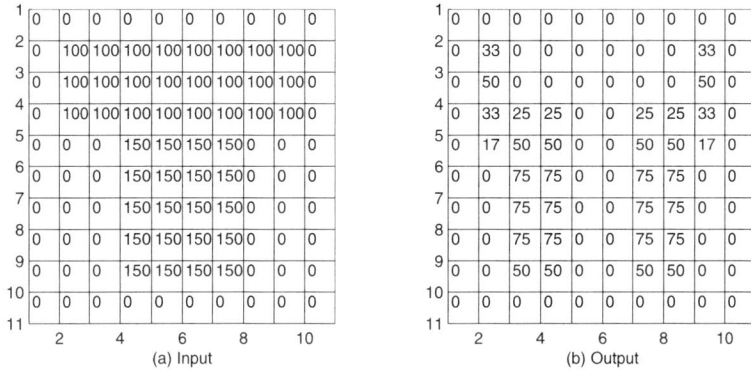

Fig. 8.7   Example of a convolution with the derivative of a convolution kernel: a) the input image and b) the result of a convolution.

8.3.2.3   *Integral of Convolutions*

Both the principles of dynamics conservation and dynamics resonance are equally valid for the integral of a convolution. In order to verify this statement, let us multiply the inverse of complex variable s to both sides of Eq. 8.13. It becomes

$$\frac{1}{s} \bullet Y(s) = \frac{1}{s} \bullet H(s) \bullet I(s). \tag{8.26}$$

Eq. 8.26 can also be rewritten as

$$\left\{ \frac{1}{s} \bullet Y(s) \right\} = H(s) \bullet \left\{ \frac{1}{s} \bullet I(s) \right\}. \tag{8.27}$$

Computation of the inverse Laplace transform of Eq. 8.27 yields

$$\int y(x) \bullet dx = h(x) * \left\{ \int I(x) \bullet dx \right\}. \tag{8.28}$$

This indicates that a linear system's response to an input's integral is equal to the integral of the system's response to that input.

Alternatively, Eq. 8.26 can also be expressed as

$$\left\{ \frac{1}{s} \bullet Y(s) \right\} = \left\{ \frac{1}{s} \bullet H(s) \right\} \bullet I(s). \tag{8.29}$$

The inverse Laplace transform of Eq.8.29 is

$$\int y(x) \bullet dx = \left\{ \int h(x) \bullet dx \right\} * I(x). \tag{8.30}$$

This indicates that the dynamics of a system (or filter) is replicated by the output, even though the input remains the same.

8.3.2.4 *Spatial Displacement of Convolutions*

Interestingly enough, the above concepts are also valid for a convolution's spatial transform (displacement). For feature detection, an important concern is the localization of detected features. Therefore, it is interesting to know whether a filter itself has a degree of freedom to adjust feature localization.

The multiplication of the term $e^{-alphas}$ to both sides of Eq. 8.13 yields

$$\{e^{-\alpha s} \bullet Y(s)\} = H(s) \bullet \{e^{-\alpha s} \bullet I(s)\} \tag{8.31}$$

or

$$\{e^{-\alpha s} \bullet Y(s)\} = \{e^{-\alpha s} \bullet H(s)\} \bullet I(s). \tag{8.32}$$

The inverse Laplace transforms of Eq. 8.31 and Eq. 8.32 are

$$y(x - \alpha) = h(x) * I(x - \alpha) \tag{8.33}$$

and

$$y(x - \alpha) = h(x - \alpha) * I(x). \tag{8.34}$$

Eq. 8.34 clearly illustrates the phenomenon of a signal feature's location shift which is caused by a filter. This phenomenon can advantageously be exploited to design filters with the ability to adjust the location of detected features in a signal.

Example 8.4   Fig. 8.8a shows an image array which the size is 10×10. In Example 8.2, if we shift the convolution kernel towards the right by 1 pixel, we will obtain

$$\{h(m, n - 1)\} = \frac{1}{6} \bullet \begin{bmatrix} 0 & 1 & 1 \\ 0 & 1 & 1 \\ 0 & 1 & 1 \end{bmatrix}.$$

If we apply Eq. 8.20, the result of the convolution between the input image and convolution kernel $\{h(m, n - 1)\}$ is shown in Fig. 8.8b. If we compare the output in Fig. 8.8b with the output in Fig. 8.6b, we can see that there is a horizontal shift. The shift is towards the left because we use Eq. 8.20 instead of Eq. 8.19 to compute the convolution. It is worth noting that the boundary of the input image array includes the first and last rows as well as the first and last columns. This is because the size of the convolution kernel is 3×3. For easy programming, pixels inside an input image's boundary are not processed.

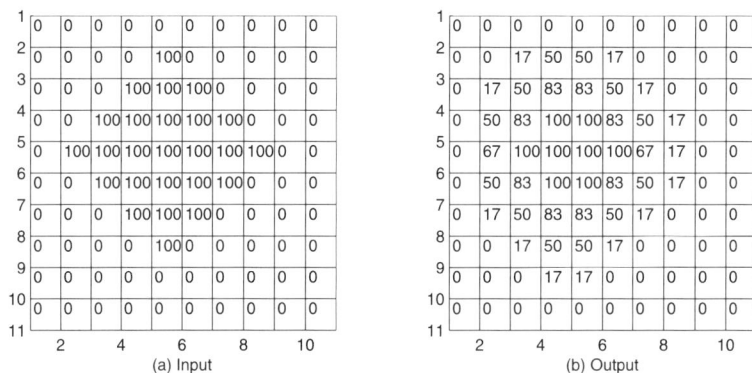

(a) Input

0	0	0	0	0	0	0	0	0	0
0	0	0	0	100	0	0	0	0	0
0	0	0	100	100	100	0	0	0	0
0	0	100	100	100	100	100	0	0	0
0	100	100	100	100	100	100	100	0	0
0	0	100	100	100	100	100	0	0	0
0	0	0	100	100	100	0	0	0	0
0	0	0	0	100	0	0	0	0	0
0	0	0	0	0	0	0	0	0	0
0	0	0	0	0	0	0	0	0	0

(b) Output

0	0	0	0	0	0	0	0	0	0
0	0	17	50	50	17	0	0	0	0
0	17	50	83	83	50	17	0	0	0
0	50	83	100	100	83	50	17	0	0
0	67	100	100	100	100	67	17	0	0
0	50	83	100	100	83	50	17	0	0
0	17	50	83	83	50	17	0	0	0
0	0	17	50	50	17	0	0	0	0
0	0	0	17	17	0	0	0	0	0
0	0	0	0	0	0	0	0	0	0

Fig. 8.8 Example of a convolution with the spatial transform of a convolution kernel: a) the input image and b) the result of the convolution.

◇◇◇◇◇◇◇◇◇◇◇◇◇◇◇◇◇◇◇

8.4   Image Feature Extraction

An input image explicitly represents the chrominance and luminance of a scene (or object). However, for manipulative and cognitive tasks, the most important attributes of a scene (or object) are the geometric features. Thus, in a visual-perception system, it is crucial to obtain the geometric representation of a scene (or object) from the chrominance and luminance information. Mathematically, an image's geometric representation is the image's alternative representation. Thus, it can also be called an *image transform* if we treat the complex process of image-feature extraction as a nonlinear image transformation.

8.4.1   *The Basics of Feature Detection*

In the frequency domain, the multiplication of complex variable s to a filter linearly enhances the high-frequency signals. In the time domain, this enhances a signal's discontinuity. Similarly, the multiplication of the inverse of complex variable s linearly attenuates the high-frequency signals which has the effect of enhancing a signal's uniformity.

As we mentioned earlier, a robot's visual-perception system is a continuous dynamic process. And, an image is the basic element of a video or image sequence. Accordingly, an image should be treated as a spatio-temporal function. However, for the sake of simplicity, let us consider an image to be a spatial function only.

8.4.1.1 Feature Definition

Mathematically, spatial function $f(x)$ in the domain of spatial variable x will exhibit three types of dynamic characteristics:

(1) Uniformity:
   $f(x) = constant$ within certain intervals of x,
(2) Continuity:
   $f(x)$ is continuously differentiable within certain intervals of x, and
(3) Discontinuity:
   $f(x)$ is not differentiable at certain locations of x, or some derivatives of $f(x)$ vanish at certain locations of x.

Since an image explicitly represents the chrominance, luminance, and texture of a scene (or object), the uniformity of these characteristics are directly related to the characteristics of the geometric surfaces in a scene. For example, if a surface is smooth and uniform in color, the corresponding region in an image plane will be uniform.

Similarly, discontinuity of these characteristics in an image depends not only on the surfaces themselves but also on the optical and geometric interference among the surfaces. For example, if light rays coming towards a surface are partly blocked by the presence of an opaque object, a shadow will form on the surface. Moreover, the reflected light rays coming off the surface and entering the optical lens of a visual sensory system can be blocked by an opaque object as well. This phenomenon is known as *occlusion*. Shadows, occlusions, and surface boundaries are the common causes for discontinuity in an image.

From the above discussions, it is clear that the presence of an object in image space is considered as the sum of an image's uniformity, continuity, and discontinuity. Since it is possible to transform an image into an alternative representation called *image transform*, we can define an image feature either in image space or in the space of its transform. However, regardless of which space is used, image features must be meaningful and useful. Formally, we can state the definition of an image feature as follows:

Definition 8.3 Any uniformity or discontinuity, which is meaningful and useful in an image (or its transform), is an image feature.

Based on this definition, the process of detecting image features is called *image feature extraction*. And, the process of analytically describing the detected image features is called *image feature description*. By default, image feature extraction is performed using intensity images.

8.4.1.2 *A Generic Procedure for Feature Detection*

In general, an image contains many types of features, including noise. In theory, it is impossible to design a filter which responds to only one type of feature and perfectly removes all the others. In electrical engineering, there is a saying about signal filters which sums up the problem. That is: "Garbage in, garbage out." If a signal contains one type of feature, it is easy to filter out the noise and enhance that feature. However, if there are multiple types of features plus noise, image-feature extraction becomes a very difficult task.

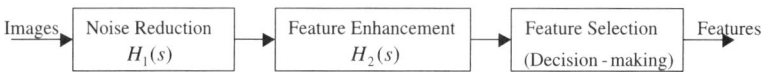

Images → | Noise Reduction $H_1(s)$ | → | Feature Enhancement $H_2(s)$ | → | Feature Selection (Decision - making) | → Features

Fig. 8.9 A Generic feature-detection procedure.

In order to cope with the difficulty caused by noise and multiple features, a feature-extraction process generally involves the following modules, as shown in Fig. 8.9:

- Noise Reduction:
  The purpose of this module is to reduce the noise level in an input image. If the image quality is good, this step is unnecessary and $H_1(s) = 1$.
- Feature Enhancement:
  It is almost impossible to directly pick up desired features from an original input image if there are multiple features. It is necessary to enhance the features of interest and, at the same time, to reduce the other types of features including noise. This is usually done with a specific filter, called a *feature detector*.

- Feature Selection:
  Whether in theory or practice, it is impossible to design a filter which can totally reduce undesired features and noise. It is, however, possible to design a filter capable of responding differently to different features. After the feature enhancement, a decision-making process is responsible for selecting the desired features.

8.4.1.3   *Criteria for Feature Detection*

"Garbage in, garbage out" generally implies that there is no ideal filter for a given type of feature. However, there are some guidelines for general filter design, which if followed, can ensure a good feature detector.

John Canny in his work on edge-detection while studying at MIT as a graduate student first stipulated three criteria for the design of a step-edge detector. We can generalize these as follows:

Criterion 1: Resistance to Noise

Canny's first criterion, called *good response*, stipulated the necessity of maximizing the signal-to-noise ratio. We can clearly explain this criterion using Laplace transform.

Let $I(s)$ denote an input image's Laplace transform. Assume that input image $I(s)$ is contaminated by white noise $N(s)$. The output from filter $H_1(s)$ (the Laplace transform of discrete filter $\{h_1(m,n)\}$) will be

$$Y_1(s) = H_1(s) \bullet I(s) + H_1(s) \bullet N(s) \qquad (8.35)$$

or

$$\frac{Y_1(s)}{I(s)} = H_1(s) + H_1(s) \bullet \frac{N(s)}{I(s)}. \qquad (8.36)$$

From Eq. 8.36, we can see that an ideal filter is one which can totally cancel out term $H_1(s) \bullet \frac{N(s)}{I(s)}$. This type of feature detector will be resistant to the presence of noise. And accordingly, it is more appropriate to call this criterion *resistance to noise*.

In theory, there is no such design of filter $H_1(s)$ which is able to cancel out term $H_1(s) \bullet \frac{N(s)}{I(s)}$ regardless of the type of noise and input. In practice, any filter which greatly reduces signal $H_1(s) \bullet \frac{N(s)}{I(s)}$ is a good filter for noise reduction.

Criterion 2: Good Accuracy for Feature Localization

Canny's second criterion is called *good localization*. As shown in Fig. 8.9, a decision-making process normally identifies the features to be the local maxima from filter H_2's output. The presence of noise may slightly shift the local maxima away from their true locations. Canny believed that a good feature detector should minimize this shift despite the presence of noise.

Again, this criterion can be better explained using the Laplace transform. Assume that the input to filter $H_2(s)$, as shown in Fig. 8.9, is $I_1(s) + N_1(s)$ with

$$\begin{cases} I_1(s) = H_1(s) \bullet I(s) \\ \\ N_1(s) = H_1(s) \bullet N(s). \end{cases} \tag{8.37}$$

The output from filter $H_2(s)$ will be

$$Y_2(s) = H_2(s) \bullet I_1(s) + H_2(s) \bullet N_1(s). \tag{8.38}$$

The results shown in Eq.8.32 and Eq.8.34 indicate two interesting facts:

- If $H_2(s) \bullet N_1(s)$ in Eq. 8.38 is a complex function having zero(s) and/or pole(s), it will cause the output shift in the time domain. This is because a complex function introduces the additive term:

$$a \bullet \cos(\omega) + j[b \bullet \sin(\omega)]$$

  or, the multiplicative term $e^{\beta s}$ to output $Y_2(s)$. A phase change in the frequency domain means a location shift in the time domain.
- If filter $H_2(s)$ incorporates multiplicative term $e^{-\alpha s}$, it will have one degree of freedom to compensate for the location shift in the time domain caused by signal $H_2(s) \bullet N_1(s)$.

Based on the above observations, it is clear that in order for a filter to satisfy the criterion of good localization, it must have the ability to cancel out signal $H_2(s) \bullet N_1(s)$. Interestingly enough, this can be achieved by incorporating term $e^{-\alpha s}$ into filter $H_2(s)$. To prove this statement, let us substitute $H_2(s)$ with $e^{-\alpha s} \bullet H_2(s)$. Then, Eq. 8.38 becomes

$$Y_2(s) = e^{-\alpha s} \bullet H_2(s) \bullet I_1(s) + N_2(s) \tag{8.39}$$

where $N_2(s) = e^{-\alpha s} \bullet H_2(s) \bullet N_1(s)$.

If we substitute term $e^{-\alpha s}$ with its Taylor series:

$$e^{\alpha s} = 1 + \frac{(-\alpha)}{1!}s + \ldots + \frac{(-\alpha)^n}{n!}s^n + \ldots ,$$

Eq. 8.39 becomes

$$\frac{Y_2(s)}{I_1(s)} = H_2(s) + N_3(s) \tag{8.40}$$

with

$$N_3(s) = N_2(s) + H_2(s) \bullet \left\{ \frac{(-\alpha)}{1!}s + \ldots + \frac{(-\alpha)^n}{n!}s^n + \ldots \right\}. \tag{8.41}$$

Obviously, a filter having good feature localization is one in which parameter α minimizes Eq. 8.41. In practice, parameter α can manually be tuned and tested with sample images.

Criterion 3: Good Selectivity of Desired Features

Canny's third criterion is called *single response*. By single response, he meant that a filter will only produce one response to a single (step) edge input if there is no noise. This criterion is applicable to the continuous step-signal but not in the discrete time domain. The following example illustrates this.

Example 8.5   Fig. 8.10a shows an image array which the size is 10×10. We choose the convolution kernel to be

$$\{h(m,n)\} = \begin{bmatrix} 0 & 1 & 0 \\ 0 & 0 & 0 \\ 0 & -1 & 0 \end{bmatrix}.$$

If we apply Eq. 8.20, the result of the convolution between the input image and convolution kernel $\{h(m,n)\}$ is shown in Fig. 8.10b. Clearly, although the input image has no noise, there are multiple responses for the horizontal edge in the middle.

◇◇◇◇◇◇◇◇◇◇◇◇◇◇◇◇◇◇

In a discrete time domain, an output's redundancy may be advantageous as it allows us to achieve sub-pixel accuracy in localizing image features.

In principle, the single-response criterion should be called *good selectivity of desired features*. This is because a feature-detection filter should, as

	2		4		6		8		10
1	0	0	0	0	0	0	0	0	0	0
2	0	100	100	100	100	100	100	100	100	0
3	0	100	100	100	100	100	100	100	100	0
4	0	100	100	100	100	100	100	100	100	0
5	0	0	0	150	150	150	150	0	0	0
6	0	0	0	150	150	150	150	0	0	0
7	0	0	0	150	150	150	150	0	0	0
8	0	0	0	150	150	150	150	0	0	0
9	0	0	0	150	150	150	150	0	0	0
10	0	0	0	0	0	0	0	0	0	0

(a) Input

	2		4		6		8		10
1	0	0	0	0	0	0	0	0	0	0
2	0	100	100	100	100	100	100	100	100	0
3	0	0	0	0	0	0	0	0	0	0
4	0	100	100	50	50	50	50	100	100	0
5	0	100	100	50	50	50	50	100	100	0
6	0	0	0	0	0	0	0	0	0	0
7	0	0	0	0	0	0	0	0	0	0
8	0	0	0	0	0	0	0	0	0	0
9	0	0	0	150	150	150	150	0	0	0
10	0	0	0	0	0	0	0	0	0	0

(b) Output

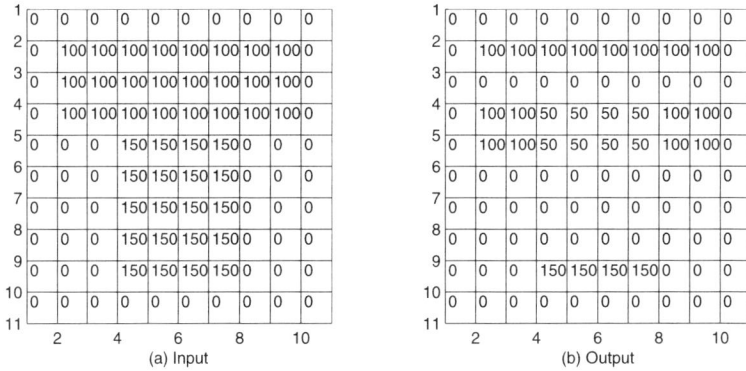

Fig. 8.10   Example of multiple responses in horizontal-edge detection.

much as possible, enhance the desired features and reduce all the others, including noise.

Example 8.6   Fig. 8.11a shows an image array which the size is 10×10. Assume that we want to detect the type of corner at location $(2,3)$. For this purpose, we design a convolution kernel as follows:

$$\{h(m,n)\} = \begin{bmatrix} -1 & -1 & -1 \\ -1 & 2 & 1 \\ -1 & 1 & 1 \end{bmatrix}.$$

The response of this filter to a uniform region will be zero because the sum of elements in $\{h(m,n)\}$ is zero.

If we apply Eq. 8.20, the result of the convolution between the input image and convolution kernel $\{h(m,n)\}$ is shown in Fig. 8.11b. We can see that there are multiple responses to this filter. However, the corner at location $(2,3)$ has the strongest output (500).

◇◇◇◇◇◇◇◇◇◇◇◇◇◇◇◇◇◇

8.4.2   *Edge Detection*

Edge is an image feature which characterizes the discontinuity of an image function in terms of chrominance or luminance. Here, we will not consider discontinuity caused by texture.

To a certain extent, it is possible to say that edges are an image's most important geometric features. Without edges, any higher-level description

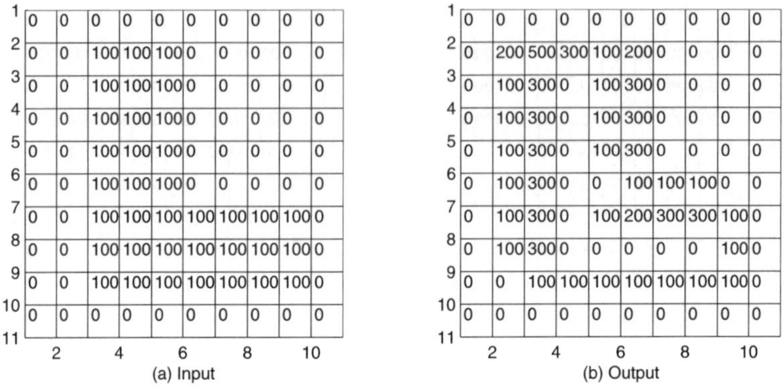

(a) Input

	2		4		6		8		10
1	0	0	0	0	0	0	0	0	0	0
2	0	0	100	100	100	0	0	0	0	0
3	0	0	100	100	100	0	0	0	0	0
4	0	0	100	100	100	0	0	0	0	0
5	0	0	100	100	100	0	0	0	0	0
6	0	0	100	100	100	0	0	0	0	0
7	0	0	100	100	100	100	100	100	100	0
8	0	0	100	100	100	100	100	100	100	0
9	0	0	100	100	100	100	100	100	100	0
10	0	0	0	0	0	0	0	0	0	0

(b) Output

	2		4		6		8		10
1	0	0	0	0	0	0	0	0	0	0
2	0	200	500	300	100	200	0	0	0	0
3	0	100	300	0	100	300	0	0	0	0
4	0	100	300	0	100	300	0	0	0	0
5	0	100	300	0	100	300	0	0	0	0
6	0	100	300	0	0	100	100	100	0	0
7	0	100	300	0	100	200	300	300	100	0
8	0	100	300	0	0	0	0	0	100	0
9	0	0	100	100	100	100	100	100	100	0
10	0	0	0	0	0	0	0	0	0	0

Fig. 8.11   Example of good selectivity in response to a desired feature.

of a scene's geometric features would be difficult to obtain.

8.4.2.1 Definition of Edges

An image is a spatio-temporal function which has two spatial variables (u, v) and one temporal variable t. If we denote $I(v, u, t)$ as an input image at time-instant t, the image's discontinuity can be measured by these first-order derivatives:

$$I_v = \frac{dI}{dv}, \quad I_u = \frac{dI}{du}, \quad \text{and} \quad I_t = \frac{dI}{dt}.$$

In fact, the local maxima of the above signals indicate the image function's discontinuity at these locations. Accordingly, we can state the definition of edge as follows:

Definition 8.4   Edges are the locations in an image array where the image function's first-order derivatives have the local maxima. Locations where the local maxima of $\left(\frac{dI}{dv}, \frac{dI}{du}\right)$ occur are called *spatial edges*. And, locations where the local maxima of $\frac{dI}{dt}$ occur are called *temporal edges*.

An edge is a geometric feature because it has a specific location within an image array. Here, let us consider spatial edges first.

8.4.2.2 Types of Edges

An image function can have two spatial, first-order derivatives: $\frac{dI}{dv}$ and $\frac{dI}{du}$. Mathematically, these two derivatives can be computed independently. In fact, $\frac{dI}{dv}$ is the image gradient in the vertical direction (column) while $\frac{dI}{du}$

is the image gradient in the horizontal direction (row). For the sake of convenience, let us analyze and detect the image function's discontinuity separately in the vertical and horizontal directions. In this way, we only need to deal with two single-variable image functions: one with spatial variable u, and the other with spatial variable v.

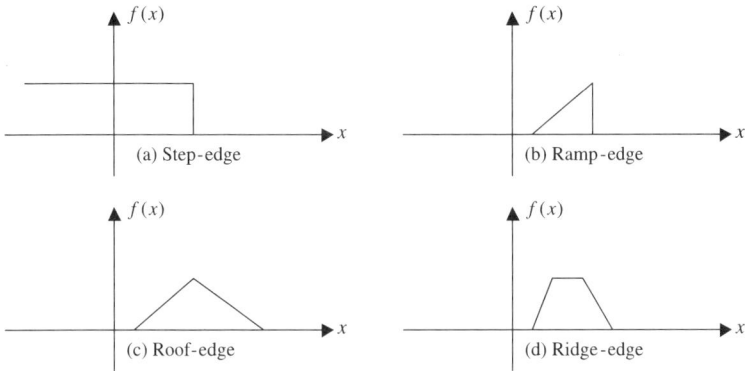

Fig. 8.12   Typical types of edge.

Let $f(x)$ denote a one-dimensional function with spatial variable x. If $f(x)$ is an image function and represents the luminance (or chrominance) of a scene (or object), the discontinuity of $f(x)$ can typically be classified into these four categories, as shown in Fig. 8.12:

- Step-type Edge:
  This edge can be caused by two different colors on an object's surface. For example, printed or hand-written characters exhibit this type of edge. Step-type edges can also be caused by: a) the outline of a shadow on an object's surface, b) impurities on the surface, or c) discontinuity in the depths of two overlapping surfaces.
- Ramp-type Edge:
  A planar surface exposed to light rays from a single point, such as a bulb, will form this type of edge at the surface's boundary.
- Roof-type Edge:
  This edge typically results from the discontinuity of the normal vectors of a single or two adjacent surfaces. For example, the edge separating two adjacent facets of a cube will form a roof-type edge in images.

- Ridge-type Edge:
  A curved surface normally exhibits a ridge-type edge. For example, a cylindrical or conical surface in certain orientations forms a ridge-type edge.

From Fig. 8.12, we can see that a roof-type edge is a combination of two ramp-type edges placed back-to-back. And, a ridge-type edge can be treated as a combination of two ramp-type edges and two step-type edges. If we denote $1(x - x_0)$ a unit-step function at location x_0, we have

$$1(x - x_0) = \begin{cases} 1 & \text{if } x \geq x_0; \\ 0 & \text{if } x < x_0. \end{cases} \qquad (8.42)$$

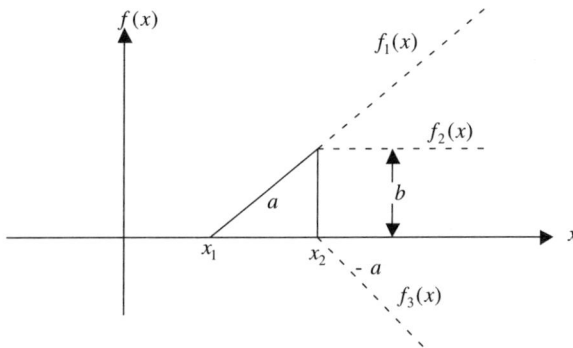

Fig. 8.13   Description of a ramp-type edge.

Refer to Fig. 8.13. If a ramp-type edge has slope a within interval (x_1, x_2), its analytical description will be the combination of $f_1(x)$, $f_2(x)$ and $f_3(x)$, as follows:

$$f(x) = f_1(x) - f_2(x) + f_3(x) \qquad (8.43)$$

with

$$\begin{cases} f_1(x) = a \bullet \int_{x_1}^{x} 1(x - x_1)dx \\[2mm] f_2(x) = b \bullet 1(x - x_2) \\[2mm] f_3(x) = -a \bullet \int_{x_2}^{x} 1(x - x_2)dx \end{cases} \qquad (8.44)$$

and $b = a \bullet (x_2 - x_1)$. Clearly, $f_2(x)$ is a step-type edge at location x_2. This leads to the following interesting observations:

- The step-type edge is the most important edge in an image.
- A good detector of step-type edges is also a good detector of other types of edges.
- The discontinuity of the ramp-edge at location x_1 can be detected with a filter which is the second-order derivative of a continuous function.

8.4.2.3 *A General Scheme for Edge Detection*

As shown in Fig. 8.9, feature detection generally consists of three steps: a) noise reduction, b) feature enhancement, and c) feature selection. Since an image's spatial dimension is two, it is convenient to detect the edges in both the image-plane's vertical and horizontal directions. For a given direction, either vertical or horizontal, an image is treated as a single-variable signal. As a result, the general scheme for edge detection includes two parallel edge-detection processes, as shown in Fig. 8.14:

- Horizontal-Edge Detection:
  The horizontal-edge detection process consists of two steps: a) noise reduction with filter $H_{11}(s)$, and b) horizontal-edge enhancement with derivative filter $H_{12}(s)$ ($H_{12}(s) = s$). Noise reduction, meaning the removal of high-frequency signals, has a smoothing effect on the edges. Since an image is a two-dimensional signal, one can choose to smooth the input image vertically, horizontally or uniformly (in both directions). To avoid the smoothing effect on the horizontal edges, the noise-reduction filter should be applied to the input image horizontally.
- Vertical-Edge Detection:
  Similarly, the vertical-edge detection process consists of two steps: a) noise reduction with filter $H_{21}(s)$, and b) vertical-edge enhancement with derivative filter $H_{22}(s)$ ($H_{22}(s) = s$). To avoid the smoothing effect on the vertical edges, the noise reduction filter should be applied to the input image vertically.
- Selection of Edges:
  If we know the image gradients in both the horizontal and vertical directions, it is possible to compute the norms of the gradients at each pixel location. Based on the norms of a gradient, it is possible to invoke a decision-making process which selects the edges as final output.

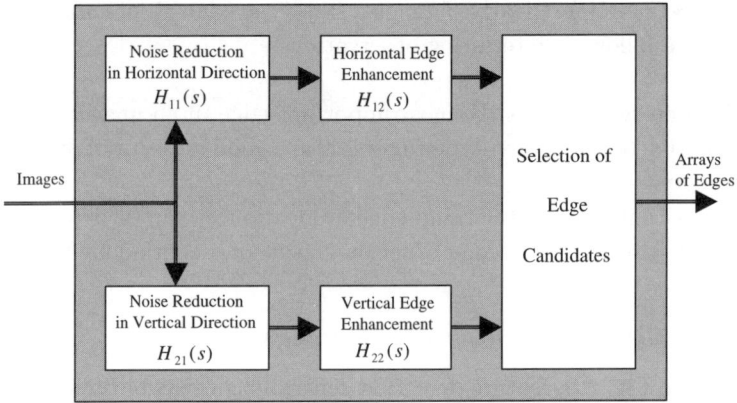

Fig. 8.14 A general scheme for performing edge detection.

8.4.2.4 *Sobel Edge Detector*

A simple and useful edge detector is the *Sobel* edge detector. If the numbers 2 in the convolution kernels are replaced by the numbers 1, the Sobel edge detector becomes the *Prewitt* edge detector. Here, we will only discuss the Sobel edge detector.

Noise Reduction in the Horizontal Direction

Sobel's convolution kernel for noise reduction in the horizontal direction is

$$\{h_{11}(m,n)\} = \begin{bmatrix} 1 & 2 & 1 \\ 1 & 2 & 1 \\ 0 & 0 & 0 \end{bmatrix}. \tag{8.45}$$

Horizontal-Edge Enhancement

In Eq.8.45, differentiating $\{h_{11}(m,n)\}$ with respect to spatial variable v in the vertical direction yields

$$\{h_{11}(m,n) * h_{12}(m,n)\} = \begin{bmatrix} 1 & 2 & 1 \\ 0 & 0 & 0 \\ -1 & -2 & -1 \end{bmatrix}. \tag{8.46}$$

(NOTE: $H_{12}(s) = s$, which represents a derivative operator in the time domain).

The kernel in Eq. 8.46 is Sobel's convolution kernel for the enhancement of an image's horizontal edges.

Horizontal-Edge Selection

Edge selection is a decision-making process which can be simple or complex. Let us denote $I_u(v, u)$ the gradient image of the enhanced horizontal-edges. A simple process for identifying location (v, u) as the horizontal-edge is to test the following conditions:

$$\begin{cases} I_u(v, u) > g_0 \\ I_u(v, u) > I_u(v - 1, u) \\ I_u(v, u) > I_u(v + 1, u). \end{cases} \tag{8.47}$$

The first inequality in Eq. 8.47 means that the first-order derivative at an edge location must be higher than gradient value g_0. This allows us to select edges of higher contrast and to remove the false edges caused by noise. The second and third inequalities mean that a horizontal-edge should be the local maximum in the vertical direction.

Noise Reduction in the Vertical Direction

Sobel's convolution kernel for noise reduction in the vertical direction is

$$\{h_{21}(m, n)\} = \begin{bmatrix} 1 & 1 & 0 \\ 2 & 2 & 0 \\ 1 & 1 & 0 \end{bmatrix}. \tag{8.48}$$

Vertical-Edge Enhancement

In Eq. 8.48, differentiating $\{h_{21}(m, n)\}$ with respect to spatial variable u in the horizontal direction yields

$$\{h_{21}(m, n) * h_{22}(m, n)\} = \begin{bmatrix} 1 & 0 & -1 \\ 2 & 0 & -2 \\ 1 & 0 & -1 \end{bmatrix}. \tag{8.49}$$

(NOTE: $H_{22}(s) = s$, which also represents a derivative operator in the time domain).

The kernel in Eq. 8.49 is Sobel's convolution kernel for the enhancement of an image's vertical-edges.

Vertical-Edge Selection

Let us denote $I_v(v,u)$ the gradient-image of enhanced vertical edges. A simple process for identifying location (v,u) as the vertical-edge is to test the following conditions:

$$\begin{cases} I_v(v,u) > g_0 \\ I_v(v,u) > I_u(v,u-1) \\ I_v(v,u) > I_u(v,u+1). \end{cases} \qquad (8.50)$$

In Eq. 8.50, the first inequality means that the first-order derivative at an edge location must be higher than gradient value g_0. The second and third inequalities mean that a vertical-edge should be the local maximum in the horizontal direction.

Edge Selection for Final Output

A simple way to combine the horizontal and vertical edges is to retain all the edge locations. For edge location (v,u), edge gradient $I_g(v,u)$ will be the norm of the corresponding horizontal and vertical gradients, that is,

$$I_g(v,u) = \sqrt{I_u^2(v,u) + I_v^2(v,u)}. \qquad (8.51)$$

Example 8.7   Fig. 8.15a shows an image array which the size is 10×10.

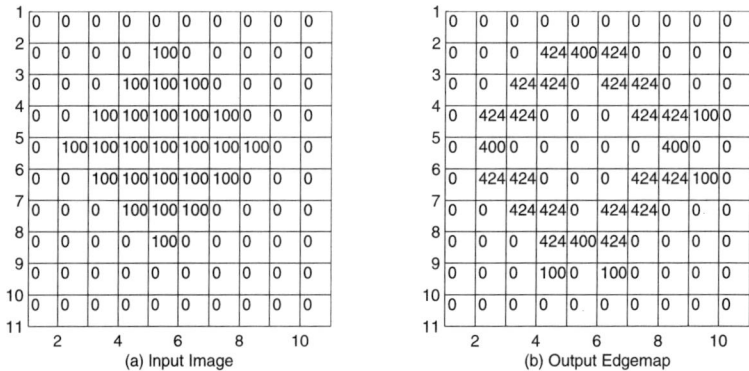

(a) Input Image

0	0	0	0	0	0	0	0	0	0
0	0	0	0	100	0	0	0	0	0
0	0	0	100	100	100	0	0	0	0
0	0	100	100	100	100	100	0	0	0
0	100	100	100	100	100	100	100	0	0
0	0	100	100	100	100	100	0	0	0
0	0	0	100	100	100	0	0	0	0
0	0	0	0	100	0	0	0	0	0
0	0	0	0	0	0	0	0	0	0
0	0	0	0	0	0	0	0	0	0

(b) Output Edgemap

0	0	0	0	0	0	0	0	0	0
0	0	0	424	400	424	0	0	0	0
0	0	424	424	0	424	424	0	0	0
0	424	424	0	0	424	424	100	0	0
0	400	0	0	0	0	400	0	0	0
0	424	424	0	0	424	424	100	0	0
0	0	424	424	0	424	424	0	0	0
0	0	0	424	400	424	0	0	0	0
0	0	0	100	0	100	0	0	0	0
0	0	0	0	0	0	0	0	0	0

Fig. 8.15   Example of edge detection using the Sobel edge detector: a) input of image array and b) output of edge-map.

Let us apply the Sobel edge detector. The intermediate results for both horizontal and vertical edges are shown in Fig. 8.16. The array of the computed gradient norms is shown in Fig. 8.15b. We can see that there are some artifacts.

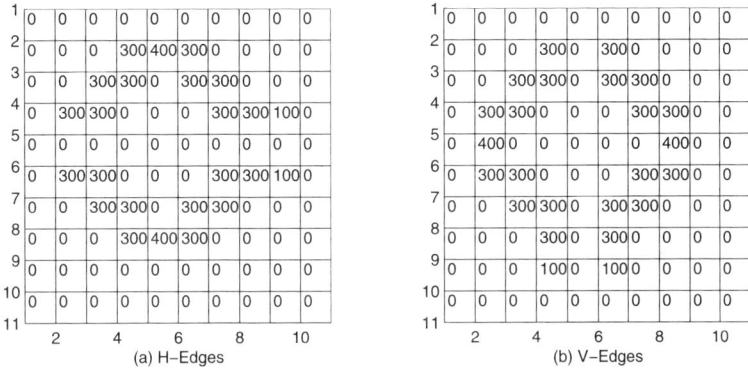

0	0	0	0	0	0	0	0	0	0
0	0	0	300	400	300	0	0	0	0
0	0	300	300	0	300	300	0	0	0
0	300	300	0	0	0	300	300	100	0
0	0	0	0	0	0	0	0	0	0
0	300	300	0	0	0	300	300	100	0
0	0	300	300	0	300	300	0	0	0
0	0	0	300	400	300	0	0	0	0
0	0	0	0	0	0	0	0	0	0
0	0	0	0	0	0	0	0	0	0

(a) H–Edges

0	0	0	0	0	0	0	0	0	0
0	0	0	300	0	300	0	0	0	0
0	0	300	300	0	300	300	0	0	0
0	300	300	0	0	0	300	300	0	0
0	400	0	0	0	0	0	400	0	0
0	300	300	0	0	0	300	300	0	0
0	0	300	300	0	300	300	0	0	0
0	0	0	300	0	300	0	0	0	0
0	0	0	100	0	100	0	0	0	0
0	0	0	0	0	0	0	0	0	0

(b) V–Edges

Fig. 8.16   Results of horizontal and vertical edges detected using the Sobel edge detector

◇◇◇◇◇◇◇◇◇◇◇◇◇◇◇◇◇◇◇

The above example illustrates that, if we want to remove undesirable artifacts, the decision-making process for edge selection is not a simple task.

In his work on edge detection, Canny proposed a hysteresis thresholding method, with two threshold values (g_1, g_2) $(g_1 < g_2)$, which allows us to remove unwanted artifacts. The idea behind this method is to perform *edge tracing*, which starts from a local maximum, the gradient-norm of which is greater than g_2. The edge-tracing process results in edge chains which are groupings of consecutive edges linked together. The tracing of an edge-chain stops when there is no more connected edge with gradient-norm greater than g_1. This process is repeated until all the local maxima, with gradient-norms greater than g_2, have been treated.

8.4.2.5   *Gaussian Edge Detector with Tunable Response and Localization*

The first Gaussian edge detector was proposed by Marr and Hildreth. Their idea did not follow the principle of dynamics resonance. Instead, they advocated the use of second-order derivatives from a Gaussian filter to perform the convolution with an input image. In this way, edges are identified at locations, where the results of the convolution make transitions across zeros. The sum of the second-order derivatives of a Gaussian function with respect to spatial variables (v, u) is called *Laplacian*, that is,

$$\nabla^2 f = \frac{\partial^2 f}{\partial v^2} + \frac{\partial^2 f}{\partial u^2} \qquad (8.52)$$

with

$$f(v, u) = \frac{1}{2\pi\sigma^2} \bullet e^{-\frac{v^2+u^2}{2\sigma^2}}.$$

The Laplacian of Gaussian (or LoG for short), as shown in Eq. 8.52, is invariant to the rotational transformation applied to function $f(v, u)$. Marr-Hildreth's edge detector is based on the detection of zero-crossings of LoG. This is why Marr-Hildreth's edge detector is also called the *zero-crossing* of LoG.

A Gaussian function is a continuously differentiable function and has the advantage of a tunable, smoothing effect for noise reduction. The larger the σ, the stronger the smoothing effect. According to the general scheme for edge detection as shown in Fig. 8.14 and the principle of dynamics resonance, a Gaussian edge detector with tunable response and localization can easily be designed as follows:

Noise Reduction in the Horizontal Direction

To smooth out the input image in the horizontal direction, we choose the following Gaussian function:

$$h_{11}(v, u) = \frac{1}{\sqrt{2\pi\sigma_{11}^2}} \bullet e^{-\frac{u^2}{2\sigma_{11}^2}}. \qquad (8.53)$$

Parameter σ_{11} controls the smoothing effect in the horizontal direction.

Example 8.8   Let us consider the following Gaussian function:

$$f(x) = \frac{1}{\sqrt{2\pi\sigma^2}} \bullet e^{-\frac{x^2}{2\sigma^2}}.$$

Now, let us choose σ to be 0.3, 0.6 and 0.9. The corresponding Gaussian curves are shown in Fig. 8.17a. We can see that the larger σ, the stronger the smoothing effect.

Since an image is an array of pixels, the continuous curve of a Gaussian function must be sampled in order to obtain a discrete convolution kernel. To perform the sampling, one has to specify the sampling range and the number of discrete values. For a Gaussian function, σ is the standard deviation of Gaussian distribution. Here, let us choose the sampling range to be $\pm 2.5\sigma$, which is very close to $\pm 3\sigma$. The number of discrete values depends on the size of the convolution kernel. Here, let us set this number to be 7. Based on these settings, the corresponding discrete convolution kernels are plotted in Fig. 8.17b.

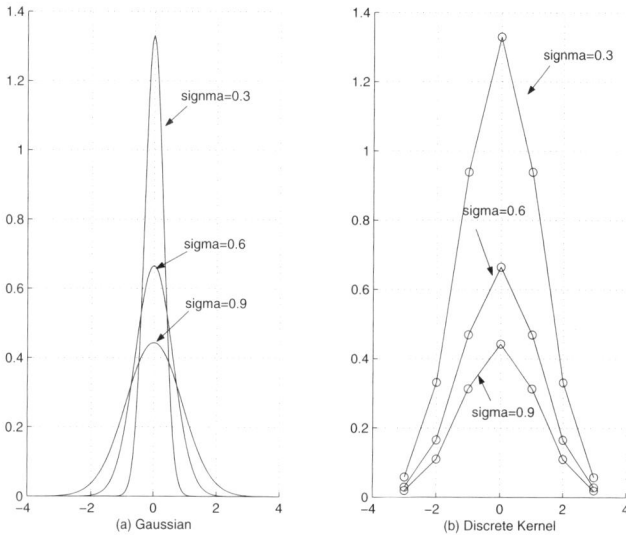

Fig. 8.17 Gaussian curves and the corresponding discrete kernels with different σ values.

◇◇◇◇◇◇◇◇◇◇◇◇◇◇◇◇◇◇◇

Horizontal-Edge Enhancement

After smoothing out the input image in the horizontal direction, it is possible to enhance the horizontal edges with the derivative of a Gaussian function, such as:

$$h_{12}(v + \alpha_v, u) = \frac{-(v + \alpha_v)}{\sqrt{2\pi} \bullet \sigma_{12}^3} \bullet e^{-\frac{(v+\alpha_v)^2}{2\sigma_{12}^2}} . \tag{8.54}$$

Parameter σ_{12} controls the amplitude of the enhanced horizontal-edges, and parameter α_v adjusts the vertical locations of horizontal-edges.

Example 8.9   Consider the following derivative of a Gaussian function:

$$g(x) = \frac{-x}{\sqrt{2\pi} \bullet \sigma^3} \bullet e^{-\frac{x^2}{2\sigma^2}}.$$

Now, let us choose σ as 0.3, 0.6 and 0.9. The corresponding curves of the Gaussian derivative are shown in Fig. 8.18a. We can see that the smaller the σ, the stronger the response of the edges. The corresponding convolution kernels, when we choose a sampling range of $\pm 2.5\sigma$ and a convolution kernel's size of 7, are plotted in Fig. 8.18b.

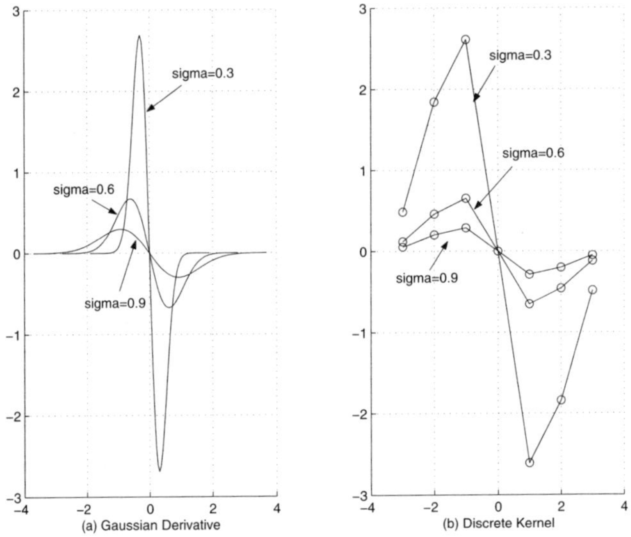

Fig. 8.18   Gaussian derivative curves and the corresponding discrete kernels with different σ values.

◇◇◇◇◇◇◇◇◇◇◇◇◇◇◇◇◇

Noise Reduction in the Vertical Direction

To detect vertical edges, we first smooth out the input image in the vertical direction with a Gaussian function:

$$h_{21}(v, u) = \frac{1}{\sqrt{2\pi\sigma_{21}^2}} \bullet e^{-\frac{v^2}{2\sigma_{21}^2}}. \tag{8.55}$$

Parameter σ_{21} controls the smoothing effect. In practice, we can set $\sigma_{21} = \sigma_{11}$.

Vertical-Edge Enhancement

Subsequently, the vertical edges can be enhanced using the following derivative of a Gaussian function:

$$h_{22}(v, u + \alpha_u) = \frac{-(u + \alpha_u)}{\sqrt{2\pi} \bullet \sigma_{22}^3} \bullet e^{-\frac{(u+\alpha_u)^2}{2\sigma_{22}^2}}. \tag{8.56}$$

Parameter σ_{22} controls the amplitude of the enhanced vertical edges, while parameter α_u adjusts the horizontal locations of vertical edges. In practice, we can set $\sigma_{22} = \sigma_{11}$.

Example 8.10 Consider the following derivative of a Gaussian function:

$$g(x) = \frac{-(x + \alpha)}{\sqrt{2\pi} \bullet \sigma^3} \bullet e^{-\frac{(x+\alpha)^2}{2\sigma^2}}.$$

Let us set σ to be 0.3, and α to be 0.0, 0.05 and 0.1. The corresponding curves of the Gaussian derivative are shown in Fig. 8.19a. The corresponding convolution kernels, if we choose a sampling range of $\pm 2.5\sigma$ and a convolution kernel's size of 7, are plotted in Fig. 8.19b. We can see that parameter α allows us to adjust the convolution kernel's shift. In other words, parameter α controls the localization of the enhanced edges. By controlling parameter α, we can easily achieve a sub-pixel accuracy regarding localization of the enhanced edges.

◇◇◇◇◇◇◇◇◇◇◇◇◇◇◇◇◇◇

8.4.2.6 *Canny and Deriche Edge Detectors*

If a function is smooth and behaves like a low-pass filter, its derivative will enhance the edges in an image. A low-pass filter in the frequency domain will have more poles than zeros. For example, the following Laplace transform:

$$\frac{s + a}{(s + a)^2 + \omega^2}$$

will behave like a low-pass filter. Not surprisingly, its inverse Laplace transform is:

$$e^{-at} \cos(\omega t).$$

Similarly, the inverse of the following Laplace transform:

$$\frac{\omega}{(s + a)^2 + \omega^2}$$

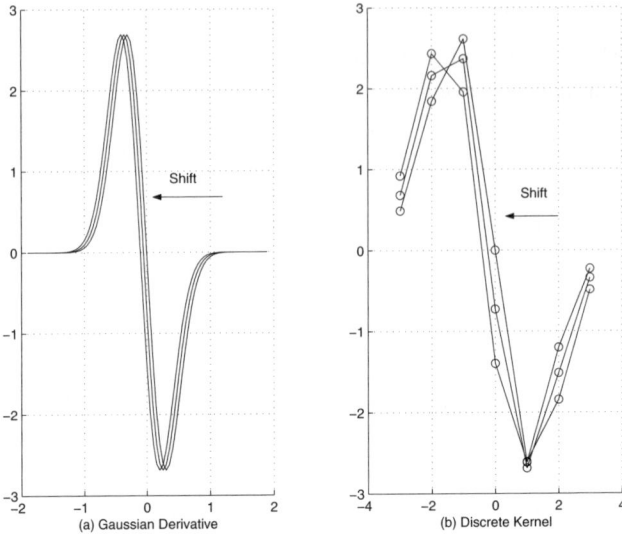

Fig. 8.19   Shift effect on the curves of a Gaussian derivative and the corresponding discrete kernels with different values of α.

is

$$e^{-at}\sin(\omega t).$$

The above observation suggests that a smooth and differentiable function's derivative can be used to enhance edge features in an image.

Nowadays, there are two widely-used edge detectors: a) the Canny method and b) the Deriche method. Here, we will not cover the details of these two methods, but instead, simply explain their functions for edge enhancement.

Canny's Function for Edge Enhancement

Canny's function for edge enhancement is as follows:

$$f(x) = \begin{cases} g(x) & \text{if } x \in [-x_0, 0] \\ -g(-x) & \text{if } x \in [0, x_0] \\ 0 & \text{otherwise} \end{cases} \qquad (8.57)$$

where

$$g(x) = [a_1 \sin(\omega x) + a_2 \cos(\omega x)]e^{-\sigma x} + [a_3 \sin(\omega x) + a_4 \cos(\omega x)]e^{\sigma x} - \lambda \qquad (8.58)$$

and $\lambda = a_2 + a_4$. Conceptually, function $g(x)$ in Eq. 8.58 is very similar to the derivative of the following function:

$$a \bullet e^{-\sigma x} \sin(\omega x) + b \bullet e^{\sigma x} \sin(\omega x).$$

Example 8.11   In Canny's function, we set

$$(a_1, \ a_2, \ a_3, \ a_4) = (0.15, \ -0.15, \ -1.0, \ -1.0)$$

and $\omega = 1.5$. Fig. 8.20a shows the curves of Canny's function in interval $[-4, 4]$ (i.e. $x_0 = 4$) when we set σ as 0.1 and 0.3 respectively. The corresponding discrete kernels, if we choose the convolution kernel's size as 7, are plotted in Fig. 8.20b.

From this example, we can see that the smaller the σ, the stronger the response of the edges.

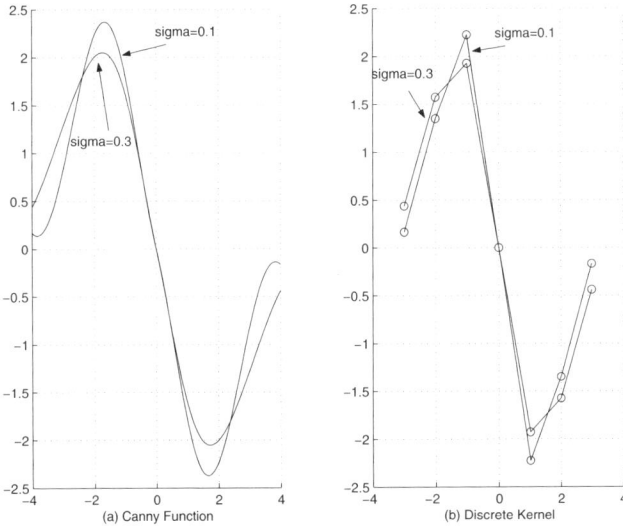

Fig. 8.20   Curves of Canny's function and the corresponding discrete kernels with different values of σ.

◇◇◇◇◇◇◇◇◇◇◇◇◇◇◇◇◇

Deriche's Function for Edge Enhancement

Deriche's function for edge enhancement is as follows:

$$f(x) = \begin{cases} g(x) & \text{if } x < 0 \\ -g(-x) & \text{if } x \geq 0 \end{cases} \tag{8.59}$$

where

$$g(x) = -[b_1 \sin(\omega x)]e^{\sigma x}. \tag{8.60}$$

Conceptually, by setting $(a_1, a_2, a_4) = (0, 0, 0)$ and $b_1 = -a_3$, Canny's function becomes Deriche's function. Since function $f(x)$ in Eq. 8.59 decays to zero when variable x approaches infinity, spatial domain $[-x_0, x_0]$ in Eq. 8.57 can be extended to infinity.

Example 8.12   In Deriche's function, we set

$$(b_1, \ \omega) = (2.0, \ 1.0).$$

Fig. 8.21a illustrates the curves of Deriche's function in interval $[-4, 4]$ when σ equals 0.5 and 1.0 respectively. The corresponding discrete kernels, if we choose a convolution kernel's size as 7, are plotted in Fig. 8.21b.

    Again, we can see that the smaller the σ, the stronger the response of the edges.

◇◇◇◇◇◇◇◇◇◇◇◇◇◇◇◇◇◇◇

8.4.2.7 *Summary of Edge-Detection Algorithm*

Refer to Fig. 8.14. The algorithm implementing the Gaussian, Canny or Deriche edge detector will consist of the following steps:

- Step 1:
  Smooth the input image horizontally with convolution kernel $\{h_{11}(m, n)\}$ computed from a Gaussian function.
- Step 2:
  Enhance the horizontal edges with convolution kernel $\{h_{12}(m, n)\}$ computed from Canny's function, Deriche's function or the derivative of a Gaussian function.
- Step 3:
  Smooth the input image vertically with convolution kernel $\{h_{21}(m, n)\}$ computed from a Gaussian function. By default, we can choose $\{h_{21}(m, n)\} = \{h_{11}(m, n)\}^t$.

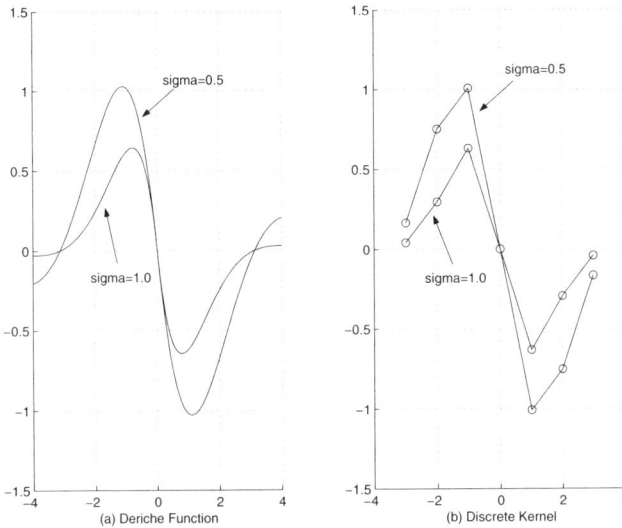

Fig. 8.21   Curves of Deriche's function and the corresponding discrete kernels with different values of σ.

- Step 4:
  Enhance the vertical edges with convolution kernel $\{h_{22}(m, n)\}$ computed from Canny's function, Deriche's function or the derivative of a Gaussian function. By default, we can choose $\{h_{22}(m, n)\} = \{h_{12}(m, n)\}^t$.
- Step 5:
  Compute the gradient norms at all pixel locations in order to obtain a *gradient image*.
- Step 6:
  Select the edges as final output using hysteresis thresholding over the gradient image. This step is also known as *nonmaximum suppression*.

Example 8.13   Fig. 8.22a shows an input image.   For the purpose of comparison, let us compute the edge gradients using the Gaussian, Canny and Deriche edge detectors.   The convolution kernels for these three edge detectors are as follows:

- Convolution Kernel for Noise Reduction:
  To reduce noise, we create a discrete convolution kernel with a Gaussian

function. By setting $\sigma = 0.3$ and the kernel size to be 5, we obtain

$$\{h_{11}(m,n)\} = \{h_{21}(m,n)\}^t = [0.0584, 0.6088, 1.3298, 0.6088, 0.0584].$$

- Discrete Convolution Kernel of a Gaussian Derivative:
  To enhance the edges in both the horizontal and vertical directions, we create a discrete convolution kernel from a Gaussian derivative function. By setting $\sigma = 0.3$, we have

$$\{h_{12}(m,n)\}^t = \{h_{22}(m,n)\} = [0.4869, 2.5368, 0, -2.5368, -0.4869].$$

- Discrete Convolution Kernel of Canny's Function:
  We can also use the discrete convolution kernel computed from Canny's function to enhance the horizontal and vertical edges. By setting $\sigma = 0.3$, $\omega = 1.5$, and the kernel size to be 5, we obtain

$$\{h_{12}(m,n)\}^t = \{h_{22}(m,n)\} = [0.4376, 2.0028, 0, -2.0028, -0.4376].$$

- Discrete Convolution Kernel of Deriche's Function:
  By setting $\sigma = 1$, $\omega = 1$, and the kernel size to be 5, the discrete convolution kernel computed from Deriche's function for the enhancement of the horizontal and vertical edges will be

$$\{h_{21}(m,n)\}^t = \{h_{22}(m,n)\} = [0.0407, 0.4860, 0, -0.4860, -0.0407].$$

The computed gradient-images are shown in Fig. 8.22b, Fig. 8.22c and Fig. 8.22d respectively. We can see that there isn't any significant difference among the three methods.

◇◇◇◇◇◇◇◇◇◇◇◇◇◇◇◇◇◇

Example 8.14   The gradient-images in the above example indicate the possible presence of edges. A final step is to make a decision on whether a pixel location is an edge or not. The hysteresis thresholding method is a better way of retaining the local maxima in a gradient-image, and removing the non-maxima. Fig. 8.23 shows the final output from a Deriche edge detector.

◇◇◇◇◇◇◇◇◇◇◇◇◇◇◇◇◇◇

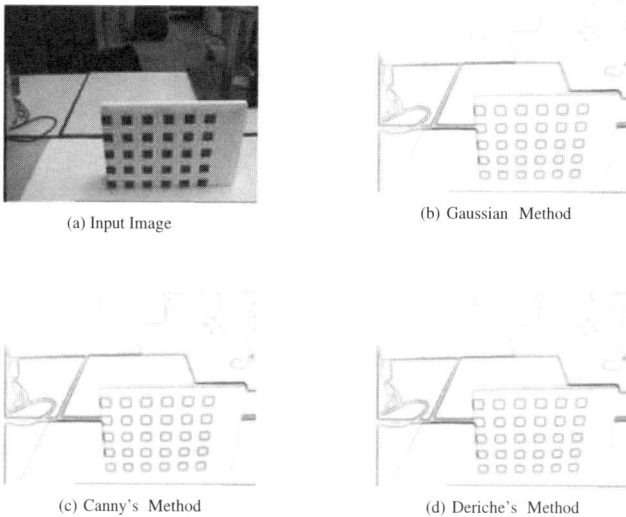

(a) Input Image

(b) Gaussian Method

(c) Canny's Method

(d) Deriche's Method

Fig. 8.22   Results of edge gradients computed from Gaussian, Canny and Deriche edge detectors.

8.4.3   *Corner Detection*

The set of edges in Example 8.14 is a geometric representation of the original input image shown in Fig. 8.22a. Theoretically, all high-level descriptions of a scene (or object) can be obtained from low-level edge features. But, in practice, it is not easy to analyze an array of edges. We will discuss this issue further in the section on feature description.

When there is an array of edges, the term *contour* can be defined as follows:

Definition 8.5   A contour is a set of interconnected edges in an image array.

Obviously, the discontinuities of curvature along a contour indicate particular points, such as corners and junctions. Because of the smoothing effect caused by an edge detector, a corner may not appear sharp. (Refer to the results in Fig. 8.23). Therefore, it is interesting to know whether it is possible to detect directly from an input image, the image feature known as a *corner* or *junction*.

Fig. 8.23   Final output of edges from a Deriche edge detector.

8.4.3.1   *Definition of Corners*

If we take the interconnected edges as input, a corner can be defined as a
location where the curvature of a contour is a local maximum. According
to this definition, corner detection directly depends on edge detection. Due
to curvature computation, the detection of corners from edges should be
considered as part of feature description. Here, we are interested in the
question of how to detect corners from an intensity image directly without
invoking any analytical description of contours.

At a time-instant, an intensity image is a two-dimensional spatial function. In a continuous space, an image function can be treated as a surface
in a spatial domain defined by the image coordinates (v, u). If an image surface can be described by a smooth and differentiable function, it is
possible to compute the curvature at each location (v, u), and along any
direction. In particular, the curvatures computed along two orthogonal directions (e.g. along the u and v axes) are called *principal curvatures*. Based
on the concept of image surface, a corner can be defined as a location where
the principal curvatures of the image surface are the local maxima.

If $f(x)$ is a single-variable function which is smooth and differentiable, the curvature of $f(x)$ at location x is

$$\kappa = \frac{\|f''(x)\|}{[\sqrt{1 + (f'(x))^2}]^3} \tag{8.61}$$

where $f'(x) = \frac{df(x)}{dx}$ and $f''(x) = \frac{d^2 f(x)}{dx^2}$. In order to compute the principal curvatures, it is necessary for the image function to be smooth and differentiable up to the second-order. This is a very stringent requirement because an image is digital and is not noise-free. And, a curvature-based corner detector is thus very sensitive to noise.

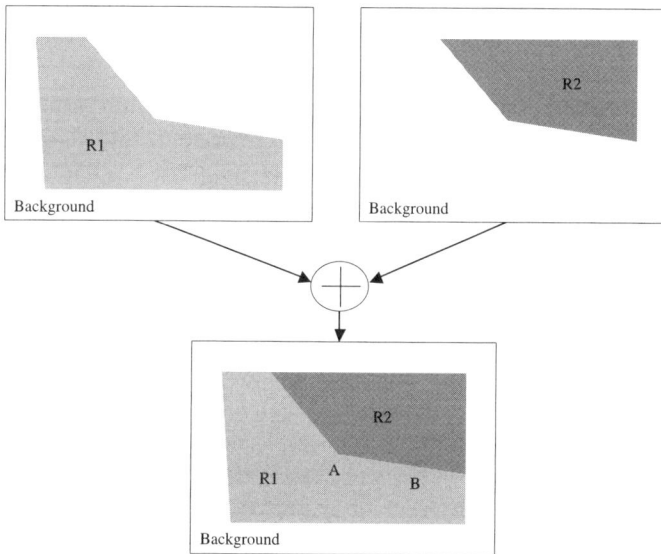

Fig. 8.24   Illustration of corner definition.

An alternative definition of corner which does not require the computation of second-order derivatives of an image function is as follows:

Definition 8.6   Location (v, u) is a corner if and only if the following conditions hold:

- Condition 1: The location (v, u) is an edge.
- Condition 2: The gradient at (v, u) is zero in at least one direction. We call this direction *zero-gradient direction*.
- Condition 3: There is no edge in the vicinity of the zero-gradient direction.

Fig. 8.24 illustrates the validity of this definition. If we examine the contour separating regions R_1 and R_2, we can see that the corner at location A satisfies all the conditions for this definition. However, while all the edges satisfy Conditions 1 and 2, a non-corner edge, such as the one at location B in Fig. 8.24 does not meet Condition 3 because a smoothed contour locally looks like a line segment.

8.4.3.2 *Zero-gradient Corner Detector*

A corner-detection algorithm depends on the definition of a corner. If a corner is considered the location where the principal curvatures are the local maxima, a corresponding corner detector must compute the curvature-related measurement.

Here, we present a simple corner-detection algorithm which uses the corner definition stated in the above section. We call this algorithm *zero-gradient corner detector*. It works as follows:

Implementation of Condition 1

As we stated in the above definition of corner, the first condition is that a corner must be an edge as well. In order to measure about the edge at a pixel location, we simply filter an input image with a set of directional convolution masks. Let $\{h_\theta(m,n)\}$ denote a convolution mask which enhances the edges in the direction of θ. For example, we can choose $\{h_0(m,n)\}$ to be:

$$\{h_\theta(m,n)\,|_{\theta=0}\} = \begin{bmatrix} 0 & 0 & 0 & 0 & 0 \\ 0 & 0 & 0 & 0 & 0 \\ 0.4869 & 2.5368 & 0 & -2.5368 & -0.4869 \\ 0 & 0 & 0 & 0 & 0 \\ 0 & 0 & 0 & 0 & 0 \end{bmatrix}. \qquad (8.62)$$

In Eq. 8.62, the matrix's row 3 is a convolution kernel computed from a Gaussian derivative function. If we rotate convolution kernel $\{h_0(m,n)\}$ with rotation angle θ, we obtain the corresponding convolution kernel for the enhancement of edges in the direction of θ. For example, we can choose θ to be 45^0, 90^0 and 135^0 in order to obtain convolution kernels $\{h_{45}(m,n)\}$, $\{h_{90}(m,n)\}$ and $\{h_{135}(m,n)\}$.

Now, let $G_\theta(v,u)$ denote the array of enhanced edges in the direction of θ. Assume that direction θ takes a value from the set $[\theta_1, \theta_2, ..., \theta_k]$. Pixel location (v, u) will be declared an edge if the following condition holds:

$$G_\theta(v, u) > g_0, \text{ for any } \theta \text{ in } [\theta_1, \theta_2, ..., \theta_k] \qquad (8.63)$$

where g_0 is a threshold value on the directional gradient.

Implementation of Condition 2

Because of noise and the discrete nature of an input image, it is rare to obtain the zero gradients perfectly across a uniform region of an image. Therefore, a zero-gradient should be interpreted as a low-gradient value. It is necessary to define a reference value below which a gradient value is judged to be low (or zero). One way to specify a reference value is to choose a default value g_1.

Let θ_{min} denote the direction in which directional gradient $G_{\theta_{min}}(v, u)$ at (v, u) is considered zero. In practice, we can implement this condition as follows:

$$G_{\theta_{min}}(v, u) < G_\theta(v, u), \quad \text{for all } \theta \text{ in } [\theta_1, \theta_2, ..., \theta_k] \text{ except } \theta_{min}. \qquad (8.64)$$

Eq. 8.64 interprets a low-gradient value as anything that is not a high-gradient value. In this way, it is not necessary to introduce threshold value g_1.

Implementation of Condition 3

As we described earlier, the third condition states that there should be no edge in the vicinity of the zero-gradient direction θ_{min}. In a discrete space, a vicinity in direction θ_{min} can be specified by two intervals of indexes: $[-i_2, -i_1]$ and $[i_1, i_2]$ $(i_1 < i_2)$. On the array of an input image, we can determine the pixel locations corresponding to intervals $[-i_2, -i_1]$ and $[i_1, i_2]$, if θ_{min} and (v, u) are known.

Let D_- and D_+ denote the sets of pixel locations corresponding to intervals $[-i_2, -i_1]$ and $[i_1, i_2]$ respectively. Then, we have

$$\begin{cases} D_- = \{(v + i \bullet \sin\theta_{min}, \ u + i \bullet \cos\theta_{min}), \ \ \forall i \in [-i_2, -i_1]\} \\[2mm] D_+ = \{(v + i \bullet \sin\theta_{min}, \ u + i \bullet \cos\theta_{min}), \ \ \forall i \in [i_1, i_2]\}. \end{cases} \qquad (8.65)$$

If we know the vicinity specified by D_- and D_+, Condition 3 can be interpreted as follows:

$$G_{\theta_\perp}(v_1, u_1) < g_0, \ \ \forall(v_1, u_1) \in D_- \text{ and } \forall(v_1, u_1) \in D_+ \qquad (8.66)$$

where $\theta_\perp = \theta_{min} + 90^0$. This means that in the vicinity of a corner candidate, we only test the gradients computed in the direction perpendicular to θ_{min}.

Summary of Zero-gradient Corner Detector

The zero-gradient corner detector does not involve any computation of curvature-related measurements. In addition, the decision-making process for verifying the three conditions is simple. The zero-gradient corner detector, if we choose two pairs of orthogonal directions: $(0^0, 90^0)$ and $(45^0, 135^0)$, is illustrated in Fig. 8.25.

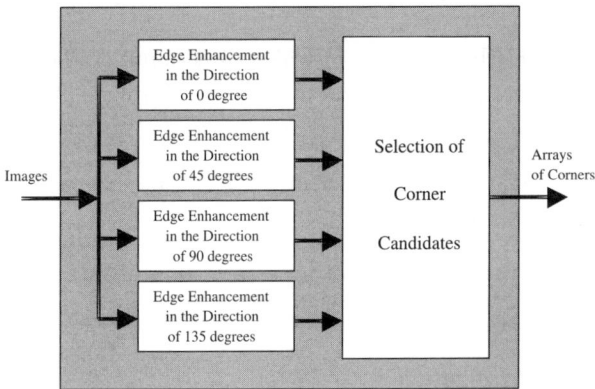

Fig. 8.25   Illustration of the zero-gradient corner detector.

The major steps in the zero-gradient corner detection algorithm are:

- Step 1: Compute the directional gradients with a set of convolution kernels.
- Step 2: Scan the image array to locate the edges (Condition 1).
- Step 3: For each edge candidate, determine the zero-gradient direction (Condition 2).
- Step 4: For each corner candidate, verify Condition 3. If confirmed, accept the candidate as a corner.
- Repeat Steps 3 and 4.

Example 8.15   The zero-gradient corner detector has three parameters: g_0 and (i_1, i_2). If we have an input image, such as is shown in Fig. 8.26a, we set $(g_0, i_1, i_2) = (90, 2, 5)$. The response of the corners is shown in

Fig. 8.26b. We can see that the result is reasonably good. At an actual corner, there may be multiple responses which are closely clustered together. A simple grouping by connectivity will cluster the multiple responses together to form a single output.

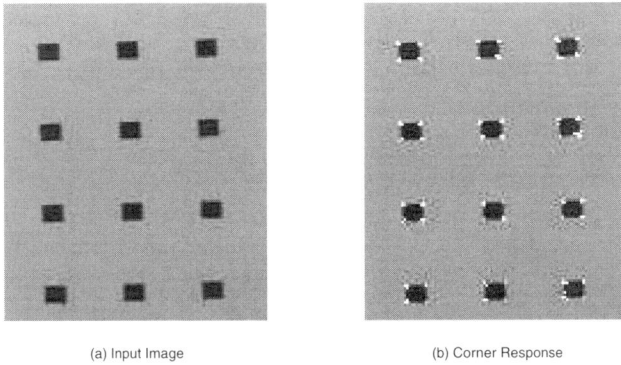

(a) Input Image   (b) Corner Response

Fig. 8.26   Results of corner detection.

◇◇◇◇◇◇◇◇◇◇◇◇◇◇◇◇◇◇

Example 8.16   Fig. 8.27 illustrates another example of corner detection. The parameters $(g_0,\ i_1,\ i_2)$ are set as $(90,\ 2,\ 5)$. The response of the corners is shown in Fig. 8.27b. The result is reasonably good. As for the multiple responses at an actual corner, they can easily be grouped together to form a single output.

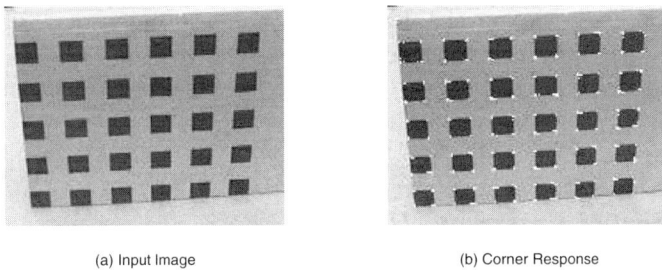

(a) Input Image   (b) Corner Response

Fig. 8.27   Results of corner detection.

◇◇◇◇◇◇◇◇◇◇◇◇◇◇◇◇◇◇

8.4.3.3   *Other Corner Detectors*

There have been many studies on corner detection. A large group of corner-detection algorithms are based on the measurement related to the curvature of contours or intensity surfaces. This category of algorithms uses the following definition of corner:

Definition 8.7   Locations where the curvatures of contours or intensity surfaces are the local maxima are considered corners.

One representative algorithm from this category, proposed independently by the teams Harris-Stephens and Forstner-Gulch, is the Plessey corner detector. Since the curvature measurement requires the second-order derivatives, it is necessary for a corner detector from this category to invoke a noise-reduction process. This makes it and all corner-detection algorithms computationally expensive.

One interesting corner detector is the SUSAN algorithm, proposed by the team Smith-Brady. This algorithm uses a circular disk to scan an input image, row-by-row and column-by-column. At each scanned location, the intensity values of the pixels within the disk are compared with the intensity value underneath the center of the disk (also known as the nucleus). The output of this circular disk at this location is the number of pixels with intensity values similar to that of the nucleus. This number is called the *Univalue Segment Assimilating Nucleus* or, *USAN* for short. A location is considered a corner if the USAN at that location is a local maximum. (In other words, it has the smallest USAN, also called SUSAN). Testing for the similarity among intensity values is a complicated issue. If a simple criterion is used, such as the difference between two intensity values, the algorithm may not be resistant to noise. If a sophisticated criterion is used, such as the exponential of the difference between two intensity values, the algorithm will be computationally expensive.

8.4.4   *Spatial Uniformity Detection*

An image is a spatio-temporal function. It not only exhibits discontinuity in both time and spatial domains but also uniformity, in terms of chrominance, luminance and texture. Therefore, uniformity detection is an interesting and important issue in image-feature extraction. We will first study spatial-uniformity detection, and then temporal-uniformity detection.

8.4.4.1 *Definition of Image Regions*

In a scene, uniform surfaces under uniform lighting conditions will create a uniform appearance in certain areas or regions on a camera's image plane. For the sake of simplicity, let us consider the case of the intensity image and the spatial domain defined by image coordinates (v, u). If $I(v, u)$ denotes an intensity image, the simplest definition of an image region is the set of pixel locations $\{(v_i, u_i), \ i = 1, 2, ...\}$ where image function $I(v, u)$ is equal to L. In other words, image region R_L having intensity value L will be

$$R_L = \{(v_i, u_i) \mid I(v_i, u_i) = L, \ \forall v_i \in [1, r_y], \ \forall u_i \in [1, r_x]\}. \qquad (8.67)$$

This way of defining an image region has two drawbacks:

- First of all, an image is not noise-free. Moreover, because of variations in lighting and viewing directions in a scene, a uniform surface does not have a uniform appearance in an image.
- Secondly, two separate surfaces in a scene may exhibit very similar appearances in a camera's image plane. Pixels having similar intensity values may not correspond to a same surface in a scene.

As noise in an image is random, we should treat image function $I(v, u)$ as a random variable. Without any *a priori* knowledge, we can reasonably assume that random variable $I(v, u)$ follows a Gaussian distribution. Thus, the intensity distribution of a uniform image-area should have a mean of μ_L and a variance of σ_L^2. Accordingly, an image region can be defined as follows:

Definition 8.8 An image region, described by intensity distribution $N(\mu_L, \sigma_L^2)$, consists of those interconnected pixel locations where the following condition holds:

$$R_L = \{(v_i, u_i) \mid \|I(v_i, u_i) - \mu_L\| < k \bullet \sigma_L, \ \forall v_i \in [1, r_y], \ \forall u_i \in [1, r_x]\} \quad (8.68)$$

with $k = 3$ (or other value).

If the input is a color image, then the image region refers to an area of a specific color. In this case, both mean μ and variance σ^2 are three-dimensional vectors.

8.4.4.2 *A General Scheme for Uniformity Detection*

According to the above definition of an image region, uniformity detection is purely a decision-making process (or nonlinear filtering). This is because no linear filter is able to enhance one region while at the same time, reducing all the other regions.

In general, there are two big challenges with using a decision-making process:

- Uncertainty:
  Decision-making is a process which associates inputs with outcomes. Possible inputs include sensory data, fact, hypothesis, context, or goal. And, possible outcomes can be an action, statement, confirmation, or rejection. If there is uncertainty in the representation (or description) of the input or outcome, decision-making is a difficult task. Uncertainty can be caused by: a) noise in the sensory data, b) incomplete sensory data, c) ambiguities in the formulation of facts, goals, hypothesis and contexts, and d) ambiguities in the description of the desired outcome.
- Redundancy:
  In our study of robot kinematics, we know that there are multiple solutions for inverse kinematics. This indicates that there are multiple pathways for associating an input with an outcome. In a robot's visual-perception system, redundancy occurs as well. For example, under different lighting conditions, certain colored surfaces (e.g. red) will appear numerically different in a camera's image plane. In this case, a similar outcome can be associated with a set of different inputs. On the other hand, sometimes there can be multiple solutions (or interpretations) for the same input. This in turn makes decision-making a difficult job. Statistical inference is a powerful tool to cope with uncertainty because it effectively deals with probability (i.e. the likelihood of an observation). When dealing with possibility (i.e. the feasibility of a solution or mapping) , it is necessary to adopt other useful tools such as a Neural Network or Fuzzy Logic.

A feature detector generally consists of three steps: a) noise reduction, b) feature enhancement, and c) feature selection. For edge detection, feature enhancement is the most important step. However, for uniformity detection, the most important step is the decision-making process of feature selection.

As shown in Fig. 8.28, a uniformity detection algorithm involves the following modules:

- Selection of Seed Regions:
  A decision-making process can be one or a combination of the following

three: a) knowledge-driven, b) experience-driven and c) expectation-driven. Generally speaking, for uniformity detection, there is no *a priori* knowledge about intensity (or chrominance) distribution $N(\mu, \sigma^2)$ of a region (or color) in a given input image. The specification of $N(\mu, \sigma^2)$ can be done either manually (by expectation) or automatically (by experience). Given an input image, a user can manually delimit a region and use it to compute the corresponding $N(\mu, \sigma^2)$. An alternative way to specify the possible regions is to subdivide an input image into an array of regular partitions. Among the partitions, those having the smallest variances, in term of luminance (or chrominance), are treated as *seed regions*. Distribution $N(\mu, \sigma^2)$ of a seed region defines a possible region to be detected from an input image.

- Storage of Seed Regions:
  If a region has a generic meaning regardless of the input images, it can be treated as knowledge. For example, a region corresponding to a red color has the generic meaning of "being red". This knowledge can be stored and reused in the future. A neural network, or associative memory, is a good mechanism for storing generic information or knowledge.
- Classification of Pixels:
  When we know a set of seed regions, the next step is to classify the pixels in an input image into groups corresponding to these seed regions. This can be done by testing the condition in Eq. 8.68 explicitly, or by using a classifier implicitly.
- Grouping of Classified Pixels:
  As we already mentioned, an input image may contain multiple regions with the same distribution $N(\mu, \sigma^2)$. It is, therefore, necessary to perform a simple process such as the grouping by connectivity to cluster the pixels belonging to the same region.

In general, there is no generic representation to codify luminance-related information. In other words, any attempt at specifying a generic representation of luminance, such as darkness, brightness or dimness, will create ambiguity.

However, this is not the case with color. Colors can be classified and codified using a generic representation or symbol (e.g. the conceptual symbols such as red, green, blue, yellow, pink, cyan, orange, etc). Obviously, the colored seed regions can be stored and reused. In the following section, we will study a general solution for detecting colored regions. By removing

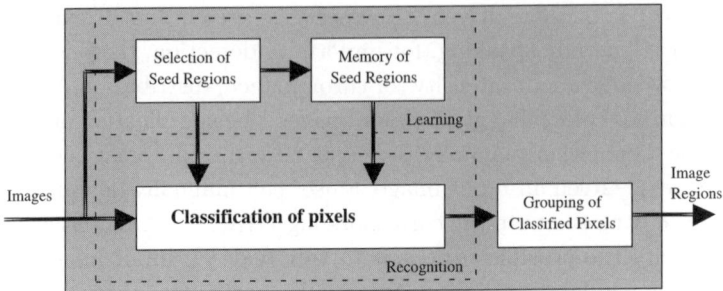

Fig. 8.28   A general scheme for uniformity detection.

the chrominance from a color, we can obtain the luminance. As a result, the solution to detecting colored regions also applies to detecting uniformity, which takes an intensity image as input.

8.4.4.3   *Color Detection*

Color is the brain's physiological and psychological interpretation of light. Physically, a color is the measurement of a light's chrominance and luminance. The chrominance characterizes the property of "pure color" (i.e. a psychological sensation), while the luminance characterizes the property of energy, or the absolute intensity (i.e. the amount of photons entering a unit area).

*L*a*b Color Space*

Color measurement by an electronic device, such as a camera, depends on the device's characteristics. Just as when we measure an object's location, we use a coordinate system as a reference. If we want an unbiased color measurement, it is necessary to have a reference. And, it is necessary to define a color space which takes into account that reference. A common reference, for color measurement, is the "white" color. So far, there isn't an ideal color space which satisfies all needs in terms of color measurement, display, and artistic visual effects. However, the most widely-adopted color space for color measurement is the *L*a*b* color space.

Let us consider a color camera and its associated image digitizer as an electronic device for color measurement. Output will be in the form of three two-dimensional arrays of pixels which correspond to the Red, Green, and Blue component colors respectively. At pixel location (v, u), the triplet

$(I_R(v, u), I_G(v, u), I_B(v, u))$ is called the *RGB tristimulus values*. If the RGB tristimulus values undergo the following transformation:

$$\begin{pmatrix} X \\ Y \\ Z \end{pmatrix} = \begin{pmatrix} 0.607 & 0.174 & 0.200 \\ 0.299 & 0.587 & 0.114 \\ 0.0 & 0.066 & 1.116 \end{pmatrix} \bullet \begin{pmatrix} I_R(v, u) \\ I_G(v, u) \\ I_B(v, u) \end{pmatrix} \qquad (8.69)$$

and

$$\begin{cases} L(v, u) = 25 \bullet \left\{ \frac{100 \bullet Y}{Y_{max}} \right\}^{1/3} - 16 \\[2ex] a(v, u) = 500 \bullet \left\{ [\frac{X}{X_{max}}]^{1/3} - [\frac{Y}{Y_{max}}]^{1/3} \right\} \qquad (8.70) \\[2ex] b(v, u) = 200 \bullet \left\{ [\frac{Y}{Y_{max}}]^{1/3} - [\frac{Z}{Z_{max}}]^{1/3} \right\} \end{cases}$$

where $(X_{max}, Y_{max}, Z_{max})$ are the results obtained using white as reference color, and triplet $\{L(v, u), a(v, u), b(v, u)\}$ are the *L*a*b tristimulus values* in L*a*b color space.

In L*a*b color space, the L axis measures the relative luminance of a color. In other words, the L value is proportional to a color's *brightness* with reference to white. The negative L axis indicates the darkness while the positive L axis measures the whiteness. However, the a and b axes measure a color's relative chromaticity. More precisely:

- The positive a axis measures the redness of a color.
- The negative a axis measure the greenness of a color.
- The positive b axis measures the yellowness of a color.
- The negative b axis measures the blueness of a color.

L*a*b color space is both numerically and perceptually uniform. This means that a numerical difference between two colors at any location in L*a*b color space causes the same sensation of color difference.

Representation of Colored Seed Regions

When we have a seed region in an RGB image as input, we first convert it using Eq. 8.69 and Eq. 8.70 into the corresponding L*a*b image. Then, we can compute the means and variances corresponding to the triplet $\{L(v, u), a(v, u), b(v, u)\}$. If μ_c and σ_c^2 denote the vectors of mean and variance, we have

$$\mu_c = \begin{pmatrix} \mu_L \\ \mu_a \\ \mu_b \end{pmatrix} \quad \text{and} \quad \sigma_c^2 = \begin{pmatrix} \sigma_L^2 \\ \sigma_a^2 \\ \sigma_b^2 \end{pmatrix}. \tag{8.71}$$

Accordingly, a colored seed region is represented by a normal distribution $N(\mu_c, \sigma_c^2)$ if the variations in its chrominance and luminance follow a Gaussian distribution.

Architecture of a RCE Neural Network

When we know the representation of the seed regions, the next question is: How do we design a memory to store seed regions and a classifier capable of recognizing image regions similar to the seed regions? A simple answer is to use an appropriate neural network which can effectively deal with both the uncertainty of input data and the redundancy of solutions.

The RCE (Restricted Coulomb Energy) neural network, developed by Reilly *et al.* in the early 1980s, is a specific design of the hyper-spherical classifier. A RCE neural network can serve as a general purpose classifier for adaptive pattern recognition. It is suitable not only for classifying separable classes, but also for recognizing non-separable classes. Moreover, an RCE neural network can serve as an active-memory mechanism and is able to perform fast and incremental learning, if necessary.

As shown in Fig. 8.29, an RCE neural network consists of three layers of neurons:

- Input Layer:
  This layer contains neuron cells that store the *feature vector* of input data (or training samples). In the case of colored seed regions, the input layer has three cells: one for mean value μ_L, one for mean value μ_a, and one for mean value μ_b.
- Prototype Layer:
  This layer contains a large number of neuron cells which are grouped into a set of clusters. Each cluster is mapped to its corresponding output. In fact, a neuron cell in a cluster represents one feasible way of mapping between cells in the input layer and corresponding cells in the output layer. Clearly, an RCE neural network is a good tool for dealing with the redundancy of solutions.
- Output Layer:
  This layer contains a number of cells, each of which represents a generic

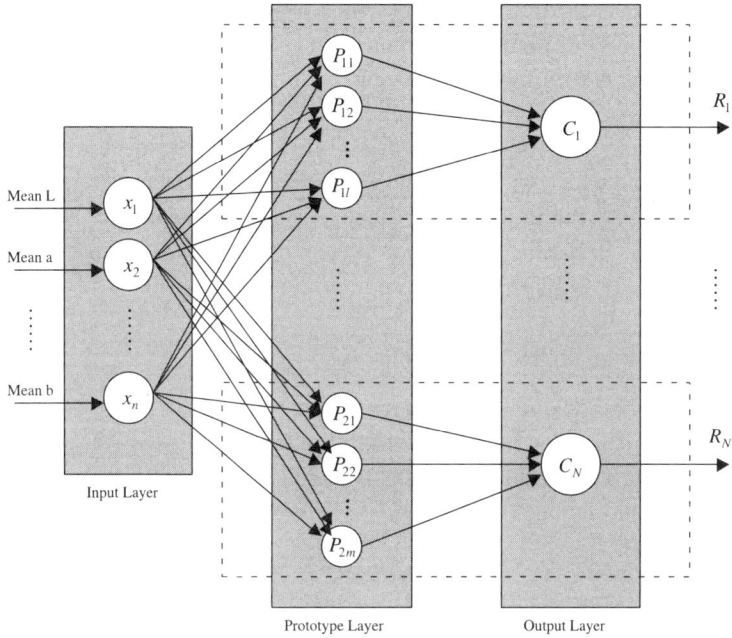

Fig. 8.29   Architecture of an RCE neural network.

class. In the case of spatial-uniformity detection in color images, a class refers to a region of a specific color.

One interesting feature of the RCE neural network is that there is no limit to the number of cells in the prototype layer. While a practical limit is imposed by the computing hardware, still, there is plenty of freedom to allocate as many cells as possible in the prototype layer to represent or memorize a class. Since it is possible for a new cluster and a new cell in the prototype layer to be allocated at any time, the RCE neural network is indeed an incremental mechanism for the incremental representation of any meaningful entity. And, this incremental representation can gradually be built through real-time interaction between a physical agent, like a humanoid robot, and its environment.

Parameters in a RCE Neural Network

In the original proposal for the RCE neural network, *prototype cells* are described by a set of five parameters:

- c: Label indicating the class to which a prototype cell belongs,
- ω: Position vector of a prototype cell in the feature space, the dimension of which is determined by the cells in the input layer,
- λ: Dimension (i.e. radius) of the hyper-spherical volume (also known as *influence field*) of a prototype cell in the feature space,
- t: Counter which indicates the number of times a prototype cell has responded to the input either during training or real-time activation,
- σ: Radial decaying coefficient of the hyper-spherical influence field which can be set to zero if there is no decaying effect.

Response from an RCE Neural Network

Assume that an RCE neural network has n cells in the input layer and N cells in the output layer. In response to input data $X = \{x_1, x_2, ..., x_n\}$, prototype cell j of class i will use a radial basis function to determine activating signal u_{ij} as follows:

$$u_{ij} = \lambda_{ij} - \|\mu_X - \omega_{ij}\| \qquad (8.72)$$

where μ_X is the mean value of input data X.

If $u_{ij} \geq 0$ and $\sigma_X < \lambda_{ij}/3$, prototype cell j of class i will respond to input feature vector X. When a prototype cell is fired, or triggered, parameter t will be increased by 1. A prototype cell in the RCE neural network has two modes of response:

(1) Fast-Response Mode:
   In this mode, prototype cell j of class i outputs value 1 to the corresponding output-layer's cell, known as the *class cell*, that is,

$$P_{ij} = \begin{cases} 1 & \text{if } u_{ij} \geq 0 \text{ and } \sigma_X < \lambda_{ij}/3; \\ 0 & \text{otherwise.} \end{cases} \qquad (8.73)$$

(2) Probability-Response Mode:
   In this mode, prototype cell j of class i outputs a probability value based on the following calculation:

$$\begin{cases} d_{ij} = \|\mu_X - \omega_{ij}\| \\ P_{ij} = e^{-\sigma_{ij} \bullet d_{ij}}. \end{cases} \qquad (8.74)$$

If we have class cell C_i ($i \in [1, N]$) in the output layer, we can first compute all the responses of the prototype cells belonging to this class cell. These responses will then be the input to the class cell. Finally, output from the class cell will be the weighted sum of inputs as follows:

$$R_i = \sum_{\forall P_{ij} \in C_i} t_{ij} \bullet P_{ij}, \quad \forall i \in [1, N]. \tag{8.75}$$

If input data triggers responses from all the class cells in the output layer, a conditional probability describing the likelihood that feature vector X belongs to class C_i, is determined as follows:

$$P(C_i|X) = \frac{R_i}{\sum_{k=1}^{N} R_k}, \quad \forall i \in [1, N]. \tag{8.76}$$

If we know the conditional probabilities $P(C_i|X)$ ($\forall i \in [1, N]$), it is easy to decide which class input data belongs to.

Training of an RCE Neural Network

Refer to the architecture of an RCE neural network, as shown in Fig. 8.29. The dimension of the input layer depends on the size of the input data. For example, the input layer will have three cells if an RCE neural network is used to memorize seed regions represented by $N(\mu_c, \sigma_c^2)$. This is because mean vector μ_c has three elements. On the other hand, the number of cells in the output layer depends on the number of seed regions, and can dynamically be increased or decreased.

When an RCE neural network is activated for the first time, there will be no cell in the prototype layer. Therefore, before an RCE neural network can be used for recognizing or classifying data, the mappings between the cells in the input layer and the cells in the output layer must be established. The process of associating the cells in the input layer to the cells in the output layer is known as *RCE neural network training*.

In order to train an RCE neural network, a set of training samples must be available. At the very least, it is necessary to have one training sample per class so that we can establish at least one pathway from input layer to output layer for each class. Let us assume that a training sample is represented by sample mean μ and sample variance σ^2. If this training sample has only one observation (one vector), the sample mean is the feature vector itself, and the sample variance is a user-defined positive value. When we use a set of training samples as input, the training process for an RCE neural network operates in the following way:

- Case 1: Creation of a New Prototype Cell:
  When we use training sample X, belonging to class i, if it does not

trigger any response (see Eq. 8.72) from existing prototype cells (if any), new prototype cell j is created with the following settings:

$$\begin{cases} c_{ij} = C_i \\ \omega_{ij} = \mu_X \\ \lambda_{ij} = 3 \bullet \sigma_X \\ t_{ij} = 1 \\ \sigma_{ij} = \sigma_X. \end{cases}$$

And, this new cell will be connected to class cell C_i which represents class i.

- Case 2: Increase in an Existing Prototype Cell's Counter t:
  When we use training sample X, belonging to class C_i, if it triggers a response from prototype cell j of the same class C_i, counter t of prototype cell j is increased by 1 as follows:

$$t_{ij} = t_{ij} + 1.$$

- Case 3 (optional): Modification of the Influence Field of an Existing Prototype Cell:
  When we use training sample X, belonging to class C_i, if it triggers a response from prototype cell j which belongs to another class C_k, the radius of the hyper-spherical influence field of prototype cell j must be decreased according to the following calculation:

$$\lambda_{kj} = \|\omega_{kj} - \mu_X\|.$$

This indicates that the radius of the hyper-spherical influence field of prototype cell j is now equal to the distance between the center of prototype cell j and the center of training sample X.

We can see that an RCE neural network has the innate mechanism to dynamically create new cells in the prototype layer. This makes it an ideal tool to incrementally represent any family of meaningful entities, such as the regions in a color image, the conceptual symbols in a natural language, etc. Most importantly, the incremental training process of an RCE neural network is very simple and has no issue of convergence.

Summary of a Color Detection Algorithm

The prototype cells in an RCE neural network perform dual roles: a) to memorize training samples and b) to classify input data with reference to stored samples. Thus, an RCE neural network can serve as both a memory and a classifier. If we treat seed regions as training samples, then an RCE neural network is an effective tool for detecting spatial uniformity in color images.

The algorithm implementing a probabilistic RCE neural network for color detection can be summarized as follows:

- Step 1: Select seed regions to train the RCE neural network.
- Step 2: Compute the means and variances of the seed regions, described by the L*a*b tristimulus values.
- Step 3: Train the probabilistic RCE neural network.
- Step 4: Scan an input image row-by-row and column-by-column.
- Step 5: At an unclassified pixel location, use the L*a*b tristimulus values as the mean value of input data and set the variance to zero. Alternatively, take a sub-image centered at the pixel location, and compute the mean and variance of the sub-image, described by the L*a*b tristimulus values.
- Step 6: Classify the scanned pixel by testing the responses from the probabilistic RCE neural network.
- Repeat Steps 4, 5 and 6 until all pixels are classified. If a pixel or its sub-image does not trigger a response from the probabilistic RCE neural network, it will be treated as belonging to an unknown class.

Example 8.17 Human-robot interaction is an important skill that a humanoid robot should possess. There are different forms of interactive communication between a humanoid robot and a human master, one of which is hand gestures. However, in order to understand the meaning behind hand gestures, it is necessary to reliably identify the hand's image which can be captured in different environments (e.g. indoor or outdoor) with simple or complex backgrounds.

A useful visual cue for quick and reliable identification of hand's images is skin color. Fig. 8.30 shows the results of hand-image segmentation based on skin-color recognition by a probabilistic RCE neural network. The classified and grouped pixels belonging to a hand's image are displayed in black. We can see that the RCE neural network works well, regardless of differences in sex or race.

◇◇◇◇◇◇◇◇◇◇◇◇◇◇◇◇◇◇◇◇

Example 8.18 In recent times, there have been many ambitious initia-

(a) Image 1   (b) Image 2

(c) Image 3   (d) Image 4

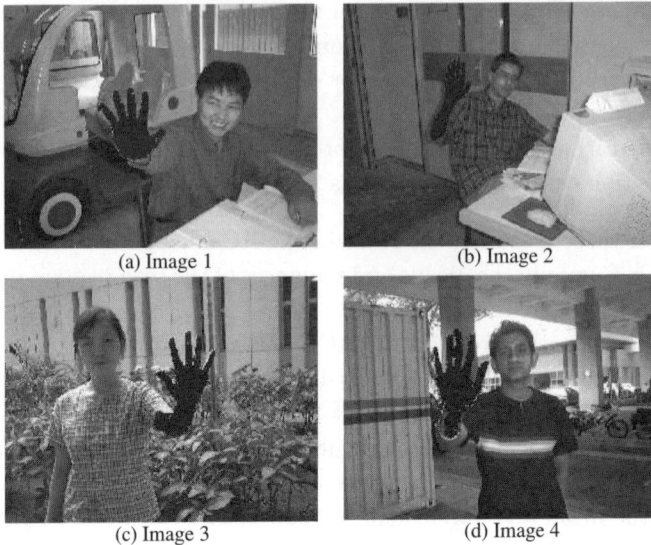

Fig. 8.30   Examples of hand-image segmentation based on skin-color identification.

tives at developing autonomous vehicles. One common way of attempting this is to develop a built-in intelligent system which is fully integrated with the vehicle. An alternative and relatively new approach is to combine a humanoid robot with an existing vehicle.

The day when a humanoid robot is capable of driving a vehicle on the road may not be far away. However, before that happens, it is necessary for the robot to be able to quickly and reliably recognize landmarks and road signs. Interestingly enough, all landmarks and road-signs are painted with specific colors. This makes it easy to identify landmarks and road-signs from color cues. Fig. 8.31 shows an example of using a probabilistic RCE neural network to identify landmarks painted in red. The classified pixels are displayed in black.

◇◇◇◇◇◇◇◇◇◇◇◇◇◇◇◇◇◇

8.4.5   *Temporal Uniformity Detection*

The property of uniformity not only exists in an image-plane's spatial domain, but also in the time domain. The latter occurs when there is relative motion between a robot's visual-sensory system and a scene. If the viewing

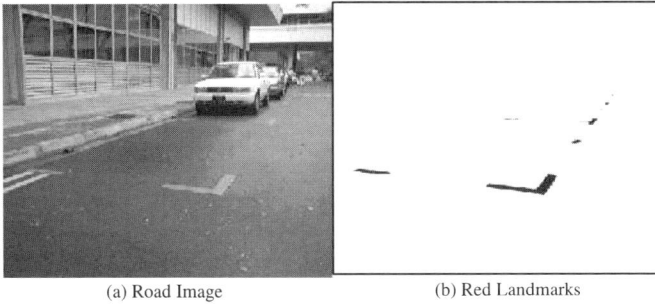

(a) Road Image   (b) Red Landmarks

Fig. 8.31   Example of a landmark segmentation based on the identification of specific colors.

angle with respect to an object in a scene does not change too much within a certain time interval, the projection of this object onto the camera's image plane will appear uniform within a sequence of images captured by the camera.

In a dynamically changing environment, it is important to be able to visually follow the temporal evolution of an object in a scene. This is commonly known as *image feature tracking*. In order to be able to perform image feature tracking, it is necessary to detect the uniformity in the image plane's time domain.

8.4.5.1   *Definition of Image Templates*

If we only have a single image, an object with a general geometry and color may not appear spatially uniform in the camera's image plane. However, if we capture a sequence of images and the viewing angle with respect to this object remains relatively constant, the projection of this object onto the image plane will exhibit temporal uniformity within a sequence of images, as shown in Fig. 8.32.

Let $I_s(v, u, t_k)$ denote the projection of an object onto the camera's image plane at time-instant t_k. If $I_s(v, u, t_k)$ appears to be constant within time interval $[t_1, t_n]$, then $I_s(v, u, t_k)$ can be treated as the time-invariant representation of an object's image. This image is commonly known as the *image template* of an object of interest. Formally, we can state the definition of an image template as follows:

Definition 8.9   An image template $I_s(v, u, t_k)$ is the time-invariant projection of an object onto the camera's image plane within a certain time

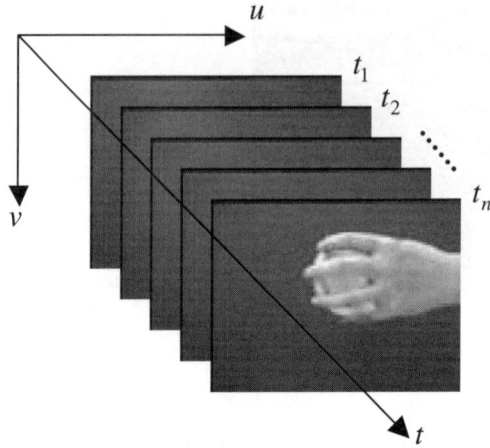

Fig. 8.32　Illustration of spatio-temporal image function describing a sequence of images.

interval.

Mathematically, this definition can be translated as

$$\triangle I_s(t) = \|I_s(v, u, t_k) - I_s(v, u, t_{k-1})\| < g_s, \quad \forall t_k \in [t_2, t_n] \qquad (8.77)$$

where g_s specifies the minimum permissible variation of an image template. Since $I_s(v, u, t_k)$ is time-invariant, it can be written more concisely as $I_s(v, u)$.

8.4.5.2　*Template Matching*

When we have a sequence of n images $I_I(v, u, t_k)$ ($k \in [1, n]$, and $n \geq 1$) and image template $I_s(v, u)$, the problem of temporal-uniformity detection is finding the locus of an image template within the image sequence. Since an image sequence is a discrete sequence, a simple way to pinpoint the locus of the image template is to determine the image-template's location in each image. A more advanced method treats the locus as a trajectory with a specific kinematic property. In this way, the temporal evolution of the locus is predictable in order to cope with uncertainty due to noise and incomplete (or missing) data.

When we have image $I_I(v, u, t_k)$ within an image sequence, the detection of temporal uniformity operates in two steps:

- Prediction of Possible Locations:

  In this step, all the possible locations of an image template in image $I_I(v, u, t_k)$ are predicted. In practice, if we know the location of the image template in the previous image $I_I(v, u, t_{k-1})$, we can define a small window $W : w_x \times w_y$ centered at this location. All the possible locations of the image template are within the window $W : w_x \times w_y$.

- Search for the Image Template's Location:

  This step is to determine the most likely location of an image template within the window $W : w_x \times w_y$. In practice, we choose a location where the sub-image is similar to the image template. If we denote $sub\{I_I(v, u, t_k) \mid (v_0, u_0)\}$ the sub-image centered at location (v_0, u_0), then the search for the image template's location is a minimization process described by

$$\min_{(v_0, u_0)} \|I_s(v, u) - sub\{I_I(v, u, t_k) \mid (v_0, u_0)\}\|, \quad \forall (v_0, u_0) \in W. \quad (8.78)$$

Example 8.19 In a visual-perception system, feature tracking is an important issue. The ability to follow an object of interest within an image sequence will enable a robot to lock its visual attention. If the template image of an object of interest is known in advance, one can apply the template matching technique directly to detect the locus of the object's template image within an image sequence. If no template image is available, we can choose some sub-images, taken from the image at time-instant t_{k-1}, as image templates.

Fig. 8.33 shows the results of temporal-uniformity detection within two consecutive images. The sub-images centered at selected points of interest are used as image templates taken at time-instant t_{k-1}. The detected locations of these image templates in the image at time-instant t_k are shown in Fig. 8.33b.

$$\diamond\diamond\diamond\diamond\diamond\diamond\diamond\diamond\diamond\diamond\diamond\diamond\diamond\diamond\diamond\diamond\diamond\diamond\diamond$$

Example 8.20 Feature correspondence in binocular vision is a crucial step towards the inference of three-dimensional geometry of a scene (or object). This is still difficult to do. However, for some distinct objects, we can treat the problem of feature correspondence in binocular vision as the problem of temporal uniformity detection if the images captured by the two cameras in binocular vision do not differ too much. In this way, the left image can be treated as an image taken at time-instant t_{k-1} and the right image as an image taken at time-instant t_k.

(a) Image at tk-1   (b) Image at tk

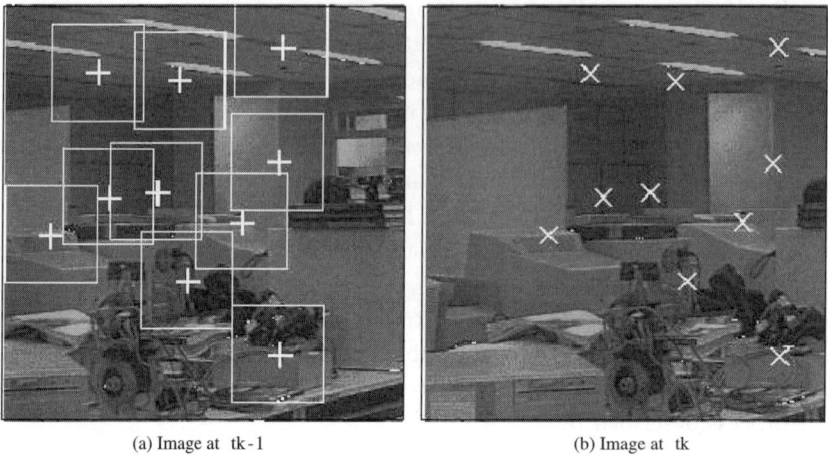

Fig. 8.33   Example of template matching between two consecutive images.

Fig. 8.34 shows an example of binocular correspondence. We can see that the objects corresponding to the computer monitor and printer can be effectively matched within the pair of images.

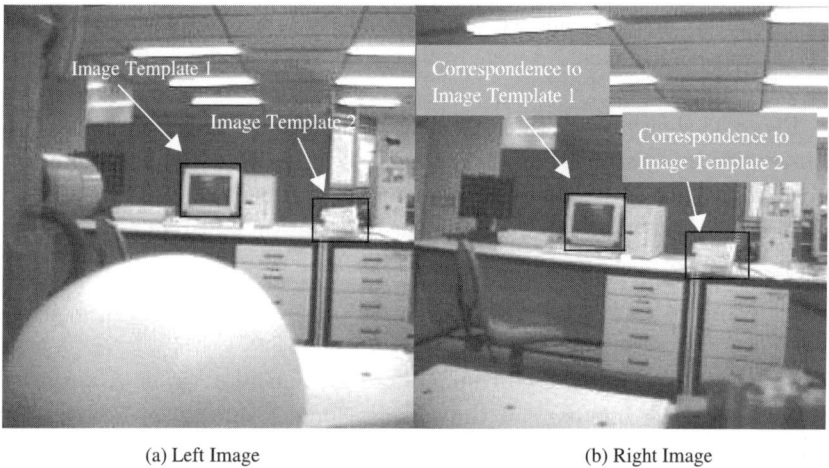

(a) Left Image   (b) Right Image

Fig. 8.34   Example of template matching between two images from binocular vision.

◇◇◇◇◇◇◇◇◇◇◇◇◇◇◇◇◇◇◇

8.4.6   *Temporal Discontinuity Detection*

When dealing with a sequence of images, one may wonder whether temporal discontinuity within an image sequence is an important issue. It is, because detection of temporal discontinuity is a quick and simple way to detect any moving object in a scene which has a stationary background.

In our study of the visual-sensory system, we learned that human vision consists of two parts: a) central vision and b) peripheral vision. One important role of peripheral vision is that it enables humans to quickly respond to the intrusion of any moving object in their field of vision. Thinking about it from an engineering point of view, it is possible that human peripheral vision makes use of the simple technique of temporal-discontinuity detection to achieve the results of quick response.

Let $I_I(t_{k-1})$ and $I_I(t_k)$ denote two consecutive images. If we consider the time interval between two consecutive time-instants to be 1, then the first-order derivative of an image function, with respect to time, can be computed from the image difference as follows:

$$\triangle I_I(t) = \| I_I(t_k) - I_I(t_{k-1}) \|. \tag{8.79}$$

Therefore, the algorithm implementing temporal-discontinuity detection will be extremely simple.

Example 8.21   In robotics, the ability of a visual-sensory system to detect a moving object in the field of vision has useful applications. For example, temporal-discontinuity detection can help a humanoid robot to choose an object of interest so that the robot's visual attention focuses on it.

Fig. 8.35 illustrates a scene with a complex background. We can see that the intrusion of an arm can be quickly and easily detected using Eq. 8.79.

$$\diamond\diamond\diamond\diamond\diamond\diamond\diamond\diamond\diamond\diamond\diamond\diamond\diamond\diamond\diamond\diamond$$

8.5   Geometric-Feature Description

A digital image is a two-dimensional array of pixels. This implies that an image array is without any unit of measurement. In fact, a pixel may correspond to any real dimension in a scene. Therefore, while it seems too early to undertake a parametric description of geometric features in an image, there are two reasons which justify it.

Image at tk   -   Image at tk-1   =   Output Image

Fig. 8.35   Example of temporal-discontinuity detection.

First of all, an image contains a large amount of information. Even after the feature extraction, the size of the data is still large. Feature description is an effective way to reduce raw data while preserving useful information. For example, in a digital image, the contour of a straight line is a list of edges. However, two end points, or a parametric equation, such as:

$$a \bullet u + b \bullet v + c = 0$$

are sufficient and accurate enough to represent a straight contour.

Secondly, a parametric description of raw data is a helpful way to infer the geometry of a scene. For example, a binocular correspondence of two circles is much simpler to handle than dealing with two lists of points. We will study the problem of binocular vision in the latter part of this chapter.

8.5.1   *The Basics of Feature Description*

The ultimate goal of feature description is to derive a compact representation of pixels belonging to the same physical entity, such as a contour or surface. In order to achieve this goal, one must address the following issues, as shown in Fig. 8.36:

- Feature Grouping:
  An image contains a large amount of raw data. After feature extraction, the results are still in the form of individual pixels, called *feature points*. In general, these feature points are unlikely to belong to a single physical entity, because a scene usually has more than one object. As a result, it is necessary to first divide the feature points into a set of groups. This process is commonly called *feature grouping*.

- Contour Splitting:
  A contour is a list of interconnected edges which may, or may not belong to a single object. Even if a contour belongs to the boundary of a surface, it may have a complicated shape which cannot be described by a single equation. Therefore, it is necessary to split a contour into a set of simple contours, each of which can be described analytically by one equation.
- Curve Fitting:
  When we have a simple contour, the purpose of curve fitting is to determine a compact representation of it. In the simplest case, a contour can always be approximated by a set of line segments.

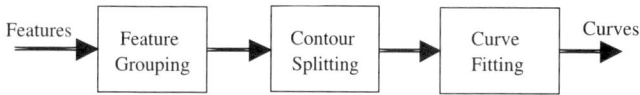

Fig. 8.36   General procedure for feature description.

8.5.2   *Feature Grouping*

Feature grouping turns out to be a very difficult process because it involves complex decision-making, as shown in Fig. 8.37. The psychological interpretation by the brain, regarding the content shown in Fig. 8.37, is that there is a rectangle overlapping four circles. However, there is no direct visual-sensory data which computationally leads to this interpretation.

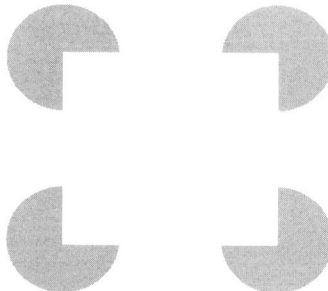

Fig. 8.37   Example of feature grouping with an invisible but expected, physical entity.

This example indicates that feature grouping may be one or a combination of the following:

- Expectation-Driven:
  Based on our visual experience, the brain is skillful at postulating a series of expectations. These expectations will eventually guide a feature-grouping process. For the example shown in Fig. 8.37, the expectation of the presence of a rectangle is indirectly verified by the visual-sensory data. Therefore, this expectation is confirmed as a valid outcome. Humans are very skillful at formulating expectations. However, even for today's robots, this is one of the most difficult things to perform well.
- Data-Driven:
  Two pixels which are close to each other are likely to belong to the same physical entity. Human vision makes use of this property to quickly and reliably partition visual-sensory data into a set of groups. This process of partitioning image features based on the property of neighborhood is commonly known as *grouping by connectivity*.

Here, we will study the data-driven process of feature grouping.

8.5.2.1 *Types of Neighborhoods*

Grouping by connectivity is a data-driven process based on a neighborhood of image features. An image is a two-dimensional array. When we have pixel location (v, u), there are three typical types of neighborhoods, as shown in Fig. 8.38:

- 4-neighborhood:
  Neighbors in the north, south, east and west are called the *4-neighborhood*. When we have pixel location (v, u), the locations of its four neighbors are

$$(v-1, u), \ (v, u+1), \ (v+1, u), \ (v, u-1)$$

- 8-neighborhood:
  An image is a regular two-dimensional array. A pixel location is generally surrounded by a set of eight neighbors, as shown in Fig. 8.38b. These eight neighbors are called the *8-neighborhood*.
- Causal Neighborhood:
  At any time-instant, an image is a two-dimensional spatial function. Therefore, when processing an image, decision-making regarding a pixel must take into account the results of its neighbors or its spatial causality. In order to achieve a deterministic outcome, it is necessary to know

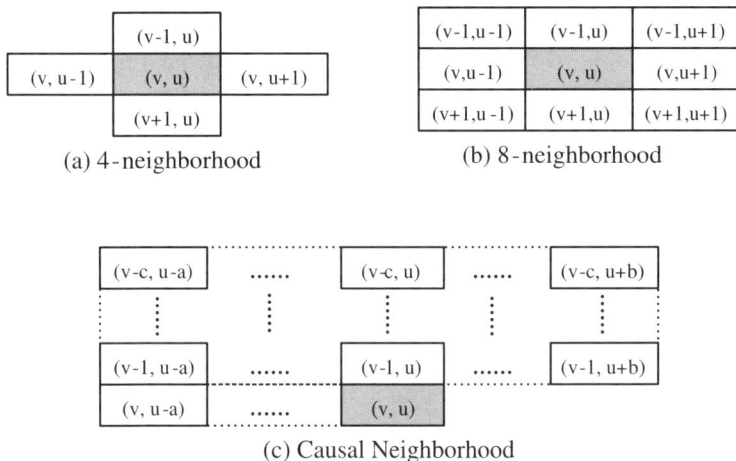

	(v-1, u)	
(v, u-1)	(v, u)	(v, u+1)
	(v+1, u)	

(a) 4-neighborhood

(v-1,u-1)	(v-1,u)	(v-1,u+1)
(v,u-1)	(v, u)	(v,u+1)
(v+1,u-1)	(v+1,u)	(v+1,u+1)

(b) 8-neighborhood

(v-c, u-a)	......	(v-c, u)	......	(v-c, u+b)
⋮	⋮	⋮	⋮	⋮
(v-1, u-a)	......	(v-1, u)	......	(v-1, u+b)
(v, u-a)	......	(v, u)		

(c) Causal Neighborhood

Fig. 8.38   Typical types of neighborhoods.

the results of decision-making with regard to the neighbors in the vicinity of a pixel. If there is mutual dependency between a pixel and its neighbors, a decision cannot be made, and the process will be deadlocked. Of all the possible neighbors of a pixel, those with known results are called *causal neighborhoods*. For example, if a decision-making process is performed by scanning an input image row-by-row (starting from the first row) and column-by-column (starting from the first column), the neighbors inside the neighborhood, as shown in Fig. 8.38c, will have already been processed when the scanning process reaches pixel location (v, u). As shown in Fig. 8.38c, the size of a causal neighborhood depends on three integer parameters: (a, b, c). For example, the neighbors of pixel location (v, u) will be

$$
N_c(v, u) = \begin{cases} \{(v_i, u_i) \mid \forall u_i \in [u - a, u + b]\} & \text{if } v_i \in [v - 1, v - c] \\ \{(v_i, u_i) \mid \forall u_i \in [u - a, u - 1]\} & \text{if } v_i = v. \end{cases}
\tag{8.80}
$$

In the following sections, we will demonstrate the usefulness of the above neighborhoods.

8.5.2.2 *Pixel Clustering*

Assume that input to feature description comes from an algorithm of spatial-uniformity detection. In this case, pixels classified as belonging to uniform region $N(\mu, \sigma^2)$ may not correspond to a single physical entity. Therefore, it is necessary to partition the pixels in order to form distinct clusters, known as *groups*.

If we know the pixels belonging to uniform distribution $N(\mu, \sigma^2)$, grouping by connectivity is a simple process which effectively groups the pixels into a set of distinct clusters. An algorithm for pixel clustering usually uses the 8-neighborhood and includes the following steps :

- Step 1:Set the group-label variable to the value of 1.
- Step 2: Scan an input image, row-by-row and column-by-column.
- Step 3: At each unlabelled pixel which belongs to uniform region $N(\mu, \sigma^2)$, check the results of its neighbors in the 8-neighborhood. If any one of them has been assigned with a group label, assign this group label to the pixel. If none of its neighbors has been assigned with a group label, assign the value of the group-label variable to the pixel, and increase the group-label variable by 1.
- Repeat Steps 2 and 3 until all pixels belonging to $N(\mu, \sigma^2)$ have been labelled.

Example 8.22   Figure 8.39a shows a small image array which contains three distinct clusters of pixels. The results of applying the pixel clustering algorithm are shown in Fig. 8.39b. We can see that these clusters are labelled as clusters 1, 2 and 3 by the algorithm.

$$\diamond\diamond\diamond\diamond\diamond\diamond\diamond\diamond\diamond\diamond\diamond\diamond\diamond\diamond\diamond\diamond\diamond$$

8.5.2.3 *Boundary Tracing*

The pixel-clustering algorithm outputs distinct regions, each of which has a label. If we know the pixels belonging to a labelled cluster (or region), we can directly compute some quantitative measurements, such as area, circumference, rectangularity, circularity and moments, which characterize the cluster.

Alternatively, we can choose to represent a labelled cluster or region by its boundary. In this case, it is necessary to identify pixels belonging to the edges of a labelled cluster (or region). A process which does this is called *boundary tracing*. A boundary-tracing algorithm normally makes use of the

Fig. 8.39 (a) Input

	2		4		6		8		10
0	0	0	0	0	0	0	0	0	0
0	200	200	0	0	0	0	0	0	0
0	200	0	0	0	200	200	200	200	0
0	200	0	0	0	200	200	200	200	0
0	200	200	0	0	0	0	0	0	0
0	0	0	0	0	200	200	0	0	0
0	0	0	0	0	200	200	0	0	0
0	0	0	0	0	0	0	200	200	0
0	0	0	0	0	0	0	200	200	0
0	0	0	0	0	0	0	0	0	0

Fig. 8.39 (b) Output

	2		4		6		8		10
0	0	0	0	0	0	0	0	0	0
0	1	1	0	0	0	0	0	0	0
0	1	0	0	0	2	2	2	2	0
0	1	0	0	0	2	2	2	2	0
0	1	1	0	0	0	0	0	0	0
0	0	0	0	0	3	3	0	0	0
0	0	0	0	0	3	3	0	0	0
0	0	0	0	0	0	0	3	3	0
0	0	0	0	0	0	0	3	3	0
0	0	0	0	0	0	0	0	0	0

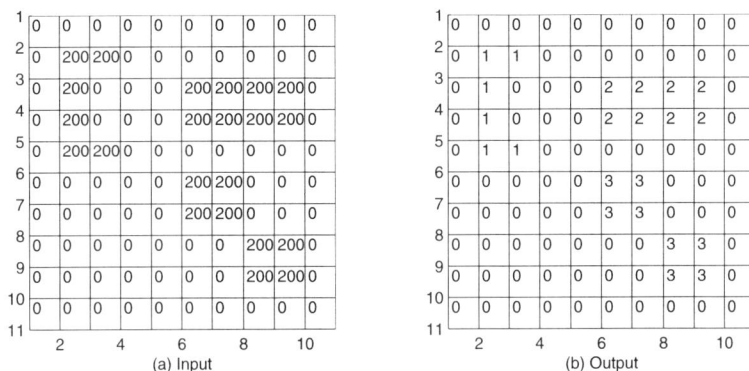

Fig. 8.39 Example of pixel clustering.

4-neighborhood and operates in the following way:

- Step 1: Scan an input image row-by-row and column-by-column.
- Step 2: At each unmarked pixel which belongs to a labelled cluster (or region), check the results of its neighbors in the 4-neighborhood. If at least one pixel does not belong to the labelled cluster or region, mark this as a boundary pixel.
- Repeat Steps 1 and 2 until all pixels belonging to the labelled cluster (or region) have been marked.

Example 8.23 Figure 8.40a shows a labelled region (label 100). The result of applying the boundary-tracing algorithm is shown in Fig. 8.40b.

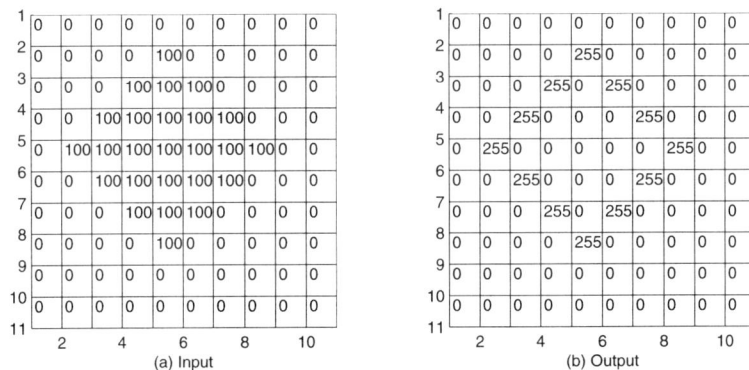

Fig. 8.40 (a) Input

	2		4		6		8		10
0	0	0	0	0	0	0	0	0	0
0	0	0	0	100	0	0	0	0	0
0	0	0	100	100	100	0	0	0	0
0	0	100	100	100	100	100	0	0	0
0	100	100	100	100	100	100	100	0	0
0	0	100	100	100	100	100	0	0	0
0	0	0	100	100	100	0	0	0	0
0	0	0	0	100	0	0	0	0	0
0	0	0	0	0	0	0	0	0	0
0	0	0	0	0	0	0	0	0	0

Fig. 8.40 (b) Output

	2		4		6		8		10
0	0	0	0	0	0	0	0	0	0
0	0	0	0	255	0	0	0	0	0
0	0	0	255	0	255	0	0	0	0
0	0	255	0	0	0	255	0	0	0
0	255	0	0	0	0	0	255	0	0
0	0	255	0	0	0	255	0	0	0
0	0	0	255	0	255	0	0	0	0
0	0	0	0	255	0	0	0	0	0
0	0	0	0	0	0	0	0	0	0
0	0	0	0	0	0	0	0	0	0

Fig. 8.40 Example of boundary tracing.

◇◇◇◇◇◇◇◇◇◇◇◇◇◇◇◇◇◇

8.5.2.4   *Edge Linking*

The purpose of edge linking is to partition edges in an image into a set of labelled groups, known as *contours*. The input to an edge-linking process is an array of edges. The output is a set of contours. However, for the ease of curve fitting, it is desirable to obtain a set of contours without any intersection nor bifurcation.

Edge linking can be treated as the process of clustering edges in an image. This can be done using the 8-neighborhood to scan the array of edges row-by-row and column-by-column. Then, the interconnected edges will be grouped into a single entity.

However, this process does not guarantee that a contour will not have any intersection or bifurcation. In order to obtain contours without any intersection or bifurcation, it is necessary to perform a second scanning on the array of edges in order to detect the junctions where a contour is broken into a subset of simple contours. Since two successive scans are performed in this method, it is called the *two-pass algorithm* for edge linking.

In summary, a two-pass algorithm for edge linking includes the following steps:

- Step 1: Set the contour-label variable to the value of 1.
- Step 2: Scan an array of edges, row-by-row and column-by-column.
- Step 3: At each unlabelled edge, check the results of its neighbors in the 8-neighborhood. If none of them has been assigned a contour-label, assign the value of the contour-label variable to the edge, and increase the contour-label variable by 1. If any one of them has been assigned a contour label, assign this contour-label to the edge.
- Repeat Steps 2 and 3 until all edges have been labelled.
- Step 4: Scan a labelled contour, edge-by-edge.
- Step 5: At each edge under the scan, check whether it is a junction point. Normally, this is not easy to do. Thus, let us assume that the thickness of a contour is one pixel. At each edge under the scan, its neighbors will be assigned the logic value "1" if they belong to the contour, and the logic value "0" if they do not. If we put the eight logic values from the eight neighbors together in a clockwise (or counterclockwise) order, we will obtain a logic byte. If the transition from "1" to "0" in the logic byte occurs more than twice, then the edge is a junction point.

- Step 6: Break the contour at the junction point if one exists.
- Repeat Steps 4, 5 and 6 until all edges on the contour are processed.

If we make use of the causal neighborhood, we can develop a more reliable edge-linking algorithm which can scan an array of edges in a single pass, and produce contours without any intersection or bifurcation. This method is called the *single-pass algorithm* for edge linking, and represents a contour with three parameters (see contour C_2 in Fig. 8.41):

- Tail:
  The tail is the starting edge of a contour. Normally, we scan an array from the top row until bottom, and from the first column until the last. Thus, the tail of a contour is the first scanned edge (or endpoint) of a contour.
- Running Head:
  During the process of edge linking, a contour grows gradually. The edge, which is added to a contour during linking, is called the actual *running head*.
- Head:
  When a contour stops growing, the running head becomes the actual head of the contour. In fact, it is the last edge (or endpoint) of a contour.

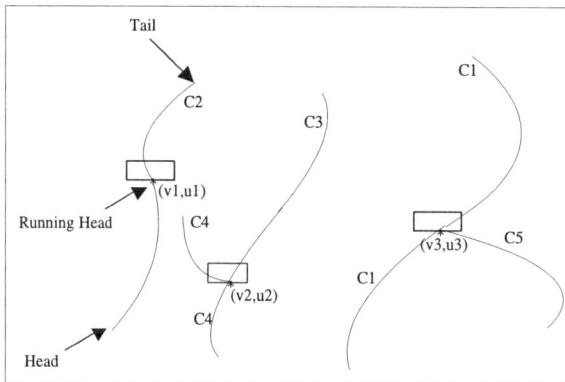

Fig. 8.41   Illustration of a single-pass algorithm for edge linking.

In Fig. 8.41, there are three clusters of interconnected edges. The one on the left-hand side is a simple contour without any intersection or bifurcation. The one in the middle has one intersection at location (v_2, u_2). The one on the right-hand side has a bifurcation at location (v_3, u_3).

Refer to Fig. 8.41. A single-pass algorithm which makes use of a causal neighborhood can be summarized, as follows:

- Step 1: Set the contour-label variable to the value of 1.
- Step 2: Scan an array of edges, row-by-row and column-by-column.
- Step 3: For an unlabelled edge at location (v, u), check the results of its neighbors in causal neighborhood $N_c(v, u)$.
- Step 4a: If there is no neighbor, assign the contour-label variable to the edge, thereby creating a new contour. Then, increase the contour-label variable by 1, and assign the edge as the tail and running head of the new contour.
- Step 4b: If there is only one running head inside $N_c(v, u)$, lengthen the contour which contains the running head to the edge, and assign the edge to be the new running head of the lengthened contour.
- Step 4c: If the running head of contour i and the tail of contour j are both present inside $N_c(v, u)$, lengthen contour i to the edge and merge these two contours at the tail of contour j. The merged contour will be assigned label i, and its running head will be the running head of contour j.
- Step 4d: If there is more than one running head inside $N_c(v, u)$, find the running head nearest to the limit on the left-hand side of $N_c(v, u)$. Lengthen the contour which contains this running head to the edge, and assign the edge to be the new running head of the lengthened contour. Subsequently, disable the other running heads inside $N_c(v, u)$. This decision rule is called the *priority-left rule*. Alternatively, we can choose the running head nearest to the limit on the right-hand side of $N_c(v, u)$. This will result in the *priority-right rule*.
- Repeat Steps 2, 3 and 4 until all edges are processed.

Assume that contour-thickness in Fig. 8.41 is one pixel. If we set (a, b, c) to be $(1, 1, 1)$ and choose the priority-left rule, it is not difficult to verify that the single-pass algorithm for edge linking will produce five contours, as shown in Fig. 8.41.

8.5.3 *Contour Splitting*

In a digital image, a contour is a list of interconnected edges. It is unlikely that all the contours extracted from an image will have a simple shape.

However, this is necessary before a curve can be fitted into a contour. If the contour is not in a simple shape, we must split it into a set of simple contours so that it can be approximated by a curve.

8.5.3.1 *Successive Linear Curves*

The strategy for splitting contours depends on the expected outcome of the curve fitting. If the intention is to obtain a set of successive linear curves (line segments), then a contour splitting algorithm can operate in the following way:

- Step 1: Arrange the edges of a contour in sequential order.
- Step 2: Compute line passing through the two endpoints of the contour. We call this line the *approximation line*. Mathematically, the equation describing an approximation line in the camera's image plane is:

$$a \bullet u + b \bullet v + c = 0.$$

- Step 3: Compute the distance from each edge to the approximation line. If the edge's location is (v_i, u_i), then its distance to the approximation line is

$$d_i = \frac{\|a \bullet u_i + b \bullet v_i + c\|}{\sqrt{a^2 + b^2}}.$$

- Step 4: Determine the edges, the distances of which to the approximation line are local maxima. We call these edges the *local-maximum edges*.
- Step 5a: Find the local-maximum edge nearest to one endpoint of the contour. If its distance to the approximation line is above threshold value d_0, split the contour at this edge.
- Step 5b: Find the local-maximum edge nearest to the other endpoint of the contour. If its distance to the approximation line is above threshold value d_0, split the contour at this edge.
- Step 6: If the contour has been split at least once, repeat the algorithm until no splitting occurs.

Example 8.24 Figure 8.42 shows an example which illustrates the procedure of splitting a contour into a set of successive linear curves. Assume that the two endpoints of the contour are A and B. In the first iteration, the local-maximum edges at C_1 and D_1 will be selected to split the contour. In the subsequent iterations, contour AC_1 will be split at C_2, contour D_1B

at D_2, and contour C_1D_1 at C_3 and D_3. The final output of the successive linear curves will be AC_2, C_2C_1, C_1C_3, C_3D_3, D_3D_1, D_1D_2, and D_2B.

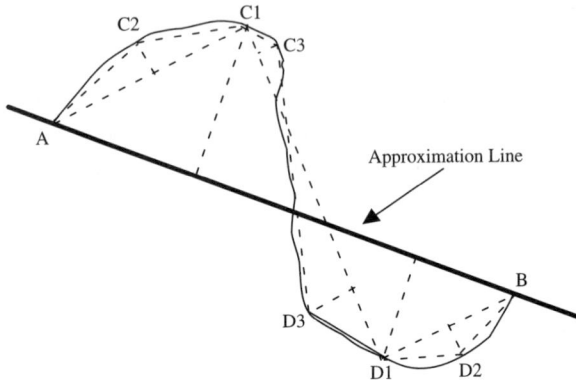

Fig. 8.42   Generation of successive linear curves by contour splitting.

◇◇◇◇◇◇◇◇◇◇◇◇◇◇◇◇◇◇◇

8.5.3.2 *Successive Nonlinear Curves*

If the desired output is a set of successive nonlinear curves (circular, parabolic or elliptic), a practical contour-splitting algorithm is as follows:

- Step 1: Arrange the edges of a contour in sequential order.
- Step 2: Compute line passing through the two endpoints of the contour. We also call this line the *approximation line*. Its equation is

$$a \bullet u + b \bullet v + c = 0.$$

- Step 3: Compute the distance from every edge to the approximation line. If the edge's location is (v_i, u_i), then its distance to the approximation line is

$$d_i = \frac{\|a \bullet u_i + b \bullet v_i + c\|}{\sqrt{a^2 + b^2}}.$$

- Step 4: Determine the edges, the distances of which to the approximation line are local maxima. We also call these edges the *local-maximum edges*.

- Step 5: Find the local-maximum edge nearest to one endpoint of the contour. Similarly, find the local-maximum edge nearest to the other endpoint of the contour. If these two edges are at the same location, take no action (Optional: Depending on the intended application, splitting can occur at this location).
- Step 6: Between the two local-maximum edges, find the edge nearest to the approximation line and split the contour at this edge.
- Step 7: If the contour has been split at least once, repeat the algorithm until no splitting occurs.

Example 8.25 Figure 8.43 shows an example which illustrates the procedure of splitting a contour into a set of successive nonlinear curves. Assume that the two endpoints of a contour are A and B. In the first iteration, the local-maximum edges at C_1 and D_1 are selected. In this example, two edges between C_1 and D_1 have a distance to the approximation line which is zero. These two points are E_1 and E_2 and the contour will be split here. In the second iteration, there is no contour splitting at AE_1 and E_1E_2. But, contour E_2B is further split at edge E_3. The final output of successive nonlinear curves is AE_1, E_1E_2, E_2E_3 and E_3B.

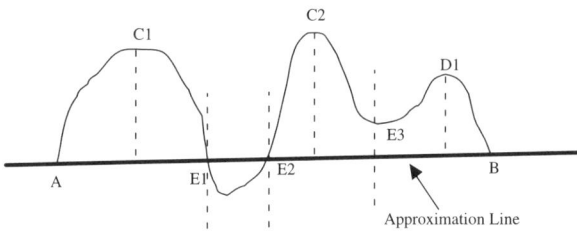

Fig. 8.43   Generation of successive conic curves by contour splitting.

◇◇◇◇◇◇◇◇◇◇◇◇◇◇◇◇◇◇◇

8.5.4   *Curve Fitting*

When we have a simple contour which can be approximated by a linear or conic curve, it is easy to estimate the parameters of the equation describing the contour. In general, the input to a curve-fitting process is a simple contour in the form of a list of edges:

$$C = \{(v_i, u_i),\ i = 1, 2, ..., n\}.$$

And, the output are the parameters of an implicit (or explicit) equation describing the contour.

8.5.4.1 Line-Segment Approximation

Let us consider image plane uv to be a continuous, two-dimensional space. A line in the uv plane can be described by the equation:

$$a \bullet u + b \bullet v + c = 0 \tag{8.81}$$

where (a, b, c) are three nonindependent parameters. Dividing Eq. 8.81 by term $\sqrt{a^2 + b^2}$ yields

$$u \bullet \cos\theta + v \bullet \sin\theta + \rho = 0 \tag{8.82}$$

with

$$\begin{cases} \cos\theta = \frac{a}{\sqrt{a^2+b^2}} \\ \sin\theta = \frac{b}{\sqrt{a^2+b^2}} \\ \rho = \frac{c}{\sqrt{a^2+b^2}}. \end{cases} \tag{8.83}$$

It is clear that parameters (θ, ρ) are independent and uniquely determine a line in the uv plane. When we have contour:

$$C = \{(v_i, u_i),\ i = 1, 2, ..., n\}$$

which can be approximated by a line, we define the following error function:

$$L = \sum_{i=1}^{n} (u_i \bullet \cos\theta + v_i \sin\theta + \rho)^2. \tag{8.84}$$

The optimal solution for ρ, which minimizes Eq. 8.84, must satisfy

$$\frac{\partial L}{\partial \rho} = 0.$$

Accordingly, we obtain

$$\rho = -\bar{u}\cos\theta - \bar{v}\sin\theta \tag{8.85}$$

with

$$\begin{cases} \bar{u} = \frac{1}{n} \bullet \sum_{i=1}^{n} u_i \\ \\ \bar{v} = \frac{1}{n} \bullet \sum_{i=1}^{n} v_i. \end{cases} \tag{8.86}$$

Substituting Eq.8.85 into Eq. 8.84 yields

$$L = \sum_{i=1}^{n} \left(\bar{u}_i \bullet \cos\theta + \bar{v}_i \sin\theta \right)^2 \tag{8.87}$$

with $\bar{u}_i = u_i - \bar{u}$ and $\bar{v}_i = v_i - \bar{v}$.

Similarly, the optimal solution for θ, which minimizes Eq. 8.87, must satisfy

$$\frac{\partial L}{\partial \theta} = 0.$$

As a result, we obtain the following equation:

$$\sum_{i=1}^{n} \left\{ (\bar{u}_i \cos\theta + \bar{v}_i \sin\theta) \bullet (-\bar{u}_i \sin\theta + \bar{v}_i \cos\theta) \right\} = 0 \tag{8.88}$$

Eq. 8.88 can also be written as

$$(B - A) \bullet \sin\theta \bullet \cos\theta + C \bullet (\cos^2\theta - \sin^2\theta) = 0. \tag{8.89}$$

with

$$\begin{cases} A = \sum_{i=1}^{n} \bar{u}_i^2 \\ \\ B = \sum_{i=1}^{n} \bar{v}_i^2 \\ \\ C = \sum_{i=1}^{n} (\bar{u}_i \bullet \bar{v}_i). \end{cases} \tag{8.90}$$

Since $2 \bullet \sin\theta \bullet \cos\theta = \sin(2\theta)$ and $\cos^2\theta - \sin^2\theta = \cos(2\theta)$, Eq. 8.89 becomes

$$\tan(2\theta) = \frac{2C}{A - B}. \tag{8.91}$$

Finally, the solution for θ will be

$$\theta = \frac{1}{2} \bullet \tan^{-1}\left(\frac{2C}{A - B} \right). \tag{8.92}$$

8.5.4.2   *Circular-Arc Approximation*

A circle (or circular arc) in the uv plane is described by the equation:

$$(u - u_0)^2 + (v - v_0)^2 = r^2 \tag{8.93}$$

where (v_0, u_0) is the center of the circle while r is the circle's radius. (u_0, v_0, r) are three independent parameters which uniquely represent a circle.

Eq. 8.93 can also be written in a parametric form as follows:

$$\begin{cases} u = u_0 + r \bullet \cos\theta \\ v = v_0 + r \bullet \sin\theta. \end{cases} \tag{8.94}$$

Now, we eliminate parameter r from Eq. 8.94. This can be done by first multiplying $\sin\theta$ to the first equation and $\cos\theta$ to the second equation, and then, computing the difference. As a result, we obtain

$$u_0 \sin\theta - v_0 \cos\theta = v\cos\theta - u\sin\theta. \tag{8.95}$$

When we have contour:

$$C = \{(v_i, u_i), \ i = 1, 2, ..., n\}$$

which can be approximated by a circular arc or circle, we can estimate the tangential angle at each edge on the contour. In this way, the raw data of a contour is represented in the form of a list of triplets as follows:

$$C = \{(v_i, u_i, \theta_i), \ i = 1, 2, ..., n\}.$$

From Eq. 8.95, we can directly establish a linear system of equations:

$$u_0 \sin\theta_i - v_0 \cos\theta_i = v_i \cos\theta_i - u_i \sin\theta_i, \ i = 1, 2, ..., n. \tag{8.96}$$

Parameters (u_0, v_0) in Eq. 8.96 can be determined by a least-square estimation. If we know (u_0, v_0), then the estimation of radius r is

$$r = \frac{1}{n} \sum_{i=1}^{n} \sqrt{(u_i - u_0)^2 + (v_i - v_0)^2}. \tag{8.97}$$

8.5.4.3   *Elliptic-Arc Approximation*

The implicit equation describing an ellipse or elliptic arc in the uv plane is

$$\frac{(u - u_0)^2}{r_u^2} + \frac{(v - v_0)^2}{r_v^2} = 1. \tag{8.98}$$

The four independent parameters of an ellipse are (u_0, v_0, r_u, r_v).

As in the case of the circle, the parametric representation of an ellipse is

$$\begin{cases} u = u_0 + r_u \cos \theta \\ v = v_0 + r_v \sin \theta. \end{cases} \tag{8.99}$$

When we have a contour in the form of a list of triplets:

$$C = \{(v_i, u_i, \theta_i), \ i = 1, 2, ..., n\},$$

we can directly establish two linear equations from Eq. 8.99 as follows:

$$\begin{cases} u_i = u_0 + r_u \bullet \cos \theta_i, \ i = 1, 2, ..., n \\ v_i = v_0 + r_v \bullet \sin \theta_i, \ i = 1, 2, ..., n. \end{cases} \tag{8.100}$$

Clearly, the four parameters (u_0, v_0, r_u, r_v) can be determined by a least-square estimation.

8.6 Geometry Measurement

There are two problems with geometric-feature grouping and description in a camera's image plane:

- First of all, a digital image is an array of pixels. The index coordinates (v, u) are without any unit of measurement. Therefore, feature description in image plane uv is qualitative, not quantitative, and all measurements done in the uv plane are relative to a projective scale. In order to obtain metric measurements, it is necessary to study the relationship between index coordinates (v, u) and coordinates (X, Y, Z) in a three-dimensional space.
- Secondly, while feature grouping in the uv plane is very elementary and it is easy to partition feature points into a set of simple contours or regions, this is insufficient for object recognition. This is because the projection of an object onto an image plane only creates a subset of simple contours and/or regions. Feature grouping at a higher level is necessary because the final outcome must be a set of contours and/or regions, each of which only corresponds to a single physical object. Without any knowledge about the real dimensions, geometries or models of image features, it is theoretically impossible to distinguish image

features which have very similar geometries/appearances but belong to
different objects.

In this section, we will first study what can be done with a single camera (monocular vision). Then we will look at the principle behind and
challenges of binocular vision.

8.6.1 *Monocular Vision*

The monocular-vision system takes images as input, and produces geometrical measurement as output. A monocular-vision system is normally
composed of: a) a single electronic camera (either color or monochrome),
b) an image digitizer (if the camera's output is analogue video), and c)
computing hardware.

The information transformation inside a monocular-vision system is
shown in Fig. 8.44. Point Q in a three-dimensional space, referenced to
frame r, is projected onto the image plane of the single camera. The image
of point Q will form image point q. Then, the electrical signals, picked up
by the imaging sensor of the camera, will be converted into an analogue
image. If the camera's output is not digital, the image digitizer will convert
analogue image into a corresponding digital image. Finally, the digital images from the image digitizer will be the input to the process of geometry
measurement in the monocular-vision system.

Assume that we are able to detect image feature q in an image array.
Then, the question is: How do we determine the coordinates of point Q
if the coordinates of its image q are known? This problem is called the
problem of *inverse projective-mapping*. In order to derive an answer, it is
necessary to know how the coordinates of image point q are related to the
coordinates of corresponding object point Q, which is referenced to frame
r. This problem is called the problem of *forward projective-mapping*.

8.6.1.1 *Forward Projective-Mapping*

When we have coordinates $(^rX,\ ^rY,\ ^rZ)$ of point Q which is referenced to
frame r, the problem of forward projective-mapping is how to determine
coordinates (u, v) of its image point q in the image array, which is referenced
to index coordinate system uv.

Assume that frame c is assigned to the camera, as shown in Fig. 8.44.
If motion transformation ^cM_r describes the transformation from reference
frame r to camera frame c, then the coordinates of point Q with respect to

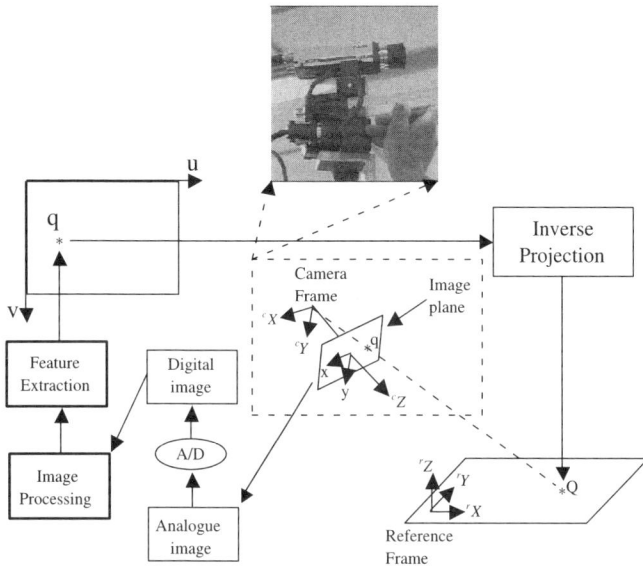

Fig. 8.44   Illustration of geometric projection in a monocular-vision system.

camera frame c are

$$
\begin{pmatrix} {}^cX \\ {}^cY \\ {}^cZ \\ 1 \end{pmatrix} = {}^cM_r \bullet \begin{pmatrix} {}^rX \\ {}^rY \\ {}^rZ \\ 1 \end{pmatrix}.
\tag{8.101}
$$

In Chapter 7, we studied the relationship between index coordinates (u, v) and the corresponding coordinates $({}^cX, {}^cY, {}^cZ)$, with respect to camera frame c. This relationship is described by projective-mapping matrix ${}^IP_c$ as follows:

$$
s \bullet \begin{pmatrix} u \\ v \\ 1 \end{pmatrix} = {}^IP_c \bullet \begin{pmatrix} {}^cX \\ {}^cY \\ {}^cZ \\ 1 \end{pmatrix}
\tag{8.102}
$$

with

$$
{}^IP_c = \begin{pmatrix} \frac{f_c}{D_x} & 0 & u_0 & 0 \\ 0 & \frac{f_c}{D_y} & v_0 & 0 \\ 0 & 0 & 1 & 0 \end{pmatrix}.
\tag{8.103}
$$

In Eq. 8.102, s is an unknown scaling factor. In Eq. 8.103, parameter f_c is the focal length of the camera, (D_x, D_y) the sampling steps of the image digitizer and (u_0, v_0) the coordinates of the optical center in index coordinate system uv.

The combination of Eq. 8.101 and Eq. 8.102 yields

$$s \bullet \begin{pmatrix} u \\ v \\ 1 \end{pmatrix} = \{^I P_c\} \bullet \{^c M_r\} \bullet \begin{pmatrix} {}^r X \\ {}^r Y \\ {}^r Z \\ 1 \end{pmatrix}. \qquad (8.104)$$

Eq. 8.104 describes *forward projective-mapping*. The coefficients inside $^I P_c$ are called the camera's *intrinsic parameters*, while the elements in $^c M_r$ are called the camera's *extrinsic parameters*. If we denote

$$H = \{^I P_c\} \bullet \{^c M_r\}, \qquad (8.105)$$

matrix H is known as the camera's *calibration matrix*. And, H is a 3×4 matrix.

Eq. 8.104 has two practical applications: a) construction of a virtual camera for vision simulation, and b) camera calibration for determining the camera's intrinsic and extrinsic parameters.

8.6.1.2 *Simulation of Monocular Vision*

For construction of a virtual camera, the specification of parameters (f_c, D_x, D_y) are not intuitive. Refer to Fig. 8.44. The most intuitive way of indirectly specifying a camera's intrinsic parameters is to define focal length f_c as well as the aperture angles in both the vertical and horizontal directions. In fact, if (r_x, r_y) are the resolution of the digital image, the size of the camera's image plane is

$$\begin{cases} w_I = r_x \bullet D_x \\ h_I = r_y \bullet D_y \end{cases} \qquad (8.106)$$

where w_I and h_I are the width and height of an image plane. If we denote θ_u the horizontal aperture-angle and θ_v the vertical aperture-angle, we have

$$\begin{cases} \tan\left(\frac{\theta_u}{2}\right) = \frac{w_I}{2f_c} \\[2mm] \tan\left(\frac{\theta_v}{2}\right) = \frac{h_I}{2f_c}. \end{cases} \qquad (8.107)$$

If we know aperture-angles (θ_u, θ_v), then the sampling steps (D_x, D_y) can be computed as follows:

$$
\begin{cases}
D_x = \frac{2f_c}{r_x} \bullet \tan\left(\frac{\theta_u}{2}\right) \\[2mm]
D_y = \frac{2f_c}{r_y} \bullet \tan\left(\frac{\theta_v}{2}\right).
\end{cases}
\tag{8.108}
$$

Example 8.26 We are going to construct a virtual camera with a focal length of 1.0cm, aperture angles of $(70^0, 70^0)$ and image resolution of $(512, 512)$. If we apply Eq.8.108, sampling steps (D_x, D_y) will be $(0.0027, 0.0027)$ (cm). From Eq. 8.103, projective-mapping matrix $^I P_c$ of the virtual camera is

$$
^I P_c = \begin{pmatrix}
365.6059 & 0.0 & 256.0 & 0.0 \\
0.0 & 365.6059 & 256.0 & 0.0 \\
0.0 & 0.0 & 1.0 & 0.0
\end{pmatrix}.
$$

(NOTE: It is assumed that the optical center is at $(256, 256)$).

◇◇◇◇◇◇◇◇◇◇◇◇◇◇◇◇◇◇◇◇◇

Example 8.27 We now position the camera we constructed in the above example at location $(200, -100, 200)$ (cm) in a scene, as shown in Fig. 8.45b. The orientation of the camera frame, with respect to the reference frame, is obtained by rotating the camera frame -100^0 about the X axis. (NOTE: The initial orientation of the camera frame coincides with the reference frame). As a result, the extrinsic parameters of the virtual camera are

$$
^c M_r = \begin{pmatrix}
1.0 & 0.0 & 0.0 & -200.0 \\
0.0 & -0.1736 & -0.9848 & 179.5967 \\
0.0 & 0.9848 & -0.1736 & 133.2104 \\
0.0 & 0.0 & 0.0 & 1.0
\end{pmatrix}
$$

and the camera's calibration matrix is

$$
H = \begin{pmatrix}
2.7446 & 1.8926 & -0.3337 & -292.9149 \\
0 & 1.4160 & -3.0366 & 748.9166 \\
0 & 0.0074 & -0.0013 & 1.0
\end{pmatrix}.
$$

(NOTE: H is normalized by the element h_{34}).

Now, let us choose three points in the reference frame: $A = (200.0, 400.0, 0.0)$ (cm), $B = (100.0, 500.0, 200.0)$ (cm), and $C =$

$(400.0, 500.0, 100.0)$ (cm). If we apply calibration matrix H, the projections of these three points onto the image plane will be: $a = (256, 332)$, $b = (194, 192)$ and $c = (376, 253)$.

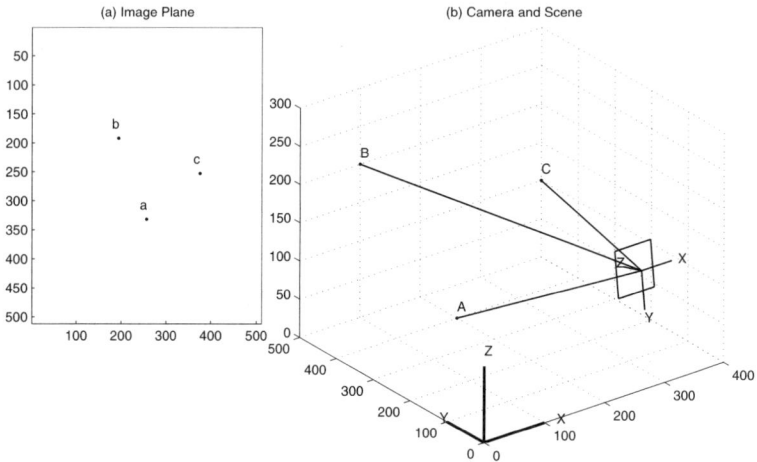

Fig. 8.45   Example of a simulated monocular vision.

◇◇◇◇◇◇◇◇◇◇◇◇◇◇◇◇◇◇

8.6.1.3   *Inverse Projective-Mapping*

Now, let us rewrite Eq. 8.104 as follows:

$$ s \bullet \begin{pmatrix} u \\ v \\ 1 \end{pmatrix} = H \bullet \begin{pmatrix} {}^r X \\ {}^r Y \\ {}^r Z \\ 1 \end{pmatrix}. \tag{8.109} $$

Then, the question is: Is it possible to determine coordinates $({}^r X, {}^r Y, {}^r Z)$, given image point (u, v) and calibration matrix H?

In Eq. 8.109, the scaling factor s is unknown. Therefore, in order to obtain a solution, it is necessary to have at least four constraints. Otherwise, there will be an infinite number of solutions for the unknowns $({}^r X, {}^r Y, {}^r Z)$ and s.

From Eq. 8.109, it is clear that one image point will impose only three constraints on the four unknowns. In order to have a unique solution, there are two possible ways to proceed: a) to increase the number of constraints, or b) to reduce the number of unknowns.

Model-based Inverse Projective-Mapping

Refer to Fig. 8.45. Assume that we have a set of three points (A, B, C), and the relative distances between them are known. Let us denote $\{({}^rX_i, {}^rY_i, {}^rZ_i), \; i = 1, 2, 3\}$ the coordinates of points (A, B, C) with reference to frame r and $\{(u_i, v_i), \; i = 1, 2, 3\}$ the coordinates of the corresponding image points. If we apply Eq. 8.109, we can establish the following system of equations:

$$s_i \bullet \begin{pmatrix} u_i \\ v_i \\ 1 \end{pmatrix} = H \bullet \begin{pmatrix} {}^rX_i \\ {}^rY_i \\ {}^rZ_i \\ 1 \end{pmatrix}, \; i = 1, 2, 3. \tag{8.110}$$

In Eq. 8.110, there are twelves unknowns and nine constraints. Thus, it is necessary to have three more constraints in order to derive a solution. Interestingly enough, the size of the triangle formed by points A, B and C provides three extra constraints. If we denote L_{12}, L_{23} and L_{31} as the lengths of AB, BC, and CA respectively, then

$$\begin{cases} L_{12} = \sqrt{({}^rX_1 - {}^rX_2)^2 + ({}^rY_1 - {}^rY_2)^2 + ({}^rZ_1 - {}^rZ_2)^2} \\ \\ L_{23} = \sqrt{({}^rX_2 - {}^rX_3)^2 + ({}^rY_2 - {}^rY_3)^2 + ({}^rZ_2 - {}^rZ_3)^2} \\ \\ L_{31} = \sqrt{({}^rX_3 - {}^rX_1)^2 + ({}^rY_3 - {}^rY_1)^2 + ({}^rZ_3 - {}^rZ_1)^2}. \end{cases} \tag{8.111}$$

Eq. 8.110 and Eq. 8.111 contain twelve constraints for the twelves unknowns. Thus, a solution exists. However, we must solve the nonlinear equations in the three constraints expressed by Eq. 8.111. This method of determining 3-D coordinates is known as *model-based inverse projective-mapping*, and the nonlinear equations must be numerically solved by a computer program.

2-D Monocular Vision

For some applications, it is likely that the coordinates, in a reference frame, are not all of equal importance. For example, the objects carried by a conveyor in a production line can be treated as two-dimensional (2-D). This is because their locations are uniquely determined by two coordinates with respect to a frame assigned to the conveyor's top surface. These two coordinates are sufficient for a robot to perform pick-and-place actions.

Refer to Eq. 8.109. If we set $^rZ = 0$, the equation becomes

$$s \bullet \begin{pmatrix} u \\ v \\ 1 \end{pmatrix} = H \bullet \begin{pmatrix} ^rX \\ ^rY \\ 0 \\ 1 \end{pmatrix}. \tag{8.112}$$

If calibration matrix H, which has been normalized by its element h_{34}, is expressed as

$$H = \begin{pmatrix} h_{11} & h_{12} & h_{13} & h_{14} \\ h_{21} & h_{22} & h_{23} & h_{24} \\ h_{31} & h_{32} & h_{33} & 1 \end{pmatrix} \tag{8.113}$$

and we define a 3×3 matrix H' as

$$H' = \begin{pmatrix} h_{11} & h_{12} & h_{14} \\ h_{21} & h_{22} & h_{24} \\ h_{31} & h_{32} & 1 \end{pmatrix}. \tag{8.114}$$

Eq. 8.112 can be rewritten as

$$s \bullet \begin{pmatrix} u \\ v \\ 1 \end{pmatrix} = H' \bullet \begin{pmatrix} ^rX \\ ^rY \\ 1 \end{pmatrix}. \tag{8.115}$$

In Eq. 8.115, H' describes the mapping from a planar surface in a scene onto a camera's image plane. As shown in Fig. 8.45, it is clear that the back-projection from image point a has a unique intersection with place $^rZ = 0$ at point A. This indicates that matrix H' has an inverse.

If we denote $D = (H')^{-1}$, Eq. 8.115 becomes

$$\rho \bullet \begin{pmatrix} ^rX \\ ^rY \\ 1 \end{pmatrix} = D \bullet \begin{pmatrix} u \\ v \\ 1 \end{pmatrix} \tag{8.116}$$

where $\rho = 1/s$.

In fact, Eq. 8.116 describes the inverse projective-mapping between a camera's image plane and the planar surface in a reference frame. This mapping can be called the *geometric principle* of 2-D monocular vision (or 2-D vision for short). Interestingly enough, inverse projective-mapping in

2-D vision is unique as long as the posture of the camera, with respect to the planar surface, remains unchanged.

The geometric principle of 2-D vision has practical applications in vision-guided walking or locomotion by a humanoid or mobile robot. We will study these applications in detail in the next chapter. Here, however, we will explore another interesting application of 2-D vision: its use in acquiring three-dimensional models of geometric objects.

Example 8.28   Nowadays, a reliable way of obtaining the 3-D coordinates of the points on an object's surface is to use a laser scanning system, commonly called a *Laser Range-Finder* or *3-D Scanner*. Another interesting approach for 3-D scanner is to use a laser projector which projects a laser plane onto an object placed on a rotating table, as shown in Fig. 8.46a.

The laser plane intersects with an object placed in front of it and this intersection creates a curved line on the object's surface. This line can easily be detected in the camera's image plane because a laser beam creates a very bright, curved line on an object. If we know matrix D which describes the inverse projective-mapping from the image plane to the laser plane, the X and Y coordinates of points on the intersection line can be computed by Eq. 8.116. If we rotate the object being scanned, a set of intersection lines can be obtained. The coordinates of the points on these lines can be computed by Eq. 8.116.

If we know rotation angle $k \bullet \triangle \theta$ which corresponds to the kth intersection line, the coordinates $\{^{r}X(k), \, ^{r}Y(k), \, ^{r}Z(k)\}$ of a point on this intersection line with reference to frame r can be determined as follows:

$$\begin{pmatrix} ^{r}X(k) \\ ^{r}Y(k) \\ ^{r}Z(k) \end{pmatrix} = \begin{pmatrix} \cos(k \bullet \theta) & 0 & \sin(k \bullet \theta) \\ 0 & 1 & 0 \\ -\sin(k \bullet \theta) & 0 & \cos(k \bullet \theta) \end{pmatrix} \bullet \begin{pmatrix} ^{l}X(k) \\ ^{l}Y(k) \\ 0 \end{pmatrix}$$

where $(^{l}X(k), \, ^{l}Y(k))$ are the coordinates of the point with reference to the frame assigned to the laser plane.

If we add all the coordinates $(^{r}X(k), \, ^{r}Y(k), \, ^{r}Z(k))$ together, the result is the set of 3-D coordinates of the scanned points on an object's surface(s). Fig. 8.46b shows one view of the result.

◇◇◇◇◇◇◇◇◇◇◇◇◇◇◇◇◇◇◇

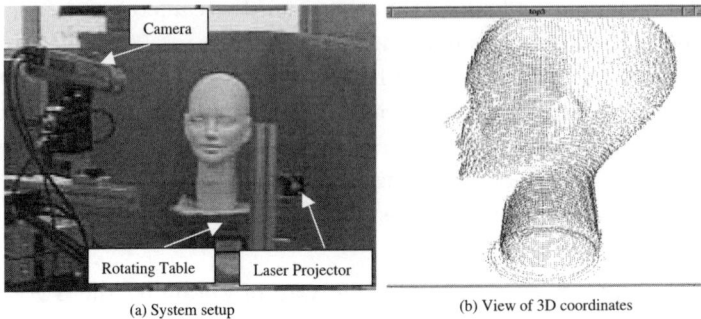

(a) System setup   (b) View of 3D coordinates

Fig. 8.46   Example of the application of 2-D monocular vision for the acquisition of a three-dimensional model of an object.

8.6.1.4   *Camera Calibration*

From our study of a camera's frame assignment in Chapter 7, we know that the frame assigned to a camera is located at the center of the camera's optical lens, and this center is not directly accessible. Therefore, for a camera randomly placed in a scene, we will not know motion-transformation matrix ^cM_r. This means, we will also not know the camera's extrinsic parameters in advance.

If a camera's extrinsic parameters are unknown, calibration matrix H will also not be known even if its intrinsic parameters are available. Without the calibration matrix, it is not possible for monocular vision to make any useful geometric measurements. Thus, it's necessary to estimate calibration matrix H using a process commonly known as *camera calibration*. The process of camera calibration can be explained as follows:

Substituting Eq. 8.113 into Eq. 8.112 yields

$$\begin{pmatrix} s \bullet u \\ s \bullet v \\ s \end{pmatrix} = \begin{pmatrix} h_{11} & h_{12} & h_{13} & h_{14} \\ h_{21} & h_{22} & h_{23} & h_{24} \\ h_{31} & h_{32} & h_{33} & 1 \end{pmatrix} \bullet \begin{pmatrix} ^rX \\ ^rY \\ ^rZ \\ 1 \end{pmatrix}. \tag{8.117}$$

If we eliminate unknown s in Eq. 8.117, we obtain

$$\begin{cases} u = \frac{^rX h_{11} + {}^rY h_{12} + {}^rZ h_{13} + h_{14}}{^rX h_{31} + {}^rY h_{32} + {}^rZ h_{33} + 1} \\ \\ v = \frac{^rX h_{21} + {}^rY h_{22} + {}^rZ h_{23} + h_{24}}{^rX h_{31} + {}^rY h_{32} + {}^rZ h_{33} + 1}. \end{cases} \tag{8.118}$$

If we define

$$A = \begin{bmatrix} {}^rX & {}^rY & {}^rZ & 0 & 0 & 0 & 0 & -({}^rX \bullet u) & -({}^rY \bullet u) & -({}^rZ \bullet u) \\ 0 & 0 & 0 & 0 & {}^rX & {}^rY & {}^rZ & -({}^rX \bullet v) & -({}^rY \bullet v) & -({}^rZ \bullet v) \end{bmatrix},$$

$$B = \begin{bmatrix} u \\ v \end{bmatrix}$$

and

$$V = \begin{bmatrix} h_{11} & h_{12} & h_{13} & h_{14} & h_{21} & h_{22} & h_{23} & h_{24} & h_{31} & h_{32} & h_{33} \end{bmatrix}^t,$$

Eq. 8.118 can be compactly written in a matrix form, as follows:

$$A \bullet V = B. \tag{8.119}$$

Eq. 8.119 describes the linear relationship among vectors A, V and B. When we have the coordinates $({}^rX, {}^rY, {}^rZ)$ of a point in a scene, and its image coordinates (u, v) are known, Eq. 8.119 imposes two constraints on the eleven unknowns in vector V. In order to derive a unique solution for the eleven unknowns in V, we must have at least eleven constraints. Since one pair of $\{(u, v), ({}^rX, {}^rY, {}^rZ)\}$ provides two constraints, it is necessary to have at least six pairs of $\{(u, v), ({}^rX, {}^rY, {}^rZ)\}$ in order to calibrate a camera.

Assume that a set of n pairs of $\{(u, v), ({}^rX, {}^rY, {}^rZ)\}$ are available ($n \geq 6$). The optimal solution which minimizes the squared error $\|A \bullet V - B\|^2$ is

$$V = (A^t A)^{-1} \bullet (A^t B). \tag{8.120}$$

Example 8.29 A common way to manually calibrate a camera is to use a calibration rig. A calibration rig can be as simple as an array of dot patterns painted on a planar surface, as shown in Fig. 8.47. In this example, a 3×3 array of dot patterns are painted on a planar surface. The distances among the adjacent dots are $(100.0mm, 100.0mm)$. Then, we place the calibration rig at three different locations along the Z axis, as shown in Fig. 8.47. The camera under calibration will output three images, which are the input to the process for camera calibration.

In each image, the dots of the calibration rig can be detected by a simple thresholding technique because the dots are in dark color on a rectangular white region. After thresholding, each dot produces a cluster of pixels. And, the center of gravity of each cluster indicates the location of

(a) At Z= 0 (mm)   (b) At Z= 200 (mm)   (c) At Z= 400 (mm)

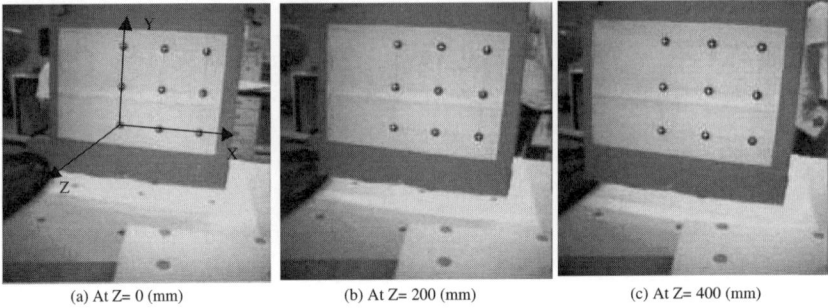

Fig. 8.47   Example of images used for camera calibration.

a dot in the image plane. As a result, we have twenty-seven (27) pairs of $\{(u,v),\ (^rX,\ ^rY,\ ^rZ)\}$, as shown in Table 8.1.

If we apply Eq. 8.120, the estimated calibration matrix of the camera is

$$H = \begin{pmatrix} 0.337599 & -0.004174 & -0.068044 & 112.500411 \\ 0.015303 & -0.367294 & -0.013599 & 110.527718 \\ -0.000209 & -0.000212 & -0.000322 & 1.0 \end{pmatrix}.$$

◇◇◇◇◇◇◇◇◇◇◇◇◇◇◇◇◇◇◇◇

8.6.1.5   *Determining the Parameters of Cameras*

Interestingly enough, when we have a camera's calibration matrix as input, we can compute its intrinsic and extrinsic parameters.

Refer to Eq. 8.104. We express motion transformation matrix ^cM_r as follows:

$$^cM_r = \begin{pmatrix} R_1 & t_x \\ R_2 & t_y \\ R_3 & t_z \\ \vec{0} & 1 \end{pmatrix} \tag{8.121}$$

with

$$\begin{cases} R_1 = \begin{pmatrix} r_{11} & r_{12} & r_{13} \end{pmatrix} \\ R_2 = \begin{pmatrix} r_{21} & r_{22} & r_{23} \end{pmatrix} \\ R_3 = \begin{pmatrix} r_{31} & r_{32} & r_{33} \end{pmatrix}. \end{cases}$$

Since R_1, R_2 and R_3 are the row vectors of the rotation matrix in ^cM_r, they are unit vectors. Moreover, they are mutually orthogonal. Accord-

Table 8.1   Camera Calibration Data

Nb	u	v	^rX	^rY	^rZ
1	116.210	38.842	0.0	200.0	0.0
2	155.200	41.075	100.0	200.0	0.0
3	195.667	44.000	200.0	200.0	0.0
4	114.500	75.500	0.0	100.0	0.0
5	152.294	78.500	100.0	100.0	0.0
6	191.891	82.027	200.0	100.0	0.0
7	112.451	110.580	0.0	0.0	0.0
8	149.500	114.500	100.0	0.0	0.0
9	187.852	118.382	200.0	0.0	0.0
10	109.814	38.512	0.0	200.0	200.0
11	151.087	41.000	100.0	200.0	200.0
12	194.400	44.140	200.0	200.0	200.0
13	107.789	77.842	0.0	100.0	200.0
14	148.128	81.128	100.0	100.0	200.0
15	190.444	84.933	200.0	100.0	200.0
16	105.806	115.055	0.0	0.0	200.0
17	145.102	119.641	100.0	0.0	200.0
18	186.075	123.950	200.0	0.0	200.0
19	102.160	38.260	0.0	200.0	400.0
20	146.358	40.679	100.0	200.0	400.0
21	193.175	44.175	200.0	200.0	400.0
22	99.808	80.362	0.0	100.0	400.0
23	143.081	84.102	100.0	100.0	400.0
24	188.500	88.500	200.0	100.0	400.0
25	97.928	120.690	0.0	0.0	400.0
26	139.837	125.674	100.0	0.0	400.0
27	184.146	130.583	200.0	0.0	400.0

ingly, we have

$$\|R_1\| = 1, \quad \|R_2\| = 1, \quad \|R_3\| = 1, \quad R_1 R_2^t = 0, \quad R_2 R_3^t = 0, \quad R_1 R_3^t = 0.$$

Then, substituting Eq. 8.121 into Eq. 8.104 yields

$$s \bullet \begin{pmatrix} u \\ v \\ 1 \end{pmatrix} = t_z \bullet \begin{pmatrix} f_x \frac{R_1}{t_z} + u_0 \frac{R_3}{t_z} & f_x \frac{t_x}{t_z} + u_0 \\ f_y \frac{R_2}{t_z} + v_0 \frac{R_3}{t_z} & f_y \frac{t_x}{t_z} + v_0 \\ \frac{R_3}{t_z} & 1 \end{pmatrix} \bullet \begin{pmatrix} ^rX \\ ^rY \\ ^rZ \\ 1 \end{pmatrix} \qquad (8.122)$$

where $f_x = f_c/D_x$ and $f_y = f_c/D_y$.

Since Eq. 8.117 and Eq. 8.122 are equal up to a scaling factor, we have

$$\begin{pmatrix} f_x \frac{R_1}{t_z} + u_0 \frac{R_3}{t_z} & f_x \frac{t_x}{t_z} + u_0 \\ f_y \frac{R_2}{t_z} + v_0 \frac{R_3}{t_z} & f_y \frac{t_x}{t_z} + v_0 \\ \frac{R_3}{t_z} & 1 \end{pmatrix} = \begin{pmatrix} h_{11} & h_{12} & h_{13} & h_{14} \\ h_{21} & h_{22} & h_{23} & h_{24} \\ h_{31} & h_{32} & h_{33} & 1 \end{pmatrix}. \tag{8.123}$$

From Eq. 8.123, we know the following equalities must hold:

$$\begin{cases} f_x \frac{R_1}{t_z} + u_0 \frac{R_3}{t_z} = \begin{pmatrix} h_{11} & h_{12} & h_{13} \end{pmatrix} \\[2mm] f_y \frac{R_2}{t_z} + v_0 \frac{R_3}{t_z} = \begin{pmatrix} h_{21} & h_{22} & h_{23} \end{pmatrix} \\[2mm] \frac{R_3}{t_z} = \begin{pmatrix} h_{31} & h_{32} & h_{33} \end{pmatrix} \\[2mm] f_x \frac{t_x}{t_z} + u_0 = h_{14} \\[2mm] f_y \frac{t_y}{t_z} + v_0 = h_{24}. \end{cases} \tag{8.124}$$

Eq. 8.124 contains a sufficient number of constraints. If a camera's calibration matrix H is known, it is possible to derive its intrinsic and extrinsic parameters as follows:

Solution for t_z

If we apply the property $\|R_3\| = 1$ to the third equality in Eq. 8.124, we can directly obtain the following solution for t_z:

$$t_z = \frac{1}{\sqrt{h_{31}^2 + h_{32}^2 + h_{33}^2}}. \tag{8.125}$$

Solution for R_3

If we know t_z, the solution for R_3 can be directly derived from the third equality in Eq. 8.124 as follows:

$$R_3 = t_z \bullet \begin{pmatrix} h_{31} & h_{32} & h_{33}. \end{pmatrix} \tag{8.126}$$

Solution for u_0

First multiply R_3^t to both sides of the first equality in Eq. 8.124. Then, apply properties $R_1 R_3^t = 0$ and $R_3 R_3^t = 1$. The result will be

$$u_0 = \begin{pmatrix} h_{11} & h_{12} & h_{13} \end{pmatrix} \bullet (t_z R_3^t). \tag{8.127}$$

Solution for v_0

Similarly, multiply R_3^t to both sides of the second equality in Eq. 8.124. Then, apply properties $R_2 R_3^t = 0$ and $R_3 R_3^t = 1$. The result will be

$$v_0 = \left(h_{21} \ h_{22} \ h_{23} \right) \bullet (t_z R_3^t). \tag{8.128}$$

Solution for f_x

From the first equality in Eq. 8.124, we have

$$f_x R_1 = t_z \left(h_{11} \ h_{12} \ h_{13} \right) - u_0 R_3.$$

Since $\|R_1\| = 1$, the computation of the norm on both sides of the above equation yields

$$f_x = \| t_z \left(h_{11} \ h_{12} \ h_{13} \right) - u_0 R_3 \|. \tag{8.129}$$

Solution for f_y

Similarly, from the second equality in Eq. 8.124, we have

$$f_y R_2 = t_z \left(h_{21} \ h_{22} \ h_{23} \right) - v_0 R_3.$$

If we compute the norm on both sides of the above equation, and apply the property $\|R_2\| = 1$, the solution for f_y will be

$$f_y = \| t_z \left(h_{21} \ h_{22} \ h_{23} \right) - v_0 R_3 \|. \tag{8.130}$$

Solution for R_1 and R_2

From the first and second equalities in Eq. 8.124, we can directly write down the solution for R_1 and R_2, as follows:

$$\begin{cases} R_1 = \frac{t_z}{f_x} \left(h_{11} \ h_{12} \ h_{13} \right) - \frac{u_0}{f_x} R_3 \\[2mm] R_2 = \frac{t_z}{f_y} \left(h_{21} \ h_{22} \ h_{23} \right) - \frac{v_0}{f_y} R_3. \end{cases} \tag{8.131}$$

Solution for t_x and t_y

Finally, the fourth and fifth equalities in Eq. 8.124 directly produce the solution for t_x and t_y as follows:

$$\begin{cases} t_x = \frac{t_z}{f_x}(h_{14} - u_0) \\ t_y = \frac{t_z}{f_y}(h_{24} - v_0). \end{cases} \qquad (8.132)$$

Example 8.30 Assume that a camera's calibration matrix is

$$H = \begin{pmatrix} 2.7446 & 1.8926 & -0.3337 & -292.9149 \\ 0 & 1.4160 & -3.0366 & 748.9166 \\ 0 & 0.0074 & -0.0013 & 1.0 \end{pmatrix}.$$

If we apply the solutions discussed above, the results of the camera's intrinsic and extrinsic parameters are as follows:

$$(f_x, f_y, u_0, v_0) = (365.5961,\ 365.5822,\ 255.9932,\ 256.0130)$$

and

$$\begin{cases} R_1 = (1.0,\ 0.0,\ 0.0) \\ R_2 = (0,\ -0.1737,\ -0.9848) \\ R_3 = (0,\ 0.9848,\ -0.1737) \\ (t_x, t_y, t_z) = (-199.9976,\ 179.5989,\ 133.2069). \end{cases}$$

The ground-truth data in this example can be found in Example 8.26 and Example 8.27. One can see that the estimated parameters are very close to the ground-truth data.

◇◇◇◇◇◇◇◇◇◇◇◇◇◇◇◇◇◇◇

Example 8.31 For vision-guided walking (or locomotion) on a man-made road, as shown in Fig. 8.48, it is necessary to estimate the robot's walking direction. Assume that the camera in a monocular vision system has been calibrated and its parameters are known. Then, the interesting question is: What is the walking direction if the robot is programmed to follow a target seen at location $q = (u, v)$ in a digital image?

If we know index coordinates (u, v) of the target, its corresponding image coordinates (x, y) will be

$$\begin{cases} x = D_x \bullet (u - u_0) \\ y = D_y \bullet (v - v_0). \end{cases}$$

If the focal length of the camera is f_c, then the vector which indicates the walking direction expressed in camera frame c, will be

$$^cV = (x, y, f_c)$$

or

$$^cV = [D_x \bullet (u - u_0), \ D_y \bullet (v - v_0), f_c].$$

When we normalize vector ^cV with scaling factor f_c, the walking direction becomes

$$^cV = \left(\frac{u - u_0}{f_x}, \ \frac{v - v_0}{f_y}, \ 1 \right).$$

This walking direction indicates the direction of the projection line passing through point a.

If the camera's extrinsic parameters in ^cM_r are known, it is easy to compute the inverse of ^cM_r. Accordingly, the walking direction with respect to reference frame r will be

$$^rV = \ (^cM_r)^{-1} \bullet^c V.$$

Clearly, the projection of vector ^rV onto the road surface indicates the robot's walking direction.

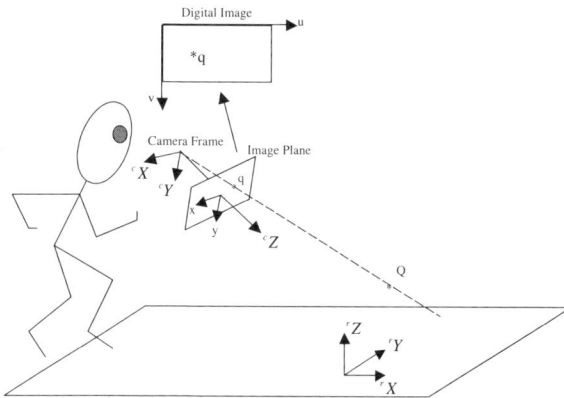

Fig. 8.48   Example of vision-guided walking.

◇◇◇◇◇◇◇◇◇◇◇◇◇◇◇◇◇◇◇

8.6.2 *Binocular Vision*

From our study of monocular vision, we know that if there is no additional knowledge about the model of an object, the coordinates of a point in a three-dimensional space cannot be determined from the forward projective-mapping of a single camera. This is because there are four unknowns, as shown in Eq. 8.109, but only three constraints.

Instead of specifying the model of an object (the relative geometry of three points), one may question whether it is possible to determine the coordinates of a point in a three-dimensional scene if the displacement vector of this point, with reference to frame r, is known. Interestingly enough, this is possible. For example, let $(^rX,\ ^rY,\ ^rZ)$ be the coordinates of point A in a scene, and $(\triangle X, \triangle Y, \triangle Z)$ be its displacement vector. From Eq. 8.109, we can establish the following two systems of equations:

$$s_1 \bullet \begin{pmatrix} u \\ v \\ 1 \end{pmatrix} = H \bullet \begin{pmatrix} ^rX \\ ^rY \\ ^rZ \\ 1 \end{pmatrix} \tag{8.133}$$

and

$$s_2 \bullet \begin{pmatrix} u + \triangle u \\ v + \triangle v \\ 1 \end{pmatrix} = H \bullet \begin{pmatrix} ^rX + \triangle X \\ ^rY + \triangle Y \\ ^rZ + \triangle Z \\ 1 \end{pmatrix}. \tag{8.134}$$

Clearly, we have five unknowns and six constraints. As a result, the coordinates of point A can be uniquely determined. This principle is called *motion stereo* or *dynamic monocular vision*. An even more challenging question is whether it is possible to determine the three-dimensional structure of a scene when the displacement vector of a scene or object is unknown. This problem is known as *structure from motion* in computer vision.

For dynamic monocular vision to work properly, there is an implicit assumption that the camera must be stationary with respect to a scene. Mathematically, the displacement of an object (or scene), with respect to a stationary camera, is equivalent to the displacement of a camera with respect to a stationary object (or scene). In other words, if an object is stationary, one can move the camera to two different locations. The coordinates of any point on the object can then be uniquely determined if the displacement vector of the camera is known.

In a dynamically changing environment, the constraint on the immobility of an object (or a camera) may not be acceptable. Mathematically, Eq. 8.133 and Eq. 8.134 illustrate that the coordinates of a point in a three-dimensional space can be fully determined if two locations and the images captured at these two locations are known in advance. Instead of moving a camera to these two locations, we can simultaneously place two cameras at these two locations. This, then, becomes the well-known *binocular vision*.

8.6.2.1 *Forward Projective-Mapping*

As shown in Fig. 8.49, there are two cameras, each of which has been assigned with a camera frame. Let us assume that frame c_1 is assigned to the left camera and frame c_2 to the right camera. In addition, we assign a common reference frame (frame r) to the scene.

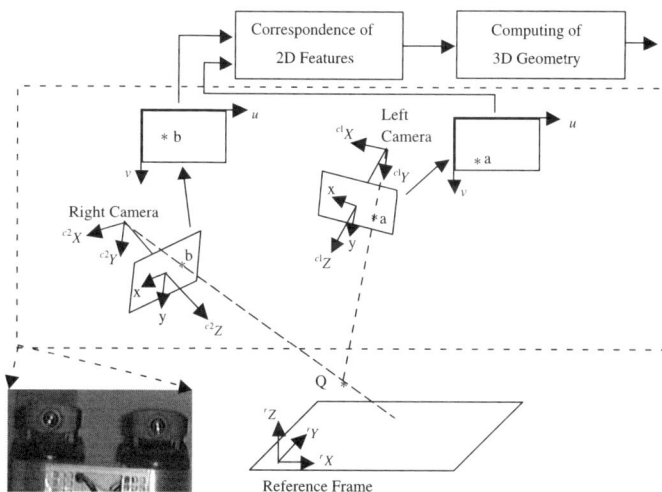

Fig. 8.49   Illustration of the geometric projection in binocular vision.

Refer to Fig. 8.49 again. It is unlikely that the two cameras used for the binocular vision will be identical. Thus, the intrinsic and extrinsic parameters of these two cameras will not be the same. Therefore, we have to calibrate these two cameras separately.

Now, let us denote H_1 the calibration matrix of the left camera, and H_2 the calibration matrix of the right camera. Consider point Q in the scene. Its coordinates are $(^rX,\ ^rY,\ ^rZ)$. The projection of point Q onto the left camera's image plane is $a = (u_1, v_1)$, and its projection onto the right camera's image plane is $b = (u_2, v_2)$. If we apply Eq. 8.109, we obtain

$$s_1 \bullet \begin{pmatrix} u_1 \\ v_1 \\ 1 \end{pmatrix} = H_1 \bullet \begin{pmatrix} {}^rX \\ {}^rY \\ {}^rZ \\ 1 \end{pmatrix} \qquad (8.135)$$

and

$$s_2 \bullet \begin{pmatrix} u_2 \\ v_2 \\ 1 \end{pmatrix} = H_2 \bullet \begin{pmatrix} {}^rX \\ {}^rY \\ {}^rZ \\ 1 \end{pmatrix}. \qquad (8.136)$$

Eq. 8.135 and Eq. 8.136 describe the forward projective-mapping in a binocular vision system.

Example 8.32 Let us construct two virtual cameras in order to form virtual binocular vision. Fig. 8.50c shows a virtual scene and two virtual cameras. The calibration matrix of the left camera is

$$H_1 = \begin{pmatrix} 2.7446 & 1.8926 & -0.3337 & -18.4574 \\ 0 & 1.4160 & -3.0366 & 748.9166 \\ 0 & 0.0074 & -0.0013 & 1.0 \end{pmatrix}$$

and the calibration matrix of the right camera is

$$H_2 = \begin{pmatrix} 2.7446 & 1.8926 & -0.3337 & -238.0234 \\ 0 & 1.4160 & -3.0366 & 748.9166 \\ 0 & 0.0074 & -0.0013 & 1.0 \end{pmatrix}.$$

When we have point A, and its coordinates in reference frame r are

$$A = (200.0,\ 500.0,\ 100.0)\ (cm),$$

then the projection of point A onto the image planes of the two virtual cameras is

$$\begin{cases} a = (u_1, v_1) = (316,\ 253) \\ b = (u_2, v_2) = (268,\ 253). \end{cases}$$

The locations of image points a and b are shown in Fig. 8.50a and Fig. 8.50b respectively.

◇◇◇◇◇◇◇◇◇◇◇◇◇◇◇◇◇◇◇

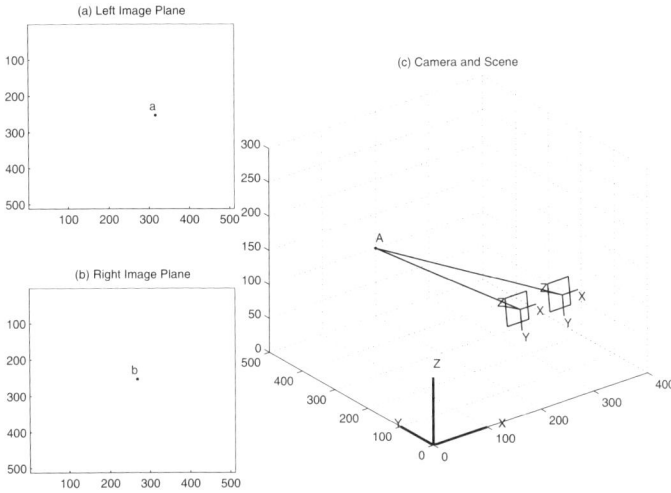

Fig. 8.50　Example of forward projective-mapping in virtual binocular vision.

8.6.2.2　*Inverse Projective-Mapping*

One primary goal we hope to achieve with a robot's visual-perception system is that it is capable of inferring the geometry of objects in a three-dimensional space. This means that inverse projective-mapping is a major concern. And, the question here is whether it is possible to determine the coordinates of a point in a three-dimensional scene from its images in binocular vision.

Refer to Fig. 8.49. Assume that the correspondence between a pair of image feature points (a, b) has been established. In other words, the index coordinates of feature points a and b are given as input. We would like to know how to determine the coordinates of point Q.

Let us denote the coordinates of feature points a and b as

- $a = (u_1, v_1)$, and
- $b = (u_2, v_2)$.

Moreover, let us represent the two calibration matrices in binocular vision as follows:

$$H_1 = \begin{pmatrix} i_{11} & i_{12} & i_{13} & i_{14} \\ i_{21} & i_{22} & i_{23} & i_{24} \\ i_{31} & i_{32} & i_{33} & 1 \end{pmatrix} \qquad (8.137)$$

and

$$H_2 = \begin{pmatrix} j_{11} & j_{12} & j_{13} & j_{14} \\ j_{21} & j_{22} & j_{23} & j_{24} \\ j_{31} & j_{32} & j_{33} & 1 \end{pmatrix}. \tag{8.138}$$

If we apply Eq. 8.135, we have

$$\begin{cases} s_1 \bullet u_1 = i_{11} \bullet {}^rX + i_{12} \bullet {}^rY + i_{13} \bullet {}^rZ + i_{14} \\ s_1 \bullet v_1 = i_{21} \bullet {}^rX + i_{22} \bullet {}^rY + i_{23} \bullet {}^rZ + i_{24} \\ s_1 = i_{31} \bullet {}^rX + i_{32} \bullet {}^rY + i_{33} \bullet {}^rZ + 1 \end{cases} \tag{8.139}$$

where $({}^rX, {}^rY, {}^rZ)$ are the coordinates of point Q. Now, let us eliminate scaling factor s_1 from Eq. 8.139. This results in

$$\begin{cases} (i_{11} - i_{31}u_1) \bullet {}^rX + (i_{12} - i_{32}u_1) \bullet {}^rY + (i_{13} - i_{33}u_1) \bullet {}^rZ = u_1 - i_{14} \\ (i_{21} - i_{31}v_1) \bullet {}^rX + (i_{22} - i_{32}v_1) \bullet {}^rY + (i_{23} - i_{33}v_1) \bullet {}^rZ = v_1 - i_{24}. \end{cases} \tag{8.140}$$

If we do the same to Eq. 8.136, we will obtain

$$\begin{cases} (j_{11} - j_{31}u_2) \bullet {}^rX + (j_{12} - j_{32}u_2) \bullet {}^rY + (j_{13} - j_{33}u_2) \bullet {}^rZ = u_2 - j_{14} \\ (j_{21} - j_{31}v_2) \bullet {}^rX + (j_{22} - j_{32}v_2) \bullet {}^rY + (j_{23} - j_{33}v_2) \bullet {}^rZ = v_2 - j_{24}. \end{cases} \tag{8.141}$$

If we define

$$A = \begin{pmatrix} i_{11} - i_{31}u_1 & i_{12} - i_{32}u_1 & i_{13} - i_{33}u_1 \\ i_{21} - i_{31}v_1 & i_{22} - i_{32}v_1 & i_{23} - i_{33}v_1 \\ j_{11} - j_{31}u_2 & j_{12} - j_{32}u_2 & j_{13} - j_{33}u_2 \\ j_{21} - j_{31}v_2 & j_{22} - j_{32}v_2 & j_{23} - j_{33}v_2 \end{pmatrix}, \tag{8.142}$$

Eq. 8.140 and Eq. 8.141 can be combined into the following matrix form:

$$A \bullet \begin{pmatrix} {}^rX \\ {}^rY \\ {}^rZ \end{pmatrix} = \begin{pmatrix} u_1 - i_{14} \\ v_1 - i_{24} \\ u_2 - j_{14} \\ v_2 - j_{24} \end{pmatrix}. \tag{8.143}$$

Accordingly, the optimal solution which minimizes the squared errors is

$$\begin{pmatrix} {}^rX \\ {}^rY \\ {}^rZ \end{pmatrix} = (A^t A)^{-1} \bullet (A^t B) \tag{8.144}$$

with

$$B = \begin{pmatrix} u_1 - i_{14} \\ v_1 - i_{24} \\ u_2 - j_{14} \\ v_2 - j_{24} \end{pmatrix}.$$ (8.145)

Eq. 8.144 describes inverse projective-mapping in a binocular-vision system. We can see that all the parameters and variables in Eq. 8.144 have values within a continuous range. For easy reference, we call it *direct 3-D reconstruction*.

Example 8.33 Let us use the same virtual binocular vision set up in Example 8.32. Assume that we have a pair of image-feature points (a, b) as follows:

$$\begin{cases} a = (u_1, v_1) = (300, \ 200) \\ b = (u_2, v_2) = (200, \ 200). \end{cases}$$

From Eq. 8.142, we obtain

$$A = \begin{pmatrix} 2.7446 & -0.3253 & 0.0574 \\ 0 & -0.0626 & -2.7759 \\ 2.7446 & 0.4140 & -0.0730 \\ 0 & -0.0626 & -2.7759 \end{pmatrix}.$$

From Eq. 8.145, we can see that vector B is:

$$B = \begin{pmatrix} 318.4574 \\ -548.9166 \\ 438.0234 \\ -548.9166 \end{pmatrix}.$$

Finally, the coordinates of the object point with respect to reference frame r assigned to the scene can be computed using Eq. 8.144. And, the results are

$$({}^r X, \ {}^r Y, \ {}^r Z) = (35.2cm, \ 195.8207cm, \ 193.33cm).$$

◇◇◇◇◇◇◇◇◇◇◇◇◇◇◇◇◇◇

Example 8.34 Figure 8.51 shows an example of the results obtained using binocular vision mounted on a mobile robot. An object (a box) is placed on the floor. The two images captured by the robot's binocular vision are displayed in Fig. 8.51a and Fig. 8.51b.

The vision process starts with edge detection and is followed by edge linking. Subsequently, simple contours are approximated by line segments. Then, a critical step is establishing correspondence between the line segments in the left image and the line segments in the right image. Here, let us assume that the characteristics of the two cameras in the binocular vision are similar. In this way, the two images are treated as a pair of consecutive images captured by a single camera, and an algorithm on the detection of temporal uniformity (template matching) is used to establish the correspondence of edges. If we know the correspondence of the edges, then the correspondence of line segments is decided by a voting technique. The final results of binocular correspondence are shown in Fig. 8.51c.

Finally, if the calibration matrices of the two cameras in binocular vision are known, the coordinates of the endpoints of the line segments with respect to frame r assigned to the scene can be computed. Fig. 8.51d shows one view of the results of 3-D line segments.

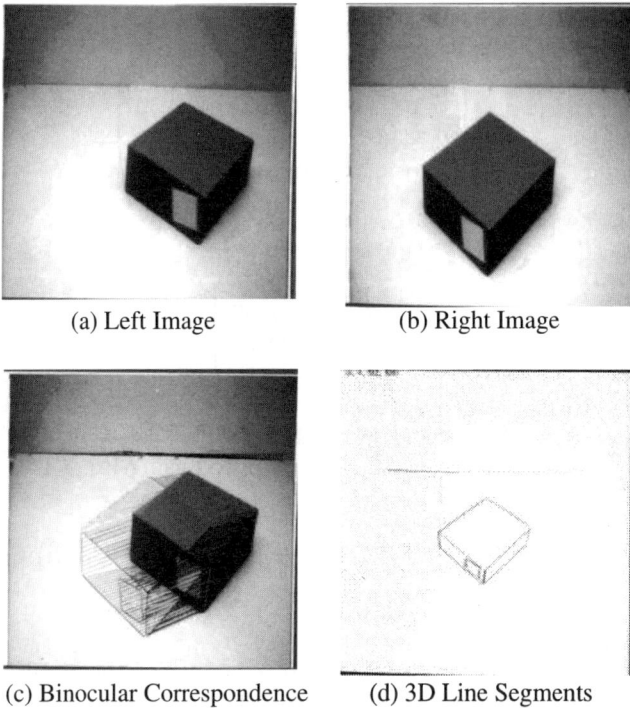

(a) Left Image   (b) Right Image

(c) Binocular Correspondence   (d) 3D Line Segments

Fig. 8.51   Example of 3-D reconstruction using binocular vision.

◇◇◇◇◇◇◇◇◇◇◇◇◇◇◇◇◇

8.6.2.3   *Unresolved Issue: Binocular Correspondence*

For feature points, inverse projective-mapping in binocular vision has a closed-form solution. (See Eq. 8.144). However, this solution depends on two necessary conditions:

- Necessary Condition 1:
  The two cameras in the binocular vision must be calibrated. In other words, the calibration matrices (H_1, H_2) must be known in advance.
- Necessary Condition 2:
  When we have image-feature point a in the left image, its corresponding point b in the right image must be known in advance. Refer to Fig. 8.49. This indicates that the determination of 3-D geometry through binocular vision is only possible if the correspondence of image-features has already been established.

The first necessary condition is easy to meet because it is not difficult to manually calibrate a camera. However, for some applications, it is still not desirable to know the calibration matrices. We will study this issue in the next chapter when we discuss the iterative approaches to robotic limb-eye coordination (or, image-space control).

The second necessary condition, known as *binocular correspondence*, is still a major problem of the visual perception system. This problem is hard to solve for two reasons:

(1) Difficulty 1:
   Images in binocular vision are digital images. When we have image-feature point a in the left image, we have to search for its corresponding point in the right image which is a two-dimensional array of pixels. If we have to verify all the possible locations in the right image, the search will be computationally prohibitive.
(2) Difficulty 2:
   The binocular vision's two cameras may not be optically and electronically identical. This means that it is unlikely that the intrinsic parameters of these two cameras will be the same. And, since these two cameras are not placed in the same location, their extrinsic parameters will not be the same either. Since the parameters will be different, these two cameras will "see" the same object differently in terms of geometry and appearance.

8.6.2.4 Continuous Epipolar-Line Constraint

A common solution which addresses the first difficulty with binocular correspondence is to use the *epipolar-line constraint*. This constraint helps limit the search for correspondence to within an interval along a line. This results in the reduction of search space for binocular correspondence. Here, we will derive the (continuous) epipolar-line constraint in a progressive manner.

Refer to Fig. 8.49. Assume that a pair of image-feature points (a, b) are projections of the same object point onto the binocular vision's two image planes. Now, let us examine what constraint is imposed onto the index coordinates of these two corresponding feature points (a, b). Let us denote the index coordinates of these two feature points as follows:

$$\begin{cases} a = (u_1, v_1) \\ b = (u_2, v_2). \end{cases}$$

As we discussed in Example 8.31, the direction of the projection line passing through feature point a in the left camera, is

$$^{c1}V_a = \left(\frac{u_1 - u_{1,0}}{f_{1,x}}, \; \frac{v_1 - v_{1,0}}{f_{1,y}}, \; 1 \right). \tag{8.146}$$

Vector $^{c1}V_a$ is expressed with respect to camera frame $c1$ and $(u_{1,0}, v_{1,0}, f_{1,x}, f_{1,y})$ are the left camera's intrinsic parameters. In fact, Eq. 8.146 can also be written in a matrix form as follows:

$$^{c1}V_a = \begin{pmatrix} \frac{1}{f_{1,x}} & 0 & -\frac{u_{1,0}}{f_{1,x}} \\ 0 & \frac{1}{f_{1,y}} & -\frac{v_{1,0}}{f_{1,y}} \\ 0 & 0 & 1 \end{pmatrix} \bullet \begin{pmatrix} u_1 \\ v_1 \\ 1 \end{pmatrix}. \tag{8.147}$$

Similarly, for feature point b in the right camera, we have the following result:

$$^{c2}V_b = \begin{pmatrix} \frac{1}{f_{2,x}} & 0 & -\frac{u_{2,0}}{f_{2,x}} \\ 0 & \frac{1}{f_{2,y}} & -\frac{v_{2,0}}{f_{2,y}} \\ 0 & 0 & 1 \end{pmatrix} \bullet \begin{pmatrix} u_2 \\ v_2 \\ 1 \end{pmatrix}. \tag{8.148}$$

where $(u_{2,0}, v_{2,0}, f_{2,x}, f_{2,y})$ are the right camera's intrinsic parameters.

If image-feature points (a, b) are projections of the same object point onto the binocular vision's two image planes, vectors $^{c1}V_a$ and $^{c2}V_b$ will intersect. Now, assume that the motion transformation between the left and right cameras is

$$^{c1}M_{c2} = \begin{pmatrix} ^{c1}R_{c2} & ^{c1}T_{c2} \\ \vec{0} & 1 \end{pmatrix} \tag{8.149}$$

where $^{c1}R_{c2}$ is the orientation of camera frame $c2$ with respect to camera frame $c1$, and $^{c1}T_{c2}$ is the origin of camera frame $c2$ with respect to camera frame $c1$.

Since $^{c2}V_b$ is a vector expressed with respect to camera frame $c2$, the same vector expressed with respect to camera frame $c1$ will be

$$^{c1}V_b = (^{c1}R_{c2}) \bullet ^{c2}V_b. \tag{8.150}$$

Mathematically, when $^{c1}T_{c2} = (0,0,0)^t$, it means that the binocular vision's two cameras coincide at the origins of their assigned frames. When the origins of the frames assigned to these two cameras are at the same location, the two projection lines passing through feature points (a, b) will be superimposed. Then, we will have

$$^{c1}V_a = ^{c1}V_b.$$

By applying Eq. 8.147, Eq. 8.148 and Eq. 8.150, the above equality becomes

$$\begin{pmatrix} u_1 \\ v_1 \\ 1 \end{pmatrix} = M_{3\times3} \bullet \begin{pmatrix} u_2 \\ v_2 \\ 1 \end{pmatrix} \tag{8.151}$$

with

$$M_{3\times3} = \begin{pmatrix} \frac{1}{f_{1,x}} & 0 & -\frac{u_{1,0}}{f_{1,x}} \\ 0 & \frac{1}{f_{1,y}} & -\frac{v_{1,0}}{f_{1,y}} \\ 0 & 0 & 1 \end{pmatrix}^{-1} \bullet (^{c1}R_{c2}) \bullet \begin{pmatrix} \frac{1}{f_{2,x}} & 0 & -\frac{u_{2,0}}{f_{2,x}} \\ 0 & \frac{1}{f_{2,y}} & -\frac{v_{2,0}}{f_{2,y}} \\ 0 & 0 & 1 \end{pmatrix}. \tag{8.152}$$

Eq. 8.151 describes computer vision's well-known *homography transform*. It states that under the pure rotation of a camera, any set of two images are related by a 3×3 transformation matrix.

Unfortunately, vector $^{c1}T_{c2}$ is not a vector of zero length because the binocular vision's two cameras cannot physically overlap in the centers of their optical lenses.

As the translational vector $^{c1}T_{c2}$ connects the origins of frames $c1$ and $c2$ together, vectors $^{c1}V_a$, $^{c1}V_b$ and $^{c1}T_{c2}$ are coplanar. The normal vector of the plane passing through these three vectors can be computed as follows:

$$^{c1}N = \left(^{c1}T_{c2}\right) \times \left(^{c1}V_a\right). \tag{8.153}$$

Accordingly, we have $\left(^{c1}N\right)^t \bullet \left(^{c1}V_b\right) = 0$. This means that the following equality holds:

$$\left[\left(^{c1}T_{c2}\right) \times \left(^{c1}V_a\right)\right]^t \bullet \left(^{c1}V_b\right) = 0. \tag{8.154}$$

If we denote $^{c1}T_{c2} = (t_x, t_y, t_z)^t$ and its skew-symmetric matrix S_T as

$$S_T = S\left(^{c1}T_{c2}\right) = \begin{pmatrix} 0 & -t_z & t_y \\ t_z & 0 & -t_x \\ -t_y & t_x & 0 \end{pmatrix},$$

Eq. 8.154 becomes

$$\left(^{c1}V_a\right)^t \bullet \left(S_T\right)^t \bullet \left(^{c1}V_b\right) = 0. \tag{8.155}$$

Substituting Eq. 8.147 and Eq. 8.148 into Eq. 8.155 yields

$$\begin{pmatrix} u_1 & v_1 & 1 \end{pmatrix} \bullet F_{3\times3} \bullet \begin{pmatrix} u_2 \\ v_2 \\ 1 \end{pmatrix} = 0 \tag{8.156}$$

with

$$F_{3\times3} = \begin{pmatrix} \frac{1}{f_{1,x}} & 0 & -\frac{u_{1,0}}{f_{1,x}} \\ 0 & \frac{1}{f_{1,y}} & -\frac{v_{1,0}}{f_{1,y}} \\ 0 & 0 & 1 \end{pmatrix}^t \bullet F_0 \bullet \begin{pmatrix} \frac{1}{f_{2,x}} & 0 & -\frac{u_{2,0}}{f_{2,x}} \\ 0 & \frac{1}{f_{2,y}} & -\frac{v_{2,0}}{f_{2,y}} \\ 0 & 0 & 1 \end{pmatrix} \tag{8.157}$$

and

$$F_0 = \left(S_T\right)^t \bullet \left(^{c1}R_{c2}\right).$$

In fact, Eq. 8.156 imposes one constraint in which matrix $F_{3\times3}$ only depends on the intrinsic parameters and relative posture $\left(^{c1}M_{c2}\right)$ of the binocular vision's two cameras. Matrix $F_{3\times3}$ is commonly called the *fundamental matrix* of binocular vision. If index coordinates (u_1, v_1) in the left image plane are given, Eq. 8.156 imposes a continuous linear constraint on index coordinates (u_2, v_2) in the right image plane. This linear constraint imposed to the index coordinates is known as the *epipolar-line constraint*.

In order to identify a possible match for the feature point at (u_1, v_1), we just need to search the pixel locations along the continuous line described by Eq. 8.156. In practice, the search interval is the intersection between the continuous epipolar line and the image plane. Although the continuous epipolar line is a one-dimensional space, the search interval will depend on the size of the image plane.

Example 8.35   Fig. 8.52 shows a virtual binocular vision. The left camera's parameters are

$$(f_{1,x}, f_{1,y}, u_{1,0}, v_{1,0}) = (365.6059,\ 365.6059,\ 256,\ 256)$$

and

$$^{c1}M_r = \begin{pmatrix} 0.9848 & -0.1710 & 0.0302 & -121.6125 \\ 0.0 & -0.1736 & -0.9848 & 179.5967 \\ 0.1736 & 0.9698 & -0.1710 & 113.8218 \\ 0 & 0 & 0 & 1.0 \end{pmatrix}.$$

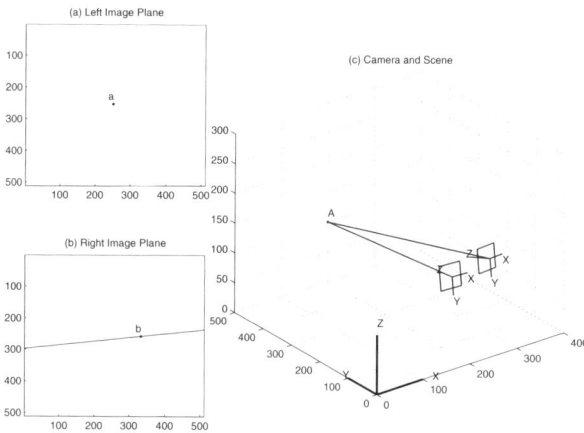

Fig. 8.52   Example of continuous epipolar-line constraint in a binocular vision system.

The right camera's parameters are

$$(f_{2,x}, f_{2,y}, u_{2,0}, v_{2,0}) = (365.6059,\ 365.6059,\ 256,\ 256)$$

and

$$^{c2}M_r = \begin{pmatrix} 0.9848 & 0.1710 & -0.0302 & -153.8321 \\ 0.0 & -0.1736 & -0.9848 & 189.4448 \\ -0.1736 & 0.9698 & -0.1710 & 164.1534 \\ 0 & 0 & 0 & 1.0 \end{pmatrix}.$$

From the extrinsic parameters of these two cameras, the motion transformation from camera frame $c2$ to camera frame $c1$ can be computed. The result is

$$^{c1}M_{c2} = \begin{pmatrix} 0.9397 & 0.0 & -0.3420 & 79.0862 \\ 0.0 & 1.0 & 0.0 & -9.8481 \\ 0.3420 & 0.0 & 0.9397 & 12.1818 \\ 0 & 0 & 0 & 1.0 \end{pmatrix}.$$

If we apply Eq. 8.157, we can obtain the fundamental matrix of binocular vision as follows:

$$F_{3\times3} = \begin{pmatrix} 0.0 & 0.0001 & -0.0045 \\ 0.0001 & 0.0 & 0.1848 \\ -0.0616 & -0.2396 & 19.0640 \end{pmatrix}.$$

When we have an image-feature point at the following location in the left image:

$$(u_1, v_1) = (252, \ 253),$$

the continuous epipolar line computed from Eq. 8.156 is

$$-0.0258 \bullet u_2 - 0.2167 \bullet v_2 + 64.6883 = 0.$$

This line is shown in Fig. 8.52c. We can see that the projection (point b) of object point A onto the right image plane is on the epipolar line.

◇◇◇◇◇◇◇◇◇◇◇◇◇◇◇◇◇◇

Example 8.36 In a four-camera vision system such as the one shown in Fig. 8.55, a set of four images is available at a time-instant. Refer to Fig. 8.53. We treat the upper-left image as the master image, and the rest the slave images. In this way, we have three pairs of images as follows:

- (upper-left image, upper-right image),
- (upper-left image, lower-left image),
- (upper-left image, lower-right image).

When we select an image feature in the master image, its match candidate in a slave image must be on an epipolar line. In this example, the selected image features are marked by "+". From Fig. 8.53, we can see that the match candidates are actually on the corresponding epipolar lines.

◇◇◇◇◇◇◇◇◇◇◇◇◇◇◇◇◇◇

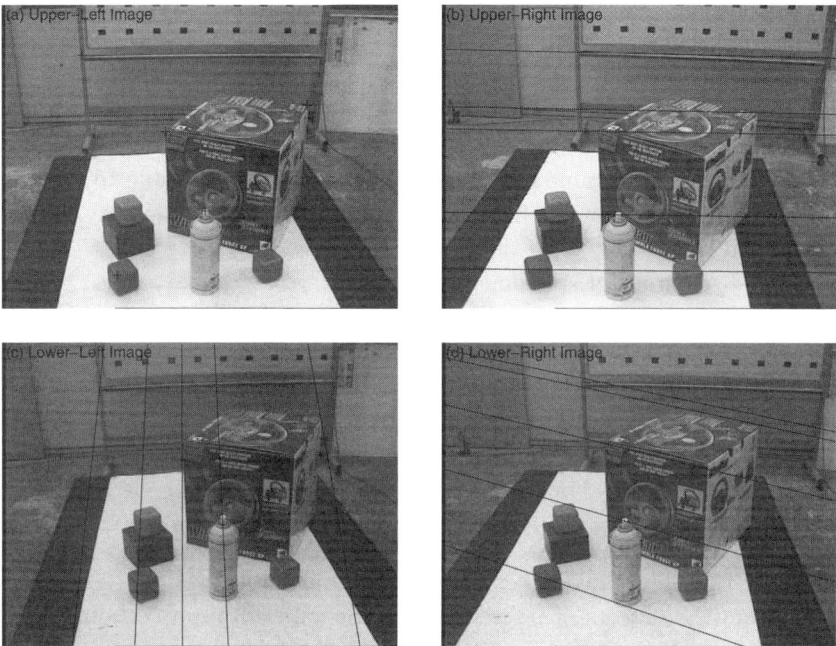

Fig. 8.53   Example of continuous epipolar-line constraint, in a four-camera vision system.

8.6.2.5   *Discrete Epipolar-Line Constraint*

Continuous epipolar-line constraint partially solves the issue of search space because a continuous epipolar line still depends on the size of the image plane. For example, when image resolution (r_x, r_y) increases, the search interval along the continuous epipolar line increases, as well. Mathematically, if the image resolution becomes infinite, the search interval along the continuous epipolar becomes infinite as well. Therefore, it is interesting to investigate whether it is possible to break the dependence between the search interval and the image resolution.

Before we study a solution, let us first examine the following two interesting observations:

- Finite Dimensions of Objects of Interest:
  As we already mentioned, one primary goal we hope to achieve with a robot's visual perception system is the ability to infer the 3-D geometry of a scene (or object) from 2-D images. Interestingly enough, for most applications, the relevant dimensions of a scene (or object) are

not infinite. For example, when a robot is walking down a road, the primary concerns are: a) where does the road go? b) are there any obstacles to avoid? A robot's visual-perception system should not be programmed to watch for objects falling from the sky, as this is not a primary concern. If the height of a robot is measured along the Z axis of the reference frame assigned to a scene, it is reasonable to assume that the Z coordinates from any object of interest should vary within a fixed range, say: $[0cm, 300cm]$.

- Finite Accuracy of a Geometry of Interest:
  The robot's visual-perception system deals with digital images. In theory, the accuracy of the results obtained from images is not infinite (i.e. the error is not infinitely small). Still, the desired accuracy of the results in a visual-perception system should depend on the intended application. In other words, we should have the freedom to specify the desired accuracy. For example, for $\triangle Z$, we can choose the Z coordinates to be within $1mm$ or within $1cm$, etc.

From the above two observations, let us assume that the Z coordinates from the objects of interest in a scene vary within a certain range: $[Z_{min}, Z_{max}]$. When inferring the 3-D geometry from 2-D images, we can specify the desired accuracy for the Z coordinates to be $\triangle Z$. In this way, all the possible values of the Z coordinates which are consistent with the specified accuracy will be

$$Z_i = Z_{min} + i \bullet \triangle Z, \quad i \in [0, n] \tag{8.158}$$

with $n \bullet \triangle Z = Z_{max}$. Clearly, n only depends on the range $[Z_{min}, Z_{max}]$ and accuracy $\triangle Z$. It is independent of the image size.

If we apply Eq. 8.158, then Eq. 8.135 and Eq. 8.136 becomes

$$s_1 \bullet \begin{pmatrix} u_1 \\ v_1 \\ 1 \end{pmatrix} = H_1 \bullet \begin{pmatrix} {}^rX \\ {}^rY \\ Z_{min} + i \bullet \triangle Z \\ 1 \end{pmatrix}, \quad i = 0, 1, 2, ...n \tag{8.159}$$

and

$$s_2 \bullet \begin{pmatrix} u_2 \\ v_2 \\ 1 \end{pmatrix} = H_2 \bullet \begin{pmatrix} {}^rX \\ {}^rY \\ Z_{min} + i \bullet \triangle Z \\ 1 \end{pmatrix}, \quad i = 0, 1, 2, ..., n. \tag{8.160}$$

Interestingly enough, index i helps us predict not only the possible co-ordinates $(^rX, {}^rY, {}^rZ)$, but also the possible binocular correspondence. This is explained as follows:

When we have a value for index i, Eq. 8.159 allows us to compute coordinates $(^rX, {}^rY, {}^rZ)$ because coordinates (u_1, v_1) have been given as input. If we know coordinates $(^rX, {}^rY, {}^rZ)$, Eq. 8.160 allows us to directly determine the location where the image feature at (u_2, v_2) in the right image, matches with the image feature at (u_1, v_1) in the left image.

Eq. 8.160 describes a discrete constraint along an epipolar line because all the predicted locations (u_2, v_2) are on a continuous epipolar line. We call this the *discrete epipolar-line constraint*. Most importantly, the dimension of search space, regarding a discrete epipolar-line constraint, depends on the value of n, and is independent of image size. This property does not hold true with a continuous epipolar-line constraint.

In summary, an algorithm which implements inverse projective-mapping and takes into account the discrete epipolar-line constraint will do the following:

- Step 1: Specify the range and accuracy of one coordinate (e.g. Z). Then, determine interval $[0, n]$ for index i.
- Step 2: Set up Eq. 8.159 and Eq. 8.160.
- Step 3: Find the value of index i within interval $[0, n]$ and compute the predicted coordinates $(^rX_i, {}^rY_i, {}^rZ_i)$ from Eq. 8.159.
- Step 4: When the predicted coordinates $(^rX_i, {}^rY_i, {}^rZ_i)$ are known, compute the predicted location (u_{2i}, v_{2i}) from Eq. 8.160.
- Step 5: Compute the difference between a sub-image centered at (u_1, v_1) in the left image, and a sub-image centered at (u_{2i}, v_{2i}) in the right image. This difference is called a *dissimilarity*.
- Repeat Steps 3, 4, and 5 for all the values of index i.
- Step 6: Choose location (u_{2j}, v_{2j}) where the dissimilarity is minimum. This dissimilarity should also be smaller than a user-specified threshold value in order to eliminate false matching. If such a location exists in the right image, it will be the match for the feature point at location (u_1, v_1) in the left image. And, coordinates $(^rX_j, {}^rY_j, {}^rZ_j)$ will be the results of the binocular vision's inverse projective-mapping.

This algorithm is fundamentally different from the one which makes use of the inverse projective-mapping described by Eq. 8.144. Here, we have the freedom to specify the range and accuracy for one coordinate. Most importantly, this algorithm simultaneously solves the issues of both binocular correspondence and determination of 3-D coordinates. Moreover, the computational time necessary for each feature point is independent of image size. For easy reference, we call this algorithm the *indirect 3-D reconstruction*.

Example 8.37   We use a four-camera vision system, as shown in Fig. 8.55. At a time-instant, a set of four images is available. Refer to Fig. 8.54. We treat the upper-left image as the master image, and the rest the slave images.

When we undertake 3-D reconstruction, we normally select image features, such as edges or corners, from the master image. If an image feature in the master image is given as input, its match candidate in a slave image must be on an epipolar line. However, if we apply the discrete epipolar line constraint, the search space for a match candidate in a slave image is a set of discrete spatial locations, as shown in Fig. 8.54.

In this example, we set the range for the Y coordinate to $[0, 300mm]$, and the desired accuracy to $10mm$. (NOTE: The Y axis is perpendicular to the table's top-surface). The selected image features in the master image are marked by "x".

$$\diamond\diamond\diamond\diamond\diamond\diamond\diamond\diamond\diamond\diamond\diamond\diamond\diamond\diamond\diamond\diamond\diamond$$

8.6.2.6   *Differential Epipolar-Line Constraint*

In literature, the second difficulty with binocular vision has been largely overlooked. All existing algorithms for binocular correspondence still require that the two cameras in the binocular vision system see the same object in as similar a manner as possible. In other words, it is necessary that the two cameras are optically and electronically similar, if not identical, so that the measurement of similarity or dissimilarity between the image features is meaningful.

Because of this stipulation, it would be inappropriate to form a binocular vision system with one monochrome camera and one color camera. However, if we could eliminate this stipulation, we could form a binocular vision system with two cameras which have very different characteristics. Here, we propose a simple solution which overcomes the second difficulty with binocular vision, and permits the use of dissimilar cameras.

Refer to Eq. 8.145. A binocular vision system's fundamental matrix $F_{3\times3}$ depends only on the intrinsic parameters and relative posture $(^{c1}M_{c2})$ of its two cameras. Now, if we assume that these intrinsic parameters and

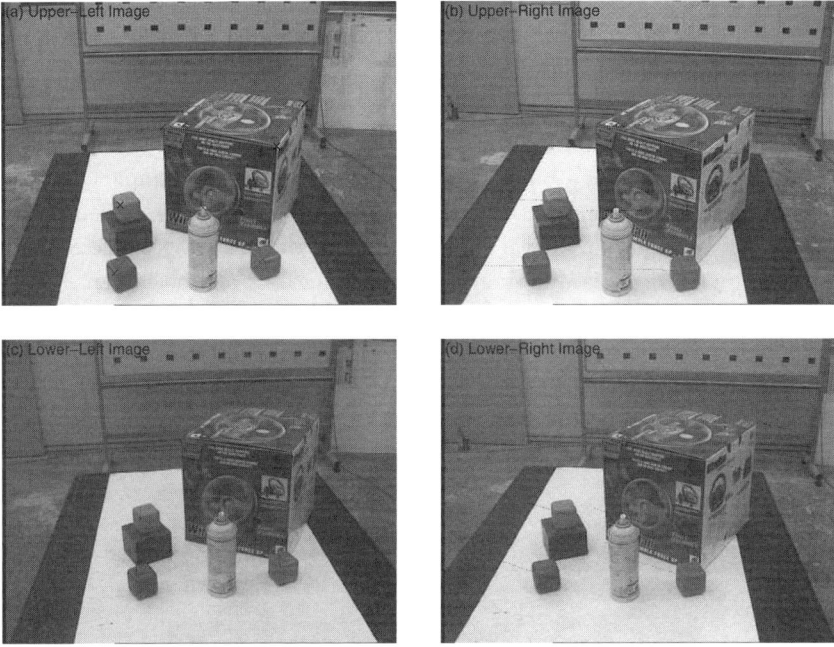

Fig. 8.54 Example of discrete epipolar-line constraint where a search space for a match candidate is a set of discrete locations.

relative posture remain unchanged, then the fundamental matrix becomes a constant matrix. And, differentiating Eq. 8.144 with respect to time allows us to obtain

$$\left(\triangle u_1 \ \triangle v_1 \ 1 \right) \bullet F_{3\times 3} \bullet \begin{pmatrix} u_2 \\ v_2 \\ 1 \end{pmatrix} + \left(u_1 \ v_1 \ 1 \right) \bullet F_{3\times 3} \bullet \begin{pmatrix} \triangle u_2 \\ \triangle v_2 \\ 1 \end{pmatrix} = 0. \quad (8.161)$$

We call Eq. 8.161 the *differential epipolar-line constraint*. It can be used in two ways:

- Dynamic Binocular Vision:
  When a binocular vision system moves along with the head of a humanoid robot (or mobile vehicle), the visual-sensory system outputs two streams of digital images: a) the left image sequence and b) the right image sequence. We can estimate the displacements of the image features within the left and right image sequences respectively using a technique for detecting temporal uniformity. Clearly, the estimation of

$(\triangle u_1, \triangle v_1)$ in the left image sequence does not depend on the characteristics of the right camera. Similarly, the estimation of $(\triangle u_2, \triangle v_2)$ is
independent of the characteristics of the left camera. As a result, the
left and right cameras can be very different.

- Composite Binocular Vision:
  Instead of forming a binocular vision system with two simple cameras,
  we can use two composite cameras, as shown in Fig. 8.55. A composite
  camera is made up of two similar cameras. In this way, the composite
  camera's output is a sequence of two images at a time instant. As a
  result, we can not only detect the image features, but also estimate their
  displacements from the output of the composite camera. Since feature
  extraction and displacement estimation are carried out independently
  using the outputs from the two composite cameras, it is not necessary
  for the (composite) cameras to be optically and electronically similar.

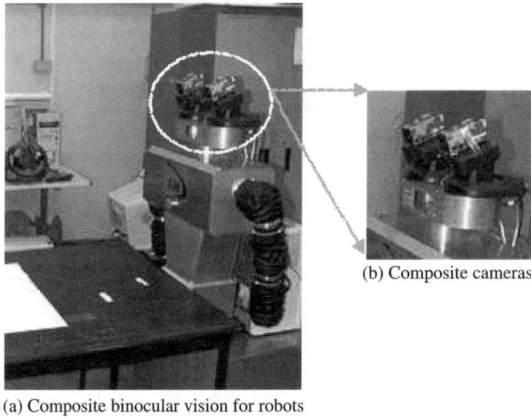

(b) Composite cameras

(a) Composite binocular vision for robots

Fig. 8.55   Illustration of a composite binocular vision system for a humanoid robot.

Interestingly enough, the differential epipolar-line constraint is unique
if and only if a scene (or an object) does not undergo a pure rotation about
the origin of camera frame $c1$ (or $c2$). This can be easily explained as
follows:

When there is an image feature at $a = (u_1, v_1)$ in the left image plane, its
possible matches in the right image plane are the projections of the object
points on the projection line passing through feature point a. Among the
possible matches, there is only one correct one. The rest are treated as false

matches.

If the scene (or object) does not undergo a pure rotation about the origin of camera frame $c1$, a pair of two consecutive images in the left camera will not be related by a homography transform. In other words, none of the false matches will reappear on the epipolar line which is determined by feature point $(u_1+\triangle u_1, v_1+\triangle v_1)$. Therefore, Eq. 8.144 and Eq. 8.161 provide the necessary and sufficient constraints for the selection of a correct match. In practice, when there is an image feature point at (u_1, v_1), the correct match is the one which verifies Eq. 8.144 and minimizes

$$E(u_1,v_1) = \begin{pmatrix} \triangle u_1 & \triangle v_1 & 1 \end{pmatrix} \bullet F_{3\times3} \bullet \begin{pmatrix} u_2 \\ v_2 \\ 1 \end{pmatrix} + \begin{pmatrix} u_1 & v_1 & 1 \end{pmatrix} \bullet F_{3\times3} \bullet \begin{pmatrix} \triangle u_2 \\ \triangle v_2 \\ 1 \end{pmatrix}.$$
(8.162)

Example 8.38 A composite binocular vision system outputs a set of four images at a time-instant. Refer to Fig. 8.56. We form two pairs of images as follows:

- (upper-left image, lower-left image),
- (upper-right image, lower-right image).

In the pair of upper-left and lower-left images, we select image features from the upper-left image. In this example, two features are selected from the locations:

$$\begin{cases} (u_a, v_a) = (160, 121) \\ (u_b, v_b) = (256, 248). \end{cases}$$

For a selected image feature, we use template matching technique and epipolar line constraint to find its match in the lower-left image. Then, the matches in the pair of upper-right and lower-right images can be easily determined by the intersections of epipolar lines. As a result, displacement vectors $(\triangle u_a, \triangle v_a)$ and $(\triangle u_b, \triangle v_b)$ can be computed directly.

By applying Eq. 8.162, the computed values of differential epipolar line constraint, with regard to the two selected image features, are

$$\begin{cases} E(u_a, v_a) = 0.174416 \\ E(u_b, v_b) = 0.599551. \end{cases}$$

Because of numerical error, differential epipolar line constraint is unlikely to be exactly equal to zero.

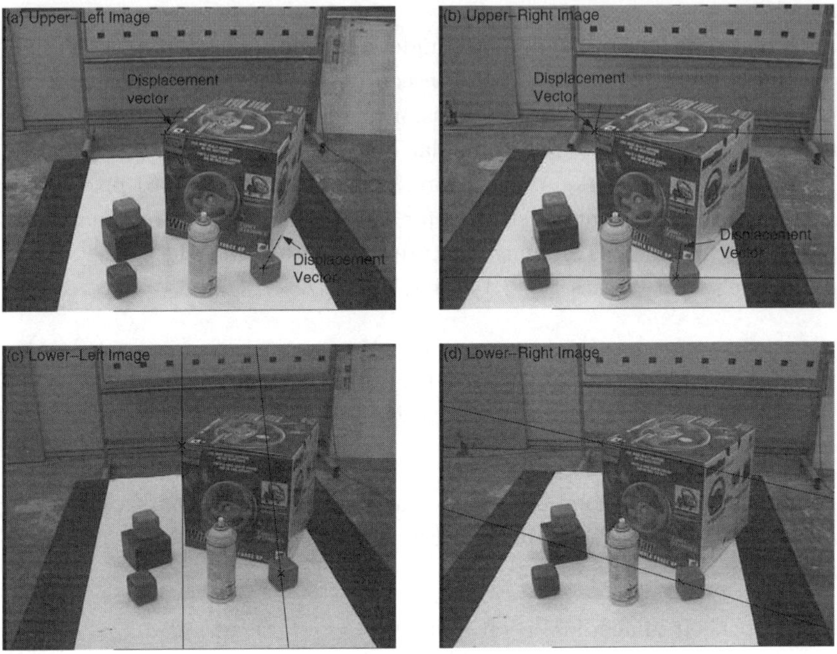

Fig. 8.56   Example of differential epipolar line constraint.

◇◇◇◇◇◇◇◇◇◇◇◇◇◇◇◇◇◇

8.7   Summary

One primary goal we hope to achieve through a robot's visual-perception system is the ability to infer the three-dimensional (3-D) geometry of a scene (or object) from two-dimensional (2-D) digital images. This is a very challenging task as digital images do not explicitly contain any geometric information about a scene (or object). Therefore, one has to go through a series of information-processing functions in order to derive geometric information from digital images. These functions include image processing, feature extraction, feature description, and geometry measurement.

In a robot's visual-perception system, there are two important categories of image-processing techniques: a) image transformation and b) image filtering. We learned that the purpose of image transformation is to obtain alternative representations of the original input image. And, the purpose of image filtering is to enhance specific dynamic characteristics in an input image. For a linear filter, we studied the principle of dynamics conservation and the principle of dynamics resonance. These two principles are clearly explained using the convolution formulae and Laplace transform.

The purpose of image-feature extraction is to detect geometric features from images. We learned that geometric features are implicitly encoded into the discontinuity, continuity and uniformity of an input image's chrominance, luminance, or texture. As a result, image-feature extraction is mainly the detection of edges, corners and uniform regions.

Interestingly enough, most edge-detection algorithms can be explained using the principle of dynamics resonance. These algorithms always start with edge enhancement followed by a process of edge selection.

With regard to corner detection, the algorithm depends on the adopted definition of corner. Here, we studied a zero-gradient algorithm for corner detection which is conceptually simple and works well with real images.

As for uniformity detection, most algorithms rely on a complex decision-making process because of the challenges caused by uncertainty and redundancy. We discussed the probabilistic RCE neural network which deals efficiently with uncertainty and redundancy, and is useful in the detection of uniform regions in color images.

The output from image-feature extraction is individual feature points. Thus, it is necessary to group these feature points into a set of clusters, each of which can be described by analytical curves. We learned that geometric-feature description involves three sequential steps: a) feature grouping, b) contour splitting, and c) curve fitting. We studied three typical examples of feature grouping which make use of three different neighborhood types: a) the 4-neighborhood, b) the 8-neighborhood, and c) the causal neighborhood.

We know that strategies for contour splitting are different, depending on the types of analytical curves to be used for feature description. And, we studied two methods of contour splitting. Subsequently, if we have a set of simple contours as input, we can choose linear, circular, or elliptic curves to approximate simple contours.

We also learned that feature description performed in the index coordinate system assigned to a digital image does not have any unit of measurement. Theoretically, the results are qualitative descriptions of the geometry of a scene (or object). In order to obtain a quantitative description, it is necessary to study the important topic of geometry measurement using either monocular or binocular vision.

We learned that the most important issue underlying geometry measurement is inverse projective-mapping. With monocular vision, we must reduce a scene's dimension (resulting in 2-D vision), or introduce extra knowledge (resulting in model-based geometry measurement). In order to measure the geometry, it is necessary to know a camera's parameters. For this reason, we also studied camera calibration. We know that if a camera's calibration matrix is known, the camera's intrinsic and extrinsic parameters can be fully determined.

Finally, we studied binocular vision. We know that the inverse projective-mapping in binocular vision has a closed form solution. However, this solution depends on two necessary conditions: a) the two cameras must be calibrated in advance, and b) the binocular correspondence of the image features must be established in advance as well. The first condition is easy to meet but the second one largely undermines binocular vision's practicality. In order to deal with the second issue, we studied two possible solutions: a) the application of a discrete epipolar-line constraint, and b) the application of a differential epipolar-line constraint. These resulted in two different algorithms for binocular vision.

8.8   Exercises

(1) Explain the process of visual perception.
(2) Is there any difference between the process behind computer graphics and that behind the visual-sensory system of a robot?
(3) Does a digital image explicitly contain any geometric information?
(4) What are the typical applications of a visual-perception system?
(5) Explain information processing in a visual-perception system.
(6) Write a sample C-program that calculates the intensity histogram of an input image.
(7) Explain the fundamental difference between image transformation and image filtering.
(8) Write a sample C-program that automatically determines the optimal threshold value v_0 in Eq. 8.9.
(9) Explain the importance of the localization of a detected feature.
(10) What are the criteria for designing a good feature-detector?
(11) If $f(x)$ represents a one-dimensional ridge-type edge, what is the exact mathematical description of $f(x)$?
(12) How do you make a feature detector possess the degree of freedom to

compensate for the shift of feature localization caused by noise?

(13) If the coefficients in Canny's function satisfy the following constraints: $a_1 \bullet a_2 < 0$ and $a_3 \bullet a_4 > 0$, what is the physical meaning behind these constraints?

(14) Explain why Deriche's function is a special case of Canny's function.

(15) What are the possible definitions of a corner feature?

(16) Explain why a boundary-tracing algorithm uses the 4-neighborhood instead of the 8-neighborhood.

(17) Explain why contour splitting is necessary before performing curve fitting.

(18) For curve fitting with circular or elliptic arcs, one problem is how to compute the tangential angle at each edge pixel of a simple contour. Propose some practical solutions to this problem.

(19) A virtual camera's focal length is 1.0 cm. Its aperture angles are $(100.0^0, 100.0^0)$. If the image resolution is $(256, 256)$, what is the projective matrix $^I P_c$? Now, if we change the focal length to 10.0cm, what is the new projective matrix $^I P_c$?

(20) Refer to Fig. 8.45. Where are locations B and C if image points b and c are inversely projected onto plane $^r Z = 0$?

(21) In a camera's calibration matrix , what is the meaning of (h_{14}, h_{24})?

(22) What is binocular vision?

(23) What is the sufficient condition for binocular vision to work properly?

(24) In Example 8.33, if $a = (400, 200)$ and $b = (200, 200)$, what are the coordinates of the object point with respect to the reference frame of the scene?

(25) In Example 8.35, compute the epipolar line corresponding to the image feature point $a = (100, 100)$.

(26) What is the epipolar-line constraint?

(27) Explain the differences between the continuous epipolar-line constraint and the discrete epipolar-line constraint.

(28) Do the two cameras in a binocular vision have to be absolutely similar in terms of optical and electronic characteristics?

(29) Project 1: Implement an edge-detection algorithm.

(30) Project 2: Implement a corner-detection algorithm.

(31) Project 3: Implement a color-segmentation algorithm.

(32) Project 4: Implement a template-matching algorithm.

(33) Project 5: Implement an edge-linking algorithm.

(34) Project 6: Implement model-based inverse projective mapping in monocular vision.

(35) Project 7: Implement solutions for determining a camera's parameters from its calibration matrix.

(36) Project 8: Implement the solution for continuous epipolar-line constraint.

(37) Project 9: Implement a binocular-vision algorithm with a discrete epipolar-line constraint.

(38) Project 10: Implement a binocular-vision algorithm with a differential epipolar-line constraint.

8.9   Bibliography

(1) Banks, S. (1990). *Signal Processing, Image Processing and Pattern Recognition*, Prentice-Hall.

(2) Canny, J. (1986). A Computational Approach to Edge Detection, *IEEE Trans. on Pattern Analysis and Machine Intelligence*, **8**.

(3) Castleman, K. R. (1996). *Digital Image Processing*, Prentice-Hall.

(4) Crane, R. (1997). *A Simplified Approach to Image Processing*, Prentice-Hall.

(5) Deriche, R. (1987). Using Canny's Criteria to Derive Optimal Edge Detector Recursively Implemented, *International Journal of Computer Vision*, **2**.

(6) Faugeras, O. (1993). *Three-Dimensional Computer Vision*, MIT Press.

(7) Fostner, W. and E. Gulch (1987). A Fast Operator for Detection and Precise Location of Distinct Points, Corners and Centers of Circular Features, *ISPRS Intercommission Workshop*.

(8) Haralick, R. M. and L. G. Shapiro (1993). *Computer and Robot Vision*, Addison-Wesley.

(9) Harris, C. G. and M. Stephens (1988). A Combined Corner and Edge Detector, *The 4th Alvey Vision Conference*.

(10) Horn, B. K. P. (1994). *Robot Vision*, MIT Press.

(11) Jain, R., Kasturi, R. and B. G. Schunck (1995). *Machine Vision*, McGraw-Hill.

(12) Marr, D. and E. Hildreth (1980). Theory of Edge Detection, *Proceedings of Royal Society, London*, **B207**.

(13) Marshal, A. D. and R. R. Martin (1992). *Computer Vision, Models and Inspection*, World Scientific.

(14) Philips, D. (1994). *Image Processing in C*, R&D Publications, Inc.

(15) Prewitt, J. M. S. (1970). Object Enhancement and Extraction, in *Picture Processing and Psychopictorics*, Academics Press.

(16) Reilly, D. L., L. N. Cooper and C. Elbaum (1982). A Neural Model for Category Learning, *Biological Cybernetics*, **45**.

(17) Roberts, L. G. (1965). Machine Perception of Three-dimensional Solids, in *Optical and Electro-optical Information Processing*, MIT Press.

(18) Smith, S. M. and J. M. Brady (1997). SUSAN - A New Approach to Low Level Image Processing, *International Journal of Computer Vision*, **23**.

(19) Specht, D. F. (1988). Probabilistic Neural Networks for Classification, Mapping or Associative Memory, *IEEE International Conference on Neural Networks*, **I**.

(20) Yakimovsky, Y. (1976). Boundary and Object Detection in Real World Images, *Journal of ACM*, **23**.

Chapter 9

Decision-Making System of Robots

9.1 Introduction

The ultimate goal of robotics research is to develop autonomous robots, which can not only be deployed in industry for better productivity, but can also co-exist in human society for better service. If a robot co-exists with humans, however, it must have the ability to develop and learn behaviors compatible with human society. Thus, an autonomous robot must also be a sociable and educable robot.

Humans have an innate mechanism for autonomously developing mental and physical abilities. What is this innate mechanism which determines the degree of autonomy in a physical system, like a human being or a robot? So far, this is still an unresolved issue which needs more research. However, as we studied in Chapter 6, we know that any complex behavior is the result of the interaction among mental and physical actors embedded in a physical body, or system.

From an engineering point of view, a physical actor is a sensing-control loop which acts on a physical kineto-dynamic chain; and, a mental actor is a perception-decision-action loop which acts on the physical actors and may also alter the mental world's internal representation(s). Mathematically, a control law is a function-based, decision-making process which may be complemented by some decision rules, such as the switching functions studied in Chapter 5. Clearly, decision making is an indispensable part of any physical or mental actor, and forms an important pillar in an autonomous system, such as a humanoid robot.

In Chapter 5, we learned that motion control can be fully automated if and only if: a) the desired output is given and b) the chosen control law makes the closed-feedback loop stable. In Chapter 5, we only addressed the second condition. In this chapter, we will address the first as we study the basics of decision making and motion planning for autonomous robots. The emphasis of our study in this chapter will be on image-guided motion planning and control.

9.2   The Basics of Decision Making

Generally speaking, the decision-making process exists everywhere. Without decision making, there would be no automated actions, tasks or even behaviors.

9.2.1   *The Key to Automated Actions*

In Chapter 5, we learned that control system is synonymous with *automatic-feedback control system*. This means that an action taken by a system can be fully automated if the desired output of the automatic-feedback control loop can be specified in advance. This requirement must be satisfied by all automated systems in manufacturing, transportation, military, etc.

However, for an automatic-feedback control system to work properly, a critical issue is the design of suitable control law(s). In particular, the control law(s) must achieve certain design specifications in terms of absolute and relative stabilities, regardless of uncertainty caused by noise, disturbance, unknown or un-modelled system dynamics, etc.

Mathematically, we can consider a control law to be a function-driven decision-making process. Depending on the complexity of a system under control, a set of control laws may be used to cope with variation caused by the system's configuration. Therefore, control laws may be complemented by some decision-making rules such as switching functions. Accordingly, decision making is the key to the automation of many industrial tasks. From this perspective, we can attempt to formally define the buzzword *automation* as follows:

Definition 9.1   Automation is the interaction between a decision-making process and an action-taking system which does not require any human intervention.

9.2.2   *The Key to Automated Behaviors*

As we discussed above, "automation" means automated actions implemented by feedback control loops. An automated action has only gained
independence in action taking. It still depends on the specification of desired output or outcome, which, for many industrial automation systems
including robots, is programmed and re-programmed by humans. Because
of this human intervention, an automated system is not an autonomous
system.

So, how do we make an automated system an autonomous system?
Perhaps, first it is necessary to define scientifically the term *autonomy*.
Autonomy is related to the state of independence. As there is no absolute
independence for any physical system on Earth, there is no absolute autonomy. As a result, autonomy is better described by a degree (or level) of
independence. Formally, we can attempt to define the term *autonomy* as
follows:

Definition 9.2   Autonomy is a characteristic which describes an automated system's degree (or level) of independence.

From Chapter 5, we know that without a specification of desired output,
an automated system will not produce any useful outcome. This indicates
that an automation system has at least one degree of dependence, which
is: the specification of desired output. Hopefully, the elimination of this
dependence will make an automated system an autonomous system. And,
the thoroughness in eliminating the degree of dependence will result in the
degree of autonomy attributed to an autonomous system. Therefore, it may
be useful to attempt to define the term *autonomous system* as follows:

Definition 9.3   An autonomous system is an automated system which
has gained at least one level of independence in specifying its desired output
without human intervention in programming or reprogramming.

It would signal tremendous progress if we could develop a robot
which could decide without human intervention the desired outcome to be
achieved. Since behavior is a sequence of ordered actions for the achievement of a desired outcome, the ability to self-specify the desired outcome
is the key to autonomous behaviors.

9.2.3 Decision-Making Processes

According to the level of sophistication in performing actions, a system can be classified into one of these categories:

- Active System:
  A system which acts on its own, regardless of desired outcome or sensory input, is called an *active system*. A system in this category does not have the ability to interact with other systems, or the outside world.
- Reactive System:
  If a system is able to respond to outside stimulus, it is called a *reactive system*. An open-loop control system is a reactive system. A reactive system can interact with the environment.
- Automated System:
  If a reactive system is equipped with a decision-making module, like a controller, it becomes an *automated system*. Au automated system acts according to desired output or outcome which has been specified in advance as input.
- Autonomous System:
  If an automated system specifies its own desired output or outcome, it is called an *autonomous system*.
- Intelligent System:
  In Chapter 1, we defined the term *intelligence* as a measurement which is inversely proportional to the effort spent achieving a predefined outcome or target. The term *effort* means energy spent in the planning and execution of a sequence of mental and physical actions. As a result, effort depends on a subjective decision made in the planning of an action sequence. Accordingly, the term *intelligent* can only be attributed to a system which encompasses a decision-making subsystem. Thus, either an automated or autonomous system can be attributed with the qualifier *intelligent* if it has the ability to achieve a predefined outcome or target in different ways.

Clearly, the enabling technology for the development of autonomous systems is decision making. Most importantly, decision-making itself is a process which has its own input and output, and must be embedded in a native system (like robot) in order to be useful. Fig. 9.1 shows a simplistic view of a generic decision-making process. The purpose here is to highlight the complexity of the input-output relationship.

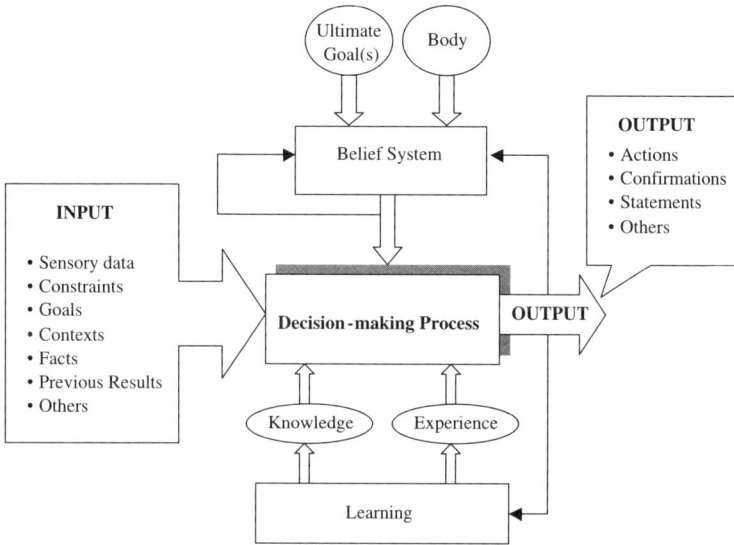

Fig. 9.1   A simplistic view of a generic decision-making process.

9.2.3.1   *Inputs of Decision-Making*

As shown in Fig. 9.1, the input to a decision-making process can be divided into two parts: a) hard input, and b) soft input. By *hard input*, we mean all data and facts which are the direct outcome of the physical systems. In contrast to hard input, *soft input* refers to all data and facts which are subject to linguistic interpretation.

Hard Input

Generally speaking, the common forms of hard input to a decision-making process include:

- Sensory Data:
  These are measurements directly obtained from sensors embedded onto physical systems.
- Constraints:
  These are models, representations and hypotheses imposed on the environment and physical systems. For example, kinematics imposes kinematic constraints on a robot's mechanism.

- Goals:
  These can be instructions received from the outside world or self-generated goals. For example, the instruction "Place the cup on the dining table" is a goal specified by others. However, the statement "I will do it in a better way" is typically a self-generated goal.
- Contexts:
  These generally refer to conditions imposed by the environment. For example, "indoor" and "outdoor" are two different contexts.
- Facts and Previous Results:
  This is knowledge and information about what a physical system can achieve. For example, the solutions and results of inverse kinematics tell us what can be done with an open kineto-dynamic chain.

Soft Input

A decision-making process must consider not only hard input but also soft input. This makes a decision-making process very different from a linear transformation or filtering process (e.g. image transformation or image filtering). So far, there is no concrete answer to the question of how to deal with the softness (also called *elasticity*) of soft input. Thus, when it comes to computing with words (the semantic level), decision making is a difficult and complicated job for a robot to perform.

As shown in Fig. 9.1, the common forms of soft input include:

- Knowledge:
  Knowledge itself has no elasticity, and is a kind of hard input. However, the interpretation of knowledge is subjective, and the application of knowledge may also introduce some approximations (i.e. softness or elasticity). For example, color has a specific range of wavelengths. However, while an imaging sensor can be very accurate, the interpretation of color in a digital image does not have a rigid range of numerical values. For another example, let us interpret the equation $W_K = \frac{1}{2}m \bullet v^2$. One can interpret it as "An increase in velocity increases the kinetic energy of a body". In this case, the statement itself has a lot of elasticity.
- Experience:
  The skill or mental syntax of a physical system can be described as the specific pattern of mapping (or association) between the input (e.g. causes) and the output (e.g. effects or actions). This pattern of mapping is not pre-established, but is acquired from experiments or trials

which result in what is called *experience*. In fact, experience cultivates the skill or mental syntax of a physical system. Accordingly, it conditions the tendency for a physical system to plan *familiar* sequences of ordered actions.

- Belief:
  Belief is the "motor" which powers an autonomous system. It is the source from which all goals and desired outcomes are formulated. Without a belief system, a human being or a physical system would not be able to self-develop and evolve. As we studied in Chapter 6, the belief system determines four types of autonomous behaviors: a) value-driven behaviors, b) expectation-driven behaviors, c) curiosity-driven behaviors and d) state-driven behaviors. How to develop a belief system for a robot is still an unresolved issue. However, if we understand the importance of the belief system, it will help us to focus our research effort in this direction. A belief system cannot exist on its own, but must be embedded in a physical system (e.g. a body). A belief system takes all the data from a body's sensory systems as input. Interestingly enough, the output from a belief system may also be taken as input to the belief system itself. When output from a belief system remains permanent, it may be designated as the *ultimate goal(s)* or *ambition(s)* of a physical system. Here, we can raise these challenging questions: What should the ultimate goal of a humanoid robot be? Could this ultimate goal be programmed once without any further reprogramming?

9.2.3.2 *Outputs of Decision-Making*

As shown in Fig. 9.1, the possible forms of output from a decision-making process include:

- Action:
  This is the most common sort of output from a decision-making process. Action can be in the form of action descriptions (e.g. what to do), motion descriptions (e.g. desired path or trajectory), or signal descriptions (e.g. voltage or current level).
- Statement:
  This is the linguistic description of facts, conclusions, conjectures, hypotheses, etc. Statements will introduce uncertainty or fuzziness depending on the lexical elasticity, or softness, of words used. For example, words with elastic meanings include: possible, probable, likely,

very, small, large, many, etc.
- Confirmation:
  This refers to statements without any ambiguity. The simplest confirmations include: "Yes", "No", "It is", "It is not", etc.

In a physical system, a set of simple decision-making processes can be nested to form a complex decision-making system. Because of the nature of soft input in a decision-making process, a robot must be equipped with the appropriate mathematical tools to handle not only certainty and redundancy but also uncertainty (including ambiguity).

9.2.4 *Difficulties in Decision-Making*

We mentioned that decision-making is a very difficult job for a robot. This is largely due to uncertainty and redundancy.

9.2.4.1 *Uncertainty*

As shown in Fig. 9.2, uncertainty can exist in input, specification of goal, and output. In general, uncertainty is the result of one or a combination of the following three causes:

- Imprecision:
  In Chapter 5, we studied the meaning of a sensor's precision. In fact, the term *precision* refers to the statistical distribution of a sensor's output $s(t)$ when its input $y(t)$ is set at a fixed value. Due to the presence of noise, the statistical distribution of a sensor's output is spread over a region, resulting in the imprecision of the sensory data.
- Incompleteness:
  Mathematically, a spatial-temporal signal from a sensor has a limited range in both space and time. Depending on the size of the spatial or temporal window, the observation of a sensor's output at a specific time-instant may not contain complete information about a physical quantity. This incompleteness of sensory data will cause uncertainty.
- Ambiguity:
  When the specification of a goal, fact or context is in the form of a statement, ambiguity can arise depending on the elasticity of the words used. For example, the goal specification, "Place part A on top of part B" contains the elastic term "on top of". Thus, this goal specification has what is called *fuzziness* or *vagueness*. If a decision-making

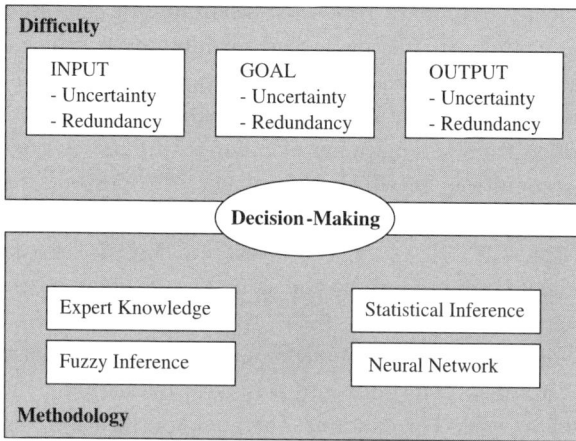

Fig. 9.2 Difficulties and methodologies in decision-making.

process is not able to completely eliminate fuzziness, ambiguity will certainly reappear in the output.

9.2.4.2 *Redundancy*

In literature, the difficulty of decision making due to redundancy has been overlooked. As shown in Fig. 9.2, redundancy may also appear in input, specification of goal, and output. In general, redundancy may come from:

- Multiple Sources of Input:
  The same physical quantity can be measured by a set of different sensors. For example, the posture of an end-effector (e.g. gripper or hand) attached to a robot can be determined from the measurements of joint angles and forward kinematics. This posture can also be measured by a robot's visual perception system. Alternatively, the same physical quantity can be derived from the same set of sensory data by using different signal-processing techniques. For example, the edges of a surface's boundary can be detected by an edge-detection algorithm. These edges can also be determined by a uniformity-detection algorithm followed by a boundary-tracing algorithm.
- Multiple Mappings:
  The pathway for associating input to output may not be unique. In some cases, there may be an infinite number of mappings from input to output. For example, take the solutions of inverse kinematics for a

kinematically-redundant robot arm manipulator (i.e. an open kineto-dynamic chain). In this case, a given input (e.g. the posture of an end-effector's frame) can be mapped into an infinite number of output (e.g. the possible sets of joint angles). Similarly, the same output may be obtained from the mapping of different inputs. A typical example is face recognition, in which the images of a person's face captured at different places and at different times by the same camera may appear different. However, the interpretation of these facial images (i.e. multiple input) must be the same (i.e. a single output).

- Semantic Overlapping:
  Lexical elasticity in statements or goal specifications creates vagueness, or fuzziness. In addition, semantic overlapping introduces redundancy as well. For example, the goal specification, "Place part A on top of part B, and center it as much as possible" has two overlapping terms: "on top of" (which includes the center) and "center it as much as possible" (which permits non-center locations).

For the most part, redundancy is different from uncertainty. Uncertainty is the source of disturbance which may cause error in the output of a decision-making process. However, redundancy is the source of information which can be used advantageously by a decision-making process (e.g. optimization, sensor fusion, etc).

9.2.5   *Methodologies in Decision-Making*

Uncertainty and redundancy cannot exist without the presence of certainty (even white noise is a mathematical abstraction). Fortunately, the whole world is dominated by a vast body of certainty. To a certain extent, we can say that the ultimate goal of science and technology is to discover the laws, principles and methodologies for the analysis, synthesis, design and control of the physical, social, economic and biological systems in the world.

9.2.5.1   *Expert Knowledge*

There is a vast amount of expert knowledge in various fields of science and technology which helps us to better understand and explain all kinds of phenomena in physics, biology, the environment and human society. Expert knowledge is a treasure for us to rely on in order to make wise decisions, efficient solutions, and new discoveries.

This chapter is devoted to the study of expert knowledge which aims at deriving efficient solutions for image-guided motion planning and control.

9.2.5.2 *Statistical Inference*

Despite uncertainty, all physical systems deployed in industry and society are stable and have deterministic behaviors. This is due to the application of expert knowledge together with powerful mathematical tools, which efficiently transform uncertainty into a specific form of certainty (e.g. probabilistic distribution, mean, variance, etc). One widely-used mathematical tool is *statistical inference* which is based on the probability theory.

We know that if input to, or output from a system, device or sensor exhibits the property of randomness, the corresponding quantity is better described by random variable X. In the probability theory, observation of a random variable is called a *sample*. Accordingly, all possible values of a random variable are called the *sample space*. In practice, the statistical inference techniques include: a) estimation of statistics & interval, b) hypothesis testing, and c) Bayes's decision rule.

Estimation of Statistics and Interval

When we have a set of n observations of random variable X:

$$\{X_i, \; i = 1, 2, ..., n\},$$

we can compute the mean:

$$\bar{X} = \frac{1}{n} \bullet \sum_{i=1}^{n} X_i \tag{9.1}$$

and the variance:

$$\sigma_X^2 = \frac{1}{n-1} \sum_{i=1}^{n} (X_i - \bar{X})^2. \tag{9.2}$$

The mean and variance are called the *statistics* of a random variable. If we know the variance of random variable X, its square root is called the *standard deviation*, which is useful for the determination of upper and lower limits. In fact, upper and lower limits define an interval that the value of a random variable is most expected to fall into. For example, the six-sigma rule defines the interval to be $[-3\sigma_X, \; 3\sigma_X]$.

Mathematically, the computation of statistics with regard to a set of observations of a random variable can be treated as a sort of transformation from uncertainty (e.g. random values) to certainty (e.g. statistics).

Hypothesis Testing

For a physical system to exhibit deterministic behaviors at the lowest level of action, it is necessary to eliminate as much as possible uncertainty at the output of a decision-making process. If we consider an output without uncertainty to have a 100% confidence level, then any output with uncertainty can better be described as a deterministic output (e.g. "Do it" or "Don't do it") associated with the confidence level γ ($\gamma \in (0,1)$). In this way, uncertainty can be eliminated by simply fixing the confidence level. This is because all outputs above a predefined confidence level are treated as deterministic outputs.

In probability, a decision or statement having uncertainty is called a *hypothesis*. A decision-making process which accepts or rejects a hypothesis is called *hypothesis testing*. Hypothesis testing can be explained as follows:

- Postulation of Hypothesis:
  Usually, a statement always has a counter statement. As a result, we normally postulate two hypotheses: a) the original statement (known as the *null hypothesis* and denoted by H_0), and b) the alternative statement (known as the *alternative hypothesis* and denoted by H_1).
- Calculation of Errors:
  When we have an input, the outcome of hypothesis testing is either H_0 or H_1. Because of uncertainty, any decision involves a certain level of risk. The risk of decision-making is better described by *Type-I* and *Type-II* errors. A Type-I error is the probability, denoted by α, of making the wrong rejection of hypothesis H_0. That is,

$$\alpha = P(\text{reject } H_0 \mid H_0 \text{ is true}). \qquad (9.3)$$

And, the Type-II error is the probability, denoted by β, of making the wrong acceptance of hypothesis H_0. That is,

$$\beta = P(\text{accept } H_0 \mid H_0 \text{ is false}).$$

Since the wrong acceptance of H_0 also means the wrong rejection of H_1, then we have

$$\beta = P(\text{reject } H_1 \mid H_1 \text{ is true}). \qquad (9.4)$$

- Decision making:

  Ideally, both Type-I and Type-II errors must be low. However, due to uncertainty, it is necessary to compromise. Depending on the applications, confidence level γ can be defined as either $\gamma = 1 - \alpha$ if we test whether to accept H_0 or, $\gamma = 1 - \beta$ if we test whether to accept H_1.

Bayes's Decision Rule

In the probability theory, the measurement of joint probability $P(AB)$ can be interpreted as the likelihood of mapping from an input (treated as event A) to a possible output (treated as event B). In a decision-making process, the input and output are correlated. Therefore, we have

$$P(AB) = P(A \mid B) \bullet P(B) = P(B \mid A) \bullet P(A) \qquad (9.5)$$

where $P(A \mid B)$ is the conditional probability of event A when event B occurs, and $P(B \mid A)$ is the conditional probability of event B when event A occurs.

The elimination of $P(AB)$ in Eq. 9.5 yields the well-known *Bayes's theorem*. That is,

$$P(B \mid A) = \frac{P(A \mid B) \bullet P(B)}{P(A)}. \qquad (9.6)$$

Under the context of decision-making, Eq. 9.6 is interpreted as the probability of making decision B when the input is A.

In practice, we frequently encounter the situation of multiple possible outputs for a given input. For example, if there are two possible outputs, H_0 and H_1, for a given input A, the application of Eq. 9.6 will yield the following Bayes's decision-making rules:

- Rule 1: Accept H_0, if

$$P(A \mid H_0) \bullet P(H_0) > P(A \mid H_1) \bullet P(H_1). \qquad (9.7)$$

- Rule 2: Accept H_1, if

$$P(A \mid H_1) \bullet P(H_1) > P(A \mid H_0) \bullet P(H_0). \qquad (9.8)$$

9.2.5.3 *Fuzzy Inference*

There are a lot of uncertainties when dealing with natural-language statements or descriptions due to the lexical elasticity of the words used, such as small, big, many, few, etc.

In order to co-exist with humans, however, a humanoid robot must become a sociable robot and, therefore, must be equipped with the necessary mathematical tools to help it interpret and understand natural languages.

In the 1970s, Lofti A. Zadeh founded a new discipline, known as Fuzzy Logic, which consists of fuzzy sets and membership functions (or possibility distribution functions).

A fuzzy set is a superset of a Boolean set. For example, in color-image segmentation, the set R (red) is a Boolean set if it is represented as follows:

$$R = \{(u_i, v_i), \ \mu_R(u_i, v_i) = 1 \mid u_i \in [1, rx], \ v_i \in [1, ry]\}$$

where (rx, ry) are image resolutions in both the horizontal and vertical directions, and μ_R is called a *membership function*.

However, a pixel can also be described by the degree of redness. In this case, the membership function μ_R can take values between 0 and 1. Accordingly, the Boolean set R becomes a fuzzy set which is represented as follows:

$$R = \{(u_i, v_i), \ 0 \le \mu_R(u_i, v_i) \le 1 \mid u_i \in [1, rx], \ v_i \in [1, ry]\}.$$

Boolean sets can not only be extended to fuzzy sets, but Boolean operators (AND, OR and NOT) can also be extended to Fuzzy operators (MIN, MAX and additive complement).

When a system's input and output can be represented by fuzzy sets, the mapping between fuzzy input and fuzzy output can be established by a process called *fuzzy inference*. In general, a fuzzy-inference process involves these steps:

- Fuzzification of Input:
  Inputs to a fuzzy-inference process are called *conditions*. Thus, fuzzy inference can be treated as a process of evaluating "if-then" rules, such as "If the neighbors are red, then the pixel itself is certainly red". Therefore, it is necessary to convert the "if" conditions into corresponding fuzzy sets.
- Evaluation of "If" Conditions:
  After the conditions in an "if-then" rule have been converted into the fuzzy sets, the next step is to apply fuzzy operators (MIN, MAX, or additive complement) to evaluate the "if" conditions. The result of this evaluation will be a single numeric number (i.e. a membership

function's value) which indicates the degree of support for the "if-then" rule.

- Evaluation of "Then" Outcomes:
  Each "if-then" rule has a "then" outcome which must be fuzzified in order to obtain the nominal fuzzy set of the "then" outcome. Subsequently, this fuzzy set is subject to evaluation by the result of the "if" conditions. One way to evaluate the nominal fuzzy set of the "then" outcome is to truncate its membership function by the result of the "if" conditions in order to obtain the actual fuzzy set for the "then" outcome.

- Fusion of "Then" Outcomes:
  A fuzzy-inference process usually involves many different "if-then" rules which lead to different outcomes. Together, these outcomes form a single output set. Thus, it is necessary to combine the fuzzy sets of the "then" outcomes, in order to obtain the fuzzy output set. A simple way of doing this is to apply the MIN fuzzy operator.

- Defuzzification:
  The purpose of defuzzification is to make a decision based on the membership function of the fuzzy output set. A simple solution for defuzzifying the output's fuzzy set is to compute the centroid of its membership function, and use the value of the centroid as the threshold to convert the output's fuzzy set into a corresponding Boolean set.

While many advances have been made in fuzzy logic, the issue of how to represent and filter out fuzziness in statements described by a natural language still remains unresolved.

9.2.5.4 *Neural Network*

In Chapter 8, we studied the probabilistic RCE neural network, and we know that its architecture is based on a multiple-layered structure composed of neurons. The neurons in the middle layer are the information-processing units. Most importantly, the neurons in the input layer are mapped to the neurons in the output layer through multiple pathways formed by the neurons in the middle layer.

Clearly, the advantages of using a computational neural network are two-fold:

- A computational neural network can effectively handle redundancy because multiple neurons are present in both the input and output layers and there are multiple pathways formed by the neurons in the middle layer(s).

- A computational neural network can effectively handle uncertainty because the neuron is an information-processing unit which can be modelled as a statistical inference engine.

Not surprisingly, human brain's neural computing network is a complex, multiple-layered neural architecture with well-established partitions, each of which is dedicated to a specific sensory-motor function.

In the following sections, we will focus on the discussions of expert knowledge underlying the decision-making for autonomous behaviors that a robot tends to achieve.

9.3   Decision Making for Autonomous Behaviors

A robot is capable of automatically executing preprogrammed motions without human intervention, because of the use of the automatic-feedback control loop. Thus, a robot is indeed an automated machine. Now, the challenging question is: How do we make a robot an autonomous machine with a certain degree or level of autonomy?

As shown in Fig. 9.3, automatic execution of motions by a robot requires the input of desired motions which must be preprogrammed. So far, all industrial robots need human intervention to program or reprogram desired motions. If we consider a goal description as the primitive input to a robot, human intervention can occur at three levels:

- Task Level:
  The aim is to translate a goal description into a sequence of ordered tasks, each of which will have its own description.
- Action Level:
  The aim is to translate a task description into a sequence of ordered actions, each of which will have its own description.
- Motion Level:
  The aim is to translate an action description into a sequence of ordered motions, each of which will be described in a form suitable as input to an automatic-feedback control loop.

Without any degree of autonomy, a robot will be dependent on human intervention at these three levels. The only way for a robot to gain a certain

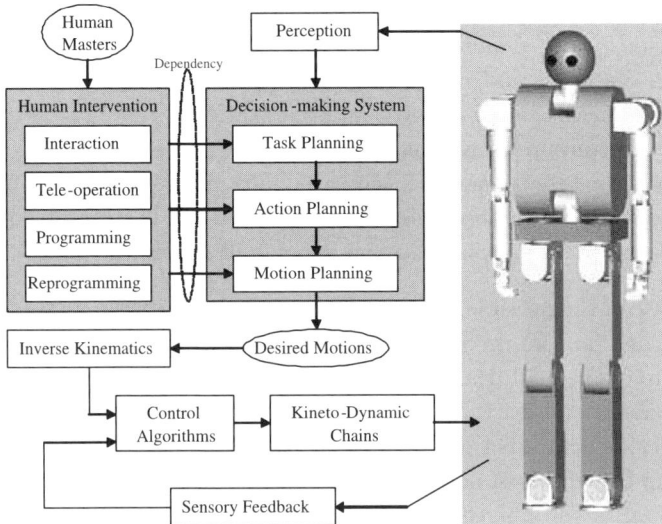

Fig. 9.3 Illustration of human intervention, dependence, and the role of a robot's decision-making system.

degree of autonomy is to develop a decision-making system which is able to self-specify desired motions from the description of actions, tasks or even goals.

9.3.1 *Task or Scenario Planning*

The aim of task or scenario planning is to translate a goal description into a sequence of ordered tasks so that the combined outcomes of these tasks result in the achievement of the goal. A scenario can be modelled by a sequence of ordered tasks. Therefore, task planning also means scenario planning.

9.3.1.1 *A Goal Description as Input*

The input to task planning is the description of goal(s). Now, the question is: What generic model represents all possible goal descriptions? There is no simple answer to this because a goal is the result of combined expectations which satisfy multiple needs or desires. These needs, or desires, can be internal (for personal consumption) and/or external (for the interest of others).

In general, a goal description must consider the following two factors:

- The Top-down Expectation:
  Obviously, a goal must have a purpose. Usually, this purpose is derived from the needs and/or desires of a physical system and a collectivity (e.g. a factory, a community, an organization).
- The Bottom-up Constraint:
  A goal must be achievable. The feasibility of a goal depends on many factors. However, the most important one is the bottom-up constraint imposed by the physical system which will realize the goal.

Because of the possible conflict or contradiction between a top-down expectation and bottom-up constraint, the goal description tends to be elastic. For example, the goal description, "Try our best to win this match", has certain semantic elasticity which is intended to accommodate the possible conflict or contradiction between the desire to win the match and the bottom-up constraint of not-being able to.

So far, it is still not very clear what the innate mechanism is that humans use to specify coherent, consistent and structured goal descriptions. Certainly, any breakthrough in this field would be a great discovery.

Example 9.1   A production line, which manufactures a specific product in large quantities, is composed of three manufacturing processes A, B and C aligned in a series. Each of these processes is fully automated by a dedicated robot, as shown in Fig. 9.4.

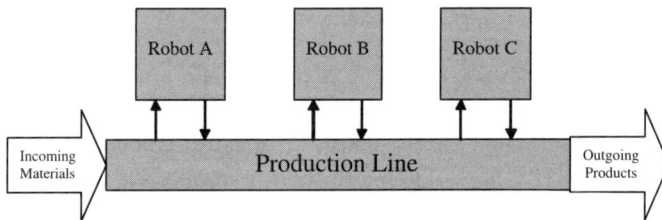

Fig. 9.4   Example of a production line with three robot-stations.

If we assume:

- the production rate using the best robot for process A is 20 units per minute,
- the production rate using the best robot for process B is 10 units per minute, and
- the production rate using the best robot for process C is 30 units per

minutes,

then a goal description such as "Manufacture the product at a rate of 5 to 10 units per minute" is a realistic and achievable goal.

◇◇◇◇◇◇◇◇◇◇◇◇◇◇◇◇◇◇◇

9.3.1.2 *A Task Sequence as Output*

Output from task planning is a sequence of ordered tasks (also known as a *task plan*). Here, the challenging question is: Will there be any general model to represent the task description? The answer to this is not simple, as all possible tasks are categorized differently. For example, in manufacturing, we have the following common tasks: a) assembly task, b) welding task, c) painting task, d) material-handling task, e) palletizing task, f) packaging task, g) inspection task, h) transportation task, etc.

It seems that as in manufacturing, every application domain has its own list of generic tasks. Accordingly, it may be useful to construct a library of generic tasks which covers as wide a range of application-domains as possible. However, this is an area in robotics where very little research has been done.

9.3.1.3 *The Task-Planning Process*

Task planning is a process which translates the goal description into a sequence of ordered tasks. Together, these ordered tasks must be able to achieve the goal. Mathematically, the task-planning process can be treated as the determination of a mapping function from the goal description to one or more possible sets of ordered tasks. The final result of a task-planning process is a selected set of ordered tasks which optimize the achievement of the goal.

Clearly, the automation of the task-planning process requires the fulfillment of the following conditions:

Necessary Condition 1: The goal description should be consistently interpretable by the computer program. Here, "consistency" means that repeated attempts at interpretation generate the same result.

Necessary Condition 2: A list of the generic tasks in the related application domains are known in advance.

Necessary Condition 3: All possible sets of ordered tasks can be validated and optimized through simulation in a virtual world. (This is somewhat similar to a human being's judgement and imagination which are performed in the mental world).

So far, these three necessary conditions constitute the major obstacles to the automation of the task-planning process in robotics. And, without the ability to automate the task-planning process, the only option is to rely on human intervention. This is why none of today's robots have gained autonomy at the task-planning level.

Example 9.2 In a production line, assume that the last station is the *palletizing station*. The goal of the palletizing station is to load the finished products, each of which has already been packaged into a box, into the pallets, and transport the filled pallets to a warehouse.

Figure 9.5 shows a proposed layout of a palletizing station where robot A is a stationary-arm manipulator, and robot B is a mobile-arm manipulator mounted onto mobile robot C.

Assume that the goal description is "Palletize the products and store them in a warehouse". If we know the proposed layout of equipment, as shown in Fig. 9.5, the results of task planning will be a sequence of ordered tasks such as:

- Task 1: Bring empty pallet to location near robot A (task assigned to conveyor B).
- Task 2: Bring products to location near robot A (task assigned to conveyor A).
- Task 3: Palletize products (task assigned to robot A).
- Task 4: Place filled pallet from conveyor B onto robot C (task assigned to robot B).
- Task 5: Transport filled pallet to warehouse (task assigned to robot C).

◇◇◇◇◇◇◇◇◇◇◇◇◇◇◇◇◇◇

9.3.2 *Action or Behavior Planning*

As with task planning, the aim of action planning is to translate a task description into a sequence of ordered actions so that the task can be performed through the execution of these ordered actions.

Since behavior can be modelled using a sequence of ordered actions, thus, action planning also refers to behavior planning.

(a) View at Time Instant T1

(b) View at Time Instant T2

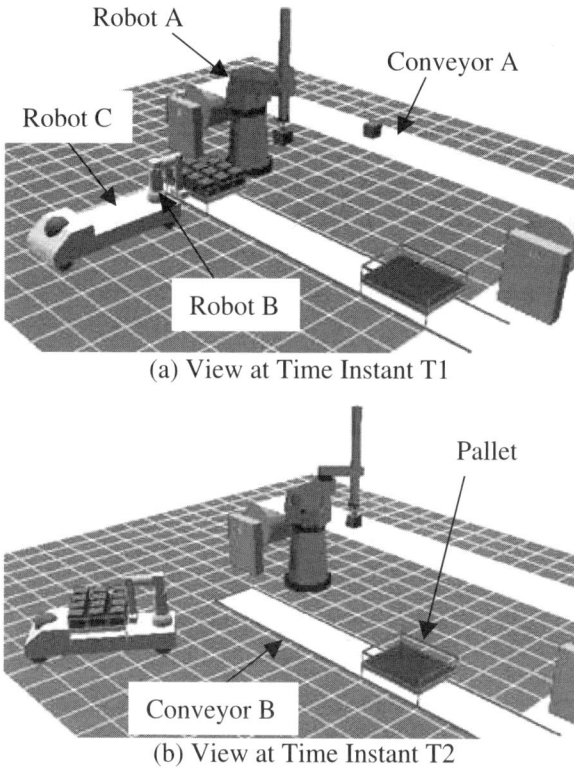

Fig. 9.5   Example of a palletizing station.

9.3.2.1   *A Task Description as Input*

The input to action planning is a task description. As we mentioned earlier, a list of generic tasks in a specific application domain can be identified in advance. However, the generic task descriptions largely depend on expert knowledge and experience.

Conceptually, a generic task which represents the common characteristics of all similar tasks is an abstracted version of a task description. So far, understanding the innate mechanism behind the human ability to abstract similar tasks into a description of generic tasks is still an object of research. Thus, the first issue to achieve the automatic formulation of task descriptions is how to abstract similar tasks into descriptions of common generic tasks.

The second issue is how to customize a generic task into a version which will effectively contribute to the achievement of a specified goal. In Example 9.2, task 1 is a customized version of the generic task, "Load and unload", within the material-handling task category. Clearly, this generic task can be customized in many different ways.

9.3.2.2 *An Action Sequence as Output*

The output from action planning is a sequence of ordered actions (also known as an *action plan*) which guarantees the effective performance of a specified task. As a task description can be treated as a customized version of a generic task, the issue becomes: How do we predefine a sequence of ordered generic actions which guarantees the performance of a generic task?

For example, the generic task of "Pick up and fix" under the assembly task category, can be accomplished with these three generic actions:

(1) Pick up an object at a location
(2) Move an object along a path, or trajectory, from one location to another
(3) Fix (e.g. screw, insert, attach, etc) an object at a location with respect to another object (or common base)

When we have the description of a specified task, the first thing to do is to determine the corresponding generic task. This can be done by an *abstraction process*. If a generic task has at least one set of predefined generic actions, a *customization process*, with regard to these generic actions, will result in an output of a sequence of ordered & customized actions. Here, an actual action is treated as the customized version of a generic action.

9.3.2.3 *The Action-Planning Process*

As with task planning, the action-planning process can also be treated as the determination of a mapping function from the task description to one or more sets of ordered actions. The final result of an action-planning process is a selected set of ordered actions which optimize the accomplishment of a specified task.

As already discussed, the task description can be treated as a customized version of a generic task. Thus, if the description of a generic task corresponding to the description of a specified task (i.e. input) can be obtained by abstraction, a sequence of ordered generic actions can easily be determined. In general, we can automate action planning if the following conditions are met:

Necessary Condition 1: The description of a specified task can be abstracted into a description of a corresponding generic task, which is known in advance.

Necessary Condition 2: One or more sets of ordered generic actions for the accomplishment of a generic task are known in advance.

Necessary Condition 3: All possible sets of ordered & customized actions can be validated and optimized through simulation in a virtual world.

In order to satisfy the first two necessary conditions, a task-action table or relational database must be available in advance. Table 9.1 shows an example of some generic task-action relations applicable to manufacturing.

Table 9.1 Example of generic task-action table.

Generic Task	Action 1	Action 2	Action 3
Assembly	pick up	move	fix
Arc Welding	attach	follow	detach
Painting	attach	follow	detach
Material Handling	load	move	unload
Palletizing	pick up	move	place
Transportation	undock	travel	dock

Example 9.3 Refer to Fig. 9.5. The results of action planning for the five tasks discussed in Example 9.2 will be:

- Action Plan for Task 1:

  (1) Load an empty pallet at a specified location on conveyor B,
  (2) Move it to a specified location near robot A.

- Action Plan for Task 2:

  (1) Load a product at a specified location on conveyor A,
  (2) Move it to a specified location near robot A.

- Action Plan for Task 3:

  (1) Pick up a product present at a specified location on conveyor A,

    (2) Move it along a specified path, or trajectory, to a location above a specified empty slot in the pallet,

    (3) Place it into the specified empty slot.

- Action Plan for Task 4:

    (1) Pick up the filled pallet from a specified location,

    (2) Move it along a specified path, or trajectory, to a specified location above robot C,

    (3) Place it at a specified location on robot C.

- Action Plan for Task 5:

    (1) Undock robot C from the present parking slot,

    (2) Travel along a specified path, or trajectory, to a specified location near a specified parking slot in the warehouse,

    (3) Dock robot C into the specified parking slot.

◇◇◇◇◇◇◇◇◇◇◇◇◇◇◇◇◇◇

9.3.3 *Motion Planning*

The purpose of motion planning is to determine feasible paths or trajectories, and their mathematical descriptions (motion descriptions) which guarantee the effective execution of a specified action. Here, motion descriptions can be in either task space or joint space. But, they must be in analytical or numerical forms so that they can be directly sent to a robot's motion control system(s), as input.

9.3.3.1 *An Action Description as Input*

The description of a specified action is the input to motion planning. Normally, action description is still in semantic form but it has less semantic elasticity than the semantic description of a task.

Depending on the intended application, it is possible to pre-establish the generic forms of action description. As shown in Table 9.1, some generic forms of action description in manufacturing include: pick up, move, follow, place, load, dock, fix, screw, insert, etc.

In robotics, action is performed by the end-effector attached to an open kineto-dynamic chain. Mathematically, the posture of an end-effector is represented by the posture of the frame assigned to it. Thus, it is important that an action's semantic description should explicitly or implicitly specify the initial and final configurations of the end-effector, which is to perform the action. This is a necessary and sufficient condition which guarantees consistent switching between the semantic description of an action and the corresponding analytical (or numerical) description of a motion.

9.3.3.2 *A Motion Sequence as Output*

The output from action planning is a sequence of ordered motions (also known as a *motion plan*), each of which is described in an analytical or numerical form. Mathematically, motion is represented by either a path (i.e. continuous or discrete spatial locations) or trajectory (i.e. a path with a time constraint).

As a path is a series of continuous or discrete spatial locations, it is always possible to use a set of parametric curves to compactly describe these locations. For example, we can choose one or a combination of the following: a) a linear curve (i.e. straight line), b) a conic curve, c) a cubic spline or other curve described by higher-order polynomials.

In practice, we can pre-establish a set of parametric representations of some generic curves and store them into a database. In this way, a set of generic motions are readily available. And, the actual motion will be treated as a customized version of a generic motion.

9.3.3.3 *The Motion-Planning Process*

Motion planning is a process which transforms a semantic description of a specified action into a sequence of analytical, or numerical descriptions of motion. As we mentioned earlier, a necessary condition which guarantees a feasible solution to motion planning, is:

> **Necessary Condition 1:** The semantic description of an action must explicitly or implicitly specify the initial and final configurations of the end-effector which is to perform the action.

Accordingly, the first issue in motion planning is to identify the initial and final configurations for the end-effector attached to an open kineto-dynamic chain. In robotics, the posture of an end-effector is represented by the frame assigned to it. This frame is known as the *end-effector frame.*

Normally, a coordinate system, or frame, uses the symbols X, Y and Z to denote the three orthogonal axes. An alternative way is to explicitly reflect the semantic meaning of the initial and final configurations in the description of an action. To do this, we can choose symbols which have physical meanings as the labels of the axes of the frames describing an end-effector's initial and final configurations.

For example, if one axis indicates the direction in which the end-effector is approaching a location, we can label this axis the "A axis". Similarly, if another axis indicates the direction in which an end-effector opens its fingers, this axis can be labelled the "O axis". If the A and O axes are orthogonal, the normal vector of the plane passing through the A and O axes can be labelled the "N axis". In this way, it makes more sense to label the frame assigned to an end-effector with the symbols AON rather than XYZ.

If we know the initial and final postures of an end-effector's frame, the second issue is how to determine a feasible path, or trajectory, which connects the initial and final postures of an end-effector's frame together. It is challenging to solve the second issue because of the following two additional requirements:

- Collision-free Path or Trajectory:
  Motion execution implies the execution of motion in a safe manner. Therefore, the path or trajectory to be followed by an end-effector's frame should not intersect with any other object in a scene or workspace. The determination of a collision-free path or trajectory is not easy because it requires advance knowledge of the geometric model of the workspace.

- Constrained Motion:
  An open kineto-dynamic chain is composed of serially-connected links which are rigid bodies. Each rigid body has its own frame and a corresponding path, or trajectory, that the frame must follow. If there is no external constraint imposed on the inner links (i.e. the links excluding the end-effector link), and the inverse kinematics of the open kineto-dynamic chain is solvable, it is relatively easy to determine the paths or trajectories for the inner links. However, for some applications, constraints may be imposed on some or all of the inner links as well. In this case, the final result is the constrained motions that the links in an open kineto-dynamic chain must follow. In robotics, planning of constrained motions is not an easy job.

Because of the requirement for a collision-free path/trajectory, given a pair of initial and final configurations, the automation of motion planning calls for another condition. That is,

Necessary Condition 2: The geometric description of a workspace is known in advance.

This necessary condition is a very stringent requirement in an uncontrolled or unstructured environment. One promising solution which would eliminate this requirement is the senor-guided (e.g. image-guided) motion planning. We will study this in the later part of this chapter.

Example 9.4 Fig. 9.6 shows an assembly station. Assume that the goal of this assembly station is to create a part by inserting object B into object A. Two important tasks in this station are:

- Task 1: Pick up object A and place it onto the conveyor.
- Task 2: Pick up object B and insert it into object A.

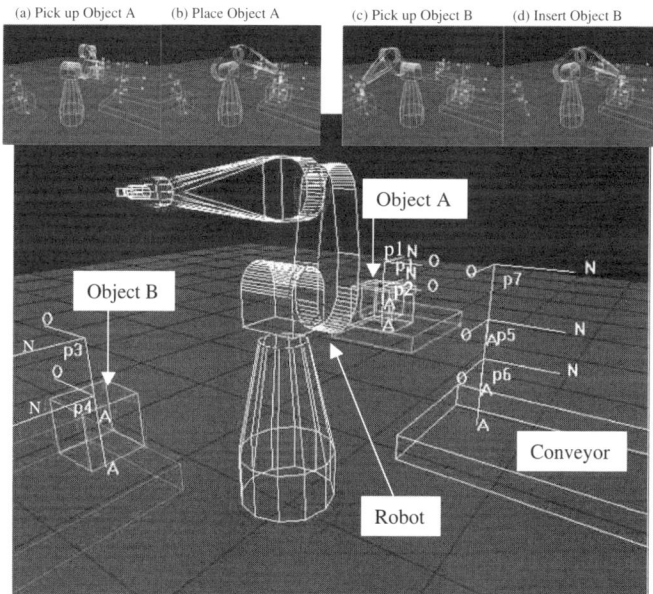

Fig. 9.6 Example of an assembly station.

Then, the results of task planning will be:

- Action Plan for Task 1:
  (1) Pick up object A at a specified location,
  (2) Move it along a specified path,

Table 9.2   Example of motion descriptions.

Motion	Initial Posture	Final Posture	Path/Trajectory
1	AON frame at p_1	AON frame at p_2	Line
2	AON frame at p_1	AON frame at p_5	Line
3	AON frame at p_5	AON frame at p_6	Line

(3) Place it at a specified location on the conveyor.

- Action Plan for Task 2:

    (1) Pick up object B at a specified location,
    (2) Move it along a specified path,
    (3) Insert it into a specified location on top of object A.

The motion descriptions for the actions in task 1 are summarized in Table 9.2.

Similarly, the motion descriptions for the actions in task 2 are summarized in Table 9.3.

Table 9.3   Example of motion descriptions.

Motion	Initial Posture	Final Posture	Path/Trajectory
1	AON frame at p_3	AON frame at p_4	Line
2	AON frame at p_3	AON frame at p_7	Line
3	AON frame at p_7	AON frame at p_6	Line

◇◇◇◇◇◇◇◇◇◇◇◇◇◇◇◇◇◇

9.3.4 *A General Framework for Automated Planning*

One of the ultimate goals of robotics research is to develop robots having a certain degree of autonomy. As we discussed above, the key to doing this depends on the ability to automate the planning processes at the task, action, or motion level. So far, we cannot answer the question: What is the innate mechanism behind the decision-making system of an autonomous robot?

However, we do know that a decision-making system which is capable of automatically generating task, action, or motion plans must possess the following elements, as shown in Fig. 9.7:

- Modelling of Generic Motions:
  The decision-making system of an autonomous robot should have a library of generic motions, each of which has a generic parametric description. When there is a library of generic motions as input, a modelling process should automatically organize these generic motions into a library of generic actions as output.

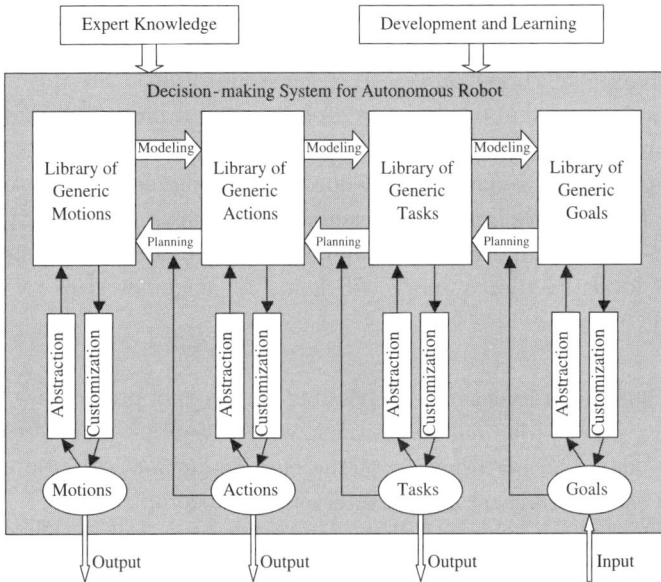

Fig. 9.7 A general framework for the automatic generation of task, action, and motion plans.

- Modelling of Generic Actions:
  Similarly, when there is a library of generic actions as input, a modelling mechanism must be in place in order to group the generic actions into a library of generic tasks.
- Modelling of Generic Tasks:
  Since a generic goal corresponds to a sequence of ordered & generic tasks, it is necessary to develop a modelling process which can automatically organize these tasks into a library of generic goals as output.
- A Semantic Analysis-Synthesis Loop:
  When there is a man-specified goal description as input, it is necessary to have an abstraction process to filter out the semantic and lexical

elasticity, and then, match the filtered description to a corresponding generic-goal description stored in the library. Abstraction is an analysis process which is likely to be performed in a closed-loop manner in order to achieve consistency and accuracy. Consistency refers to the stable mapping between a man-specified goal and its corresponding generic goal. Accuracy is the degree of semantic conformity between a man-specified goal and its corresponding generic goal. Within the closed-loop of the semantic analysis-synthesis, the result of semantic analysis (i.e. abstraction) can be re-mapped back to its original form through a semantic-synthesis process (i.e. customization). In this way, the customized version of a generic description can be compared with the original description in order to decide whether the closed-loop of the semantic analysis-synthesis was successful or should be repeated. If we want to eliminate human intervention at the goal, task or action level, the closed-loop of the semantic analysis-synthesis must be deployed at these levels so that a robot will gain full autonomy (up to the goal level).

With human intervention, it is possible to construct preliminary versions of libraries which store generic motions, generic actions, generic tasks and generic goals. The performance of the closed-loop for semantic analysis-synthesis can also rely on human intervention to complete.

It is clear that expert knowledge and experience will enable a human master to program a robot as an automated machine. However, the challenging question here is: Is it possible to develop an innate mechanism for a robot to automatically acquire knowledge and skills, in order to perform the modelling, planning and semantic analysis-synthesis processes inside its decision-making system? Again, this is an interesting research topic in the area of development and learning.

9.4 The Basics of Motion Planning

A first and necessary step towards the development of an autonomous robot is to investigate principles and methodologies for automating the motion-planning process. This is because, without the mental ability to automatically undertake motion planning, a robot will never be able to achieve any degree of autonomy at action or task level.

As we studied earlier, the purpose of motion planning is to identify the initial and final configurations of an end-effector from the description of a given action first. Then, it is necessary to determine a feasible path or trajectory connecting these two configurations together while at the same time, satisfying some constraints imposed by the workspace and physical system (e.g. a robot).

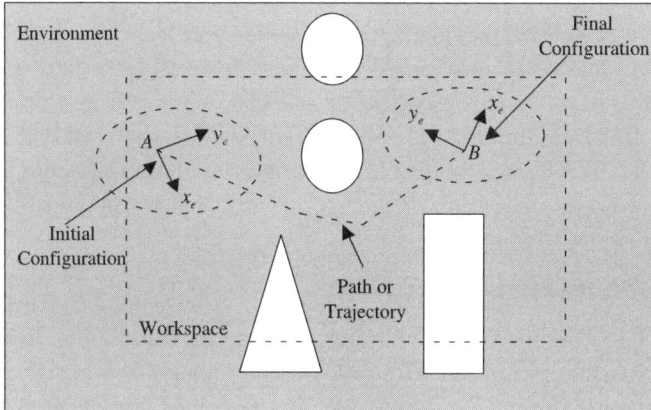

Fig. 9.8   Illustration of the motion-planning problem.

Figure 9.8 illustrates the problem of motion planning. For the sake of simplicity, let us assume that the description of a given action explicitly specifies the initial and final configurations of the end-effector attached to a robot's open kineto-dynamic chain. The critical issue, then, is how to determine a feasible path or trajectory which will bring the end-effector's frame from the initial configuration to the final configuration. By default, the frame assigned to an end-effector is denoted as frame e.

9.4.1   *Path and Trajectory*

As shown in Fig. 9.8, the output from motion planning is a feasible path or trajectory. Formally, with regard to motion planning, the definition of path is as follows:

Definition 9.4   Path is a series of continuous, or discrete, spatial locations which connect the initial and final configurations of an end-effector.

Mathematically, a path can be described by a spatial curve such as a linear, conic or spline curve. The spatial locations of a path only describe the static configurations of an end-effector. They do not impose any dynamic constraints in terms of velocity and acceleration. And, for some applications, these may be the most important constraints. Therefore, it is necessary to associate time to a planned path. This results in what is called a *trajectory*. A formal definition of trajectory can be stated as follows:

Definition 9.5   Trajectory is a path with a time constraint.

In order to determine velocity and acceleration from the description of a path and time constraint, it is convenient to consider a path as a one-dimensional space. In this way, the end-effector is treated as a point which travels in this one-dimensional space. If L denotes the travelled distance with respect to reference point O_s (the origin of the initial configuration), the velocity and acceleration of the end-effector will be $\frac{dL}{dt}$ and $\frac{d^2L}{dt^2}$.

9.4.2   *Motion-Planning Strategy*

If neither the workspace nor the robot impose any constraints, it is simple to plan a feasible path, or trajectory, which connects the initial and final configurations. However, motion planning becomes complicated when it is necessary to consider the geometric constraints imposed by a workspace and the kinematic (or dynamic) constraints imposed by a robot. Obviously, different motion-planning strategies achieve different levels of effectiveness depending on the constraints.

9.4.2.1   *Forward Planning*

A common motion-planning strategy is to determine a feasible path or trajectory which connects the initial configuration to the final configuration. Subsequently, the planned path or trajectory is directly executed by a robot's end-effector. We call this strategy *forward planning*.

Fig. 9.9 illustrates forward-motion planning. The search for a feasible path or trajectory starts from the initial configuration. Among all the feasible paths or trajectories, we choose one which satisfies certain criterion (e.g. shortest distance).

However, the forward-planning strategy has two drawbacks:

- If the final configuration is tightly constrained by the presence of other objects (or obstacles), the forward search for a feasible path may be computationally expensive. In certain cases, one may even fail to find a solution. A typical example which illustrates the inefficiency of the forward-planning strategy is the problem of planning a feasible path, or

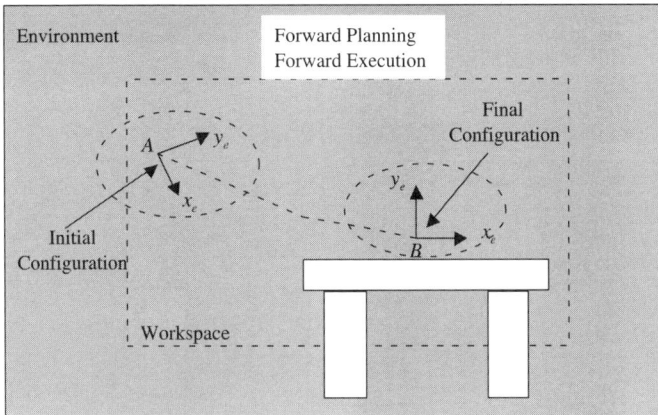

Fig. 9.9   Illustration of the forward-planning strategy.

trajectory, for the parallel parking of a car-like vehicle. This problem, by the way, can be easily solved by the inverse-planning strategy (see Example 9.5 below).

- Nowadays, image-guided motion planning is a very popular research topic, and researchers aim at building future robots capable of imitating human behaviors, such as hand-eye coordination. In Chapter 8, we studied the visual-perception system of robots. We know that an image is an array of pixels, as well as the perspective projection of a three-dimensional scene onto the camera's image plane. Due to the digital nature of an image, inverse projective-mapping is not very accurate when the actual coordinates of a point in the scene are far away from the camera's image plane. Thus, hand-eye coordination implemented with the forward planning strategy will perform poorly if the initial configuration of the hand is closer to the eyes than the final configuration. This problem can be easily solved, however, if we adopt the inverse-planning strategy. We will study hand-eye coordination in greater detail in the later part of this chapter.

9.4.2.2 *Backward Planning*

An alternative motion-planning strategy is to determine a feasible path or trajectory which connects the final configuration to the initial configuration. Subsequently, the planned path or trajectory is executed in reverse by a robot's end-effector. We call this strategy *backward or inverse planning*.

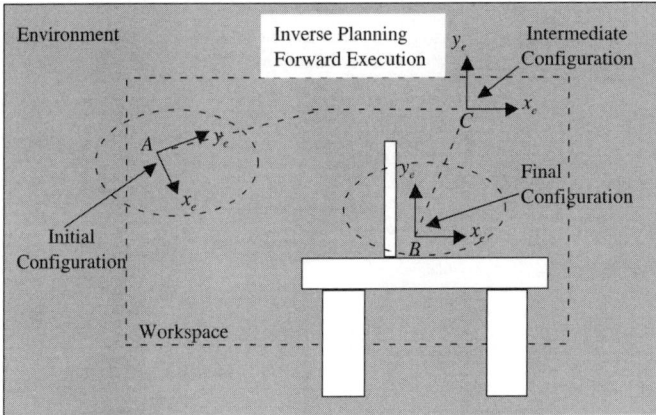

Fig. 9.10   Illustration of the inverse-planning strategy.

Fig. 9.9 illustrates inverse-motion planning in task space. The key idea behind backward planning in task space is to determine an intermediate configuration which can easily be reached from the final configuration. Normally, we can assume that the intermediate configuration is not so constrained by the workspace. Thus, it is easy to determine a path or trajectory connecting the intermediate configuration to the initial configuration. If the initial configuration is highly constrained, one can iteratively apply both forward and backward planning strategies.

Example 9.5   Under the initiative of intelligent transportation, there has been a lot of studies about the development of smart vehicle which can automatically execute maneuvers, such as road following, car following, obstacle avoidance and parking. But, among the parking maneuvers, parallel parking is undoubtedly the most difficult. This is an interesting issue in robotics because future humanoid robots will certainly have the skill to drive car-like vehicles.

Figure 9.11 illustrates the problem of parallel parking, and shows the simulation results obtained using the inverse-planning strategy. If we treat the road surface in the proximity of the car-like vehicle as a planar surface, the posture of the car-like vehicle, with respect to a reference frame, is fully determined by three variables: a) the coordinates (x, y), and b) the orientation θ. For a car-like vehicle, these three variables are not independent. In fact, the first-order derivatives of these three variables depend on one

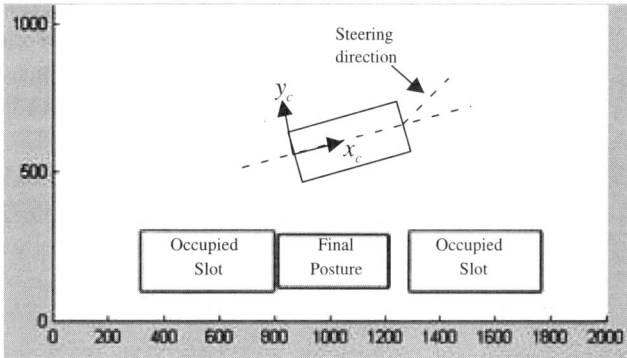

(a) Final Configuration and Geometric Model

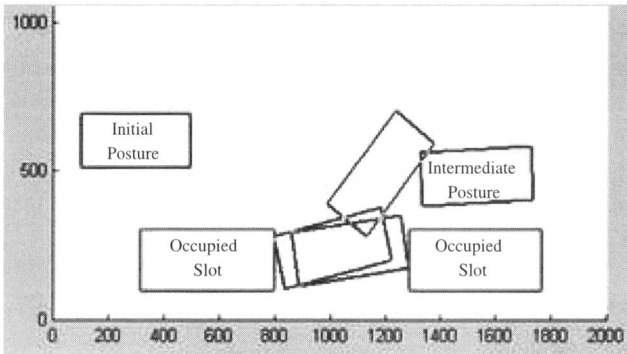

(b) Intermediate and Initial Configurations

Fig. 9.11   Simulation results obtained using backward planning to parallel park a car-like vehicle or mobile robot. The scale of the workspace is in millimeters.

another. This means that a constraint in the following form exists:

$$g(\dot{x}, \dot{y}, \dot{\theta}) = 0.$$

However, there is no constraint in the form of

$$f(x, y, \theta) = 0.$$

Thus, a car-like vehicle (or mobile robot) is governed by a nonholonomic constraint. With regard to a nonholonomic robot, a feasible path must be one in which the maximum curvature does not exceed a limit. This requirement makes forward-motion planning for a nonholonomic robot a difficult job. But, it is manageable if one uses the inverse-motion planning strategy.

As shown in Fig. 9.11a, the final posture of a car-like vehicle is between two occupied parking slots. An important controllable variable in maneuvering a car-like vehicle is steering angle α. Normally, steering angle α is within a limited range $[-\alpha_0, \alpha_0]$ ($\alpha_0 > 0$) which can be quantized into a set of discrete values.

If we denote α_f the steering angle for forward driving, and α_r the steering angle for reverse driving, the set $(\alpha_{f,i}, \alpha_{r,i})$ specifies steering angles for one cycle of forward-reverse or reverse-forward driving. When there is a set of $(\alpha_{f,i}, \alpha_{r,i})$, we can compute the number of forward-reverse or reverse-forward driving cycles which will bring the final posture to a free zone in the workspace. We call this number the *drive-out cycle number*. If there is no solution, the drive-out cycle number is set to infinity.

Interestingly enough, for all the possible sets $(\alpha_{f,i}, \alpha_{r,i})$ ($i = 1, 2, ...$), we can establish a reference table which is indexed by i and stores the drive-out numbers. In fact, this reference table memorizes all feasible solutions for determining intermediate postures. Figure 9.11b shows one solution which brings the final posture to an intermediate posture in one cycle. Once an intermediate posture is obtained, it is easy to plan a feasible path connecting the intermediate posture to the initial posture.

$$\diamond\diamond\diamond\diamond\diamond\diamond\diamond\diamond\diamond\diamond\diamond\diamond\diamond\diamond\diamond\diamond$$

9.4.2.3 *Formation Planning*

By default, motion planning means the determination of a path or trajectory which will make a frame reach a final configuration from an initial configuration. The frame under consideration can either be the end-effector's frame in an open kineto-dynamic chain or a frame assigned to a mobile robot (or vehicle). For some applications, we have trouble planning paths or trajectories for multiple frames (which can physically or virtually constrain each other). Two typical scenarios are:

- Open Kineto-dynamic Chains in Highly Constrained Workspace:
  In general, an open kineto-dynamic chain has n ($n > 1$) rigid links which connect in a series. The last link in the series is the end-effector link. When dealing with kinematic analysis, a frame is assigned to each link. In this way, we simply consider the kinematic constraints among a set of n frames. In a less-constrained workspace, it is sufficient to plan only the path or trajectory for the end-effector's frame (frame n). Then, the paths or trajectories of the remaining $n - 1$ frames will be

determined by a solution of inverse kinematics. However, in a highly-constrained workspace, this strategy does not work, and it is necessary to plan the path or trajectory for each individual frame.

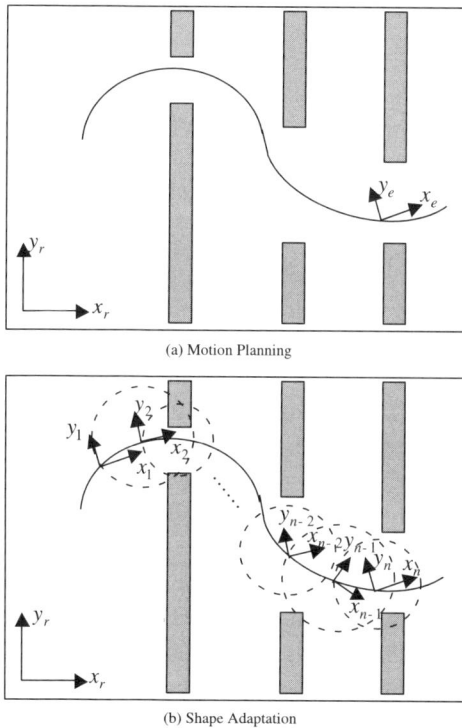

- Formation of Mobile Robots:
  A mobile robot is a rigid body and can be treated as a link. Thus, a fleet of mobile robots (or vehicles) can be modelled as an open-kinematic chain formed by a set of rigid bodies which are virtually connected in a series. The mobile robot at the head of the fleet is treated as the end-effector link. If a frame is assigned to each rigid body (i.e. mobile robot), n mobile robots will have n frames. If the mobile robots in a fleet must form a certain shape or pattern, it is necessary to plan the path or trajectory for each individual frame.

(a) Motion Planning

(b) Shape Adaptation

Fig. 9.12   Illustration of formation planning, and two subproblems.

Planning paths or trajectories for a set of n frames is known as *formation planning*. As shown in Fig. 9.12, the problem of formation planning can be divided into two subproblems: a) motion planning for frame n, and b) shape adaptation for the remaining $n - 1$ frames.

If we know the current posture of frame n and its path or trajectory, shape adaptation can be operated in the following way (see Fig. 9.12b):

- Step 1: Initialize index i to value n.
- Step 2: Draw a circle (or sphere, if in three-dimensional task space) centered at the origin of frame i. The radius of the circle will be the distance between the origin of frame i and the origin of frame $i - 1$.
- Step 3: Determine the posture of frame $i - 1$ which must be located on the circle (or sphere, if in three-dimensional task space) centered at frame i. One simple way to do this is to locate frame $i - 1$ at an intersection point between the circle and the path of frame n.
- Step 4: Decrease index i by 1. Repeat Steps 2 and 3 until $i = 2$.

Since shape adaptation is not a difficult issue, the central problem of formation planning is how to plan a feasible path or trajectory for the end effector's frame in an open kineto-dynamic chain. Clearly, this problem is the same as that of motion planning (for a single frame).

9.5   Motion Planning in Task Space

Motion planning in task space is a problem which can be stated as follows: When we have the initial and final configurations of an end-effector's frame, and we know the geometric model of the workspace as well as the (kinematic) constraint of the robot, what is a feasible path or trajectory which will make the end-effector's frame reach the final configuration from the initial configuration?

There is no simple solution to the problem of motion planning in task space due to the complexity of the geometric constraint imposed by the workspace, and the kinematic/dynamic constraint imposed by the robot. Here, we present a solution which assumes that the initial and final configurations are not highly constrained. If this is not the case, one can apply the inverse-planning strategy to turn a highly-constrained configuration into a less-constrained intermediate configuration.

9.5.1 *Planning of Collision-Free Paths*

If the initial and final configuration are not highly constrained, motion planning becomes a problem of determining a collision-free path in the workspace. A practical solution to this is to first construct a path map in the workspace similar to a city's road map, then select a pathway connecting the initial configuration to the final configuration. For the sake of clarity, we will consider a two-dimensional task space and explain the solution step-by-step.

9.5.1.1 *A Discrete and Normalized Workspace*

If we know the initial and final configurations, we can normalize the workspace, as shown in Fig. 9.13. One solution to normalize a workspace is to choose the line passing through the origins of the initial and final configurations as the horizontal axis. Then, the normalized workspace will be a rectangular window (or a box in 3-D workspace) which is symmetrical about the horizontal axis and centered at the mid-point between the two origins. The dimension of the normalized workspace is to be specified by the user.

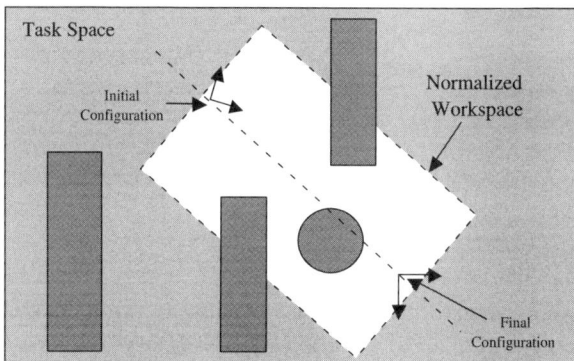

Fig. 9.13   Illustration of a normalized workspace.

The geometry of an object in a normalized workspace can be simple or complex. A simple way to avoid the complexity of describing the geometry of objects present in a normalized workspace is to quantize the workspace into a matrix of cells (or voxels in 3-D workspace). Thus, a binary number is assigned to each cell to indicate whether it is occupied or free. For example, the value "1" can be assigned to an occupied cell, while the value "0" can be assigned to a free cell.

Example 9.6   Figure 9.14 shows an example of a discrete, normalized workspace. The scale is chosen as 100×100, and the workspace is quantized into a matrix of 10×10 cells. The cells colored in grey are occupied.

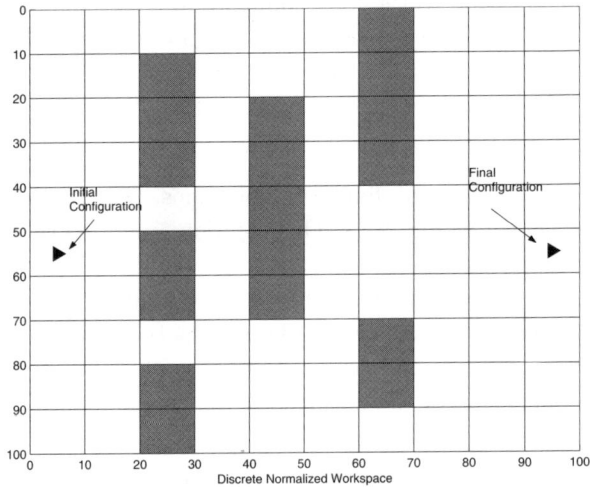

Fig. 9.14   Example of a discrete, normalized workspace.

◇◇◇◇◇◇◇◇◇◇◇◇◇◇◇◇◇

Seed Cells of Paths

When we have a normalized workspace as input, the next step is to construct a path map similar to the road map of a city. This can be done in two steps: a) vertical grouping to determine the seed cells of a path, and b) horizontal linking to construct the path map.

Vertical grouping is operated within each column (or vertical plane if in a 3-D matrix of voxels). The idea is to group consecutive free cells together to form a cluster. The cell at the center of this cluster is called the *seed cell* of a path.

Assume that vertical grouping is done by scanning a column from the first row to the last. If $r_{0,j}$ is the row index of the first cell in cluster j, and $r_{1,j}$ the last cell, the row index of the seed cell corresponding to cluster j

will be

$$r_j = r_{0,j} + (r_{1,j} - r_{0,j})/2. \tag{9.9}$$

Alternatively, it is also possible to choose the cell at one end of a cluster to be the seed cell.

Example 9.7 If we consider the discrete, normalized workspace in Example 9.6 as input, the seed cells of the path obtained by vertical grouping are shown in Fig. 9.15. The cells marked "x" are the seed cells of the path.

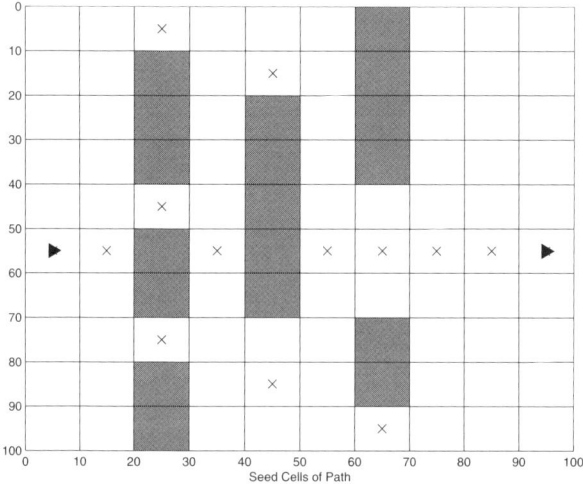

Fig. 9.15 Example of seed cells of a path.

◇◇◇◇◇◇◇◇◇◇◇◇◇◇◇◇◇◇

9.5.1.2 *A Path Map*

If we know the seed cells of paths in the columns, it is easy to construct a path map which connects the seed cells together. This can be achieved by a horizontal-linking process which finds the free paths and connects the seed cells in column c to the seed cells in column $c + 1$.

Refer to Fig. 9.16. When constructing a path map, it is necessary to consider the following two cases:

- Case 1:
  A pair of seed cells are not in the same row. One is in column c and the

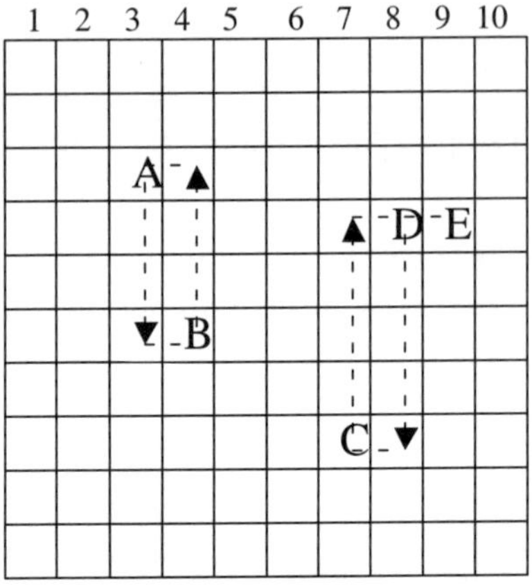

Fig. 9.16   Illustration of the horizontal linking of seed cells.

other is in column $c + 1$. For example, seed cells A and B in Fig. 9.16 are not in the same row. In this case, we first test to see if there is a free path in column c which connects seed cell A to seed cell B. If not, then we test to see if there is a free path in column $c+1$ which connects seed cell A to seed cell B. If not, seed cell A will not have a direct path to seed cell B.

- Case 2:
  A pair of seed cells are in the same row, one in column c and the other in column $c + 1$. For example, seed cells D and E in Fig. 9.16 are in the same row. In this case, these two seed cells have a direct horizontal path.

Example 9.8   If we apply the horizontal-linking process to Example 9.7, we obtain a path map, as shown in Fig. 9.17. In this example, there are multiple feasible paths which connect the initial configuration to the final configuration.

◇◇◇◇◇◇◇◇◇◇◇◇◇◇◇◇◇◇

9.5.1.3   *A Collision-Free Path*

In a path map, we call the cells on the paths the *path cells*. As shown in Fig. 9.17, a path map obtained by the horizontal-linking process consists of:

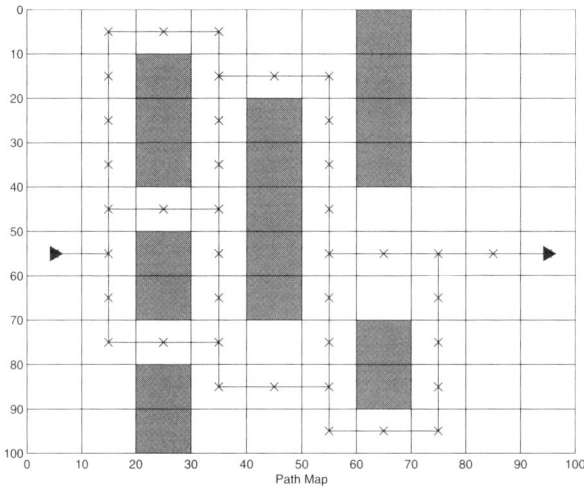

Fig. 9.17 Example of a path map.

- Junctions:
  A junction is a path cell where a bifurcation or intersection occurs.
- Terminals:
  A terminal is a seed cell which has only one path cell as its neighbor.
- Branches:
  A branch is a simple path where there is no bifurcation or intersection. The branches in a path map can be classified into three categories: a) double-ended branch, b) single-ended branch, and c) in-loop branch. A double-ended branch is one which links two terminals together, while a single-ended branch is one which links a terminal to a junction. All branches in the closed paths are called *loop-branches*. A loop-branch connects two junctions together.

In order to select a feasible path which connects the initial configuration to the final configuration, it is possible to apply the following algorithm:

- Step 1: Eliminate all double-ended branches
- Step 2: If the initial configuration is not on a path, connect it to the nearest single-ended or in-loop branch. We call this the *initial branch*.
- Step 3: If the final configuration is not on a path, connect it to the nearest single-ended or in-loop branch. We call this the *final branch*.
- Step 4: Eliminate all remaining single-ended branches.

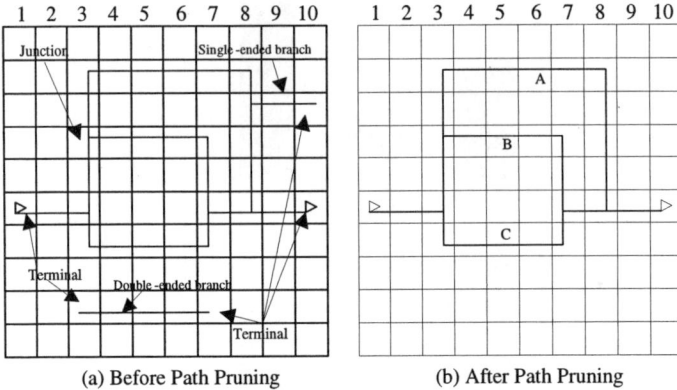

(a) Before Path Pruning   (b) After Path Pruning

Fig. 9.18   Illustration of path pruning.

- Step 5: Choose a subset of in-loop branches to form a feasible path connecting the initial and final branches.

The operation for eliminating double-ended and single-ended branches is called *path pruning*. (See Fig. 9.18). After path pruning, there will be three feasible paths: A, B and C, which connect the initial configuration to the final configuration. In practice, we can choose the one which has the shortest distance as output.

9.5.1.4   *Constraints of Robots*

Here, the output of a feasible path connecting the initial and final configurations is a polyline. Interestingly enough, the angle between two consecutive line segments in the polyline is 90^0. In other words, a feasible path from this motion-planning algorithm contains only straight angles, if any. These straight angles can be approximated by circular arcs in order to take into account the kinematic and dynamic constraints imposed by a robot.

When a mobile robot or a robot's end-effector follows a path, it is necessary to consider two types of constraints:

- Kinematic Constraint:
  This is an explicit constraint imposed on the robot's mechanism. If the input and output motions of a robot's mechanism can be described by both kinematics and motion (or differential) kinematics, the constraint imposed on the mechanism is a *holonomic constraint*. But, if

the input-output motion relationship is only governed by motion (or differential) kinematics, the constraint is a *nonholonomic constraint*. A car-like mobile robot is governed by a nonholonomic constraint. A nonholonomic constraint implies that the curvature of the robot's path should have a maximum limit.

- Dynamic Constraint:
  This is an implicit constraint imposed on the robot's inertia. When a robot is in motion, the change in its linear or angular momentum requires effort (force or torque). If a path can be locally approximated by a circular arc, this means that centrifugal acceleration exists even if the robot is moving at a constant velocity along a path. In order to stay on the planned trajectory (path with a time constraint), the robot must expend extra effort to overcome the centrifugal acceleration. Thus, the dynamic constraint implies that the radius of a circular arc should have a minimum limit.

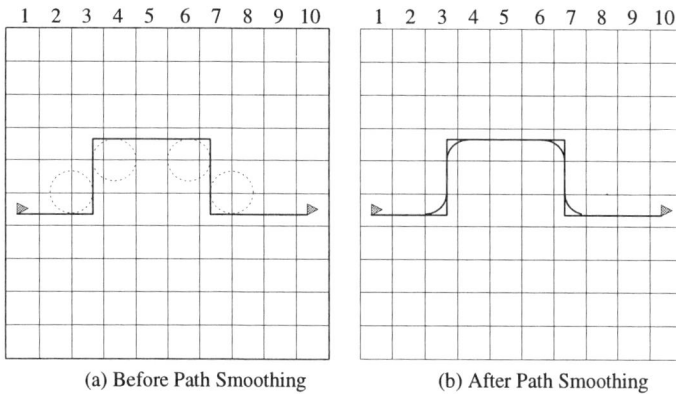

(a) Before Path Smoothing   (b) After Path Smoothing

Fig. 9.19   Illustration of path smoothing.

Since the maximum curvature limit is compatible with the requirement for the minimum limit of a circular arc's radius, it is simple to consider only the minimum-radius constraint . Mathematically, turning at a single point (e.g. corner) is equivalent to movement along a circular arc, the radius of which is zero. Clearly, all the corners on a feasible path must be approximated by the corresponding circular arcs which must have radii greater than the minimum limit. This operation is called *path smoothing*.

Figure 9.19 illustrates the path-smoothing process. The purpose here is to approximate the corners on a feasible path using circular arcs. After path smoothing, the final output of a feasible path satisfies the requirement for a minimum radius. Depending on the applications, an alternative solution is to approximate the corners using spline curves (e.g. cubic spline).

9.5.2 Motion Description

The output from motion planning is a path which is represented in the form of a list of line segments, circular arcs, or spline curves. These line segments, circular arcs, or spline curves are called *path curves*, as shown in Fig. 9.20.

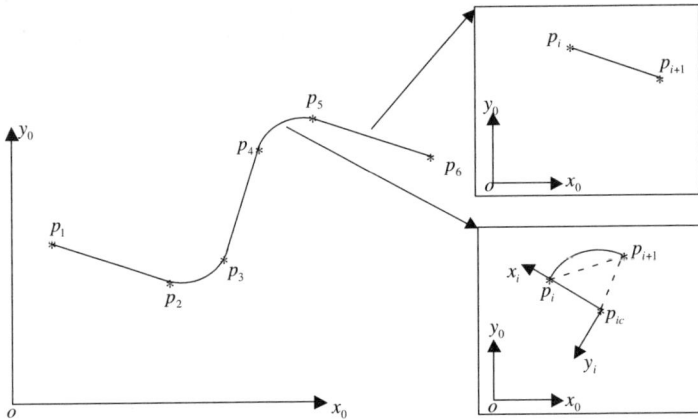

Fig. 9.20   Parametric representation of path curves.

The points which connect two consecutive path curves together are called the *key points* (or *knots*, in computer graphics). In general, the knots of a path can be represented as

$$P = \{p_i, \ i = 1, 2, ..., n\} \tag{9.10}$$

where $p_i = (^0x, \ ^0y)^t$ in 2-D workspace or $p_i = (^0X, \ ^0Y, \ ^0Z)^t$ in 3-D workspace.

And, a path curve can be concisely described by: a) implicit equation ($f(x, y) = 0$), b) explicit equation ($y = f(x)$), or c) parametric equation ($x = f(t)$ and $y = g(t)$). As a path is to be followed by a robot at a certain velocity or acceleration, it is more convenient to choose a parametric representation. In this way, a time constraint can easily be added to the path description in order to obtain the corresponding *trajectory* description.

9.5.2.1 *Linear Curves*

Refer to Fig. 9.20. Let λ denote the travelled distance along a path by a mobile robot or a robot's end-effector. If two consecutive knots p_i and p_{i+1} are connected by a line segment (linear curve), the parametric representation of this line segment is

$$p_i(\lambda_i) = p_i + \frac{p_{i+1} - p_i}{\|p_{i+1} - p_i\|} \bullet \lambda_i \qquad (9.11)$$

where $\|p_{i+1} - p_i\|$ is the length between the two knots p_i and p_{i+1}.

And, one can easily verify the following results:

$$p_i(\lambda_i) = \begin{cases} p_i & \text{if } \lambda_i = 0 \\ p_{i+1} & \text{if } \lambda_i = \|p_{i+1} - p_i\|. \end{cases} \qquad (9.12)$$

Thus, the physical meaning of Eq. 9.11 is that point $p_i(\lambda_i)$ will travel from p_i to p_{i+1} along a straight line if λ_i varies from 0 to $\|p_{i+1} - p_i\|$ and λ_i measures the travelled distance.

When we have two consecutive knots and we use a cubic spline to interpolate these two knots, there will be no parametric representation in terms of the parametric variable λ. However, a common solution is to approximate a cubic spline using a polyline having a sufficient number of line segments. In this way, Eq. 9.11 can be applied to represent the line segments which approximate a cubic spline.

9.5.2.2 *Circular Curves*

Refer to Fig. 9.20 again. Assume that the path curve connecting two consecutive knots p_i and p_{i+1} is a circular arc. Here, we denote the circular arc's center and radius as p_{ic} and r_i respectively. In order to obtain a parametric representation of the circular arc, we need to assign a frame to it. One way of doing this is to choose p_{ic} as the origin, and line $p_{ic}p_i$ as the x axis.

If a frame assigned to a circular arc connecting two consecutive knots p_i and p_{i+1} is denoted as frame i, then the orientation of frame i, with respect to frame 0 (reference frame), can be determined from the direction of line $p_{ic}p_i$. Let us denote ^0R_i the orientation of frame i with respect to frame 0.

In frame i, the circular arc is centered at the origin. Thus, its parametric representation is

$$\begin{cases} {}^i x = r_i \bullet \cos\theta \\ {}^i y = r_i \bullet \sin\theta \end{cases} \tag{9.13}$$

or

$$ {}^i p = \begin{pmatrix} r_i \bullet \cos\theta \\ r_i \bullet \sin\theta \end{pmatrix}. \tag{9.14}$$

In Eq. 9.14, parametric variable θ is a rotation angle about the origin. If λ_i denotes the distance travelled by a point on the circular arc, then

$$\theta = \frac{\lambda_i}{r_i}. \tag{9.15}$$

Substituting Eq. 9.15 into Eq. 9.14 yields

$$ {}^i p(\lambda_i) = \begin{pmatrix} r_i \bullet \cos\left(\frac{\lambda_i}{r_i}\right) \\ r_i \bullet \sin\left(\frac{\lambda_i}{r_i}\right) \end{pmatrix}. \tag{9.16}$$

According to the Cosine Theorem governing a triangle, a point travelling along a circular arc will reach point p_{i+1} after a rotation angle of

$$\beta = \cos^{-1}\left(\frac{\|p_{i+1} - p_i\|^2 - 2 \bullet r_i^2}{2 \bullet r_i^2} \right). \tag{9.17}$$

Thus, one can easily verify the following results:

$$ {}^i p(\lambda_i) = \begin{cases} {}^i p_i & \text{if } \lambda_i = 0; \\ {}^i p_{i+1} & \text{if } \lambda_i = r_i \bullet \beta. \end{cases} \tag{9.18}$$

And, the physical meaning of Eq. 9.16 is that point ${}^i p(\lambda_i)$ will travel from ${}^i p_i$ to ${}^i p_{i+1}$ along a circular arc if λ_i varies from 0 to $r_i \bullet \beta$, and λ_i measures the travelled distance.

If we know the position and orientation of frame i with respect to frame 0, the coordinates of point ${}^i p(\lambda_i)$ in frame 0 will be

$$p_i(\lambda_i) = p_{ic} + ({}^0 R_i) \bullet ({}^i p(\lambda_i)). \tag{9.19}$$

Substituting Eq. 9.16 into Eq. 9.19 yields

$$p_i(\lambda_i) = p_{ic} + ({}^0 R_i) \bullet \begin{pmatrix} r_i \bullet \cos\left(\frac{\lambda_i}{r}\right) \\ r_i \bullet \sin\left(\frac{\lambda_i}{r}\right) \end{pmatrix}. \tag{9.20}$$

Eq. 9.20 is the parametric representation of a circular arc.

9.5.2.3 *Paths*

Refer to Fig. 9.20. A path consists of a set of path curves connected in a series. If a path curve can be described by a parametric equation, such as Eq. 9.11 or Eq. 9.20, the next question is: Can we represent a path in an analytical form?

Consider a path with n knots:

$$\{p_i, \quad i = 1, 2, ..., n\}.$$

If $p_i(\lambda_i)$ is a point on the path curve between two consecutive knots p_i and p_{i+1}, the relative position of $p_i(\lambda_i)$ with respect to p_i will be

$$\triangle p_i(\lambda_i) = p_i(\lambda_i) - p_i. \tag{9.21}$$

Now, let L_i denote the length between two consecutive knots p_i and p_{i+1}. As we discussed above, if a path curve is a linear curve, then

$$L_i = \|p_{i+1} - p_i\|.$$

However, if a path curve is a circular curve, then

$$L_i = r_i \bullet \beta$$

with

$$\beta = \cos^{-1}\left(\frac{\|p_{i+1} - p_i\|^2 - 2 \bullet r_i^2}{2 \bullet r_i^2}\right).$$

If λ is the total distance travelled measured from p_1 by point $p_i(\lambda_i)$, then we have

$$\lambda_i = \lambda - \sum_{j=1}^{i-1} L_j. \tag{9.22}$$

Substituting Eq. 9.22 into Eq. 9.21 yields

$$\triangle p_i(\lambda - \sum_{j=1}^{i-1} L_j) = p_i(\lambda - \sum_{j=1}^{i-1} l_j) - p_i. \tag{9.23}$$

If a path curve is a linear curve, from Eq. 9.11, we can obtain

$$\triangle p_i(\lambda - \sum_{j=1}^{i-1} L_j) = \frac{\lambda - \sum_{j=1}^{i-1} L_j}{\|p_{i+1} - p_i\|} \bullet (p_{i+1} - p_i). \tag{9.24}$$

For a circular curve, the application of Eq. 9.20 will result in

$$\triangle p_i(\lambda - \sum_{j=1}^{i-1} L_j) = p_{ic} - p_i + ({}^0 R_i) \bullet \begin{pmatrix} r_i \bullet \cos\left(\frac{\lambda - \sum_{j=1}^{i-1} L_j}{r}\right) \\ r_i \bullet \sin\left(\frac{\lambda - \sum_{j=1}^{i-1} L_j}{r}\right) \end{pmatrix}. \quad (9.25)$$

If we define

$$\triangle p_i(\lambda) = \begin{cases} 0 & \text{if } \lambda \le \sum_{j=1}^{i-1} L_j, \\ \triangle p_i(\lambda - \sum_{j=1}^{i-1} L_j) & \text{if } \sum_{j=1}^{i-1} L_j < \lambda \le \sum_{j=1}^{i} L_j, \\ L_i & \text{if } \lambda > \sum_{j=1}^{i} L_j, \end{cases} \quad (9.26)$$

then the position of a point on the path, after travelling distance λ, will be

$$p(\lambda) = \sum_{i=1}^{n-1} \triangle p_i(\lambda). \quad (9.27)$$

Eq. 9.27 is the analytical representation of a path described by a series of (linear and/or circular) path curves.

9.5.2.4 *Trajectories*

Eq. 9.27 is a closed-form description of the spatial locations on a path. The parametric variable λ is the distance travelled by a mobile robot or a robot's end-effector. Therefore, it is easy to impose a time constraint on λ so that the velocity and acceleration along the planned path can be specified.

Let us consider parametric variable λ to be a function of time t. Then, its first- and second-order derivatives are: $\frac{d\lambda}{dt}$ and $\frac{d^2\lambda}{dt^2}$. From Eq. 9.27, it is easy to determine the velocity and acceleration of point $p(\lambda)$ as follows:

$$\begin{cases} \frac{dp(\lambda)}{dt} = \frac{\partial p(\lambda)}{\partial \lambda} \bullet \frac{d\lambda}{dt} \\ \frac{d^2 p(\lambda)}{dt^2} = \frac{\partial^2 p(\lambda)}{\partial \lambda^2} \bullet \left(\frac{d\lambda}{dt}\right)^2 + \frac{\partial p(\lambda)}{\partial \lambda} \bullet \frac{d^2\lambda}{dt^2}. \end{cases} \quad (9.28)$$

Clearly, if we impose a time constraint on λ, the velocity and acceleration along a path will be fully determined by Eq. 9.28. In theory, there are many choices for function $\lambda(t)$ as long as it is differentiable up to the second order. In practice, a common solution is to choose a function for the first-order derivative of $\lambda(t)$. This function is called a *velocity profile*.

Let us denote:

$$\begin{cases} v(t) = \frac{d\lambda(t)}{dt} \\ d(t) = \lambda(t). \end{cases}$$

The widely-adopted velocity profile $v(t)$ is the one shown in Fig. 9.21. This velocity profile is known as a *trapezoid*.

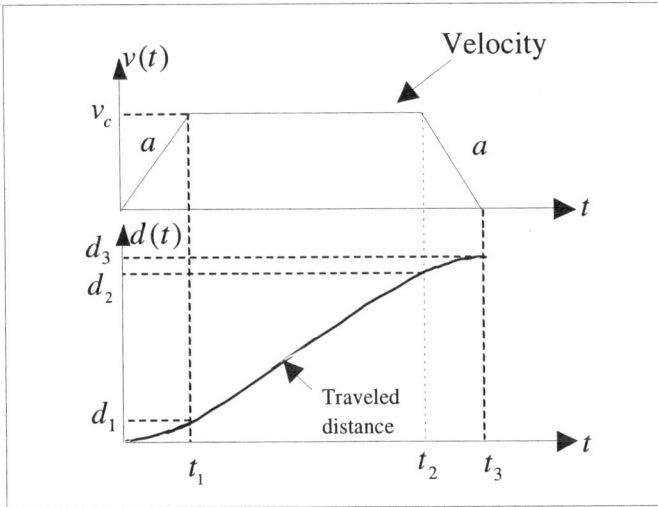

Fig. 9.21   Trapezoidal velocity profile.

A trapezoidal velocity profile has six parameters:

- L: total traveled distance,
- a: acceleration (or deceleration),
- v_c: cruising speed,
- (t_1, t_2): cruising time-interval,
- t_3: stop time.

Usually, the parameters (L, a, v_c) in a trapezoid are specified by the user as input. The other three parameters must be determined as follows:

- Solution for t_1:
  A point travelling on a path will reach cruising speed at time-instant t_1, then

$$v_c = a \bullet t_1.$$

As a result,

$$t_1 = \frac{\pi \, v_c}{a}.$$

- Solution for t_2:

  The total distance traveled is

$$L = 2 \bullet \int_0^{t_1} a \bullet t \bullet dt + \int_{t_1}^{t_2} v_c \bullet dt.$$

And, the solution for t_2 is

$$t_2 = t_1 + \frac{L - a \bullet t_1^2}{v_c}.$$

- Solution for t_3:

  From Fig. 9.21, it is easy to see that

$$t_3 = t_1 + t_2.$$

If all six parameters are known, a trapezoidal velocity profile can be concisely described as follows:

$$d(t) = \begin{cases} \int_0^t a \bullet t \bullet dt & \text{if } t \in [0, t_1]; \\[2mm] \frac{1}{2} a \bullet t_1^2 + \int_{t_1}^t v_c \bullet dt & \text{if } t \in [t_1, t_2]; \\[2mm] \frac{1}{2} a \bullet t_1^2 + v_c \bullet (t_2 - t_1) - \int_{t_2}^t a \bullet t \bullet dt & \text{if } t \in [t_2, t_3]. \end{cases} \quad (9.29)$$

9.5.2.5 *Interpolation of Orientations*

For a nonholonomic robot, the change in the orientation of the robot's assigned frame is coupled with the motion of the frame's origin. In this case, planning the frame's angular motion is not a concern. However, for a holonomic robot, the angular motion of its end-effector's frame is considered as independent to the linear motion at the frame's origin. In this case, it is necessary to plan the motion for the orientation of the end-effector's frame.

If an application does not specify any requirement for the orientation of an end-effector's frame, it is easiest to keep it at a fixed orientation. But, if the orientations of the initial and final configurations are different, it is necessary to plan the motion of orientation. And, a common way to do this is to linearly interpolate the relative orientation of the final configuration with respect to the initial configuration.

(a) Knots (b) Path/Trajectory

(c) Equivalent Axis of Rotation

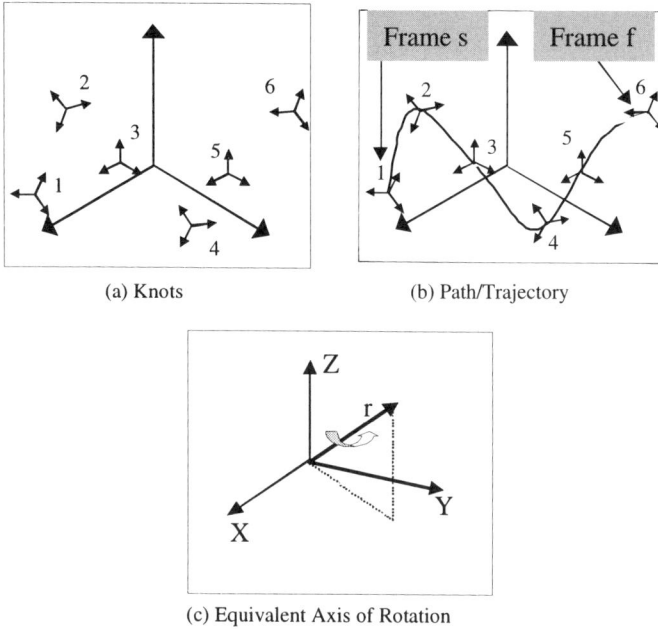

Fig. 9.22   Determination of independent motion of orientation.

Refer to Fig. 9.22b. Assume that the initial configuration is denoted by frame s, and the final configuration by frame f. Let ${}^sR_f$ be the relative orientation of frame f with respect to frame s. In Chapter 2, we learned that a rotation matrix can be represented by the equivalent axis of rotation r and the equivalent rotation angle θ. r and θ can be determined by Eq. 2.72 and Eq. 2.73 respectively.

Now, assume that θ is divided into n equal intervals. Then, the equivalent rotation angle at interval i will be

$$\theta_i = i \bullet \frac{\theta}{n}, \ i = 1, 2, ..., n. \tag{9.30}$$

Refer to Fig. 9.22c. If $r = (r_x, r_y, r_z)^t$ and β denotes the angle between r and the Z axis, then we have

$$\begin{cases} \sin\beta = \dfrac{\sqrt{r_x^2 + r_y^2}}{\sqrt{r_x^2 + r_y^2 + r_z^2}} \\ \cos\beta = \dfrac{r_z}{\sqrt{r_x^2 + r_y^2 + r_z^2}}. \end{cases} \tag{9.31}$$

Similarly, if α denotes the angle between the X axis and the plane passing through the Z axis and r, we have

$$\begin{cases} \sin \alpha = \frac{r_y}{\sqrt{r_x^2 + r_y^2}} \\ \cos \alpha = \frac{r_x}{\sqrt{r_x^2 + r_y^2}}. \end{cases} \qquad (9.32)$$

Let ${}^s R_i$ denote the intermediate orientation corresponding to θ_i. Then, the rotation matrix ${}^s R_i$ $(i = 1, 2, ..., n)$ can be calculated as follows:

$${}^s R_i = R_z(\alpha) \bullet R_y(\beta) \bullet R_x(\theta_i) \bullet R_y(-\beta) \bullet R_z(-\alpha) \qquad (9.33)$$

with

$$R_z(\alpha) = \begin{pmatrix} \cos(\alpha) & -\sin(\alpha) & 0 \\ \sin(\alpha) & \cos(\alpha) & 0 \\ 0 & 0 & 1 \end{pmatrix},$$

$$R_y(\beta) = \begin{pmatrix} \cos(\beta) & 0 & \sin(\beta) \\ 0 & 1 & 0 \\ -\sin(\beta) & 0 & \cos(\beta) \end{pmatrix}$$

and

$$R_x(\theta_i) = \begin{pmatrix} 1 & 0 & 0 \\ 0 & \cos(\theta_i) & -\sin(\theta_i) \\ 0 & \sin(\theta_i) & \cos(\theta_i) \end{pmatrix}.$$

The physical meaning of Eq. 9.33 is as follows: First, rotate the equivalent axis r to coincide with the X axis by rotation $R_z(-\alpha)$ followed by rotation $R_y(-\beta)$. Then, perform a rotation about the equivalent axis r using a θ_i angle (i.e. $R_x(\theta_i)$). Finally, bring the equivalent axis r back to its initial orientation with rotation $R_y(\beta)$, followed by rotation $R_z(\alpha)$. And, the results of interpolation

$$\{{}^s R_i, \ i = 1, 2, ...n\}$$

represent the intermediate orientations of an end-effector's frame when it travels along a planned path/trajectory.

9.6 Image-Guided Motion Planning and Control

Today's robot is an automated machine which has well-developed physical abilities to perform actions. However, the automatic execution of motion requires programming and reprogramming, which is done manually by human masters. The only way for a robot to gain a certain level of autonomy is if we can develop its mental ability to perform motion, action or even task planning. We learned that the final goal of motion, action, and task planning is to specify desired motions for automatic motion-control loops. Without the ability to self-specify a desired outcome, a robot will never be autonomous.

In general, a robot's mental ability depends on its perception-decision-action loop. Among the possible perception systems, vision is undoubtedly the most important one because it is the best system for identifying physical quantities such as motion and geometry. Therefore, a necessary step towards the development of an autonomous robot (e.g. humanoid robot) is the investigation of the loop composed of: a) visual perception, b) decision making and c) motion generating.

Research on the visual guidance of robotic systems dates back to the early 1970s. However, it has only been since the late 1980s that extensive work in this field has started to emerge. All these early works focused on vision as a feedback sensor in a motion-control loop. This framework, commonly known as *visual servoing*, has two limitations:

- Vision cannot outperform motion sensors with regard to sensory feedback in an articulated machine because the input to vision (i.e. image) is the projection of task space, not joint space.
- The sensory-motor mapping from the visual-sensory system to the actuators is not direct or indispensable. For example, without vision, a robot can still move on its own, just as a blind person can.

An alternative framework is to treat vision as an "on-line motion planner" or an "action (execution) controller", which is placed at a higher level than a "motion (execution) controller". The purpose of an "on-line motion planner" is to automatically specify the desired motions based on information from a visual-perception system. This framework is clearly illustrated in Fig. 9.23. If we use terminology from control engineering, an autonomous robot should have a hierarchy of controllers in the following order:

- Motion Controller (action taking):
  This is a closed-feedback control loop which performs physical actions at the lowest level. The input to motion-control loop is called the *desired motion*.

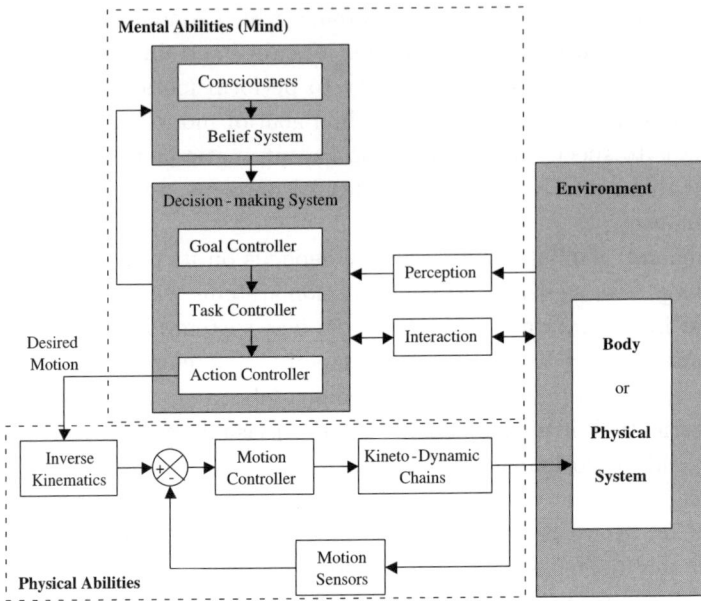

Fig. 9.23   Hierarchy of controllers in an autonomous robot.

- Action Controller (motion planning):
  This controller encompasses the motion-control loop. Its role is to automatically specify the desired motions which effectively perform a given action (input). The action controller is the decision-making unit of the *action-execution loop*.
- Task Controller (action planning):
  A task can be represented by a sequence of ordered actions. A task controller encompasses the action-execution loop. It automatically translates a given task (input) into a sequence of ordered actions (output). Similarly, the task controller is the decision-making unit of the *task-execution loop*.
- Goal Controller (task planning):
  A goal can be achieved by performing a sequence of ordered tasks (or a scenario/strategy for short). If a robot gains autonomy at the level of task execution, it should have the mental ability to automatically plan a sequence of ordered tasks (output) which will lead to the accomplishment of a given goal (input).

As we already mentioned, a necessary step towards the development of an autonomous robot is the investigation of the principles underlying the robot's action controller. Undoubtedly, image-guided motion planning and control have paramount importance because they constitute the general framework within which it is possible to automate the motion planning and control for some generic actions, such as:

- Image-guided manipulation (hand-eye coordination),
- Image-guided positioning (head-eye coordination),
- Image-guided locomotion (leg-eye coordination).

9.6.1 *Hand-Eye Coordination*

In Chapter 1, we learned that "manufacturing" means to "make things by hand" directly or indirectly. And, we all know that it would be a difficult, if not impossible job, if the motion of our hands was not under the visual guidance of our eyes. Clearly, the coordination between hand and eye plays an important role in daily activities, such as typing, writing, playing, handling, manipulating, etc.

9.6.1.1 *Input*

Figure 9.24 illustrates hand-eye coordination with a binocular vision system and a robotic hand-arm. Frames $c1$ and $c2$ are assigned to the binocular vision's two cameras. The base link of the robotic hand-arm is assigned frame 0, while its end-effector is assigned frame e.

Assume that the generic-action's description is to go from the initial configuration at location P to the final configuration at location Q. Obviously, input to the robotic hand-eye coordination should include:

- Initial Configuration:
  This is the initial posture of a frame attached to a robot's end-effector, or hand. Under the context of hand-eye coordination, the initial posture is explicitly specified in image space but not in task space. Thus, it must be visible to the binocular vision system. If not, it is necessary to define an intermediate posture to replace the invisible initial posture. For some applications, an action may not explicitly specify the initial posture. In this case, a default initial posture can be easily defined by a robotic arm because its forward kinematics is usually known in advance.

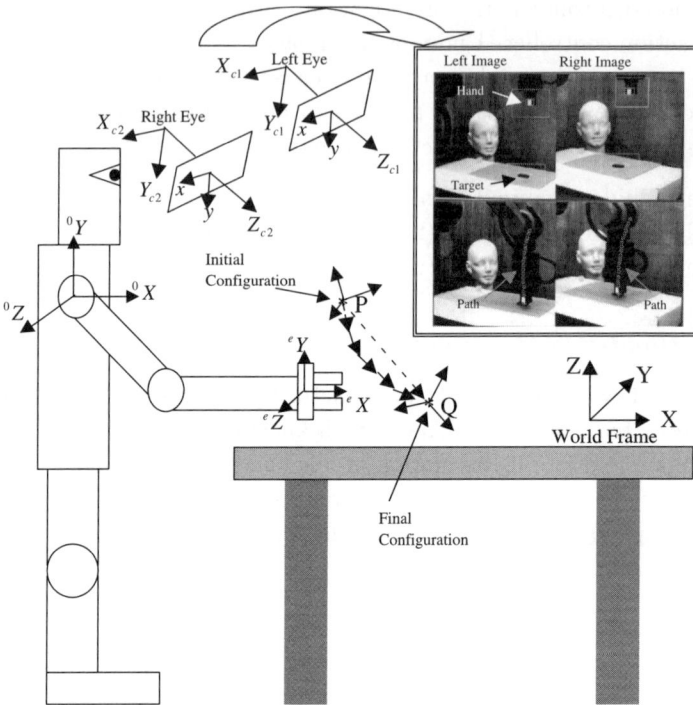

Fig. 9.24   Illustration of robotic hand-eye coordination.

- Final Configuration:

  This is the final posture which a robot's hand or end-effector must reach. Similarly, the specification of final posture is done in image space but not in task space. As a result, the final posture must be visible to a robot's binocular vision system.

For the sake of simplicity, we call the projection of the initial configuration on an image plane the *hand*, and the projection of the final configuration on an image plane the *target*. Since the main purpose here is to automatically plan a feasible path/trajectory from images, we will not consider issues related to image-feature extraction and correspondence. Thus, input to the robotic hand-eye coordination may also include:

- Images of Hand and Target:

  The detection of a hand and target in a robot's binocular vision needs to be automatic. Moreover, a robot's visual-perception system must automatically handle the tracking and binocular correspondence with regard to the hand and target in image space.

- Calibration Matrices (H_1, H_2):

  If the calibration matrices (H_1, H_2) of the binocular vision's two cameras are known in advance, a feasible path in task space can be directly planned from input specified in image space. If these two matrices are unknown, image-guided motion planning will act as an action controller. In this case, a perception-planning-control loop will be formed, which automatically brings a robot's hand, or end-effector, from the initial configuration to the final configuration.

9.6.1.2 *Output*

Hand-eye coordination is a decision-making process at the action-controller level. From a control point of view, under the context of hand-eye coordination, a given action's description is the desired output of a robot's action controller. In general, an action can be either a generic action or a customized version. However, the description of desired action should define the initial and final configurations of a robot's hand or end-effector.

Accordingly, the primary goal we hope to achieve in robotic hand-eye coordination is the robot's ability to automatically generate, from images, a feasible path/trajectory which brings the robot's hand, or end-effector, from the initial configuration to the final configuration.

However, in practice, the hand-eye coordination's decision-making process may result in one or a combination of the following two outcomes:

- Feasible Path/Trajectory:

  If the calibration matrices (H_1, H_2) are known, image-guided motion planning automatically generates a feasible path in task space, based on the input specified in image space. In addition, a robot's action controller for hand-eye coordination can also monitor the motion controller's execution of the planned path/trajectory (see Fig. 9.23).
- Control Law of the Action-Controller:

  If the calibration matrices (H_1, H_2) are unknown, image-guided motion planning acts as an action controller which generates the desired displacements in task space and monitors the motion controller's execution of the planned displacements. This forms a perception-planning-control loop (see Fig. 9.23) which operates in an iterative manner within the action-execution loop.

9.6.1.3   *A Closed-form Solution*

Refer to Fig. 9.24. Let $(^0X_P,\ ^0Y_P,\ ^0Z_P)$ be the coordinates of point P with respect to frame 0. If the image coordinates on the left camera's image plane are $(^{c1}u_p,\ ^{c1}v_p)$, according to the forward projective-mapping studied in Chapter 8, we have

$$s_1 \bullet \begin{pmatrix} ^{c1}u_p \\ ^{c1}v_p \\ 1 \end{pmatrix} = H_1 \bullet \begin{pmatrix} ^0X_P \\ ^0Y_P \\ ^0Z_P \\ 1 \end{pmatrix} \tag{9.34}$$

where s_1 is an unknown scaling factor.

Similarly, if $(^{c2}u_p,\ ^{c2}v_p)$ are the image coordinates of point P on the right camera's image plane, the following equation holds:

$$s_2 \bullet \begin{pmatrix} ^{c2}u_p \\ ^{c2}v_p \\ 1 \end{pmatrix} = H_2 \bullet \begin{pmatrix} ^0X_P \\ ^0Y_P \\ ^0Z_P \\ 1 \end{pmatrix} \tag{9.35}$$

where s_2 is another unknown scaling factor.

If the calibration matrices (H_1, H_2) are known, and the binocular correspondence of point P's images has been established in advance, the coordinates $(^0X_P,\ ^0Y_P,\ ^0Z_P)$ of point P can be directly estimated from Eq. 9.34 and Eq. 9.35.

In a similar way, the coordinates $(^0X_Q,\ ^0Y_Q,\ ^0Z_Q)$ of point Q can also be directly computed if the binocular correspondence of point Q's images has been established in advance.

If we know the coordinates of point P and point Q, it is easy to plan a feasible path connecting point P to point Q. A simple way to do this if there is no obstacle between point P and point Q is to use a straight line.

9.6.1.4   *An Iterative Approach*

The closed-form solution for robotic hand-eye coordination has two limitations:

- Lack of Flexibility:
  The closed-form solution is based on the metric reconstruction of 3-D coordinates. As a result, it is necessary to know the calibration matrices (H_1, H_2) in advance. This imposes a constraint on the intrinsic

and extrinsic parameters of a binocular vision's two cameras. In other words, the camera parameters should remain unchanged or the binocular vision must be recalibrated. Clearly, this is not human-like hand-eye coordination because human vision is qualitative (not metric).

- Lack of Error Compensation:
  The noises from images and calibration contaminate the 3-D coordinates of points A and B. As the closed-form solution operates in an open-loop manner (a single step of perception-planning), there is no feedback mechanism to compensate for the uncertainty, if any, caused by noise.

Interestingly enough, an alternative solution to robotic hand-eye coordination is the *iterative approach* which is based on a robot's *qualitative binocular-vision* and is similar to human vision. This method can be explained in the following progressive manner.

Qualitative Binocular-Vision

Consider the 3-D coordinates (X, Y, Z) of point P in a reference frame which can be any user-specified coordinate system. If $(^{c1}u,\ ^{c1}v)$ are point P's image coordinates on the left camera's image plane in a robot's binocular vision, we have

$$s_1 \bullet \begin{pmatrix} ^{c1}u \\ ^{c1}v \\ 1 \end{pmatrix} = H_1 \bullet \begin{pmatrix} X \\ Y \\ Z \\ 1 \end{pmatrix} \tag{9.36}$$

where s_1 is an unknown scaling factor. Now, let us purposely drop out the unknown scaling factor s_1 in Eq. 9.36. Accordingly, we obtain:

$$\begin{pmatrix} ^{c1}u \\ ^{c1}v \\ 1 \end{pmatrix} = H_1 \bullet \begin{pmatrix} X \\ Y \\ Z \\ 1 \end{pmatrix}. \tag{9.37}$$

The removal of the third row in Eq. 9.37 yields

$$\begin{pmatrix} ^{c1}u \\ ^{c1}v \end{pmatrix} = B_1 \bullet \begin{pmatrix} X \\ Y \\ Z \\ 1 \end{pmatrix} \tag{9.38}$$

with

$$B_1 = \begin{pmatrix} ^{c1}b_{11} & ^{c1}b_{12} & ^{c1}b_{13} & ^{c1}b_{14} \\ ^{c1}b_{21} & ^{c1}b_{22} & ^{c1}b_{23} & ^{c1}b_{24} \end{pmatrix}.$$

Mathematically, Eq. 9.38 is called the *affine transformation*. But here, we call it *qualitative projective-mapping* by a camera.

In a similar way, with regard to the binocular vision's right camera, we can obtain the following qualitative projective-mapping:

$$\begin{pmatrix} ^{c2}u \\ ^{c2}v \end{pmatrix} = B_2 \bullet \begin{pmatrix} X \\ Y \\ Z \\ 1 \end{pmatrix} \tag{9.39}$$

with

$$B_2 = \begin{pmatrix} ^{c2}b_{11} & ^{c2}b_{12} & ^{c2}b_{13} & ^{c2}b_{14} \\ ^{c2}b_{21} & ^{c2}b_{22} & ^{c2}b_{23} & ^{c2}b_{24} \end{pmatrix}.$$

Binocular vision, governed by the qualitative projective-mappings described by Eq. 9.38 and Eq. 9.39, is called *qualitative binocular-vision*.

Calibration of Qualitative Binocular-Vision

Matrices (B_1, B_2) in Eq. 9.38 and Eq. 9.39 are the *calibration matrices* in a robot's qualitative binocular-vision. Interestingly enough, the following properties can be easily verified:

- Property 1:
  The first column vectors in both B_1 and B_2 are projections of unit vector $(1, 0, 0)^t$ in a reference frame onto the two cameras' image planes.
- Property 2:
  The second column vectors in both B_1 and B_2 are projections of unit vector $(0, 1, 0)^t$ in a reference frame onto the two cameras' image planes.
- Property 3:
  The third column vectors in both B_1 and B_2 are projections of unit vector $(0, 0, 1)^t$ in a reference frame onto the two cameras' image planes.
- Property 4:
  The fourth column vectors in both B_1 and B_2 are image coordinates of origin $(0, 0, 0)^t$ of a reference frame onto the two cameras' image planes.

Because of the above properties, it is easy to determine (B_1, B_2). In practice, it is possible to choose an orthogonal coordinate system and project it onto the binocular vision's image planes. Two possible scenarios include:

- If the robot's binocular vision has been calibrated in advance, it is easy to define a virtual coordinate system and project it onto the image planes by using (H_1, H_2).
- If the robot's binocular vision has not been calibrated, it is possible to move the robot's end-effector into three orthogonal directions to define an orthogonal coordinate system. The projection of this orthogonal coordinate system will be captured by the actual (uncalibrated) cameras in the binocular vision. Clearly, a robot's qualitative binocular-vision does not require calibration matrices (H_1, H_2), if a robotic hand-arm is present.

Let us consider the base vectors and the origin of a randomly chosen coordinate system which is visible from the robot's binocular vision. Then, the images of points

$$\{(1,0,0),\ (0,1,0),\ (0,0,1)\}$$

and origin $(0,0,0)$, captured by the actual cameras, directly form matrices (B_1, B_2) as follows:

- If $(^{c1}u_x,\ {}^{c1}v_x)$, $(^{c1}u_y,\ {}^{c1}v_y)$ and $(^{c1}u_z,\ {}^{c1}v_z)$ are the image coordinates of points

$$\{(1,0,0),\ (0,1,0),\ (0,0,1)\}$$

which are projected onto the left camera's image plane, then

$$B_1 = \begin{pmatrix} {}^{c1}u_x - {}^{c1}u_o & {}^{c1}u_y - {}^{c1}u_o & {}^{c1}u_z - {}^{c1}u_o & {}^{c1}u_o \\ {}^{c1}v_x - {}^{c1}v_o & {}^{c1}v_y - {}^{c1}v_o & {}^{c1}v_z - {}^{c1}v_o & {}^{c1}v_o \end{pmatrix} \tag{9.40}$$

where $(^{c1}u_o,\ {}^{c1}v_o)$ are the image coordinates of origin $(0,0,0)$ which is projected onto the left camera's image plane.

- If $(^{c2}u_x,\ {}^{c2}v_x)$, $(^{c2}u_y,\ {}^{c2}v_y)$ and $(^{c2}u_z,\ {}^{c2}v_z)$ are the image coordinates of points

$$\{(1,0,0),\ (0,1,0),\ (0,0,1)\}$$

which are projected onto the right camera's image plane, then

$$B_2 = \begin{pmatrix} {}^{c2}u_x - {}^{c2}u_o & {}^{c2}u_y - {}^{c2}u_o & {}^{c2}u_z - {}^{c2}u_o & {}^{c2}u_o \\ {}^{c2}v_x - {}^{c2}v_o & {}^{c2}v_y - {}^{c2}v_o & {}^{c2}v_z - {}^{c2}v_o & {}^{c2}v_o \end{pmatrix} \tag{9.41}$$

where $(^{c2}u_o, \ ^{c2}v_o)$ are the image coordinates of origin $(0,0,0)$ which is projected onto the right camera's image plane.

Example 9.9   Let us construct a virtual binocular-vision, as shown in Fig. 9.25. The left camera's calibration matrix is

$$H_1 = \begin{pmatrix} 3.5538 & 1.6320 & -0.2878 & -134.6303 \\ 0.3906 & 1.6235 & -3.5479 & 832.8808 \\ 0.0015 & 0.0085 & -0.0015 & 1.0 \end{pmatrix}$$

and the right camera's calibration matrix is

$$H_2 = \begin{pmatrix} 1.9226 & 1.8934 & -0.3339 & -86.6181 \\ -0.2708 & 1.1257 & -2.4601 & 677.9354 \\ -0.0011 & 0.0059 & -0.0010 & 1.0 \end{pmatrix}.$$

Now, let us place an orthogonal coordinate system, parallel to the reference frame, at $(50cm, 100cm, 100cm)$. The projections of the base vectors and the origin are shown in Fig. 9.25. For display purposes, the base vectors' projections are amplified by a factor of 50. However, the actual calibration matrices of qualitative binocular-vision are

$$B_1 = \begin{pmatrix} 1.8400 & 0.3600 & -0.0800 & 100.0000 \\ -0.1000 & -17.5000 & -1.7600 & 371.0000 \end{pmatrix}$$

and

$$B_2 = \begin{pmatrix} 1.4800 & 0.7000 & -0.1400 & 115.0000 \\ 0.1000 & -0.6000 & -1.5000 & 370.0000 \end{pmatrix}.$$

◇◇◇◇◇◇◇◇◇◇◇◇◇◇◇◇◇◇

Image-guided Motion-planning Function

In qualitative binocular-vision, (B_1, B_2) fully describe the projections of a configuration (coordinate system) onto image planes. Thus, motion planning in task space becomes equivalent to motion planning in image space if inverse projective-mapping is available.

In image space, we can always determine displacement vectors $(\triangle^{c1}u, \ \triangle^{c1}v)$ and $(\triangle^{c2}u, \ \triangle^{c2}v)$ which move towards the target. Then, the question is: What should the corresponding displacement vector $(\triangle X, \triangle Y, \triangle Z)$ in task space be, which moves a robot's hand from the initial configuration to the final configuration?

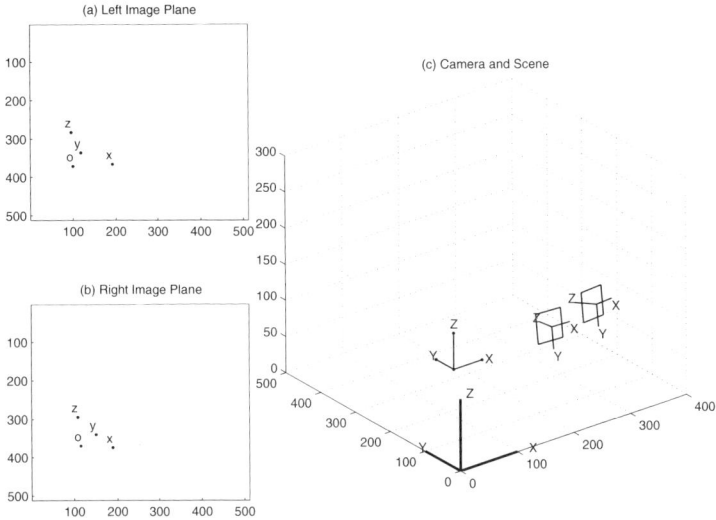

Fig. 9.25   Example of calibration of qualitative binocular-vision.

As B_1 is a constant matrix, differentiating Eq. 9.38 with respect to time will yield the following difference equation:

$$\begin{pmatrix} \triangle^{c1}u \\ \triangle^{c1}v \end{pmatrix} = C_1 \bullet \begin{pmatrix} \triangle X \\ \triangle Y \\ \triangle Z \end{pmatrix} \qquad (9.42)$$

with

$$C_1 = \begin{pmatrix} {}^{c1}u_x - {}^{c1}u_o & {}^{c1}u_y - {}^{c1}u_o & {}^{c1}u_z - {}^{c1}u_o \\ {}^{c1}v_x - {}^{c1}v_o & {}^{c1}v_y - {}^{c1}v_o & {}^{c1}v_z - {}^{c1}v_o \end{pmatrix}.$$

Similarly, differentiating Eq. 9.39 with respect to time allows us to obtain the difference equation with regard to the right camera, that is,

$$\begin{pmatrix} \triangle^{c2}u \\ \triangle^{c2}v \end{pmatrix} = C_2 \bullet \begin{pmatrix} \triangle X \\ \triangle Y \\ \triangle Z \end{pmatrix} \qquad (9.43)$$

with

$$C_2 = \begin{pmatrix} {}^{c2}u_x - {}^{c2}u_o & {}^{c2}u_y - {}^{c2}u_o & {}^{c2}u_z - {}^{c2}u_o \\ {}^{c2}v_x - {}^{c2}v_o & {}^{c2}v_y - {}^{c2}v_o & {}^{c2}v_z - {}^{c2}v_o \end{pmatrix}.$$

Eq. 9.42 and Eq. 9.43 impose four constraints on these three unknowns:

$$(\triangle X, \ \triangle Y, \ \triangle Z).$$

If we define

$$\triangle I = \left(\triangle^{c1}u, \ \triangle^{c1}v, \ \triangle^{c2}u, \ \triangle^{c2}v\right)^t \tag{9.44}$$

as the displacement vector in image space, and

$$\triangle P = (\triangle X, \ \triangle Y, \ \triangle Z)^t \tag{9.45}$$

as the displacement vector in task space, the least-square estimation of $\triangle P$ will be

$$\triangle P = A \bullet \triangle I \tag{9.46}$$

with

$$A = (C^t \bullet C)^{-1} \bullet C^t \tag{9.47}$$

and

$$C = \begin{pmatrix} {}^{c1}u_x - {}^{c1}u_o & {}^{c1}u_y - {}^{c1}u_o & {}^{c1}u_z - {}^{c1}u_o \\ {}^{c1}v_x - {}^{c1}v_o & {}^{c1}v_y - {}^{c1}v_o & {}^{c1}v_z - {}^{c1}v_o \\ {}^{c2}u_x - {}^{c2}u_o & {}^{c2}u_y - {}^{c2}u_o & {}^{c2}u_z - {}^{c2}u_o \\ {}^{c2}v_x - {}^{c2}v_o & {}^{c2}v_y - {}^{c2}v_o & {}^{c2}v_z - {}^{c2}v_o \end{pmatrix}. \tag{9.48}$$

As C is a constant matrix, A will also be a constant matrix. This indicates that mapping from image space to task space can be done by a constant matrix. Clearly, this is the simplest result that one could expect to obtain. Here, we call matrix A the *image-to-task mapping matrix*.

If we consider $\triangle I$ the error signal to a robot's action controller, A will be the proportional-control gain matrix, and $\triangle P$ the control action which acts on a robot's motion-controller, as shown in Fig. 9.23. In practice, we can introduce one extra scalar proportional-gain g ($0 < g < 1$) so that the transient response of a robot's action-controller can be manually tuned. Accordingly, Eq. 9.46 becomes

$$\triangle P = g \bullet A \bullet \triangle I. \tag{9.49}$$

Alternatively, one can also choose g to be a gain matrix as follows:

$$g = \begin{pmatrix} g_x & 0 & 0 \\ 0 & g_y & 0 \\ 0 & 0 & g_z \end{pmatrix}.$$

If the calibration matrices (H_1, H_2) are known, Eq. 9.49 can be called the *image-guided motion-planning function*. Otherwise, it can serve as the control law in an image-guided action controller.

Determination of Displacement Vectors in Image Space

When we have the initial and final configurations of an action, we know their projections onto the binocular vision's image planes. We denote $(^{c1}u_s, \, ^{c1}v_s)$ and $(^{c2}u_s, \, ^{c2}v_s)$ the projections of the initial configuration's origin. Similarly, let $(^{c1}u_f, \, ^{c1}v_f)$ and $(^{c2}u_f, \, ^{c2}v_f)$ denote the projections of the final configuration's origin. Then, error signal $\triangle I$ can be defined in two ways:

- Forward-Planning Strategy:
  In this case, we define the error signal to the action controller to be

$$\triangle I = \begin{pmatrix} ^{c1}u_f - \, ^{c1}u_s \\ ^{c1}v_f - \, ^{c1}v_s \\ ^{c2}u_f - \, ^{c2}u_s \\ ^{c2}v_f - \, ^{c2}v_s. \end{pmatrix}$$

  And, the computed $\triangle P$ at each iteration will update the position of the initial configuration. Thus, the initial configuration moves towards the final configuration under the control law described by Eq. 9.49.
- Inverse-Planning Strategy:
  In this case, we define the error signal to the action controller to be

$$\triangle I = \begin{pmatrix} ^{c1}u_s - \, ^{c1}u_f \\ ^{c1}v_s - \, ^{c1}v_f \\ ^{c2}u_s - \, ^{c2}u_f \\ ^{c2}v_s - \, ^{c2}v_f. \end{pmatrix}$$

  Then, then computed $\triangle P$ at each iteration will update the position of the final configuration. This means that the final configuration moves towards the initial configuration under the control law described by Eq. 9.49.

Example 9.10 Refer to Example 9.9. By applying Eq. 9.48, matrix C is

$$C = \begin{pmatrix} 1.8400 & 0.3600 & -0.0800 \\ -0.1000 & -0.7000 & -1.7600 \\ 1.4800 & 0.7000 & -0.1400 \\ 0.1000 & -0.6000 & -1.5000 \end{pmatrix}.$$

From Eq. 9.47, the calculated A matrix will be

$$A = \begin{pmatrix} 0.9106 & -0.1053 & -0.4716 & 0.1190 \\ -1.7710 & 0.2074 & 2.2399 & -0.3580 \\ 0.7016 & -0.4114 & -0.8907 & -0.1383 \end{pmatrix}.$$

Now, let us choose g to be 0.3. And, let us place an initial configuration at $(50cm, 100cm, 100cm)$ and a final configuration at $(150cm, 200cm, 200cm)$.

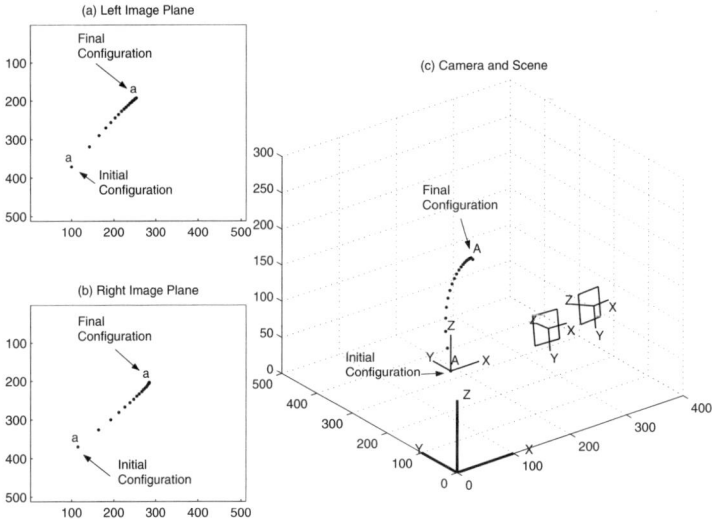

Fig. 9.26 Example of image-guided motion-planning using Eq. 9.47 and forward-planning strategy.

If we adopt the forward-planning strategy, the application of Eq. 9.49 will result in the path shown in Fig. 9.26. The initial configuration will reach location $(150.6003cm, 205.8964cm, 200.4105cm)$ after 19 iterations.

However, if we adopt the inverse-planning strategy, the application of Eq. 9.49 will result in the path shown in Fig. 9.26. The final configuration will reach location $(50.5214cm, 99.1274cm, 100.9902cm)$ after 17 iterations.

This example clearly indicates that the inverse-planning strategy produces better results if the initial configuration is closer to the cameras than the final configuration.

◇◇◇◇◇◇◇◇◇◇◇◇◇◇◇◇◇◇

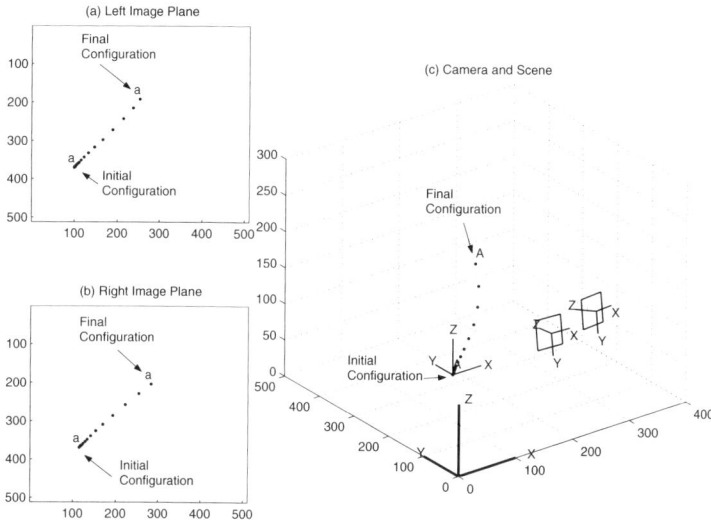

Fig. 9.27 Example of image-guided motion-planning, using Eq. 9.47 and inverse-planning strategy.

Stability of Image-guided Action Controllers

Refer to Fig. 9.23. A robot's action-controller encompasses its motion-controller. If a robot's motion-control loop is stable, then the question is whether the nested action/motion-control loop remains stable.

For the sake of simplicity, let us treat the robot's motion-control loop as a plant, which is described by transfer function $G_m(s)$. From Eq. 9.49, we know that a robot's action-controller uses a proportional-control law. Accordingly, let us denote K the proportional-control gain (or gain matrix) in the action controller. In this way, the closed-loop transfer function of the action-controller will look as follows :

$$G_a(s) = \frac{K \bullet G_m(s)}{1 + K \bullet G_m(s)}. \tag{9.50}$$

Now, let us increase proportional-control gain K (or the determinant of the gain matrix K) to infinity. As a result, Eq. 9.50 will become

$$G_a(s) \simeq 1. \tag{9.51}$$

Eq. 9.51 indicates that the nested action/motion-control loop is absolutely stable if and only if the motion-control loop itself is absolutely stable.

In fact, the proportional-control gain K in the action controller will only affect relative stability (i.e. the transient response).

Robustness of Image-guided Action Controllers

The following two examples illustrate the robustness of an image-guided action controller or motion-planning function.

Example 9.11   Refer to Example 9.10. Let us now rotate the left camera about the Y axis by -10^0. And, let us keep matrix A as it is before:

$$A = \begin{pmatrix} 0.9106 & -0.1053 & -0.4716 & 0.1190 \\ -1.7710 & 0.2074 & 2.2399 & -0.3580 \\ 0.7016 & -0.4114 & -0.8907 & -0.1383 \end{pmatrix}.$$

The planned paths before and after the rotation are shown in Fig. 9.28. We can see that the two paths in task space are very close to each other.

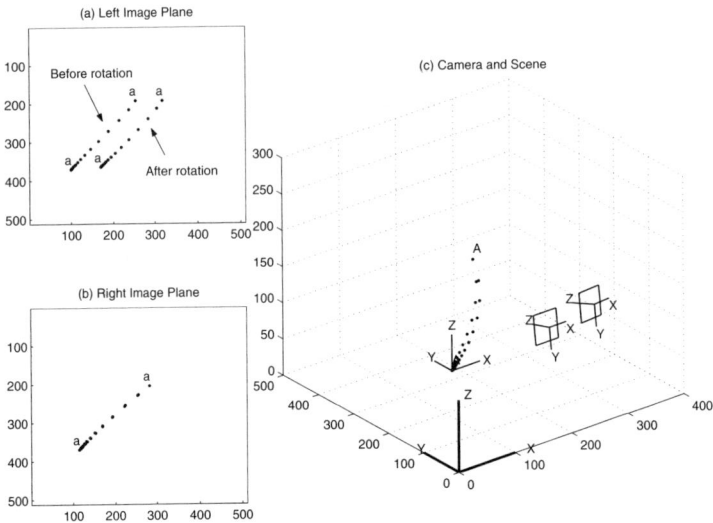

Fig. 9.28   Example of robustness with respect to change in a camera's parameters.

◇◇◇◇◇◇◇◇◇◇◇◇◇◇◇◇◇

Example 9.12   Refer to Example 9.10 again. Let us now move the final configuration to $(250cm, 300cm, 300cm)$. The distance between the initial and final configurations is about $346cm$. In robotics, this is considered to be a large working range.

The planned paths resulting from the application of both forward- and inverse-planning strategies are shown in Fig. 9.29. We can see that the inverse-planning strategy produces much better results.

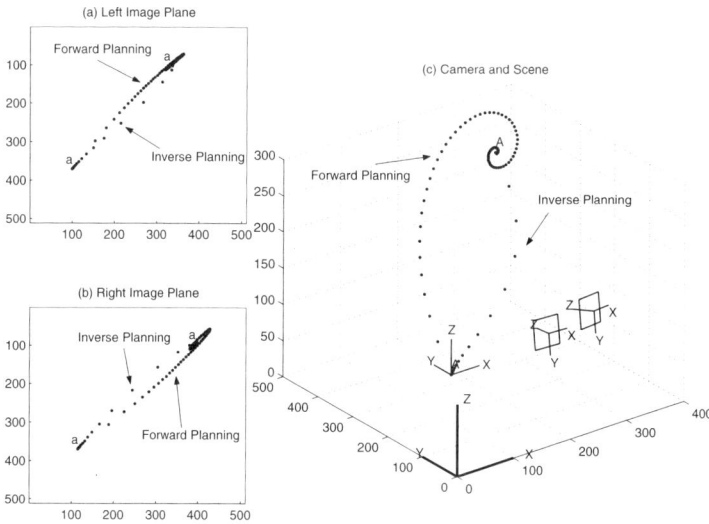

Fig. 9.29   Example of robustness with respect to a change in working range.

◇◇◇◇◇◇◇◇◇◇◇◇◇◇◇◇◇◇◇

Orientation Planning

By default, motion planning for a path/trajectory is separated from motion planning for orientation. As a set of three points in task space fully defines the position and orientation of a frame, it is possible to apply Eq. 9.49 to automate motion planning for orientation.

Example 9.13   Refer to Example 9.10. In the initial configuration, we choose three points

$$\begin{cases} A = (50cm, 100cm, 100cm) \\ B = (50cm, 100cm, 150cm) \\ C = (100cm, 100cm, 100cm). \end{cases}$$

When these points reach the final configuration, their coordinates are

$$\begin{cases} A = (50cm, 200cm, 200cm) \\ B = (50cm, 200cm, 250cm) \\ C = (100cm, 200cm, 200cm). \end{cases}$$

The planned paths for points A, B and C resulting from the application of inverse-planning strategy are shown in Fig. 9.30. From the coordinates of these planned spatial locations, it is easy to determine the position and orientation of intermediate configurations.

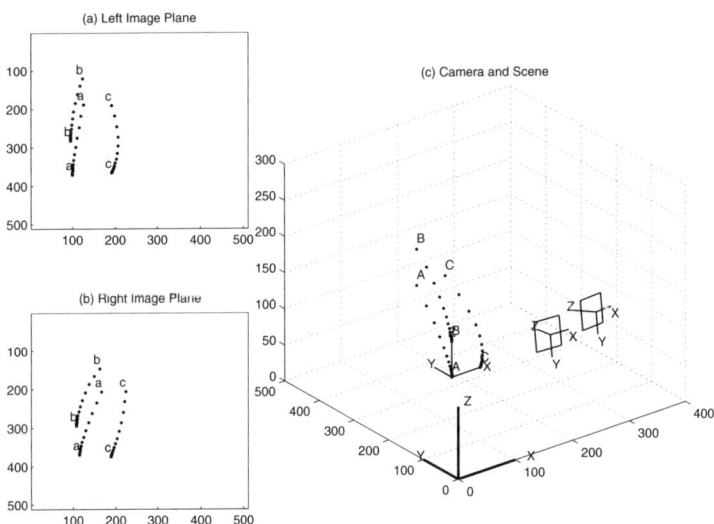

Fig. 9.30   Example of image-guided motion planning for a set of three points.

◇◇◇◇◇◇◇◇◇◇◇◇◇◇◇◇◇◇

9.6.2   *Head-Eye Coordination*

Hand-eye coordination is probably the most important behavior that human beings perform daily. It is indispensable for autonomous manipulation. Similarly, another important type of image-guided behavior is head-eye coordination. The main objective of head-eye coordination is to automatically plan a feasible path/trajectory which will move a robot's head from its initial configuration (posture) to a reference configuration (posture). Normally, a robot's head is supported by its body. Therefore, head-eye coordination is indispensable for autonomous positioning such as driving a car on a road, landing a helicopter on a helipad, generating images for augmented-reality, etc.

9.6.2.1 *Input*

Figure 9.31 shows an actual example of robotic head-eye coordination. We can see, in Fig. 9.31a, that monocular vision is mounted on an arm manipulator, and this combination imitates a robot's body-head mechanism. For the sake of description, let us call the assembly of a robot's vision mounted on top of its body-head mechanism a *vision-head*.

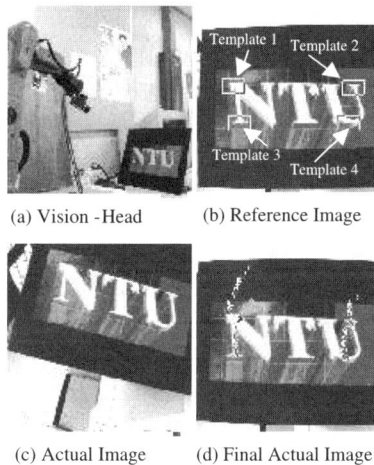

(a) Vision -Head   (b) Reference Image

(c) Actual Image   (d) Final Actual Image

Fig. 9.31   Example of robotic head-eye coordination with monocular vision.

At a reference configuration (also called a final configuration), a robot's vision-head captures an image of a specified reference object and stores it as a *reference image* (Fig. 9.31b). If a robot's vision-head is shifted from the reference configuration for whatever reason, the question is: Using the reference image (Fig. 9.31b) and the actual image (Fig. 9.31c) of a reference object, is it possible to determine a feasible path/trajectory which brings a robot's vision-head back to the reference configuration, so that the actual image (Fig. 9.31d) is equal to the reference image?

According to the problem statement, input to the robotic head-eye coordination includes:

- Reference Object:
  The aim of head-eye coordination is to position a frame assigned to a robot's vision-head at a final configuration relative to a reference object (e.g. a configuration in front of a door). Therefore, the reference object must be defined in advance. In practice, the reference object can be a door, a road, a helipad, or any physical object in a scene.
- Reference Image:
  As we mentioned earlier, a reference image is an image captured by a robot's vision-head at a reference configuration (or final configuration). For repeated actions, it is not difficult to capture and store a reference image. However, there are applications in which reference images are not available in advance. Thus, this input should not be absolutely necessary.
- Actual Image of Reference Object:
  The issue here is how to automatically plan a feasible path for a robot's vision-head. We assume that the detection and tracking of the reference object's images can be automatically done by the robot's visual-perception system.

Head-eye coordination with monocular vision in three-dimensional task space is difficult when there is no *a priori* knowledge about a reference object. So far, none of the existing solutions developed within the framework of monocular vision can outperform solutions based on binocular vision. Most importantly, binocular vision provides much richer information than monocular vision.

As human vision is binocular and humanoid robots are usually equipped with binocular vision, it is more interesting to study robotic head-eye coordination with binocular vision, as illustrated by Fig. 9.32.

Within the framework of binocular vision, additional input to the robotic head-eye coordination may include:

- Calibration matrices (H_1, H_2) of the two cameras in a robot's binocular vision.
- Motion-transformation matrix $^{c1}M_{c2}$ (or $^{c2}M_{c1}$) between the two frames assigned to the binocular vision's two cameras (NOTE: Motion-transformation matrix $^{c1}M_{c2}$ or $^{c2}M_{c1}$ can easily be estimated from (H_1, H_2)).

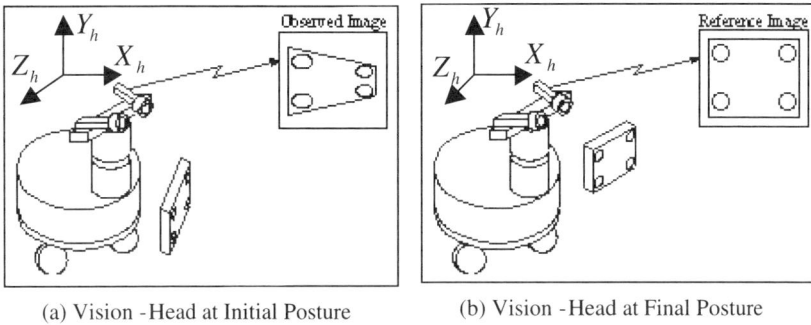

(a) Vision -Head at Initial Posture   (b) Vision -Head at Final Posture

Fig. 9.32   Illustration of robotic head-eye coordination with binocular vision.

9.6.2.2 *Output*

As with hand-eye coordination, head-eye coordination is also a decision-making process. Under the context of the robotic head-eye coordination, the desired output of the robot's action-controller is an action description, such as "Position the vision-head at the reference configuration in front of the reference object". Thus, the primary goal of head-eye coordination is to automatically generate a feasible path/trajectory which will bring a robot's vision-head from its current (initial) configuration to the reference configuration.

Since the binocular vision's two cameras will be assigned two frames, it is possible to use one of these frames for the robot's vision-head as well. In this way, we can reduce the number of frames under consideration. In practice, we can assume that the frame assigned to the left camera also serves as the vision-head's frame. Figure 9.33 illustrates the frame assignment when a robot's vision-head is at its initial and reference configurations. Here, frame $c1$ and $c2$ are assigned to the two cameras when the robot's vision-head is at its actual configuration. And, frames $c3$ and $c4$ are assigned to the two cameras when the robot's vision-head is at its reference configuration. All these frames can be called the *camera's configurations* or, *views* for short.

Accordingly, two possible outputs from the robotic head-eye coordination are:

- Motion-Transformation Matrix ${}^{c1}M_{c3}$:
  This matrix describes the reference configuration (frame $c3$) with respect to frame $c1$. If we know ${}^{c1}M_{c3}$ or its inverse matrix, it is easy to choose a feasible path/trajectory which will move frame $c1$ to frame

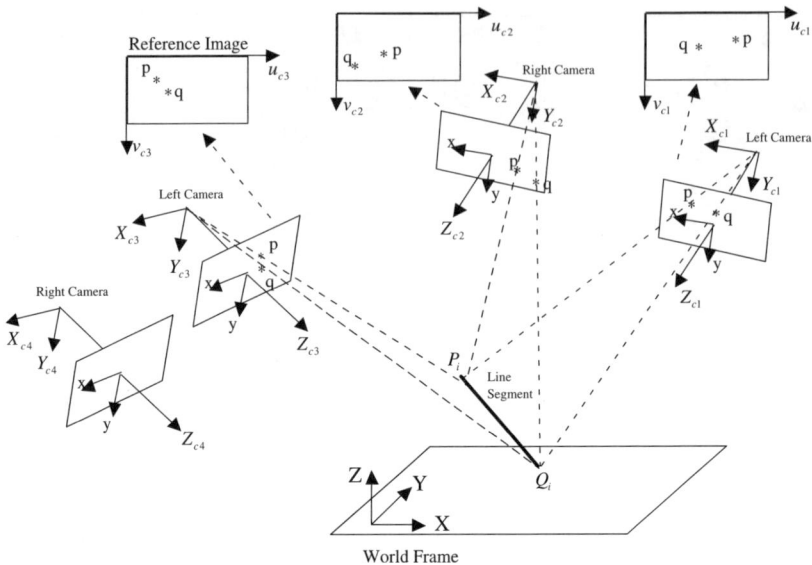

Fig. 9.33   Illustration of camera frames/configurations in robotic head-eye coordination with binocular vision.

$c3$. For example, one can choose a straight line connecting the origins of frames $c1$ and $c3$.

- Feasible Path/Trajectory:
  If $^{c1}M_{c3}$ cannot be obtained directly, it is possible to iteratively plan a feasible path/trajectory in task space, based on information from images which will move frame $c1$ to frame $c3$.

9.6.2.3   *A Closed-form Solution*

The objective of a closed-form solution is to estimate motion-transformation matrix $^{c1}M_{c3}$ or $^{c3}M_{c1}$ from the reference image (captured at configuration $c3$), and two actual images (captured at configurations $c1$ and $c2$).

Refer to Fig. 9.33. Assume that a reference object contains at least three line segments:

$$\{(P_i, Q_i),\ i = 1, 2, 3, ...\}$$

where P_i and Q_i are the two end-points of line segment i.

If calibration matrices (H_1, H_2) are known, an easy solution is to estimate the end-points $\{(P_i, Q_i),\ i = 1, 2, 3, ...\}$ in the binocular views $(c1, c2)$

and $(c3, c4)$ first, and then estimate motion-transformation matrix $^{c1}M_{c3}$ or $^{c3}M_{c1}$ from the correspondence of the line segments between views $c1$ and $c3$. This solution is called the *four-view method*. However, it has two drawbacks:

- It requires two reference images. As a result, it is not applicable to head-eye coordination with monocular vision.
- It is not a minimum-condition solution. In fact, just three views are sufficient to estimate motion-transformation matrix $^{c1}M_{c3}$ or $^{c3}M_{c1}$.

Because of the above drawbacks, the four-view method is not effective in either theory or practice. Here, we study the minimum-condition solution which requires only three views and is called the *three-view method*. There are two advantages to the three-view method:

- It is applicable to head-eye coordination with monocular vision because views $c1$ and $c2$ can be obtained by moving the camera from its current configuration to an intermediate configuration. As the kinematics of a robot's body-head mechanism is usually known in advance, the motion-transformation matrix between views $c1$ and $c2$ is automatically known. If we take advantage of our knowledge of the kinematics, head-eye coordination with monocular vision can be treated as a special case of head-eye coordination with binocular vision.
- It is computationally efficient because there is one image less than in the four-view method.

Now, let us introduce the three-view method step-by-step.

Equation of Line Segments in Image Space

Refer to Fig. 9.33 again. Let us consider the image plane in frame $c3$ (i.e. a reference frame). Points p and q are the projections of points P_i and Q_i onto the image plane in view $c3$. As the image plane is a two-dimensional space, a line passing through points p and q can be described in the following general form:

$$a \bullet u + b \bullet v + c = 0 \qquad (9.52)$$

where (a, b, c) are the coefficients in the equation describing a line in image space.

Equation of Line Segments in Task Space

In Chapter 8, we learned that the index coordinates (u, v) are related to the real-image coordinates (x, y) by

$$\begin{cases} u = u_0 + \frac{x}{D_x} \\ v = v_0 + \frac{y}{D_y}. \end{cases} \tag{9.53}$$

According to a camera's perspective projection, we have

$$\begin{cases} x = f \bullet \frac{c3X}{c3Z} \\ y = f \bullet \frac{c3Y}{c3Z}. \end{cases}$$

Thus, Eq. 9.53 can also be expressed as follows:

$$\begin{cases} u = u_0 + f_x \frac{c3X}{c3Z} \\ v = v_0 + f_y \frac{c3Y}{c3Z} \end{cases} \tag{9.54}$$

where $f_x = f/D_x$ and $f_y = f/D_y$.

Substituting Eq. 9.54 into Eq. 9.52 yields

$$a' \bullet \{^{c3}X\} + b' \bullet \{^{c3}Y\} + c' \bullet \{^{c3}Z\} = 0 \tag{9.55}$$

where

$$\begin{cases} a' = a \bullet f_x \\ b' = b \bullet f_y \\ c' = a \bullet u_0 + b \bullet v_0 + c. \end{cases}$$

If we define

$$^{c3}L_i = (a', \ b', \ c')^t$$

and

$$^{c3}P_i = (^{c3}X, \ ^{c3}Y, \ ^{c3}Z)^t,$$

then Eq. 9.55 can be rewritten in the following concise form:

$$\{^{c3}L_i^t\} \bullet \{^{c3}P_i\} = 0. \tag{9.56}$$

Eq. 9.56 is also valid for end-point Q_i on line segment i. As a result, we have

$$
\begin{cases}
\{^{c3}L_i^t\} \bullet \{^{c3}P_i\} = 0 \\
\\
\{^{c3}L_i^t\} \bullet \{^{c3}Q_i\} = 0
\end{cases}
\tag{9.57}
$$

In fact, Eq. 9.57 is the image-constrained representation of the line segment, in task space. This is because $^{c3}L_i$ is the parameter vector computed from the image and the camera's intrinsic parameters, $(u_0, \ v_0, \ f_x, \ f_y)$.

Equations of Deterministic Head-Eye Coordination

We denote:

- $(^{c1}P_i, \ ^{c1}Q_i)$ the end-points of line segment i in frame $c1$,
- $^{c3}R_{c1}$ the orientation of frame $c1$ with respect to frame $c3$ (i.e. the rotation matrix from frame $c1$ to frame $c3$),
- $^{c3}T_{c1}$ the origin of frame $c1$ with respect to frame $c3$ (i.e. the translation vector from frame $c1$ to frame $c3$).

According to the motion transformation between frames, we have the following relationship between coordinates $(^{c1}P_i, \ ^{c1}Q_i)$ and $(^{c3}P_i, \ ^{c3}Q_i)$:

$$
\begin{cases}
^{c3}P_i = \{^{c3}R_{c1}\} \bullet \{^{c1}P_i\} + \{^{c3}T_{c1}\} \\
\\
^{c3}Q_i = \{^{c3}R_{c1}\} \bullet \{^{c1}Q_i\} + \{^{c3}T_{c1}\}.
\end{cases}
\tag{9.58}
$$

By applying Eq. 9.58 to Eq. 9.57, we obtain

$$
\begin{cases}
\{^{c3}L_i^t\} \bullet \{^{c3}R_{c1}\} \bullet \{^{c1}P_i\} + \{^{c3}L_i\} \bullet \{^{c3}T_{c1}\} = 0 \\
\\
\{^{c3}L_i^t\} \bullet \{^{c3}R_{c1}\} \bullet \{^{c1}Q_i\} + \{^{c3}L_i\} \bullet \{^{c3}T_{c1}\} = 0.
\end{cases}
\tag{9.59}
$$

The subtraction of the two equations in Eq. 9.59 yields

$$
\{^{c3}L_i\} \bullet \{^{c3}R_{c1}\} \bullet \{^{c1}P_i - \ ^{c1}Q_i\} = 0.
\tag{9.60}
$$

Finally, when we have line segment i, we can establish the following system of equations:

$$
\begin{cases}
\{^{c3}L_i^t\} \bullet \{^{c3}R_{c1}\} \bullet \{^{c1}P_i - \ ^{c1}Q_i\} = 0 \\
\\
\{^{c3}L_i^t\} \bullet \{^{c3}R_{c1}\} \bullet \{^{c1}P_i\} + \{^{c3}L_i^t\} \bullet \{^{c3}T_{c1}\} = 0.
\end{cases}
\tag{9.61}
$$

From Eq. 9.61, we can make the following remarks:

- A line segment will impose two constraints on $(^{c3}R_{c1}, \ ^{c3}T_{c1})$.
- Rotation matrix $^{c3}R_{c1}$ can be independently estimated by using the first equation in Eq. 9.61.
- As a rotation matrix has only three independent parameters, a set of three line segments is sufficient to fully determine rotation matrix $^{c3}R_{c1}$ if these three line segments are not in the same plane. However, one must solve a system of three nonlinear equations.
- In practice, it is possible to treat the nine (9) elements inside $^{c3}R_{c1}$ as nine unknowns. In this way, there will be twelve unknowns in Eq. 9.61 because translation vector $^{c3}T_{c1}$ has three unknowns. Consequently, a set of six line segments is sufficient to fully estimate these twelve unknowns by solving a system of twelve linear equations:

$$\begin{cases} \{^{c3}L_i^t\} \bullet \{^{c3}R_{c1}\} \bullet \{^{c1}P_i - \ ^{c1}Q_i\} = 0 \\ \\ \{^{c3}L_i^t\} \bullet \{^{c3}R_{c1}\} \bullet \{^{c1}P_i\} + \{^{c3}L_i^t\} \bullet \{^{c3}T_{c1}\} = 0 \end{cases} \quad i = 1, 2, ..., 6.$$

$$(9.62)$$

Here, we call Eq. 9.62 the *equations of deterministic head-eye coordination*.

Example 9.14   A binocular vision system is mounted on a mobile platform, similar to the one illustrated in Fig. 9.32. At two different configurations, two pairs of stereo images are captured. In total, there are four images. But, we only need to use three images to validate the three-view method of head-eye coordination. The input images are shown in Fig. 9.34.

When we move a mobile platform from one configuration to another, it is impossible to know the motion transformation between the two configurations. This is because the kinematics of a mobile platform cannot be known exactly due to the possible slippage of the wheels. Therefore, it is important to be able to estimate the motion transformation from images. In this example, the binocular vision has been calibrated and thus the calibration matrices (H_1, H_2) are known in advance. From Fig. 9.34, we can see that the reference object contains more than three line segments. In fact, all its edges can be approximated by line segments.

Fig. 9.35 shows the results of our experiment. Let us first detect the edge pixels in the input images. (See Fig. 9.35a, Fig. 9.35b and Fig. 9.35c). Then, let us approximate the edge pixels by line segments and establish the binocular correspondence of the line segments in the binocular views $(c1, c2)$ (Fig. 9.35d), and $(c1, c3)$ (Fig. 9.35e).

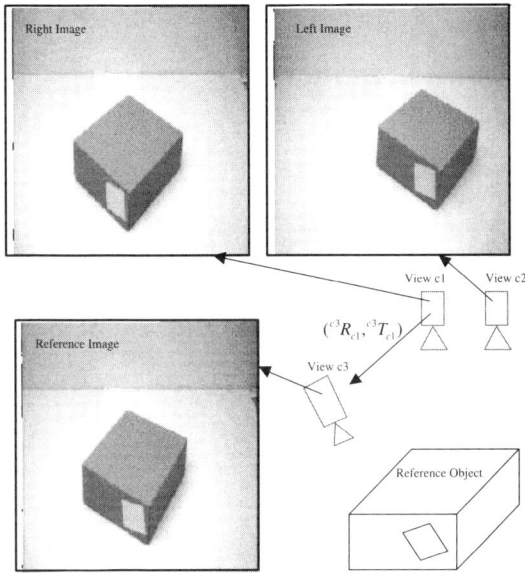

Fig. 9.34 Input images in robotic head-eye coordination with binocular vision.

The application of Eq. 9.62 yields the estimated rotation matrix $^{c3}R_{c1}$ as follows:

$$^{c3}R_{c1} = \begin{pmatrix} 0.980591 & 0.107180 & -0.164174 \\ -0.104814 & 0.994225 & 0.023033 \\ 0.165694 & -0.005378 & 0.986162 \end{pmatrix}$$

and the estimated translation vector is

$$^{c3}T_{c1} = (22.871370cm, -3.478399cm, 1.009764cm)^t.$$

In order to visually judge the correctness of the estimated solution $(^{c3}R_{c1},\ ^{c3}T_{c1})$, let us transform the 3-D line segments from frame $c1$ to frame $c3$ (reference frame), and project the 3-D line segments back to the camera's image plane at frame $c3$. The superimposition of the 2-D line segments, estimated from the reference image, and the projected 3-D line segments is shown in Fig. 9.35f. We can see that the superimposition is quite accurate.

◇◇◇◇◇◇◇◇◇◇◇◇◇◇◇◇◇◇

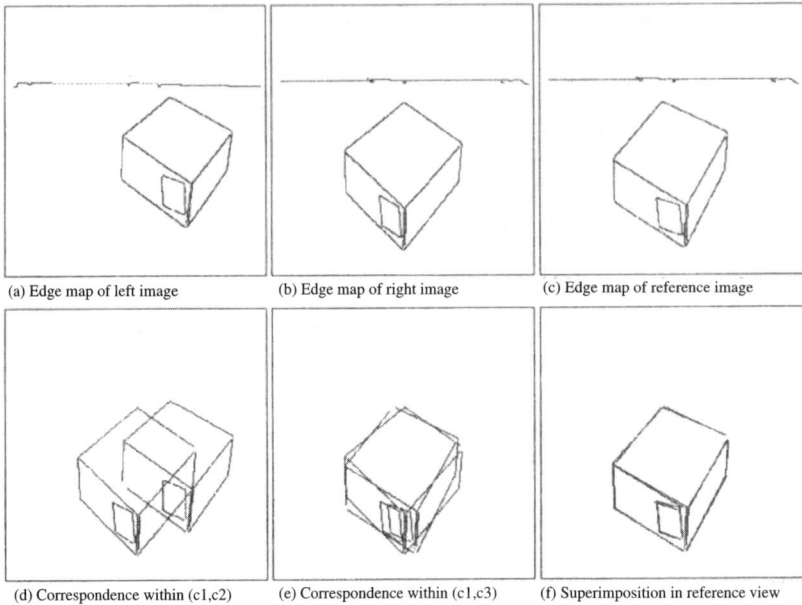

(a) Edge map of left image   (b) Edge map of right image   (c) Edge map of reference image

(d) Correspondence within (c1,c2)   (e) Correspondence within (c1,c3)   (f) Superimposition in reference view

Fig. 9.35   Results of a closed-form solution for robotic head-eye coordination with binocular vision.

Example 9.15   The closed-form solution for robotic head-eye coordination can also be applied to solve a critical issue in *augmented reality*. In engineering visualization and medical applications, it is necessary to generate what is called *look-through images* (also known as *augmented images*). For example, in surgery, a doctor can only see what is visible in his/her field of vision. However, if the doctor can see look-through images, it will make operations easier.

As shown in Fig. 9.36, look-through images are obtained by superimposing virtual images onto actual images. The virtual images contain other relevant information which are absent or invisible in actual images. In order to make look-through images realistic, the common areas in virtual and actual images must be perfectly superimposed. This indicates that a critical issue in generating look-through images is to position the virtual binocular-vision at a configuration with respect to the virtual scene which is exactly the same as the configuration of the actual binocular vision with respect to the actual scene. This is a difficult problem because the actual binocular-vision is mounted on the head of a user (e.g. surgeon) who may

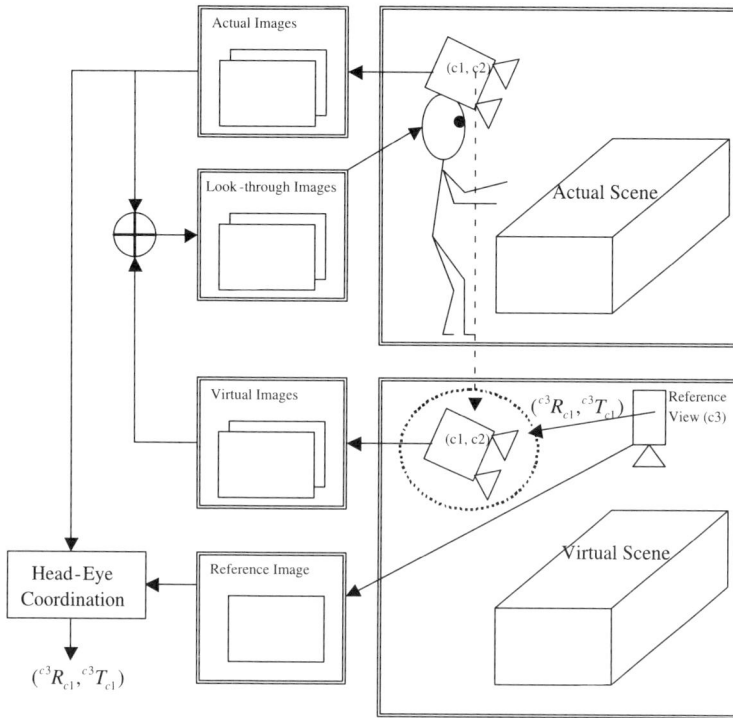

Fig. 9.36 Illustration of how the problem of positioning virtual binocular vision, in a virtual scene, can be treated as a problem of head-eye coordination.

change the head's posture depending on various needs.

Interestingly enough, the problem of positioning virtual binocular-vision in a virtual scene can be treated as a problem of head-eye coordination. In this case, the reference image is captured by a virtual camera at a reference view (frame $c3$), as shown in Fig. 9.36. The two actual images and virtual reference image are the input to the three-view method of head-eye coordination. The output is motion transformation $(^{c3}R_{c1}, \ ^{c3}T_{c1})$ which describes the posture of the virtual binocular vision with respect to the reference view (frame $c3$).

◇◇◇◇◇◇◇◇◇◇◇◇◇◇◇◇◇◇◇◇

9.6.2.4 *Iterative Approaches*

There are two limitations to closed-form solution(s) of head-eye coordination because of the following two necessary conditions:

(1) The two calibration matrices (H_1, H_2) must be known in advance in order to compute the end-points $({}^{c1}P_i, {}^{c1}Q_i)$ on the 3-D line segments in frame $c1$. We know that any change in the camera's parameters requires the recalibration of the camera. As a result, a closed-form solution is not robust with regard to changes in a camera's parameters.

(2) The reference image must be available in advance. A reference image is an image captured at a reference view in front of a reference object. For some applications, we may not have any reference image even though a reference object is present in a scene, because we may not be able to position the camera at a reference view in advance in order to capture the reference image. And, even if reference images are available, it is possible that they are already obsolete for whatever reason (e.g. change of reference object).

Because of the above limitations, it is interesting to investigate the possible application of qualitative binocular-vision to robotic head-eye coordination. Interestingly enough, it is possible because we can place the frame, assigned to the robot's vision-head, in front of its binocular vision and treat it as the initial configuration. In this way, it is possible to plan a feasible path/trajectory which brings the reference view back to the frame that is assigned to the robot's vision-head. Finally, the planned path/trajectory can be executed in reverse by the robot.

Mathematically, a set of three points fully defines a frame in three-dimensional space as long as these three points are not on the same line. Let us choose three noncollinear points (A, B, C) to represent the frame assigned to the robot's vision-head, as shown in Fig. 9.37. In this way, head-eye coordination becomes a problem of planning a feasible path/trajectory which brings a set of at least three noncollinear points to a reference view. As the robot's vision-head moves together with these three points, the final result is the positioning of the robot's vision-head at a reference view.

In order to apply the principle of qualitative binocular-vision, we simply attach an orthogonal coordinate-system to each of these three points (A, B, C), as shown in Fig. 9.37. In this way, each point has its own image-to-task mapping matrix. Then, from Eq. 9.49, we can establish the following system of equations for motion planning under the context of head-eye

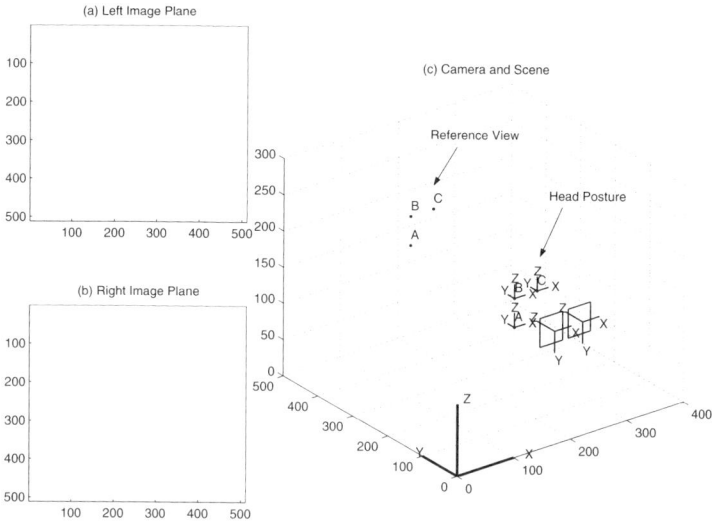

Fig. 9.37 Illustration of the representation of a head frame using a set of three non-collinear points (A, B, C).

coordination:

$$\begin{cases} \triangle P_A = g_A \bullet A_A \bullet \triangle I_A \\ \triangle P_B = g_B \bullet A_B \bullet \triangle I_B \\ \triangle P_C = g_C \bullet A_C \bullet \triangle I_C \end{cases} \tag{9.63}$$

where

- (g_A, g_B, g_C) are three proportional gains (or gain matrices),
- (A_A, A_B, A_C) are image-to-task mapping matrices corresponding to points (A, B, C),
- $(\triangle P_A, \triangle P_B, \triangle P_C)$ are three displacement vectors in task space which act on points (A, B, C),
- $(\triangle I_A, \triangle I_B, \triangle I_C)$ are three displacement vectors in image space observed by the robot's binocular vision.

Example 9.16   For the simulation setup shown in Fig. 9.37, the calibration matrices (H_1, H_2) of the binocular vision's two cameras are

$$H_1 = \begin{pmatrix} 3.6561 & 2.5600 & 0.0000 & -146.1665 \\ 0 & 2.5600 & -3.6561 & 987.2118 \\ 0 & 0.0100 & 0.0000 & 1.0000 \end{pmatrix}$$

and

$$H_2 = \begin{pmatrix} 3.6561 & 2.5600 & 0.0000 & -328.9694 \\ 0 & 2.5600 & -3.6561 & 987.2118 \\ 0 & 0.0100 & 0.0000 & 1.0000 \end{pmatrix}.$$

(NOTE: The two image planes are parallel to one another).

In the reference frame, a set of three non-collinear points are chosen to represent the posture of the robot's vision-head. The coordinates of these three points are

$$\begin{cases} A = (100cm, 0cm, 180cm) \\ B = (100cm, 0cm, 220cm) \\ C = (140cm, 0cm, 220cm). \end{cases}$$

By attaching an orthogonal coordinate-system to each of these points, it is possible to obtain the corresponding image-to-task mapping matrices. They are

$$A_A = \begin{pmatrix} 0.3344 & 0.0000 & -0.0650 & 0.0000 \\ -0.6781 & 0.0000 & 0.6874 & 0.0000 \\ 0.1115 & -0.1370 & -0.1130 & -0.1370 \end{pmatrix},$$

$$A_B = \begin{pmatrix} 0.3344 & -0.0000 & -0.0650 & -0.0000 \\ -0.6781 & -0.0000 & 0.6874 & -0.0000 \\ -0.1115 & -0.1370 & 0.1130 & -0.1370 \end{pmatrix}$$

and

$$A_C = \begin{pmatrix} 0.1061 & 0 & 0.1679 & 0 \\ -0.6452 & 0 & 0.6452 & 0 \\ -0.1061 & -0.1370 & 0.1061 & -0.1370 \end{pmatrix}.$$

Now, we set

$$(g_A, g_B, g_C) = (0.3, 0.3, 0.3)$$

and choose three points in reference view (see Fig. 9.38) with the coordinates

$$\begin{cases} A = (100cm, 300cm, 210cm) \\ B = (100cm, 300cm, 250cm) \\ C = (140cm, 300cm, 250cm). \end{cases}$$

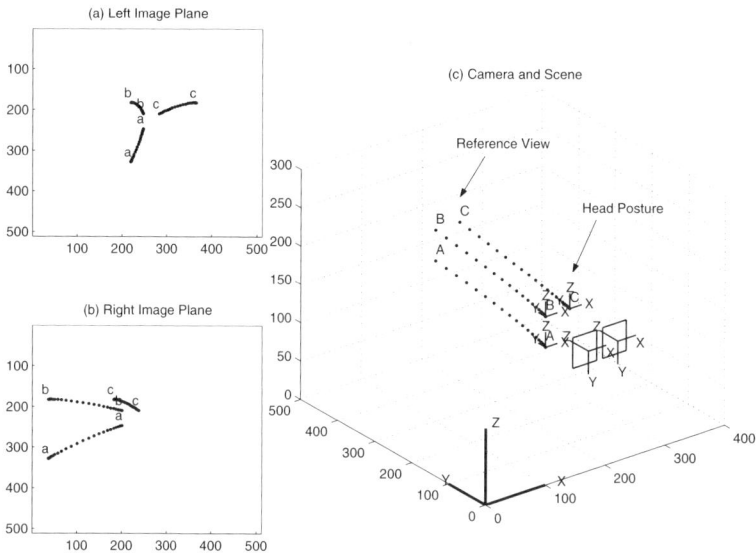

Fig. 9.38   Example of results obtained from the iterative solution to head-eye coordination.

The application of Eq. 9.63 yields the planned paths shown in Fig. 9.38. We can see that these paths almost follow a straight line.

◇◇◇◇◇◇◇◇◇◇◇◇◇◇◇◇◇◇

Example 9.17   Head-eye coordination provides an effective solution to a range of image-guided positioning applications, such as missile guidance, autonomous landings of aircraft, etc.

Figure 9.39 illustrates the autonomous landing of a helicopter. We mount the binocular vision system at the bottom of the helicopter's body. And, they form a vision-head together. At the base of the helicopter's supporting legs, we place a rectangular landing guide, the four corners of which will represent the frame that is assigned to the helicopter's vision-head. The desired action here is to land the helicopter on top of the helipad. Clearly, this action can be translated into the problem of positioning the landing guide at the reference configuration represented by the helipad.

Figure 9.40 shows the simulation setup. For display purposes, the binocular vision's two cameras are indicated by two small triangles. The frames of the vision-head and the reference are represented by four corner points. In this setup, the Z axis is perpendicular to the ground.

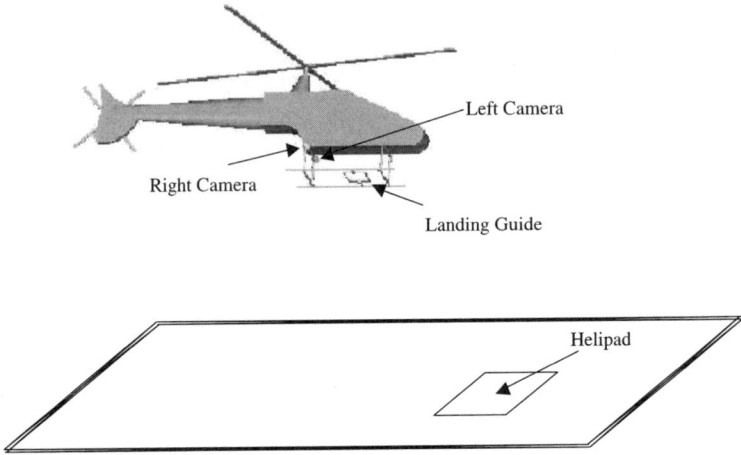

Fig. 9.39   Illustration of how an autonomous landing of a helicopter is seen as a problem of head-eye coordination.

For the autonomous landing of aircraft, it is desirable to approach the landing area slowly in order to avoid sudden impact with the ground. Interestingly enough, it is easy to perform this slow approach by choosing a small proportional-gain for the $\triangle Z$ component in Eq. 9.63.

In this example, we choose proportional-gain matrix g to be

$$g = \begin{pmatrix} 0.5 & 0 & 0 \\ 0 & 0.5 & 0 \\ 0 & 0 & 0.2 \end{pmatrix}.$$

The planned and executed path is shown in Fig. 9.41. We can see that the helicopter approaches the helipad smoothly. During landing, the images of the landing guide remain constant. However, the images of the helipad follow the path which first moves down and then converges towards the images of the landing guide.

◇◇◇◇◇◇◇◇◇◇◇◇◇◇◇◇◇◇◇

9.6.3   *Leg-Eye Coordination*

The aim of leg-eye coordination is to automate the generation of a feasible path/trajectory for the generic action, commonly known as *image-guided*

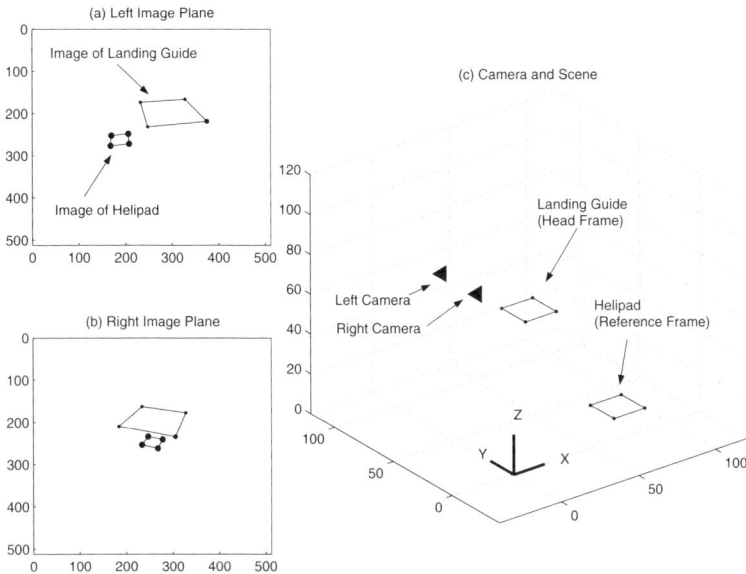

Fig. 9.40   Simulation setup for testing an autonomous landing of a helicopter.

locomotion. However, in special cases, leg-eye coordination is equivalent to hand-eye coordination (e.g. when a robot kicks a ball with its foot).

For the sake of simplicity, let us study leg-eye coordination dedicated to generic actions, such as walking, running, cycling, driving, etc. As shown in Fig. 9.42, image-guided locomotion has two subproblems:

- Image-guided Road Following:
  For example, the first robot in Fig. 9.42 must determine a feasible path/trajectory which will keep its body on the road.
- Image-guided Target Following:
  In Fig. 9.42, the second robot can treat the first one as a moving target. Thus, the problem becomes how to generate a feasible path/trajectory which will position a robot (e.g. the second robot) at a relative configuration with respect to the moving target (e.g. the first robot).

9.6.3.1   *Image-Guided Road Following*

In the proximity of a robot, the road surface can be approximated by a plane. As the plane is a two-dimensional space, there is a deterministic relationship between the image plane and the two-dimensional plane in

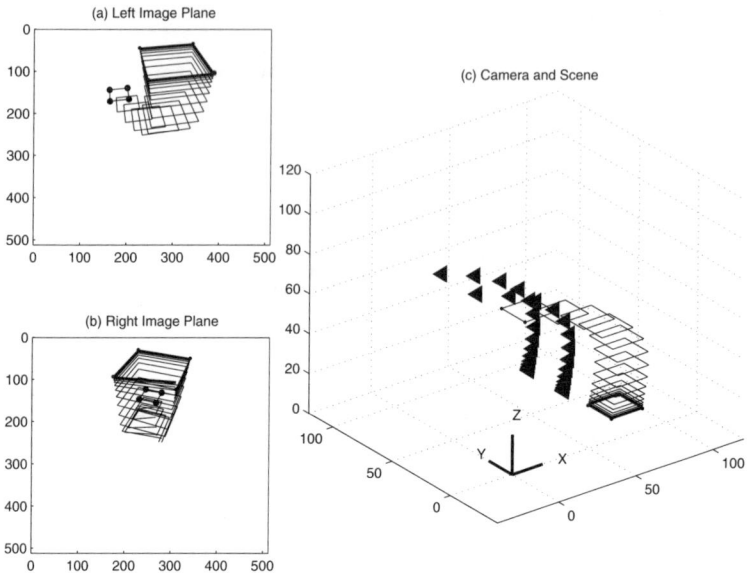

Fig. 9.41   Simulation results of the autonomous landing of a helicopter.

task space (see Chapter 8).

As illustrated in Fig. 9.43, if D denotes the inverse projective-mapping from the image space to a two-dimensional plane, we have

$$
s \bullet \begin{pmatrix} X \\ Y \\ 1 \end{pmatrix} = D \bullet \begin{pmatrix} u \\ v \\ 1 \end{pmatrix}
\tag{9.64}
$$

where (X, Y) are the coordinates of a point in road space and (u, v) the index coordinates of its image.

According to Eq. 9.64, the automatic generation of a feasible path/trajectory for image-guided road following involves the following steps:

- Step 1: Determine the inverse projective-mapping matrix D using calibration.
- Step 2: Detect road boundaries in image space using color identification of landmarks (or edge detection followed by edge linking).
- Step 3: Back-project road boundaries from image space to road space using Eq. 9.64.

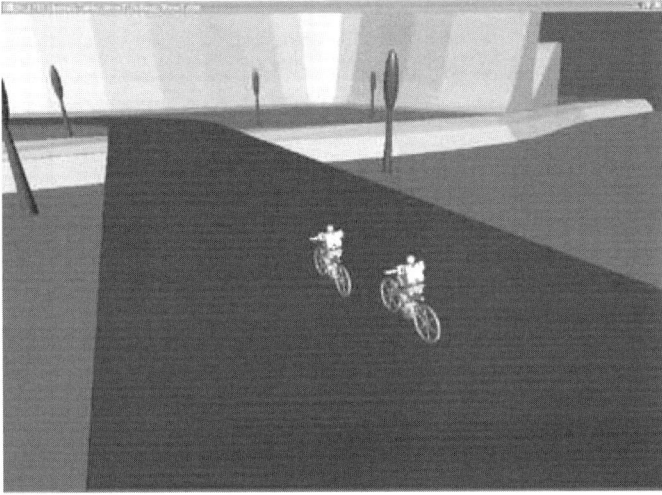

Fig. 9.42   Illustration of image-guided locomotion.

- Step 4: Plan a feasible path that is close to the central line of the road.
- Step 5: Generate a trajectory according to a time constraint.

Example 9.18   Parking is a special type of road following. In a parking maneuver, the initial and final configurations are available. In fact, the initial configuration is the actual posture of the robot (or vehicle), and the final configuration is determined by the parking slot available.

Figure 9.44 shows the results of image-guided autonomous parking. The process starts with color identification of the input image (Fig. 9.44a). In this example, landmarks are painted in red. Thus, we first identify the red pixels (Fig. 9.44b). Then, we filter out the small clusters of red pixels. For a cluster corresponding to a landmark, a curve-fitting process results in two angled line segments (Fig. 9.44c).

Since the inverse projective-mapping matrix D is known in advance (through calibration), the angled line-segments can be back-projected onto the road surface (Fig. 9.44d). In road space, the initial and final configurations of the robot can be represented by two rectangles. Finally, a path connecting these two configurations can be planned using a fifth-order polynomial function which allows us to impose a limit on maximum curvature.

◇◇◇◇◇◇◇◇◇◇◇◇◇◇◇◇◇◇

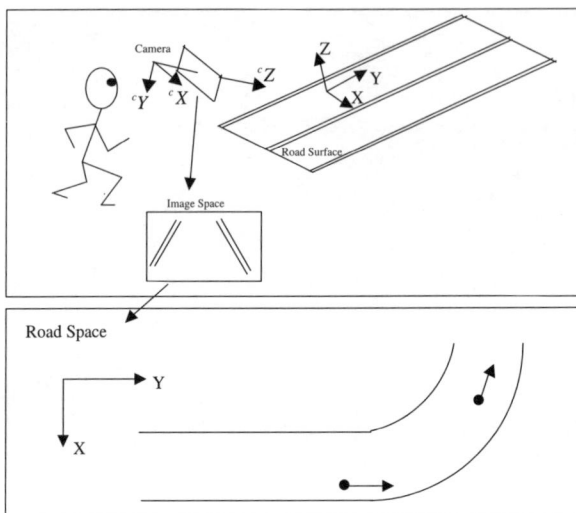

Fig. 9.43	Illustration of image-guided road following.

### 9.6.3.2	*Image-Guided Target Following*

Target following is a common scenario in transportation. When we drive, we often follow a car at a safe distance. When a team of robots move together in formation, one robot must follow another at a certain distance. Target following, therefore, is an important behavior.

As shown in Fig. 9.45, the problem of target following can be translated into a problem of head-eye coordination. In fact, we can assign a frame to a target, called a *target frame*. Interestingly enough, in image-guided target following, the frame assigned to a vision-head called a *head frame* for short, can be anywhere. Thus, we choose to put it in front of the robot's body. For example, if a safe distance is about 6 meters, we can place the head frame 6 meters in front of the robot's body. In this way, the problem of target following becomes how to automatically generate a feasible path/trajectory which ensures that the head frame coincides with the target frame. Clearly, this problem can be solved using the iterative approach to head-eye coordination.

Example 9.19	Figure 9.46 shows a simulation of image-guided target following. The calibration matrices (H_1, H_2) of the binocular vision's two

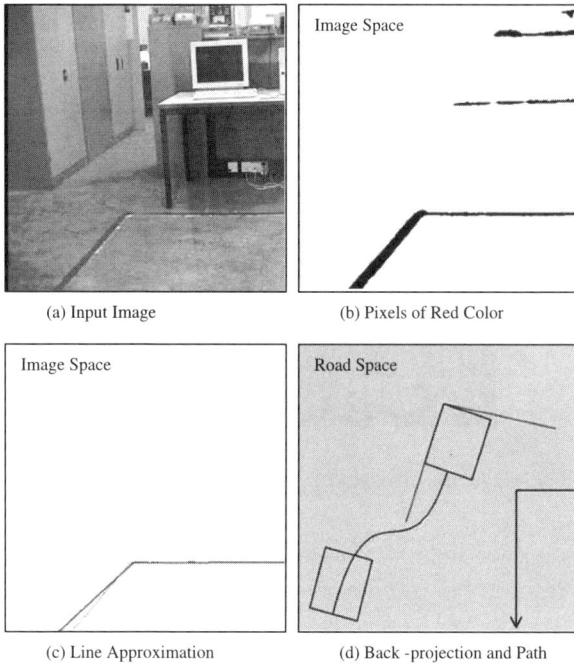

(a) Input Image

(b) Pixels of Red Color

(c) Line Approximation

(d) Back -projection and Path

Fig. 9.44   Example of image-guided parking.

cameras are

$$H_1 = 10^3 \bullet \begin{pmatrix} 0.0096 & 0.0026 & 0.0000 & -0.7949 \\ 0 & 0.0026 & -0.0096 & 2.1668 \\ 0 & 0.0000 & 0.0000 & 0.0010 \end{pmatrix}$$

and

$$H_2 = 10^3 \bullet \begin{pmatrix} 0.0096 & 0.0026 & 0.0000 & -1.2726 \\ 0 & 0.0026 & -0.0096 & 2.1668 \\ 0 & 0.0000 & 0.0000 & 0.0010 \end{pmatrix}.$$

In fact, these two camera frames are located respectively at

$$(110cm, \; -100cm, \; 200cm)$$

and

$$(160cm, \; -100cm, \; 200cm).$$

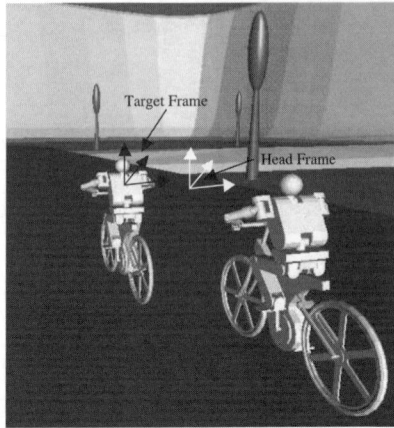

Fig. 9.45   Illustration of image-guided target following.

Now, we choose three points to represent the head frame. In the reference frame (world frame), the coordinates of these three points are

$$\begin{cases} A = (100cm, \ 500cm, \ 180cm) \\ B = (100cm, \ 500cm, \ 220cm) \\ C = (140cm, \ 500cm, \ 220cm). \end{cases}$$

At each of these three points, an orthogonal coordinate-system, which is parallel to the reference frame, is attached. As a result, the corresponding image-to-task mapping matrices are

$$A_A = \begin{pmatrix} 0.8333 & 0 & -0.2083 & 0 \\ -6.6667 & 0 & 6.6667 & 0 \\ 0.2083 & -0.3125 & -0.2083 & -0.3125 \end{pmatrix},$$

$$A_B = \begin{pmatrix} 0.8333 & 0 & -0.2083 & 0 \\ -6.6667 & 0 & 6.6667 & 0 \\ -0.2083 & -0.3125 & 0.2083 & -0.3125 \end{pmatrix}$$

and

$$A_C = \begin{pmatrix} 0.2083 & 0 & 0.4167 & 0 \\ -6.6667 & 0 & 6.6667 & 0 \\ -0.2083 & -0.3125 & 0.2083 & -0.3125 \end{pmatrix}.$$

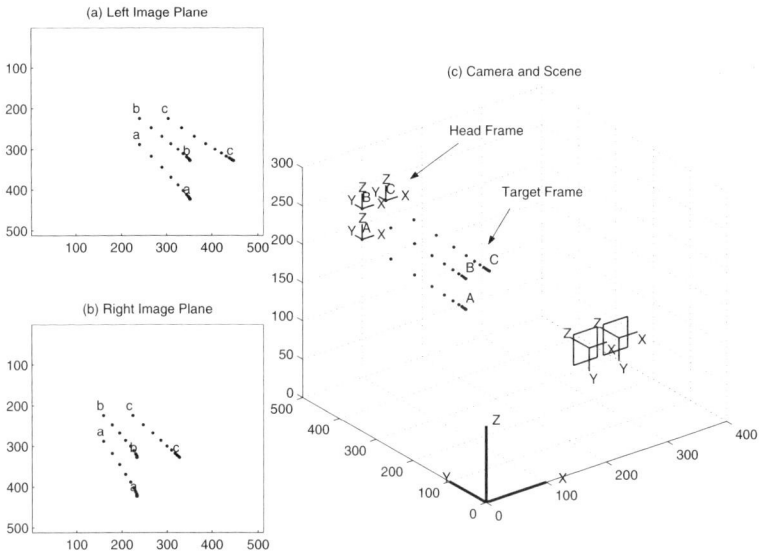

Fig. 9.46   Example of image-guided target following.

Similarly, the target frame is also represented by a set of three points. At a time-instant, assume that the coordinates of the three points representing the target frame in the reference frame are

$$\begin{cases} A = (150cm, \ 300cm, \ 130cm) \\ B = (150cm, \ 300cm, \ 170cm) \\ C = (190cm, \ 300cm, \ 170cm). \end{cases}$$

The above settings simulate the situation when one robot follows too closely behind another. The generated path/trajectory must increase the relative distance so that the head frame will coincide with the target frame.

Now, we choose the proportional gains (g_A, g_B, g_C) to be $(0.2, 0.2, 0.2)$. The application of Eq. 9.63 yields the paths shown in Fig. 9.46c. Again, the paths almost follow a straight line.

◇◇◇◇◇◇◇◇◇◇◇◇◇◇◇◇◇◇

9.7   Summary

In this chapter, we learned that the ultimate goal of robotics research is to develop autonomous robots which can be deployed in industry and society.

We also learned that a robot's autonomy can be achieved at three levels: a) action-execution level, b) task-execution level, and c) goal-execution level. And, a robot's autonomy depends on its decision-making system.

Decision making is a complex process. This is because of uncertainty and redundancy which may be present at both the input and output of a decision-making process. We now know that possible methods for dealing with uncertainty and redundancy include: a) application of expert knowledge, b) statistical inference, c) fuzzy inference, and d) application of neural networks.

Under the motion-centric theme in robotics, the primary concern of a robot's decision-making system is to automatically break down the description of a desired goal into detailed descriptions of motions, in terms of feasible paths/trajectories. For this reason, we studied the topics of: a) task planning, b) action planning, and c) motion planning.

We know that the objective of task planning is to break down the description of a desired goal (input) into detailed descriptions of a sequence of desired tasks. So far, this level of decision-making still requires human intervention.

Similarly, the objective of action planning is to break down the description of a desired task into detailed descriptions of a sequence of desired actions. Again, this level of decision-making requires human intervention.

As for motion planning, the objective is to identify the initial and final configurations of a robot's end-effector from the action description first, and then plan a feasible path/trajectory which will move the robot's end-effector from its initial configuration to the final configuration.

Subsequently, we studied motion planning in task space which requires *a priori* knowledge of the geometric model of the scene. We know that a feasible path can be easily obtained from a path map constructed in a discrete, normalized workspace if the initial and final configurations are given as input.

In order to gain autonomy at the motion-execution level, it is necessary to automate motion planning. Otherwise, autonomy attained at higher levels has no use because motion planning still requires human intervention. For this reason, we studied the important topic of image-guided motion planning in task space. And, we learned about three typical scenarios: a) hand-eye coordination, b) head-eye coordination, and c) leg-eye coordination.

Hand-eye coordination is synonymous with image-guided manipulation. We learned a closed-form solution, which is based on 3-D reconstruction.

However, the most important method of hand-eye coordination is the iterative approach, which is based on qualitative binocular-vision. We studied an image-guided motion-planning function. We know that there is a constant mapping matrix called an image-to-task mapping matrix which directly maps paths in image space to corresponding paths in task space. This mapping is robust, stable, and effective.

Head-eye coordination is synonymous with image-guided positioning. We studied a closed-form solution which makes use of three views. This solution is also applicable to solving the interesting problem of augmented reality. By placing the head frame (assigned to a vision-head) in front of the binocular vision, the image-guided motion-planning function, derived from hand-eye coordination, is also applicable to solving the problem of head-eye coordination. This illustrates the strong similarity between qualitative binocular-vision and human vision under the context of visual guidance.

Leg-eye coordination is synonymous with image-guided locomotion. There are two subproblems under this topic: a) image-guided road following and b) image-guided target following. We learned that the first problem can be easily solved using inverse projective-mapping of 2-D vision, while the second can be treated as a problem of head-eye coordination.

A robot's mental ability to automate motion planning is a decisive step towards autonomy. Future research in robotics will certainly look into the issues of automatic planning at task and action levels.

9.8   Exercises

(1) What is automation?
(2) What is autonomy?
(3) What is an autonomous system?
(4) What is an automated system?
(5) What is an intelligent system?
(6) What are the possible inputs to a decision-making process?
(7) What are the possible outputs from a decision-making process?
(8) What are the possible difficulties in decision making?
(9) What is the necessary condition for a robot to become autonomous?
(10) What are the possible levels of autonomy that a robot is likely to reach?
(11) What are the possible generic actions relevant to the generic task of "walking"?
(12) What are the difficulties in automating a task-planning process?

(13) What are the difficulties in automating an action-planning process?

(14) What are the difficulties in automating a motion-planning process?

(15) What is a path?

(16) What is a trajectory?

(17) Explain the physical meaning of Eq. 9.9.

(18) Explain how to use a cubic spline to represent a path.

(19) What is a holonomic constraint?

(20) What is a nonholonomic constraint?

(21) Why must a feasible path satisfy the constraint of maximum limit imposed to its curvature?

(22) What is the hand-eye coordination problem statement?

(23) What is the head-eye coordination problem statement?

(24) What is the leg-eye coordination problem statement?

(25) Derive an image-guided motion planning function.

(26) In Example 9.14, what is the equivalent rotation axis of $^{c3}R_{c1}$? And, what is the equivalent rotation angle?

(27) Project 1: Implement a motion-planning algorithm in task space.

(28) Project 2: Implement the iterative approach for hand-eye coordination.

(29) Project 3: Implement the three-view method for robotic head-eye coordination.

(30) Project 4: Implement the iterative approach for head-eye coordination.

(31) Project 5: Implement a simulation of a flying vehicle making an autonomous landing.

(32) Project 6: Implement a simulation of a robotic hand catching a moving target.

9.9   Bibliography

(1) Balch, T. and R. C. Arkin (1999). Behavior-based Formation Control for Multi-robot Team, *IEEE Trans. on Robotics and Automation*, **14**.

(2) Brady, M., J. M. Hollerbach, T. L. Johnson, T. Lozano-Perez and M. T. Mason (1982). *Robot Motion: Planning and Control*, MIT Press.

(3) Cameron, S. and P. Probert (1994). *Advanced Guided Vehicles*, World Scientific.

(4) Desai, J. P., J. P. Ostrowski and V. Kumar (2001). Modelling and Control of Formations of Nonholonomic Mobile Robots, *IEEE Trans. on Robotics and Automation*, **17**, 6.

(5) Dubois, D. and H. Prade (1988). *Possibility Theory*, Plenum Press.

(6) Hashimoto, K. (1993). *Visual Servoing*, World Scientific.

(7) Hosoda, K and M. Asada (1994). Versatile Visual Servoing without Knowledge of True Jacobian, *IEEE International Conference on Intelligent Robots and Systems*.

(8) Latombe, J. C. (1991). *Robot Motion Planning*, Kluwer Academic.

(9) Meystel, A. (1991). *Autonomous Mobile Robots*, World Scientific.

(10) Mezouer, Y. and F. Chaumette (2000). Path Planning in Image Space for Robust Visual Servoing, *IEEE International Conference on Robotics and Automation*.

(11) Walpole, R. E., R. H. Meyers and S. L. Meyers (1998). *Probability and Statistics for Engineers and Scientists*, Prentice-Hall.

Chapter 10

Prospects

Towards a Sociable Robot

An industrial robot is strongly characterized by its ability as a reprogrammable machine. This makes it a machine tool for manufacturing. However, motivation behind the development of the humanoid robot should go beyond the scope of simply making another machine tool for manufacturing. The ultimate goal of humanoid-robot research should be to develop *sociable robots* which may or may not be shaped like humans.

A sociable creature means an entity which is able to acquire and exhibit coherent social behaviors without reprogramming. A human is a sociable creature, as most of our social behaviors are learned through interactions with others rather than from genetic programming and reprogramming. A robot will become sociable if it evolves to the stage where it can be deployed anywhere after a one-time programming (i.e. the birth of the robot). We call this property *Program-Once and Run-Everywhere* (or PORE, for short). A sociable robot will be a PORE robot.

Obviously, a sociable or PORE robot must be *educable*. In this way, a less-experienced robot can learn from a more-experienced robot, or human master, without being reprogrammed. A robot will become educable if and only if it has an innate learning mechanism. Accordingly, a robot with an innate learning mechanism will not depend on reprogramming. How to develop an innate learning principle for educable robots is still an unresolved issue. And, as long as it is unresolved, there is no point in speculating on sociable robots.

An Energy-efficient Mechanism

From a mechanical point of view, the human body has an innate mechanism which evolves quantitatively but not qualitatively. Unless an external cause alters this mechanism, it remains qualitatively the same. Similarly, we can say that today's robots have an innate mechanism with stable kinematic and dynamic properties.

However, the human biomechanical system is highly energy-efficient. Humans can easily lift an object heavier than the weight of the arm. In general, this is not true of today's industrial or humanoid robot.

Intelligent Control

From a dynamics point of view, the purpose of control is to achieve a desired dynamic behavior by altering the intrinsic dynamics of a system under control with a set of externally-added control and sensing elements. It is challenging to design a good controller for a complex dynamic system like a robot. There are two difficulties. The first one is uncertainty. A robot is a dynamic system with a variable configuration which largely contributes to the variation in the robot's dynamic property. The second difficulty is redundancy. When there is a task as input, there may be multiple feasible solutions with regard to the robot's configuration and the control strategy of its controller.

In control engineering, the methodology of robust control aims at dealing with uncertainty in a control system, while the commonly-called *intelligent control* copes with the issue of redundancy. Mathematically, the goals of robust control and intelligent control are related to the principle of control optimization. Then, the question is: What is the innate principle for control optimization, given the innate structure (i.e. the feedback-control loop) of a control system? In other words, could a robot's control system be programmed once, and run everywhere without reprogramming?

Behavior Development and Learning

The human brain is a complex neural computing network with well-established partitions and pathways of mappings among the partitions of nerve cells. A human brain contains about 100 billion nerve cells (i.e. neurons) and 1000-5000 billion glial cells (i.e. glia). At birth, almost all the neurons are present. However, the glial cells form as the brain grows. It is the neurons, not the glial cells, which are in charge of all the sensory-data processing, decision-making, and sensory-motor controls.

In addition, the transmission speed of neural information (i.e. potential signal or pulse) among the neurons is in the range of 0.5 meters/s to 120 meters/s. Therefore, from an engineering point of view, a human brain's neural computing network is a stable and innate system. Then, the question is: What would a robot's stable and innate information system be which could function continuously during its lifetime?

A distributed & networked platform of microprocessors is a kind of modular system which is almost infinitely expandable. However, computing hardware or architecture without the support of appropriate software (i.e. an operating system) has very limited usage (if any). The question here is: When will a distributed real-time operating system, supporting a cluster of networked heterogeneous microprocessors, become a reality?

A sociable robot should not only learn, but also develop its own mental and physical capabilities. The most important part of a robot's information system is the information process running inside it. It is still not clear what information process is supporting the development and learning of a robot's behaviors. In other words, the question is: What is the innate mechanism behind autonomous mapping among sensors, actuators, and decision-making modules?

Real-time Visual Perception

A human receives a large amount of information from the visual-sensory system. The brain is able to respond in a timely manner to visual-sensory input no matter what the size of the visual information. The quality and timely responsiveness of the human visual-perception system significantly contributes to the degree of autonomy that we enjoy.

For a sociable robot to act and interact autonomously, it is clear that the robot's visual-perception system must produce coherent, consistent, and timely results. With the quality and timely responsiveness of the visual-perception system, the next question is: Can all motion-centric actions be controlled in image space without any reprogramming (e.g. without human intervention in motion planning)?

In addition, human vision performs well in image-understanding and object-recognition. We can effortlessly build a virtual representation of the real world although the result is highly qualitative. So far, it is not clear what the innate principle underlying the powerful visual-perception process of human vision is. Could a sociable robot perceive and describe a scene or object in a timely manner in the near future ?

Emotion and Language

A human being is a complex decision-making system which exhibits multiple behaviors. Our thinking, learning, perception, and action are strongly dependant on a unique entity known as a *belief system*. The dynamism of our belief system is usually reflected through our expressions of emotion. If a robot is sociable, it should also have the ability to express emotions.

Interestingly enough, a belief system is not genetically preprogrammed. A belief system is gradually built up based on outcomes of actions, behaviors, and linguistic input from others (e.g. what others say). Then, the questions are: What is the innate principle underlying a belief system? Is a natural language the basic building block of a belief system (e.g. the statement of goals, the statement of outcomes, the statement of values, the statement of rules, the statement of expectations, etc)?

Intelligence and Consciousness

We are able to consciously, and intelligently, coordinate behaviors which may be simple or very elaborate. The questions are: What is human consciousness? Is it a kind of resonance created by the multiple perception-decision-action loops concurrently running inside our brain? Could consciousness be imitated by an artificial information system?

Promising Future

Robotics is a fascinating field. There are many challenges, but also many opportunities waiting ahead.

Index